ECONOMIC PROBLEMS OF
AGRICULTURE IN INDUSTRIAL SOCIETIES

Other International Economic Association symposia

ECONOMIC PROBLEMS OF AGRICULTURE IN INDUSTRIAL SOCIETIES

Proceedings of a Conference
held by the International Economic Association

EDITED BY

UGO PAPI

AND

CHARLES NUNN

MACMILLAN
London Melbourne · Toronto
ST MARTIN'S PRESS
New York
1969

54348

© The International Economic Association 1969

Published by
MACMILLAN AND CO LTD
Little Essex Street London WC2
and also at Bombay Calcutta and Madras
Macmillan South Africa (Publishers) Pty Ltd Johannesburg
The Macmillan Company of Australia Pty Ltd Melbourne
The Macmillan Company of Canada Ltd Toronto
St Martin's Press Inc New York
Gill and Macmillan Ltd Dublin

Library of Congress catalog card no. 69–11372

Printed in Great Britain by
R. & R. CLARK LTD
Edinburgh

54348

CONTENTS

Contents

vi

Contents

Contents

ACKNOWLEDGEMENTS

THE International Economic Association wishes to thank all those who contributed to the success of the Conference recorded in this volume. It is indebted to UNESCO and to the Ford Foundation whose financial aid enabled the work to be undertaken and to the FAO for allowing the meetings to take place on its premises. The facilities provided and the helpfulness of the FAO staff contributed greatly to the smooth working of the Conference and were highly appreciated.

But the success of such a Conference depends on the quality of the papers and of the discussion. The Association wishes to record its gratitude to all those who contributed, and especially to Professor G. U. Papi, who was the architect of the programme and who did so much to make participants welcome to Rome — the evening reception in the magnificent setting of the Castello Sant' Angelo will long be remembered — and to Mr. Charles Nunn, who has devoted so much time and talent to editing the volume and recording the discussion.

LIST OF PARTICIPANTS

Mr. G. R. Allen, Director, Food & Agricultural Development, W. R. Grace & Co., New York, U.S.A.

Professor J. Ashton, Dept. of Agricultural Economics, The University, Newcastle upon Tyne, U.K.

Professor H. Astrand, Director, Swedish Agricultural Credit Association, Stockholm, Sweden.

Professor R. Badouin, Faculté de Droit, Université de Montpellier, France.

Professor M. Bandini, Rome, Italy.

Lic. F. Belaunzaran, Mexico 10 D.F., Mexico.

Mr. D. R. Bergmann, Institut National de la Recherche Agronomique, Paris, France.

Professor R. Bićanić, Faculty of Law, University of Zagreb, Yugoslavia.

Professor C. E. Bishop, University of North Carolina, U.S.A.

Dr. G. Blau, FAO.

Professor K. O. Campbell, Dept. of Agricultural Economics, University of Sydney, Australia.

Dr. Q. Camus, Jr., Philippine Economic Society, Manila, Philippines.

Professor V. M. Dandekar, Gokhale Institute of Politics and Economics, Poona, India.

Professor D. Delivanis, University of Salonika, Greece.

Dr. D. Dimitriu, Bucarest, Roumania.

Professor L. Dupriez, Centre de Recherches Économiques, Louvain, Belgium.

Professor H. C. Eastman, Canadian Political Science Association, Toronto, Canada.

Mr. Zyong-Zin Em, Vienna, Austria.

Mr. R. P. H. de Farcy, Vanves, Seine, France.

Professor L. Fauvel, Secretary General, IEA, Paris, France.

Dr. Fleischhauer, Frankfurt-am-Main, Germany.

Dr. J. Flek, Prague, Czechoslovakia.

Professor K. A. Fox, Dept. of Economics and Sociology, Iowa State University, U.S.A.

Professor N. Georgescu-Roegen, Dept. of Economics and Business Administration, Vanderbilt University, U.S.A.

Professor G. M. Gerhardsen, Norges Handelshøyskole, Bergen, Norway.

Professor O. Gulbrandsen, Landbrukshögskolan, Institutionen för Lantbruketsmarknadslära, Sweden.

Professor G. Haberler, Harvard University, Cambridge, U.S.A.

Professor B. Haley, Stanford University, U.S.A.

xi

List of Participants

Professor D. E. Hathaway, Dept. of Agricultural Economics, Michigan State University, U.S.A.

Professor S. Hoos, Berkeley, California, U.S.A.

Dr. R. Hsia, Economic Dept., University of Hong Kong, Hong Kong.

Dr. E. H. Jacoby, FAO.

Professor E. James, Faculté de Droit, Université de Paris, France.

Professor Gale Johnson, Dean, Division of the Social Sciences, University of Chicago, U.S.A.

Professor Glenn L. Johnson, Dept. of Agricultural Economics, Michigan State University, U.S.A.

Professor W. A. Jöhr, St. Gallen, Switzerland.

Professor E. Knoellinger, Åbo, Finland.

Mr. J. Le Bihan, Unité de Recherches Économiques (INRA) de l'École Supérieure des Industries Agricoles et Alimentaires au Cerdia, France.

Professor G. Leduc, Faculté de Droit, Université de Paris, France.

Professor Edward Lipinski, Warsaw, Poland.

Dr. Ernst E. Lipinsky, Bonn, Germany.

Professor E. Lundberg, Stockholm, Sweden.

Professor V. A. Martinov, Association of Soviet Economic Scientific Institutions, Moscow, U.S.S.R.

Professor R. Mossé, Grenoble, France.

Professor C. Mouton, Économie Rurale, Institut National Agronomique, Paris, France.

Professor Y. Mundlak, The Hebrew University of Jerusalem, Faculty of Agriculture, Rehovot, Israel.

Dr. C. Murgescu, Director, Institut des Recherches Économiques, Bucarest, Roumania.

Professor Myun-Suck Lee, Bank of Korea, Frankfurt-am-Main, Germany.

Professor F. Neumark, Frankfurt-am-Main, Germany.

Professor W. H. Nicholls, Vanderbilt University, U.S.A.

Mr. C. C. Nunn, University of Alberta, Edmunton, Canada.

Professor A. Nussbaumer, Vienna, Austria.

Professor K. Obolenski, All Union Research Institute in Agricultural Economics, Association of Soviet Economic Scientific Institutions, Moscow, U.S.S.R.

Dr. C.-E. Odhner, Stockholm, Sweden.

Mr. H. Ogura, Kyoto, Japan.

Dr. E. M. Ojala, Director, Commodities Division, FAO, Rome, Italy.

Professor G. U. Papi, Rome, Italy.

Professor D. Patinkin, The Eliezer Kaplan School of Economics and Social Sciences, The Hebrew University, Jerusalem, Israel.

Professor K. N. Plotnikov, Association of Soviet Economic Scientific Institutions, Moscow, U.S.S.R.

Professor M. Pohorille, Warsaw, Poland.

List of Participants

Professor U. Renborg, Lantbrukshögskolan Institutionen för Lantbrukets företagsekonomi, Uppsala, Sweden.

Professor E. A. G. Robinson, Cambridge University, Cambridge, U.K.

Mr. Robinson, FAO.

Professor V. W. Ruttan, International Rice Research Institute, Manila, Philippines.

Dr. V. Stipetić, Zagreb, Yugoslavia.

Professor J. Tepicht, Instytut Ekonomiki Rolnej, Warsaw, Poland.

Professor V. Travaglini, Rome, Italy.

Dr. I. Vajda, Hungarian Economic Association, Budapest, Hungary.

Professor J. Valarché, Fribourg, Switzerland.

H.E. Sir Ronald Walker, Australian Ambassador, Paris, France.

Ing. S. Zemborain, Buenos Aires, Argentine.

*Professor L. Komló, Budapest, Hungary.

*Professor H. Priebe, Frankfurt-am-Main, Germany.

*Dr. J. H. Richter, McLean, Virginia, U.S.A.

OBSERVERS

Mr. P. Baudin, European Economic Community, Brussels.

Professor T. Dams, European Economic Community, Brussels.

Dr. H.-B. Krohn, European Economic Community, Brussels.

Mr. L. G. Rabot, European Economic Community, Brussels.

Dr. V. Sorenson, Agriculture Directorate, O.E.C.D., Paris.

Mr. M. Tracy, Agriculture Directorate, O.E.C.D., Paris.

Dr. A. M. Weisblat, Agricultural Development Council, University of Philippines, Laguna, Philippines.

Dr. William C. Pendleton, The Ford Foundation, New York.

* Authors of papers who were not able to be present at Rome or to take part in the discussions of the Conference.

INTRODUCTION

BY

G. UGO PAPI

AND

CHARLES NUNN

THE Conference whose proceedings make up this volume was attended by contributors from many different countries and so from different systems of government and traditions of enquiry ; this accounts for much of the interest of the discussions that took place and for the comparisons, between different papers and discussions, which this volume affords. But there are, of course, many features that are common to the approaches adopted by the contributors, and it may serve the modest purpose of introducing the volume to such readers as have not yet been able to pick out the salient features of economists' enquiry into agricultural problems, to make a few remarks on some aspects of the common ground. Only after this will we proceed to draw out some of the major topics running through the Conference and to offer some comparisons between the different contributions.

I. INTRODUCTORY REMARKS

As may be expected from the volume's title, many of the contributions give directly onto questions of policy and there are none which fail to make explicit reference to the way in which they might be of significance to policy-makers. Nor is it surprising that treatment of similar problems occurs in different papers and discussions since by far the largest part of the Conference centred upon the problems of industrial societies. What, however, may not be so immediately evident or expected are, first, that we can find a large measure of agreement on what are the important contemporary lines of enquiry and, second, the view implicit in much of the work that the agricultural sector is marked off from other areas of economic activity by such singularities as to call for a quite distinct set of principles for policy guidance and an unusually large transformation of his methods on the part of the economist who would play a part in this sector.

Introduction

Some evidence on the extent of the Conference consensus on what are the important lines of enquiry will be found at a number of points in this introduction, and particularly in the second part, where we comment directly upon the groupings of papers and discussions. It is, however, the specific nature of the agricultural sector as seen by the economist which is our point of departure.

The special treatment of agriculture and the provision of special institutions to administer and enquire into that sector would seem to be co-extensive with the development of modern government. To account even superficially for this state of affairs would involve us in a vast and intriguing line of historical enquiry which is not called for here. However, we can characterize, in the most general terms, the factors which impel the trend. There are ways of administering and analysing agricultural activity which align it with other productive and social behaviour and there are ways of treating it which emphasise its idiosyncracies. Professor Georgescu-Roegen's paper provided us with a striking example of the argument for special analytical treatment. His paper would persuade us to reconceive the agricultural productive system by giving proper analytical weight to the 'inevitability' of agricultural production imposing 'some idleness on both capital and labour over the production period and complete idleness on every fund factor during the rest of the year'. 'This alone', he continues, 'would suffice as proof that industry and agriculture are governed by different laws', and Professor Georgescu-Roegen concludes his paper with policy suggestions, the power of which derives from the terms of his analysis. For the theoretical economist these analytical distinctions are of great importance and suggest that agriculture, by its very nature, requires a special box of tools ; contemporary work in the application of mathematical programming techniques in agricultural economics would seem to support this. However, at the present time, the main differentiation of agriculture from other activities is based on more pragmatic than analytical grounds ; that is, in terms of special problems and institutions. Clearly, then, it is to the conditions of production and the activities of governments that we must draw attention to indicate why agriculture receives special treatment.

It is a mark of modern governments and policy-makers, that to function efficiently — or, indeed, to function at all within modern ideological conditions — they see themselves as being obliged to define in material terms what constitutes benefits for the members of their societies and to provide the conditions in which the proper flows of these benefits are achieved. Under these circumstances it is not hard to see why they devote special attention to the agricultural sector. For, from this point of view, the sector plays a crucial role : it is the provider of unusually strategic benefits ; it constitutes a powerful and especially fastidious part of the body to be benefited ; while, at the same time, it is relatively wayward in its responsiveness to government action or other external stimuli. The need for the special treatment of agriculture has become even more striking in recent times since the preoccupation with problems of economic growth and what constitutes benefit and ever more exhaustive discussions of how the increased flow of benefits is to be assured.

Thus it is not surprising that, in spite of the diversity of political and academic backgrounds of those who took part in the Conference, they referred to institutional structures and types of enquiry that were in many respects similar. The contemporary institutional sociology of the sector may be characterized by the universal provision of independent ministries for the sector, by the activities of the FAO, by the proliferation of research facilities devoted to technical problems within agriculture, and by the growth of agricultural economics as a distinct line of enquiry, as well as by a number of other institutional provisions for agricultural sectors. We are thus faced by highly comparable frameworks which reflect a common strategy among economist policy-makers. Typical of the problems which these institutions treat as central to modern agricultural development are the consolidation of agricultural productive units ; the problems of rural labour mobility and the need for improved vertical communications ; these problems are endemic in agriculture in the developed world and, under different forms, in less-developed regions. While other sectors display analagous problems, the capacity for self-sufficient and concerted behaviour of the agricultural sector renders these difficulties peculiarly intractable. It is these sorts of issues rather than a body of special theory over and above the common theoretical fare of general economists which underly the papers in this volume.

The similarities in institutions and problems between countries help to show up the common traits of most work in this field. But these frameworks are not static or immutable. There are, it is true, conditions which have never been absent from agricultural sectors in societies where other sectors are developing and which, from contemporary points of view, constitute chronic 'problems' demanding continuous institutional provision. The special costs associated with

Introduction

factor movements out of the agricultural sector are the simplest example. But neither the external form nor the internal composition of the institutions that bring focus to our field of problems are 'given' in any simple sense. It is one of the benefits of the Conference that it casts light upon the changes that are under way or that are suggested as being desirable. The origins and characters of these sorts of institutions are conditioned by special political interests, and their growth and development cannot be relied upon to bear the correct continuous relation to the reality of the areas they administer or study : the case of an ill-adjusted ministerial alignment treated by Mr. Komló in his paper on the structural problems of a centrally planned economy digesting the impact of technical improvement and growth provides an explicit instance of this point.

It is in the field of international co-operation and research that the largest shifts in institutions and frameworks have occurred in recent years. As in so many other fields, it has been borne in upon policy-makers in the agricultural world and upon the societies within which they work, that the institutions and interests typical of the nation state are insufficient to realise the full benefits that the world agricultural sector can provide. It is, then, the modifications in institutions and policies at an international level which most require clarification and about which those that formulate policy are firmly united in their interest even if divergent in their views. For it is a further distinguishing characteristic of the agricultural world that while, to a peculiar degree, it shows a great diversity of conditions of production between countries and regions, it displays fundamental international and inter-regional interdependencies that lie at the heart of economic development and co-operation throughout the world. Even if the view is taken that it is not institutions but the free play of enlightened self interest that will release the economic benefits that world agriculture can yield, it will be recognized that supra-national institutions are essential both to ensure that world market behaviour is sufficiently orderly, secure and well-informed and to provide the counter-balance to national interests and institutions which are likely to take a cramped and relatively isolationist view.

This volume offers a number of contributions concerned with these matters : the discussions in Parts I and II are particularly helpful. The major topics dealt with are : the international impact of the Common Agricultural Policy of the European Economic Community ; questions of national output surpluses and their disposal regarded as an international problem ; the role of world

Introduction

trade in agricultural products as part of a potential strategy of development of poor regions and as furthering a long-run exploitation of international resources. We will provide, in a moment, some guidance on the references to these questions as they occur in different parts of the volume. But first we would emphasize the peculiar difficulties in the field of international co-operation in agricultural development and the large areas for comparative studies which these problems present for greater exploration.

Three of the major circumstances with which policy-makers must contend in their approach to long-term international rearrangements are : the strategic importance of the products of agriculture in a less than perfectly secure world ; the peculiar instabilities to which its markets are subject ; and the contemporary rapid advance of organizational and technical changes which bring in their wake radical revaluations of the factors in the sector. In the light of these difficulties, a peasant-like unresponsiveness on the part of particular governments to the incentives of an international free-trade system may be as soundly based as that of the peasant himself. We have to add to these influences the consideration that the time period over which governments are typically found wanting is shorter than the period within which the benefits of major changes in the agricultural sector can usually be realised. The pecuniary and non-pecuniary costs of change, which in the agricultural sector are commonly very high, are likely to be incurred well before the rather uncertain benefits of the changes, aimed at a more rational use of international resources, have been fully realized ; and almost certainly before they have been so distributed as to over-compensate the factor owners of the agricultural sector. These constraints and complexities must be seen against the background of world trade as a whole, of the present contrived system of international prices and of inhibited international factor movements ; and they must also be situated in the context of the ambiguities in any analysis of comparative advantages in a world of rapidly applied new techniques. For further introduction to these problems, we would refer the reader to Professor Robinson's paper and the discussion arising from his suggestions on the possible future shape of world trade.

A Conference of this span could not be expected even fully to survey the present attempts at international co-operation or the increase in common application to the issues brought up by the last paragraph. The discussions of the ramifications of the European Economic Community's policies — varying from international labour

Introduction

migrations and international quality requirements raised by Professor Delivanis through to the problems concerning the application of an immediately effective price policy alongside a relatively slow-working structural policy — would provide something of a microcosm within which the problems of the world at large can be grasped. However, it is worth emphasizing that at a number of points in the conference regret was expressed that in two important areas the major issues had not been centrally confronted. First ; there is the question of the sorts of institutional requirements and types of negotiations that could best treat the problems of greater international co-operation. This would provide occasion for the discussion of some of the concrete proposals put forward by the institutions which exist at present, and for assessment of the relations (both as research bodies and as quasi-executive organs) between such institutions as UNCTAD, the FAO and those conducting the bilateral and multi-lateral trade and aid activities of the different nations. A few papers touch upon this point (for example those of Dr. Martinov and Dr. Ojala), but it was not the central concern of any paper or discussion. Second ; there is the question of a review of current technical changes in agriculture, of the difficulties of applying them in different parts of the world (this was touched on at a number of points in the Conference ; for example, by Professor Ashton after Professor Delivanis' paper), and of the impact such changes would have on the comparative advantages between regions and countries. Much of the difficulty in knowing how to discuss possible international changes is rooted in ignorance of the consequences of applying the best possible techniques and skills in different parts of the world, or of the relative costs of acquiring and develop-ing those skills and techniques. The discussion of feed-lot systems and their possible impact upon areas where extensive grazing is practised (to be found in the discussion to Chapter 1 starting with Professor Campbell's question), is a case in point. It is plain that these issues would provide matter for at least one further IEA symposium. But further enquiry into these two areas must surely form the heart of any search for solutions to to the most acute problems in the international agricultural world ; that is, the set of problems that centre on the under-developed conditions of so much of the world with its reliance upon unreliable markets and structures devised in a different age, and the set of problems that centre on the costly over-production (at least with respect to unsupported prices) of much of the relatively advanced world?

Introduction

II. PAPERS AND DISCUSSIONS

The papers and discussions on international issues form Parts I and II of the volume. The discussion in these parts show the areas of greatest common concern. In the first discussion, following Dr. Ojala's paper, Professor Lundberg suggests that it may be 'more unrealistic to believe in quick acceleration of agricultural production in new countries than to hope for a change of attitude in industrial countries so that they become prepared to accept more manufactured imports, in exchange for openings for their agricultural products in the markets of developing countries'. On the other hand, Dr. Ojala's paper, which is mainly devoted to sketching the likely trends in world agricultural production and demand up to 1975, concludes that, 'if the development effort in the agriculture of the low-income countries fails to improve significantly on past rates of increase in marketed supplies of food, the dark colours of today's picture will remain dark'. There is a difference in emphasis here which seems to indicate different priorities in what we should be most concerned with and which seem likely to lead to different short-run policy suggestions.

The first emphasis is taken by Dr. Lundberg, by Professor Robinson in his paper and in his final comments to the discussion of his paper, by Professor Haley in the discussion of Professor Hathaway's paper, and — in the guise of criticisms of the international attitudes implicit in the Common Agricultural Policy of the EEC — by Professor Gale Johnson and Mr. Allen in the discussion to Professor Bandini's paper. This emphasis would have us concentrate on the long-run economic forces and the major shifts in international trade that would surely solve the problems referred to at the end of the last paragraph of Part I of this Introduction.

The second emphasis is that taken by Dr. Ojala, by the policymakers within the EEC (see particularly Professor Bandini's paper and the last part of the discussion after Professor Mouton's paper), and, implicitly by some of the contributions in Part II where the trade prospects of poor countries is discussed. This emphasis focuses our attention on the immediate prospect of agricultural development in regions that have — or at least see themselves as having — short-run problems of such magnitude that they cannot at this juncture afford to wait upon the major shifts. It is in this context that the absence of the sort of work we referred to at the end of our general remarks above is most felt. Without a full assessment of the effects of the measures taken and proposed under UNCTAD, GATT, the moves towards regional common markets, etc., the area of discussion open to try to reconcile the different emphases is small. It is not remarkable that it is on these international issues that the Conference displays the greatest divergences. Repeatedly there seems to be a dividing line, often running through the individual economists themselves. On the one hand, there is the economist engrossed by policy problems in the present constrained circumstances who emphasises the structural and developmental problems of today (see, for example, the last page of Dr. Martinov's paper where he underlines the technically complex nature of the trade-growth industries, and Professor Papi's points in the discussion after Chapter 2). On the other hand, there is the general economist looking in the long run with Professor Gale Johnson (in the discussion after Chapter 2) towards 'a concept of desired output based on optimum use of world resources'.

The questions raised by the Common Agricultural Policy of the European Economic Community are treated as being of fundamental importance by all participants ; but the division indicated in the last paragraph is strikingly displayed. The review of the history and philosophy of European agricultural policy presented in Professor Bandini's paper and the defence of the Common Agricultural Policy by Dr. Krohn and Dr. Dams in the discussion after Professor Mouton's paper are faced by the criticisms which are found in the discussions in Part I and in Professor Mouton's paper on the trend towards self-sufficiency in the EEC. There are three main features of these criticisms : first, that the Common Agricultural Policy must, in the long view, be retrogressive with respect to the development of world trade (not least for the industrial exports of the Community itself) ; second — a particular aspect of the first — that the policy represents an undue deference to the interests of intra-European agriculture from which the agricultural interests of other advanced countries must suffer ; and, third, that the confidence placed by the policy-makers in the results of a structural policy designed to strengthen the European agricultural leviathan before it is fully launched into the deep waters of international competition, is to be treated with suspicion. These questions on the directions taken by the European Economic Community recurred in different forms at other points in the Conference. Thus, in the discussion of Professor Priebe's paper on the performance of a sample of German farms subsidised by the provision of cheap capital, it was remarked that the

Introduction

'middle European' view of the durability of the family farm contrasted with views in other advanced countries largely because the
agriculture in the latter countries had had to face market forces from
which the agriculture in the former was relatively protected.

However, as Dr. Krohn points out, there will be a misrepresen-
tation of the impact of EEC policies in the context of world trade if
comparison is not made between the level of protection that would
have existed had there been no common market and the levels of
protection that in fact attend it. This suggests to us the further
question as to whether such a common market in the present world
context is a step in the direction of greater international co-operation
and rationalization, or whether it is a local 'bloc' situation which will
hamper the sort of co-operation that long-run economic forces might
otherwise bring about. The answer will, of course, depend on the
policies adopted by the common market ; but the sort of international
negotiations and researches which have brought the European
Community to its present stage of successful development and which
are bringing about inter-regional and inter-sectoral reallocations
which it is unlikely would have occurred without these negotiations,
demonstrate the viability and suggest the wider use of these methods.
The first part of Professor Delivanis' paper on the reasons which may
lead a developing country to seek association with the European
Community, and much of the discussion to that paper, are relevant in
this context. We would also point out that discussion of Professor
Nicholls' paper on the U.S. South as an under-developed region
casts further light on the difficulties raised, and the benefits provided,
by a common market consisting of relatively advanced economies for
an under-developed or developing region.

The contributions concerning the repercussions on developing
countries of present policies in industrial countries are grouped in
Part II. The central issue dealt with here is the question of the effect
on international trade of the agricultural protection and subsidy
policies practised by industrial countries. There are marked differences of emphasis to be found between the papers of Dr. Richter and
Dr. Martinov and the contributions of the discussants. Dr. Richter's
paper is mainly limited to an examination of questions of direct competition between tradeable outputs of the two groups of countries ; it
finds that the effect of protection in industrial countries is mainly on
other *developed* countries but has not greatly intensified the trade
problem of *developing* countries. Dr. Blau, in the discussion following

Introduction

Chapter 8, points out that studies which examine the many substitution possibilities suggest that a large part of the trade of developing countries is affected by competition from advanced countries. Both Dr. Martinov, in his Marxist historical review of trade patterns, and Professor Dandekar in the first part of his paper, regard the economic and foreign-policy behaviour of advanced countries as impinging seriously upon the trade opportunities for the agriculture of developing regions. Part of the differences again turns on the views held of long-run developmental and inter-sectoral possibilities. If it is thought that technical improvements along with specialization in the production of select agricultural commodities in developing areas will lead to the lowering of costs relative to the costs of the same outputs in developed countries, then protective policies in advanced areas may be read as closing the door on trade opportunities for developing areas in that they both reduce any advantage that might be achieved and suggest that any attempt to overcome the barrier that this protection presents will be answered by an increase in its level. These arguments are not inconsistent with a static review of direct contemporary competition which offers little evidence of these protection policies affecting current trading possibilities for developing areas. It was in the discussion after these papers that some speakers developed the point that we have been stressing ; that so much of the problem turns on whether or not the appropriate skills, technology and managerial ability are introduced in developing countries, so that their competitive position is improved and so that the flexibility which is increasingly achieved in the agriculture of rich countries, is achieved in poorer areas.

Parts III to V consist of papers dealing with contemporary problems in developed agriculture. There is a striking consensus here on what are the main problems and on what are the basic moving components behind these problems. It is this aspect of the contributions that we will stress in the following introductory paragraphs, leaving to the reader to find the diverse forms in which the problems present themselves and the variety of policy instruments and strategies which different regions and systems of government afford. As our introductory paragraphs suggest, we are dealing with a sector which may show unresponsive behaviour to the free play of market forces or great vulnerability to those forces or, in many cases, both, according to its own stage of development and to the level of development of the rest of the economy in which it operates. These characteristics can constitute such serious impediments in a developing economy and can give rise to such social distress and economic inhibition within the agricultural sector itself, that capitalist governments have typically intervened to supplement or modify the market forces, and the governments of centrally planned economies have adopted especially powerful methods of incentive or coercion. The present condition of developed agriculture is the outcome of the interplay between these special features, the consequent government intervention and the impact of recent developments in techniques and tastes. To the contemporary economist, the central problems arising from this state of affairs may be grouped under three heads : first, the pursuit of the optimum scale of the farming unit ; second, the attainment of structures of control and communication which will best digest the technology now available to the sector and the demands for improved final products that higher incomes render effective ; and third, the achievements of a mobility of factors sufficiently great to bring about a proper degree of specialization and concentration. These three issues first arise in Part III which is concerned with nature of the family farm, and we have found it most suitable for our introductory purposes to discuss them in that context.

The problem of achieving changes in the scale of productive units is intimately related to questions concerning the role of the family farm ; so, the prejudices and arguments put forward in favour of the family farm are relevant to the problem. Some notion of the family farm, however hazy, typically lies deeply implanted in any patriot's view of his identity, even if his contacts with the countryside are such that he might more rationally regard it as a series of irksome though necessary gaps in urban civilization. As Professor Glen Johnson reminds us, the concept is part of the individual's mythology about the 'good life' with its reputed charms of independence and cleanliness. These dreams are the reflection of a number of real and important features that distinguish the social and political views and solidarities of the farmer from those of entrepreneurs in other sectors. In this world the standard problems of method faced by the social scientist are unusually acute : the first problem is to avoid analytical confusion between issues of productivity and social and political issues ; and the second is to translate analytical results into terms that are consistent with the continuity of social, political and economic phenomena. (Professor Ruttan's paper on the problems of productivity and equity in land reform legislation places us in that context where these difficulties are most dramatically exemplified.)

Introduction

What is the family farm? The definitions offered include, with varying emphases, the two dimensions of size and degree of independence. The choice of the appropriate measure of these dimensions makes for further difficulties. Is size to be measured in terms of value added, full time labour force, capital invested, acreage or by some other measure? The first part of the discussion to Chapters 10–12 illustrates this problem of size criteria in the context of figures on agricultural performance in the United States. The independence criterion is also ambiguous. On the one hand there is the possible degree of peasant-like insulation from market forces varying from that of a rich peasant farmer with diversified production, through to the cases discussed by Professor Nussbaumer where the farmer can rely upon his woodlands as a source of injections of capital. On the other hand, there is the question of relative dependence on other sectors and funds for stability in marketing conditions, for capital requirements and for information and aid on techniques and quality requirements. Whether size or independence is taken as the predominant criterion, what sort of measurement should be used will depend on the particular comparison being made, on the type of farming in question, on the structure of the economy in which it operates, on the sort of government intervention that exists and on the sort of government intervention that is expected in the future.

The first three papers in Part III and the related discussions furnish instances of these points. Professor Renborg expects that, in Sweden at least, farms will essentially remain family units in the future ; but he points out that in some lines of production acreage increases are to be expected in the achievement of greater specialization. In such cases, the opportunities for specialization and the technology available will determine at what size — measured either by units of factors of production or by an output measurement — the returns to increased scale of the productive unit will be constant or falling. On the influence of government intervention, Professor Glen Johnson's paper remarks on the way that government support compensates for over-investment by family farms in immobile resources and for the capital losses which result from overproduction. Government thus creates conditions that inhibit the development of larger units. Furthermore, as contributions to the discussion underlined, the demands for land for non-farming uses and the expected government support action capitalized into high land prices, both operate as deterrents to increasing the size of consolidated units. There is also the *expectation* that governments will favour family farms in their efforts to maximise rural support, and this acts both to inhibit larger units from farming and to encourage the production of capital equipment for agriculture which is of a size and type suitable for family farm units.

In spite of the analytical ambiguities, with respect to scale and independence, that surround the concept, the discussion of the family farm focuses attention on the second of our central problems in the development of contemporary farming : the attainment of structures appropriate to modern technology. The family farm proposes a particular sort of management unit in agriculture. The management unit need not, in principle, be constrained by the size of the productive unit or by the preservation of its family character. This is shown by the increase in the number of cases of a single management farming two or more units, and by planned projects in developing agriculture which centralize the management of similar units in order to economize on scarce management skills. Such a trend, however, seems likely to move slowly in view of the alternative opportunities for management skills and the complexity of natural features which mean that the same techniques and complementary activities may only be appropriate over a narrow area. It is in regions where the case for monoculture is unanswerable and in instances where new technology leads to large increases in the returns to scale that the case for larger management units is strongest.

It is with these larger and vertically more comprehensive systems of management that Part IV of the volume is largely concerned. The arguments for such systems are presented in the papers by Professor Le Bihan and Professor Valarché. Professor Le Bihan assesses the impact on European agriculture of trends of this sort ; he makes comparison between the European research efforts and those in the United States and he distinguishes between the different sorts of activity — research and diffusion — and the different sorts of institutions that pursue these activities. The consequences upon the management and risk-bearing structure of farming is the main topic taken up in the discussion after those two papers. Mr. Allen points out that the effect of the changes upon the family farm or other micro units will vary with the technical change involved. While the case of the broiler industry studied by Professor Valarché is one where the improvements are largely embodied in the feed and breeding sectors and where processing and marketing methods are best conducted centrally, this does not apply in many other cases and in these the economies of scale may not be 'internalized' to the same degree.

However, in these areas, the units, or at least the interconnected systems within which the improvements are realized, are likely to become larger, so that the high fixed costs which the research and development of innovations entail may be spread over a larger output. It is in the nature of such changes that the micro-unit has the area reduced within which its own decisions are effective.

The papers by Professor Komló and Professor Pohorille show how the issues raised in the last paragraph are requiring important changes in planning methods in centrally planned economies. Professor Komló discusses the problems of highly collectivized Hungarian agriculture where lack of co-ordination in central planning and where the existence of a food-processing structure which had been located and built for an agriculture of a non-specialized structure, hold back the developments necessary for the introduction of new techniques and for the satisfaction of more fastidious markets. Professor Pohorille touches on the potential use of the government contract agencies, through which prices and total output balance are achieved, as sources of technical information and innovation diffusion.

The third of our three central problems in contemporary developed agriculture concerns the mobility of factors. With the need for technical changes, for integration and for size adjustments, the rate of movement of factors into and out of agriculture becomes crucial. The low scrap value of capital instruments, the difficulties in land consolidation and the specificity of skills and tasks of agricultural labour, all act to impede the movement of factors out of agriculture. These same influences bar the entry of factors better adapted to new demands and new techniques. Moreover, government support policies, unless conducted with expensive and rather inquisitorial subtlety, are liable to reduce factor mobility even further by ensuring a socially tolerable return to the factors already engaged in farming. Professor Glen Johnson develops a further argument in this context on the overcommitment of resources within sub-optimal structures.

It is to the problems of the labour market that most attention is directed. The welfare problem is here most acute and political interests most seriously touched. Also, it is often a prerequisite of mobility in land and of proper adjustment in capital resources. The status of the family farm is central to this problem also. The attractions of ownership or of secure occupancy may outweigh the low *per capita* income for a family, particularly where that income can be supplemented by part-time work in other sectors. In effect this means that the family will tolerate incomes below those that a large-

Introduction

scale entrepreneur is obliged to pay if he is to attract labour from other activities. Two sides of the problem are set out in different national contexts in the papers by Professor Bishop and Dr. Flek. The former discusses the disequilibrium in the agricultural labour market with particular emphasis upon the factors that render the small farmer and the unskilled farm labourer immobile ; while the latter discusses the Czechoslovakian case where a successful post-war campaign to shift labour into industry has resulted in an unbalanced age and skill composition in the rural population. Both Professor Astrand and Professor Gulbrandsen are concerned with this topic in their papers on Swedish agricultural policy. The paper of the former, in particular, relates labour mobility to the question of maintaining income objectives for those remaining in agriculture.

Our chosen route through some of the topics in this volume has enabled us to indicate very little of the wealth of material to be found there. Our comments on how the family farm features in the volume are quite incomplete without commenting on the references (to be found, for example, in Professor Tepicht's paper and in the discussion to Professor Pohorille's paper) to the persistence of private-plot farming, to the nature of collective farming and its relation to old peasant structures, and to the secondary market in some centrally planned economies. However, the most serious ommission is the absence of comment on the price aspect of both international and national agricultural sectors. Our excuse, to be shared presumably with a number of the contributors, must be that such are the special features of this sector and so considerable are the interventions of governments that many *current* prices in agriculture are to be regarded less as determinants of, and indicators for, long-run changes than as by-products of short-term and highly regulated phenomena. A major factor in the interest of this volume is that it contains contributions that try to come to grips with the long-term issues.

PART I

AGRICULTURE IN THE WORLD ECONOMY

Chapter 1

AGRICULTURE IN THE WORLD OF 1975
GENERAL PICTURE OF TRENDS [1]

BY

E. M. OJALA

Director, Commodities Division, FAO, Rome

I. INTRODUCTION

THE long-term character of the basic problems of world agriculture probably justifies a working assumption that the situation in 1975 will be mainly the result of factors and trends that can be currently identified. This is the approach that will be adopted in this paper. Nevertheless, it may be necessary to speculate on the advent of new policies, and to attempt a judgement as to their possible effects on agriculture. Within the limited scope of a few pages, one cannot do more than try to sketch the main lines of likely changes.

I should add that FAO is currently embarking on a far-reaching programme of studies and consultations with a view to preparing an Indicative World Plan for Agricultural Development. This Indicative Plan, which is to reach out to 1975 and tentatively to 1985, is to be based on the concerted preparation of two types of studies : on the one hand, production, demand and trade projections for individual commodities on a world-wide basis ; and on the other, a series of subregional studies of agricultural development prospects in less developed areas, based on country data and national plans. This work is only in its early stages, and anything said below should not be taken to prejudge any issues or anticipate in any way the conclusions and recommendations of the Indicative Plan.

II. POPULATION

A first thing to look at is the relative size of world agriculture in 1975. This can be assessed in two ways — by reference to the popula-

[1] Acknowledgement is due to a number of FAO staff colleagues for valuable comments made on an early draft of this paper. However, the views expressed are not necessarily those of the Organization.

TABLE I

POPULATION AND GROSS DOMESTIC PRODUCT, AGRICULTURAL AND NON-AGRICULTURAL, BY MAIN ECONOMIC REGIONS, 1960 AND PROJECTED 1975

	1960			Projected 1975			Rate of projected growth 1960–75 % per year compound		
	Total	Agri-cultural	Non-Agri-cultural	Total	Agri-cultural	Non-Agri-cultural	Total	Agri-cultural	Non-Agri-cultural
Population (million)									
World	3,005	1,596	1,409	4,000	1,894	2,106	1·9	1·2	2·7
Developed countries	651	121	530	781	96	685	1·2	–1·6	1·7
Developing countries	1,322	902	420	1,913	1,157	756	2·5	1·7	4·0
Centrally planned countries	1,032	573	459	1,306	641	665	1·6	0·7	2·5
GDP ($ billion at constant prices)									
World	1,381	203	1,178	2,898	322	2,576	5·1	3·1	5·4
Developed countries	920	61	859	1,913	82	1,831	5·0	2·0	5·2
Developing countries	170	61	109	362	111	251	5·2	4·1	5·7
Centrally planned countries	291	81	211	623	129	494	5·2	3·2	5·8

Note—Table prepared by Dr. D. Basu, FAO, Rome. The selection of the population and GDP growth rates, total and agricultural, is the key to the projections.

Sources : *UN World Economic Survey 1963*, p. 19, Table 2–1; *UN Yearbook of National Accounts, 1963*; *UN Studies in Long-Term Economic Projections for the World Economy*; *UN Provisional Report on World Population Prospects*, 1964; *FAO Production Yearbook*; *OECD Statistics of National Accounts 1950–61*, 1964; Rosenstein-Rodan, "International Aid for Underdeveloped Countries," *The Review of Economics and Statistics*, May 1961, No. 2; Records available in FAO.

tion directly dependent on the sector, and to the product generated. Both are indicated in Table I, by estimates for 1975 in comparison with the situation in 1960. The basic world data for the table have been derived mainly from United Nations and FAO sources. Breakdowns between agricultural and non-agricultural sectors have been attempted but, in view of the nature of the basic statistics, they should not be regarded as more than indicative. The key to the projections lies in the assumed rates of growth for the total economy and for the agricultural sector, at the world level, and for the three major geographical zones. Agricultural product in the developing countries is assumed to grow at a rate which will enable demand for food to be met at the assumed level of overall economic growth. Alternative assumptions could be used.

It seems likely that the absolute size of world agriculture in terms of population directly supported will increase to about 1·9 billion people by 1975, approximately 19 per cent greater than in 1960. The total world population may reach 4 billion, an increase of about one-third over 1960. However, the total non-agricultural population depending upon the market for their agricultural supplies will increase at a faster rate, from 1·4 to over 2·1 billion people, or about 50 per cent. Nevertheless, most of the world's agricultural producers will still be working mainly for the direct subsistence of their own families. Although nearly one-fifth larger in absolute terms, the relative size of world agricultural population will fall from about 53 per cent of total population in 1960 to perhaps just under 48 per cent in 1975 (Table II). These global indications mask the significance of the trends, which differ markedly according to the stage of economic development reached.

In the developing regions of the world (comprising some two-thirds of the human race if China is included) incomes are low, diets are inadequate and underemployment of the agricultural population is common. These are the regions of explosive population increase. In 1960 an estimated 68 per cent of their people were on the land. With optimum emphasis on industrialization and assuming an 80 per cent increase in urban population, the developing economies by 1975 will nevertheless have to support some 30 per cent more people on the land. This will intensify the rural employment problem, especially where land resources are scarce. With their agricultural population still as high as 60 per cent of the total, their economies will remain predominantly agricultural. The relatively high rate of urbanization will mean a strong growth of commercial demand for basic foods.

5

TABLE II
PERCENTAGE DISTRIBUTION OF POPULATION AND GROSS DOMESTIC PRODUCT AGRICULTURAL AND NON-AGRICULTURAL SECTORS

	1960			1975		
	% Total	% Agricultural	% Non-Agricultural	% Total	% Agricultural	% Non-Agricultural
Population						
World	100	53·1	46·9	100	47·4	52·6
Developed countries	100	18·5	81·5	100	12·3	87·7
Developing countries	100	68·2	31·8	100	60·5	39·5
Centrally planned countries	100	55·5	44·5	100	49·1	50·9
GDP						
World	100	14·7	85·3	100	11·1	88·9
Developed countries	100	6·6	93·4	100	4·3	95·7
Developing countries	100	35·9	64·1	100	30·7	69·3
Centrally planned countries	100	27·7	72·3	100	20·7	79·3

TABLE III
SECTORAL PRODUCT PER HEAD OF POPULATION IN AGRICULTURAL AND NON-AGRICULTURAL SECTORS

	1960			1975		
	Overall	Agricultural	Non-Agricultural	Overall	Agricultural	Non-Agricultural
GDP per caput (in U.S. $)						
World	460	127	836	724	170	1,223
Developed countries	1,409	506	1,614	2,449	854	2,673
Developing countries	128	68	259	189	96	332
Centrally planned countries	282	141	459	477	201	743

This is a favourable factor for the expansion of agriculture, provided their agricultural and commercial institutions can be geared to the supply and distribution of food to meet this increased demand largely from domestic resources. The presently rising trend of net food imports into these regions does not encourage an optimistic view. There is an opportunity, if not an imperative, for the developing countries to harmonize production plans in order to exploit complementarities and promote trade, including trade in foods, among themselves.

In the developed market economies, the U.S.S.R. and the centrally planned countries of Europe, which together make up about one-third of the world's people, similar population movements will occur. However, such movements continuing from different levels will have a completely different significance, and the problems for the agricultural sector though difficult will also be of a different character. Where industrial development has been under way for many decades, levels of income and of food consumption are high. The dynamic shift of employment from agriculture to other sectors has already proceeded far, to the point where only 10–15 per cent of the population is engaged in farming. The rise in total population, at rates no more than half that expected in the developing regions, will pose no problems of food production for the modern, capital-intensive agricultural sectors of these countries. In fact, in some of those more advanced economically, there are already problems of over-supply of agricultural products, which are likely to be intensified by 1975. Thus the downward pressure on farm prices and incomes may be expected to continue.

The decline in the proportion of the population dependent on agriculture in the developed regions as a whole will continue rapidly from 18·5 per cent in 1960 to around 12·3 per cent in 1975. Extremely low levels are likely to be reached in several more countries by the end of the period. In nearly all the countries of this group there will be a decline in absolute numbers as well. Only in the less industrialized parts of Southern Europe are the numbers still growing, and the turn-down may well occur before 1975. In Europe structural rigidities in agriculture will retain more people in the sector than will be technologically necessary, and governments will develop policies and measures to speed their exit. The income gap will remain, but the basic necessities of food and work will not be lacking, as they could be in large areas of Asia, Africa and Latin America.

7

III. AGRICULTURAL PRODUCT AND INCOME

Assumptions regarding the gross product of the world economy in 1975 and the aggregate net product of the agricultural sector are much more speculative than projections of population. Some recent United Nations studies in long-term economic projections are available for the world economy, broken down into three zones, namely : the developed market economies, the developing countries and the centrally planned economies.

In Table I, an annual economic growth rate of 5 per cent has been assumed for developed countries as a group over the period 1960–1975. For centrally planned economies and developing countries as separate groups, a rate of 5·2 per cent anually has been assumed. These are largely based on the United Nations studies. The breakdown as between agriculture and other sectors in 1960 has been derived in FAO from United Nations national accounts data, and for 1975 it results from the differential agricultural growth rates assumed for the three zones.

For the developing countries the rate of growth taken is in line with the minimum target rate of 5 per cent set to be reached by 1970 at the end of the UN Development Decade. Special efforts will be needed, particularly in agriculture, during the remainder of the 1960's and in the 1970's to ensure that their economic growth over the period 1960–75 will average 5·2 per cent annually, as assumed. This, like the assumed agricultural growth rate of 4·1 per cent annually, is a target rather than a projection.

These assumptions would represent an average expansion of the world economy of 5·1 per cent annually over the period 1960 to 1975. World agriculture in terms of gross domestic product would have to grow at about 3·1 per cent annually on the average, while the product of the non-agricultural sectors would grow much more rapidly, at about 5·4 per cent annually. In 1960 agriculture constituted about one-seventh of the world economy in terms of share in the sum of gross domestic products. By 1975, while producing some 60 per cent more, the share of agriculture may fall off to about one-ninth of the aggregate world gross domestic product, in view of the more rapid expansion of the non-agricultural sector.

These are the continuation of well-established long-term trends associated with economic progress. These changes are interesting enough at the global level. However, since they have been going on fairly rapidly for a century or more in the developed countries, and are only beginning in most developing countries, it is necessary to look at these two major groups separately.

In the developed market economies as a group, the share of the gross domestic product generated directly by agriculture is currently only about 6 per cent. By 1975 this proportion will have fallen to the very low figure of around 4 per cent, reflecting mainly the great preponderance of non-agricultural goods and services in the patterns of consumption prevailing at the high income levels enjoyed in these countries.

In the developing countries on the other hand (excluding centrally planned economies), agriculture accounts currently for about 35 per cent of aggregate gross domestic product. This figure is lower than in the past. If economic development proceeds at the assumed rates in this zone, the share of agriculture will fall further, perhaps to around 30 per cent by 1975. The key importance of agriculture in these economies will thus be little diminished by that time, especially when it is borne in mind that many of the new industries will be based on agricultural raw materials. In a few developing countries at very low income levels, half or more of the national product will still be coming from agriculture.

The centrally planned economies include the great land masses of the U.S.S.R. and Mainland China, which produce nearly 40 per cent of world gross product from agriculture. Since these two vast areas are at very different stages of economic development, the share of agriculture in their aggregated economies lies intermediate between the values observed and projected for developed and developing countries, but probably closer to the latter.

In 1960 the agricultural sectors of developed and developing zones each contributed about 30 per cent of world agricultural product. On the assumptions made, agriculture in the developing countries would expand to around 35 per cent by 1975, while agriculture in the developed countries growing less rapidly would fall off in proportion to contribute about 25 per cent of world agricultural product.

The combined effect of the population and product assumptions on agricultural incomes may now be considered. By dividing the gross product aggregates in Table I by the corresponding population totals one can make a crude first approach at estimates of average sectoral product per head of the agricultural and non-agricultural populations in the various geographical zones, as at 1960 with projections to 1975. The results of this calculation are shown in Table III, and must be interpreted carefully.

9

It may be noted that on the assumptions made, average product or income per head of the world population would rise by 58 per cent, from $460 in 1960 to about $720 in 1975. In developed countries and the centrally planned economies, the average increase would be around 70 per cent. Because of the faster growth of population in the developing world, their ratio of increase in product per head, from $128 to $189, would be less, about 48 per cent.

Average agricultural product per head of the agricultural population would rise from about $500 to $850 in the developed countries, maintaining the same ratio of about 32 per cent to average non-agricultural product per head of the non-agricultural population. In the developing world the corresponding rise would be from $68 to $96. This would constitute a relative improvement in product per head in agriculture, reflecting the assumed speedup in agricultural growth rate in this zone to the target of 4·1 per cent.

The gap between the figures for agricultural and non-agricultural sectors within each zone overstates the difference in average income levels per person between the two sectors, more particularly in the developed countries. Total income per head of the agricultural population will be higher than the figures shown because of the non-agricultural earnings of the farm families, and the incomes per head of the non-agricultural population will be correspondingly lower than shown. It is not possible to make the corrections for this factor. However, these, and other smaller corrections needed, would not show any great improvement in the relative income position of the agricultural population. In developed countries through agricultural support prices and subsidies it is possible for governments to reduce this income disadvantage to some extent. The much greater scope for income transfers to farmers from the non-agricultural sector in developed than in developing countries is well illustrated by Table III.

In this broad approach it is not possible to be precise about food consumption levels in 1975. In the high income developed countries where nutritional levels have been satisfactory, little change may occur by 1975 on a *per caput* basis, except possibly a slight increase in animal protein intake in some marginal areas. In the medium income developed countries such as those of Mediterranean Europe, the quality of the diets would be greatly improved by 1975. In the developing regions, *if demand for food were met at the income levels assumed for* 1975, a significant improvement would result in the food consumption of the populations. Their average calorie intake would rise to the minimum levels considered necessary for normal health and activity, but areas of insufficiency would remain. There would also be gains in average intake of animal protein, but developing country diets in this respect would still fall short of the calculated requirements.

IV. IMPORT DEMAND AND TRADE

It has generally been found useful to divide agricultural output into four broad groups in discussions of demand and trade prospects, since these vary so widely commodity by commodity. These groups are the following: temperate-zone foodstuffs (*e.g.* grains, dairy products, meat, eggs), world trade, 1959–61 = $12·1 billion; tropical products (*e.g.* coffee, cocoa, tea, bananas, spices), world trade, 1959–61 = $4·1 billion; temperate/tropical products (*e.g.* rice, sugar, fats and oils, citrus, tobacco), world trade, 1959–61 = $8·6 billion; agricultural raw materials (*e.g.* cotton, wool, jute, hard fibres, rubber, hides and skins, forest products), world trade, 1959–61 = $14·3 billion.

Temperate-zone Foodstuffs. Rapid technological advance and resulting increases in yields and output in both exporting and importing countries, coinciding with policies of agricultural protection in importing countries and price support measures in some of the exporting countries, have had the combined effect of raising supplies of temperate-zone grains to levels in excess of the absorptive capacities of the traditional high income markets, where income elasticities for staple commodities are generally low, if not — as for cereals for human consumption — negative. The demand for meat, especially beef, still enjoys a high income elasticity in developed countries and may well continue strong to 1975. In recent years beef production in Western Europe and North America has not kept pace with demand, and these regions will probably continue to provide good import markets in coming years. Reflecting rising labour costs, the rate of increase in dairy production in Western Europe and North America has slowed down. Surplus problems have accordingly eased, but at present it still seems unlikely that the net import demand of these regions for dairy products will rise since their dairy production is highly protected.

The tendency for urban populations in developing countries to switch consumption from tropical grains to wheat and flour seems likely to be strengthened by 1975, because of the lag in the production and export supplies of rice and the availability of wheat on concessional terms from the United States. Developed countries are not likely to expand their commercial sales of temperate-zone products to developing countries, and any major increase in this flow is likely to be on an aid basis. Europe may well have joined North America as an important supplier of 'surplus' grains before 1975. It seems not unreasonable to expect that by then the developed countries will have worked out an international agreement to eliminate the conflicts in national policies that lie at the root of the lack of balance in the commercial trade in temperate-zone grains.

The earnings from trade in temperate-zone foodstuffs flow mainly to developed countries, although some developing countries such as Argentina and Mexico are also important beneficiaries.

Tropical Products. For tropical products, originating only in developing countries, the demand situation in the dominant high income markets is akin to that for temperate-zone commodities. *Per caput* consumption is already high in these regions. Although income elasticities of demand are still such that scope remains for some increases as incomes rise, it seems hardly likely that the growth in demand, even on optimistic assumptions, can lead to increased export earnings from these products to anything like the extent required to finance the import needs of the exporting countries' development plans. Allowing for population increases, the most that can be hoped for at current price levels is an increase in the value of imports into the major markets of the order of some 2 per cent per annum. In a total trade of some $4 billion per annum, this implies an increase of some $80 million per year, to be spread over some 40 exporting countries.

Since demand for these products is price inelastic in the main markets, the total elimination of fiscal charges in importing countries would not greatly change the situation. The most hopeful prospects seem to lie in the centrally planned economies of the U.S.S.R. and Eastern Europe, where *per caput* consumption of these products is still well below that in other countries at comparable income levels.

Some of these commodities are already in surplus, and the producing countries cannot afford to carry long-term stocks. Hence the problem of price fluctuations remains grave. Failing radical changes, a sustained improvement in the situation of exporters does not appear likely, a fact of vital importance since export earnings from this group of commodities account for nearly 30 per cent of the value of the agricultural exports of all the developing countries taken together.

On the other hand, since these products originate only in developing countries and have no competition in developed countries, they offer scope for some international management of prices in the interests of producer countries, provided burdensome surpluses can be avoided.

Temperate/Tropical Products. These commodities are produced in both the temperate-zone high income areas of the world and in the tropical and semi-tropical low income regions. Because of the present high levels of consumption and the competition from domestic products, both natural and in some cases synthetic, the growth of import requirements of the high income group of countries is likely to be limited. In the United States the production potential for some of these commodities is already under control, and output could be greatly and rapidly increased to meet demand, at home or overseas.

In developing countries, on the other hand, demand is rising rapidly and production increases can be achieved only gradually. For low-income exporters, therefore, the best prospects for long-term increases in trade are likely to be found in planned exchanges among themselves. For rice the major part of world production, consumption and trade already takes place in developing regions. The general picture should be modified for citrus, demand for which in most high income, non-producing countries may remain buoyant where not limited by policy measures.

Agricultural Raw Materials. This group of commodities constitutes an important source of export income for developing countries. Unfortunately, they face increasing competition from synthetics and other substitutes, as well as replacement by new techniques in a large number of end-uses. The rising competition from synthetics will tend increasingly to set limits to prices, and indeed will probably further depress them, as well as restrict the markets available to the natural products. The level of *per caput* utilization of these products in many of the main markets is already high. Moreover, the rise in output of certain of these products in developed and centrally planned producing countries will probably continue. Thus there appears to be little prospect of a dynamic increase in export earnings from most of these commodities.

Exceptions may, however, be made in the important cases of cotton and forestry products. For cotton, demand in developing regions is likely to grow strongly, but it is by no means certain that this demand will be supplied from other developing countries, unless special trade arrangements are made.

Unlike foodstuffs and most other agricultural products, demand for many wood products continues to grow vigorously in the developed countries, with income elasticities in Europe well above unity for paper and paper-board and between 3 and 5 for wood-based panel products. Here is an important potential export market for developing countries, to the extent of possibly an extra one billion dollars annually by 1975. However, the potential of these industries for import saving is even greater, as the demand for processed wood products is growing faster in developing countries, entailing significant imports. To enable this export-earning and import-saving potential to be achieved, developing countries will have to develop wood-processing industries, and tariff structures in developed countries which discriminate in favour of raw material imports, may have to be revised.

During the 1950's the world fish catch increased at a faster rate (5·6 per cent annually) than agricultural production, developing countries and centrally planned countries contributing most of the increase. World trade in fish and fish products grew even faster, partly through greater movement of frozen fish between developed countries, but mainly through the growth of the fishmeal industry in Peru and Chile. The fishmeal trade is very narrowly based, and it is doubtful whether it will continue to expand as rapidly as in the past.

V. PRODUCTION PROBLEMS AND POLICIES

This picture of agriculture in the world of 1975 is based on the assumptions that this sector will grow at the average rate of 2 per cent per annum in the developed countries and about 4 per cent in the developing countries, with an intermediate assumption for the group of centrally planned economies. During the period 1958–63 gross agricultural product expanded faster in the developing than in the developed countries — at 2·5 per cent in the former group and 2·2 per cent in the latter. The projections thus assume a widening of this differential — a slight reduction in the agricultural growth rate in the developed zone and a marked acceleration of the rates in developing countries. The broad implications of these assumptions must now be examined.

There is no doubt that the technological revolution of the past twenty years will continue in the developed countries, whose potential rate of agricultural growth is well above 2 per cent annually. Policy

in the more industrialized importing countries and the grain producing countries of this group will increasingly be directed towards
limiting the increase in agricultural production to the available
commercial outlets and some assessment, which may or may not be
organized internationally, of the prospective requirements of developing countries for commodity aid. If domestic agriculture in developing countries can be improved fast enough to meet the effective
demand for food in those regions, as largely assumed in Table I, these
developed countries will need to be holding back their agricultural
potential strongly by 1975.

The main problem of developed agriculture, if held down to a
growth rate of something like 2 per cent, will be how to ensure
satisfactory income levels for farm producers. Measures will be
intensified for moving unneeded resources of people and in some
cases land out of agriculture. The technological revolution will be
accompanied by changes in the structure of agriculture, in farm
organization and in management. The trend towards larger farms,
greater technical efficiency and capital-intensive methods of production, will continue. The process of vertical integration will
become more widespread. Family farmers will be increasingly
threatened by highly capitalized 'no-land' livestock raising, and will
develop new or strengthened forms of economic association in order
to compete.

The year 1975 could show a great improvement in policies and
measures for influencing agricultural production. On the other
hand, in the absence of international agreements, the emergence of
widespread surpluses of temperate zone grains could threaten the
stability of developed agriculture and heighten the interest in disposing of agricultural surpluses in the most useful ways. The
developed primary exporting countries may be able to avoid the
limitations or support of their production during this period by
negotiating access to or a share in the increase of demand in developed
importing countries for meat and dairy products, and by increasing
their sales to newly expanding markets in Japan and some developing
countries.

In the developing world as a whole, a 4 per cent agricultural growth
rate will only be feasible if the commodity pattern of production
undergoes a marked shift towards food products for domestic or
regional markets. Over the last ten years in developing regions
(except the Far East) the output of non-food products mainly for
export has increased faster than that of food products. It is evident

from the state of the international markets, which are predominantly in developed countries, that this situation cannot continue. Coffee, cocoa, sugar and cotton are in world surplus, tea is largely a static market and rubber and fibres are suffering severely from competition by synthetics and other substitutes.

The constraints of the world market will thus lead to greater concentration in developing countries on policies and measures for stimulating the production of basic foods and livestock products for the market. For these the demand is internal or regional, and is currently outstripping supply, especially in the case of basic foods. An agricultural growth rate of 4 per cent is by no means unreasonable. A number of developing countries have equalled or exceeded this figure over relatively long periods, *e.g.* Mexico, Thailand. Nevertheless, there can be no illusions about the new policies and special efforts needed if this accelerated rate is to be achieved over the next ten years for the developing countries as a group.

There is a hopeful indication that governments are increasingly recognizing the significance of the institutional impediments to agricultural progress. In addition to physical investments, the new policies are likely to place greater emphasis on improving the agrarian and marketing structures, establishing rural networks of communications and economic institutions, and on the stabilization of prices to producers of food crops. These measures are well known but not easy to implement. They take time. Yet without progress in these directions other measures for the supply and widespread distribution of physical inputs such as improved seeds, pesticides and fertilizers, can hardly be expected to achieve great results. The problem of food production is so urgent that unconventional methods may have to be tried to supplement the orthodox. These would include the concentration initially of needed inputs, advice and institutional services in favourable zones, rather than spreading total resources over the whole country so thinly that no break-through can be achieved. The 'package programmes' of India and Pakistan are akin to this. The direct intervention of the state in large-scale food production through state farms is being tried in a number of countries. Other governments may be willing to apply the plantation approach, licensing private corporations with access to capital and technical knowledge to lease land where it is available and produce basic foods or the market, adding provisions for the training of local personnel, production of improved planting material or livestock for distribution to local farmers or whatever service is appropriate. There may also be scope in some areas for the vertical integration of marketing and production. 'Islands' of special effort, whether package programmes, state farms or food plantations should, of course, be fully integrated in national plans for setting the whole agricultural sector alight, with the torch of production for the market at stable prices.

Employment policy will also play a larger part in developing countries in the selection of agricultural development projects and measures, and of methods of construction and implementation. A minimum amount of foreign exchange, technical assistance and domestic capital will be a critical need. While economizing in this direction, governments will give greater weight to the selection of labour-intensive projects and methods of work. By 1975 experimentation with methods of organization, inducement and training may have led to the adoption of national policies of utilizing rural manpower to the optimum extent in agricultural and rural development programmes.

VI. DEVELOPMENT ISSUES

Since changes in population trends could have only limited effects by 1975, a key development issue is whether or not an agricultural growth rate of at least 4 per cent annually can be achieved soon in the developing regions. If this goal can be reached, the base will be laid for satisfactory overall economic growth rates in these regions, based on a relatively progressive and prospering agriculture. The developing world would thus be on the way to self-sustaining growth. With a rising trend in the mass purchasing power of the rural populations, the industrialization and transformation of the developing economies would be stimulated and supported. Dependent on their steady growth, their purchases from developed countries and centrally planned economies would increase in volume and complexity, thus contributing to the prosperity of the latter and facilitating the adjustment of the more industrialized economies to the requirements of economic advancement in the world at large.

Responsibility rests mainly with the developing countries themselves, but the developed countries also have an essential part to play. Assistance from developed countries, in the form of technical assistance, capital aid or trade, is virtually a prerequisite for a satisfactory rate of progress in the developing countries. This is accepted by developed countries. Their net outflow of long-term funds to developing countries and multilateral agencies increased from 0·6 per cent of their combined gross domestic product in 1956–59 to 0·7 per cent in 1960–61, according to United Nations sources. However, this flow has since been barely maintained.

The need for assistance in trade matters has been much more widely recognized since the 1964 UN Conference on Trade and Development. Here new guidelines for international trade behaviour linked with the development needs of the developing countries were propounded. Thus the developing countries seek better conditions of access for their traditional exports in the markets of the developed countries, and — much more important — for their new exports of processed and manufactured products, to which they look for escape from the demand limitations of the market for raw materials. This means, in effect, a shift of increases in certain simpler types of manufacturing capacity from developed and some centrally planned economies to developing countries, a shift which has already started in certain cases. In the GATT a new Chapter on Trade and Development has been drawn up, and under the Kennedy Round of tariff negotiations ways are being sought to give trade advantages to the developing countries which could enable them to accelerate and diversify their export earnings and economic growth. Since many of the changes sought, and in some cases agreed in principle, touch vital interests of the developed countries, progress towards their implementation in the world economy is likely to be slow and regulated.

In addition to trade and processing policies, there will be increasing pressure during the next ten years for related production adjustment policies to be developed for a number of agricultural commodities. In the case of commodities which can be produced in temperate or sub-tropical as well as tropical zones, agriculture in the developed and centrally planned economies is often in competition with production in developing countries, as in the case of sugar, tobacco, cotton and oilseeds. Tobacco is a labour-intensive crop and in this respect is better adapted to the conditions of agriculture in developing than in developed countries. Sugar can be produced from cane in the tropics at lower cost than beet sugar in the temperate zones. In these and other cases, structural and other obstacles render the expansion of output a slower process in developing than in developed countries, so that the latter tend to capture a large share of the increases in demand. The implications of agricultural policies in the developed and centrally planned economies for the development of export agriculture in low-income countries will increasingly be held up by governments for international scrutiny. Livestock products can be produced at lower cost in the primary exporting countries of the southern hemisphere than in the importing developed countries. This important conflict arises mainly among more developed countries, who have initiated some moves to resolve it by agreed understandings on the sharing of large import markets, including market growth in some cases, among traditional exporters. A similar approach may be pressed in future consultations leading to wider agricultural production adjustments affecting products exported by developing countries. Such adjustments, if agreed, will take place gradually, because of the fixed investment in current production patterns, the farm income problem in most developed countries and the need to reconcile the interests of the various parties.

Pressures will also rise, for similar reasons, for production adjustments affecting tropical products, especially coffee, cocoa and tea. These will be the subject of discussions and negotiations among developing countries. New producers may well argue that traditional exporters, having benefited from the market for many years, are now better fitted to make progress towards the industrial transformation of their economies. They may urge that the limited annual increases in world demand be shared progressively with economies at an earlier stage of development which can fairly easily adapt themselves to export production from tree crops but not yet to industrialization.

It will be hard to reconcile these interests. If all continents and the world as a whole are enjoying economic progress, co-operative arrangements and mutual concessions will be more quickly arrived at. Development, especially of the developing regions, will increasingly be the context of international consultations on these trade and production problems. These consultations may lead to additional and more effective international agreements on agricultural commodities, which may include provision on national production policies and surplus disposal, as well as questions of prices, quotas and stocks. Arrangements for economic co-operation among developing countries themselves may be expected to gain in strength as they increasingly collaborate in commodity or regional groupings to harmonize production and processing plans in order to make the most of trade with each other and with the industrially advanced world.

New decisions will have to be taken in the next few years, at both national and international levels, if agriculture in the world of 1975 is to look very different from what it does today. The pressure of world population on food supplies will by then have increased by one third as against 1960. If the development effort in the agriculture of the low-income countries fails to improve significantly on past rates of increase in marketed supplies of food, the dark colours of today's picture will remain dark. These are some of the issues which the FAO Freedom from Hunger Campaign has done much to bring to the attention of public opinion throughout the world.

DISCUSSION OF DR. OJALA'S PAPER

Professor Lundberg said that Dr. Ojala painted a picture of increasing population pressure, of improvements in the standard of living in developing countries and of the creation of problems for individual governments because of serious dis-equilibrating tendencies. Thus Dr. Ojala posed problems which will recur frequently throughout the Conference, but it is important to have a preliminary presentation of the different tendencies in world agriculture. We are necessarily talking in abstractions, but, at this stage, this approach is useful.

Dr. Ojala stated his working assumption that the situation in 1975 will be the result of factors and trends that can be identified at present. Professor Lundberg said there are three categories of trends which Dr. Ojala required us to keep in mind. First the trends which can be extrapolated from past and present situations ; for example, population. Second, there are trends based on economic assumptions ; for example, the rate of growth of GNP. Third, trends which refer to targets in the future ; for example, the acceleration of growth of agricultural production in new countries. The trends of types two and three are susceptible to policy measures.

Professor Lundberg said that in the case of population trends, Dr. Ojala gave a number of projections which seem quite plausible and which imply severe problems of population pressure. With regard to these population trends, Professor Lundberg said that we have had sad experience of the forecasting of demographic trends, and have been surprised by accelerations of population growth. Mr. Ojala reached the conclusion that in new countries there will be a significant improvement in diets. Forewarned by past experience one has to ask whether this will not have a powerful effect on population rates of growth. Professor Lundberg said another question arose from the fact that Mr. Ojala talked of an optimum increase in industrialization. What does this optimum mean ? Why should not the rate of industrialization be greater than he projected ? A higher rate of industrialization implies a higher rate of urbanization. Is this the major obstacle, and, if not, what are the obstacles to a quicker rate ?

Turning to the second category of trends, Professor Lundberg said that population growth is closely related to trends in production, consumption and trade. The assumption of a 4·1 per cent growth rate in agricultural production in poor countries is the crux of the whole problem. It means an acceleration from 2·5 per cent to 4·1 per cent, which may be thought of as a large increase in relation to the past, or as a small one in relation to the potential. But this growth rate does imply a change in the distribution of agriculture in the world, in that the share of agriculture of new countries in world agricultural production will rise from 30 per cent to 35 per cent, while that of advanced countries' agriculture will fall. All this suggests an important shift in the distribution of labour and trade between the developing countries and the rest of the world. At the same time, according to Dr. Ojala, the income gap will continue to increase and the *per capita* income in industrial countries will go up by 70 per cent while in new countries it will only rise by 50 per cent.

Professor Lundberg said it is necessary to look at the inter-relationship between these trends and those in other parts of the economy. Dr. Ojala had emphasized the importance of an increase in agricultural production in new countries in order to reach a stage of self-sustained growth. Agriculture is seen as a bottle-neck, and is taken as strategic to development. However, an increase of 4·1 per cent in supply will currently be overtaken by a corresponding increase of effective demand in these countries. What will this mean for other sectors ? Hunger and destitution in these countries will remain. Is there not the serious risk that when urbanization occurs on the scale envisaged, the income elasticity of demand for food products will be at least one ? At the same time, the paradox remains that industrialized countries will be struggling with surplus policies to keep down production.

How will the production acceleration in the new countries be brought about, and what are the basic essentials ? On this topic there are a host of questions. Dr. Ojala enumerated a programme for infrastructural development, but what are the real incentives on individual farms in a world with surpluses ? If there are difficulties in accelerating growth of agricultural production in these countries, may there not be alternative ways of achieving the goal ? To what period does the 4·1 per cent refer ? When will it be reached ? What difficulties are envisaged in trying to reach it ? And the main question : cannot the inelasticities of food supplies in new countries be changed by better international division of labour ?

Professor Lundberg said that when Dr. Ojala discussed, with some pessimism, the gains and prospects of agricultural trade for the developing countries, the analysis is complicated by the fact that we do not know what the real costs and prices are. In all countries there are double price systems facing a set of world price levels which are themselves of doubtful significance.

On the specific point about forest products, Professor Lundberg said that he was not so optimistic as Dr. Ojala. European and United States' potential for low-cost expansion in these products is impressive, so that the development of big export markets in Europe do not seem likely.

Returning to the general question of the international division of labour, Professor Lundberg asked whether the problem of inelasticity of supply of food in the new countries may not be improved by more intensive world trade. There is the paradox that industrialized countries are holding back production in fields where they seem to have the comparative advantage. Why should they not simply increase production and supply new countries with these agricultural products ? This perhaps is the naivety of a general economist, but it is not a greater naivety than that which suggests that new countries can find markets in industrial products. Dr. Ojala seems over-pessimistic about the attitude of industrialized countries in this respect.

Professor Lundberg said that he would stress, as a general economist, that the *diversification of exports* is important for new countries, in leading towards stability. He asked whether it is not more unrealistic to believe in quick acceleration of agricultural production in new countries, than to hope for a change of attitude in industrial countries so that they become prepared to accept more manufactured imports, in exchange for openings for their agricultural products in the markets of developing countries.

Finally, Professor Lundberg said that he would prefer to change the terminal date that the study referred to from 1975 to 1980.

Professor Nussbaumer noted that Dr. Ojala had said that the 5·2 per cent growth rate for developing countries is a target figure, and he asked how realistic are the other growth rates given by Dr. Ojala. Certainly a growth rate of 5 per cent for developed countries is high in view of recent experience in many European countries. A growth rate of 4·5 per cent seems to be more realistic.

Professor Nussbaumer noted the importance of the part played by agricultural production in national income and in the formation of growth rates. He asked whether the figures given by Dr. Ojala are on the basis of current prices or on the basis of prices of some past period. Does the projected low share of agriculture in the national income of developed countries in 1975, estimated in Dr. Ojala's paper at 4·3 per cent, indicate that figures have been calculated on the basis of past prices ? Otherwise agricultural prices rising faster than industrial prices in free markets suggest that a declining physical share of agricultural product in gross domestic product might be offset, to a certain extent, by a more favourable development in prices.

Professor Nussbaumer agreed with Dr. Ojala concerning the good prospects for beef production. We may expect a higher turnover of capital in these products, a change towards a higher quality of meat and in many countries a more labour-intensive process of production relative to other agricultural products ; this could be advantageous to countries with large

22

labour reserves. This, however, left unanswered the question of countries with a high level of employment and high wages in industry.

Professor Nussbaumer said that he did not agree with Dr. Ojala's optimistic view of the prospects in wood production. In wood production it is important to distinguish between the different qualities of wood, and to note the heavy costs of transportation and labour.

Professor Delivanis stressed how dangerous it is to rely on averages, even on a world scale. Professor Delivanis questioned Dr. Ojala's optimistic view that institutional obstacles are now being better understood by governments keen to increase agricultural output. Governments simply permit price increases of agricultural products while they ought instead to be inducing farmers to reduce costs and to improve quality, processing and presentation. Experience shows that demand frequently increases after such improvements, even when the market is considered to be already saturated with the goods concerned.

Dr. Stipetic asked whether the FAO planning group is simply extrapolating the rates of agricultural production growth from past trends or demonstrating wishful thinking about world agriculture. On what arguments is this wishful thinking based ? The projected rates of growth for developing countries given by Dr. Ojala are much higher than in the past. If this is based on expected technical improvements, we must consider what obstacles there are to the diffusion of technological change. Technological changes come, in large measure, from temperate zones and they cannot be automatically transplanted to the tropical zone. Obviously, education and scientific research are the basis of increased progress in agriculture in developing countries, but education is not discussed in the paper. Dr. Stipetic contrasted the case of the kibutzim in Israel, where the level of general education is very high, so that new technological advances can be easily read about and digested, with the ordinary case in the developing world, where the farm workers are illiterate peasants. The rates of growth of agricultural production differ, and must differ, considerably in two such different environments.

Professor Campbell said that he wished to ask about an apparent inconsistency in Dr. Ojala's paper. In one place (page 15) he had written : 'Family farmers will be increasingly threatened by a highly capitalised "no land" livestock raising and will develop new or strengthened forms of economic association in order to compete.' This seems inconsistent with the passage (page 19) : 'Livestock products can be produced at lower cost in the primary exporting countries of the southern hemisphere than in the importing developed countries.' Professor Campbell said that these countries of the southern hemisphere raise livestock under range conditions and he thought that they would continue to have a substantial comparative advantage in so doing. Can Dr. Ojala explain this inconsistency ?

Mr. Allen said that he was also interested in Professor Campbell's question and that he took the opposite point of view to that of Professors

Nussbaumer and Campbell. 'No land' methods of raising cattle represent a serious threat to the grass ranching systems of Australia and Latin America. In Europe we are used to producing meat from grass lands but this is an outdated technique. He disagreed with Professor Nussbaumer that beef production is likely to remain labour-intensive with its elasticity of supply consequently being exceptionally low in comparison with other farm products. The North American feedlot is a much more economic method of converting grain to meat and it is a highly capital-intensive form of agriculture. The most economic feedlot can take at least 5,000 head at a time with an annual turnover of more than 12,000 head of cattle. In addition, we may be on the threshold of important new advances and of higher meat/grain conversion rates in feeds. Mr. Allen expected to see the grain feedlot method growing in Europe. Already there is 'barley beef' in England as a consequence of the rapidly increasing grain yield. Thus, he would expect the elasticity of supply of beef to be much higher than is implied by the FAO projections mentioned in Dr. Ojala's paper. It should be added that the recent shortage of beef was due to a simultaneous conjunction of temporary factors, including heavy drought in Argentina, a very cold winter in Europe and a down-turn of the cattle cycle in the United States of America.

Professor Obolenski said that he considered Dr. Ojala's paper thorough and interesting but, at the same time, he thought that some of the conclusions are dubious. Professor Obolenski thought that the statements on the rate of development of agriculture in socialist countries are unduly pessimistic.

Although he had not at his disposal all the relevant figures on agriculture in socialist countries, Professor Obolenski said that, in the Soviet Union, measures undertaken for the development of agriculture give every reason to expect much higher rates of growth in development. For example, there is the plan for capital investments in 1966–70 which provides for investment in agriculture of 71 billion roubles. He said that the production of technical equipment for agriculture is to be doubled, that the annual production of artificial fertilizers will be more than doubled and that measures have been either planned or adopted to increase the role of material incentives.

Professor Obolenski said he believed that in evaluating the probable future rates of development of agriculture it is necessary to take into consideration, not only the trends of the past few years, but also efforts and measures which are being made in certain countries for strengthening the material and technical basis of agriculture.

Dr. Ojala in reply said that the growth rates used certainly contain an element of 'target'. While it is true that overall growth rates in developed countries have been around 4 to 4·5 per cent in the recent past it seemed useful to add an aspirational element in looking ahead. It would be quite possible to recalculate the tables with different assumptions about growth rates. Indeed, by selecting some alternative growth rates for agricultural and non-agricultural sectors in the different zones, it should be possible to throw more light on the implications for world agriculture.

Regarding questions raised as to the target growth rate assumed for agricultural output in developing countries, Dr. Ojala said that he hoped that the economists present from developing countries would give their views. The 4 per cent growth rate in agricultural production that he had assumed is not without precedent; several developing countries have already achieved it. The governments of developing countries have shown evidence of wanting to break from past trends and of being prepared to take some of the necessary policy measures including agrarian reform and marketing improvements. Dr. Ojala agreed that incentives to the farmer are a key element. Here also there is evidence that governments of developing countries are groping towards the right policies; for example, price stabilization policies for producers, though very difficult to implement in a developing country, are increasingly being tried.

Dr. Ojala agreed that technological advance and education are obviously important, but he had not intended that his paper should be exhaustive. He argued that the technological advance already achieved in developed countries means that the developing countries do not have to repeat all the basic research even though they have to adapt the results to their own conditions. On the question of livestock raising, Dr. Ojala said that he agreed with Mr. Allen. Capital-intensive methods are becoming established and while they still require land for the growing of grain feeds, they can pose serious competition for low cost grass-land livestock production in the southern hemisphere countries, especially as the latter have to face access problems in the markets of many importing countries.

In reply to the question by Dr. Stipetic, Dr. Ojala said that the FAO Indicative World Plan for Agricultural Development would not be based purely on extrapolations. It would be an attempt, on the basis of detailed studies of the past and the present, to indicate the changes which would be needed in developing countries to achieve growth rates in agriculture consistent with the overall economic growth rates assumed.

Referring to Professor Lundberg's suggestion that the developing countries should industrialize so as to buy food from the advanced countries by exporting manufactures, Dr. Ojala said that this solution is not satisfactory for developing countries in general. In most of them, 60–70 per cent of the workers are engaged in agriculture. If this sector is neglected the incomes of these workers will remain at depressed levels, and then agriculture could not provide the increase in mass purchasing power and savings that is needed to stimulate and finance the transformation of the economy.

Replying to Professor Nussbaumer's question, Dr. Ojala confirmed that the projections had been based on constant 1960 prices. Thus, any increases in real agricultural prices in the developed countries would raise the projected share of agriculture in the national income.

Chapter 2

THE DESIRABLE LEVEL OF AGRICULTURE IN THE ADVANCED INDUSTRIAL ECONOMIES

BY

E. A. G. ROBINSON

Cambridge University, England

I. INTRODUCTORY

THE desirable level of agriculture in the advanced industrial economies is not a simple concept which can be established independently
of the general environment of the particular economy. Some of the
advanced economies, such as Canada or the United States, even if
now considerably industrialized, have a predominantly agricultural
background of history and rich agricultural resources. And the
plains of Western Europe, even if now one of the world's greatest
centres of industry, led in the development of the modern economy
largely because they represented one of the richest agricultural areas
of the temperate zone.

If one is to ask what is the desirable level of agriculture in such
countries, one must in effect ask where their comparative advantages
now lie, and are likely in future to lie, as between meeting their own
consumption needs, including those for food, directly, and using
their resources indirectly to buy their food needs by the export and
sale of manufactures.

At the turn of the century anyone writing a study of this theme
would have emphasized on the one hand the great gains from international specialization and exchange, and on the other hand the problems of diminishing returns in agriculture and of rising real marginal
costs as more resources were devoted to the land and attempts were
made to increase yields per acre. He would have been confirmed in
this by the cheapening of the real costs of international transport with
the steamship and refrigeration, by the low costs of new world
agriculture, by the wide discrepancy between net product per worker
in agriculture and in industry, by the wide differences between

26

productivity in industry in the advanced and in the backward countries.

A writer approaching the same theme today feels very much more cautious regarding the trends. Fertilizers, pesticides, better seeds have increased the yields of crops ; improvements of breeds have contributed similar gains to animal husbandry and dairy-farming. The mechanization of agriculture has greatly increased the horse-power available per worker and his product ; at the same time it has released land, the scarce factor, from providing for the supply of horse-power in the most literal sense, to providing food either directly for human consumption or for feeding animals for human consumption.

Thus the point of serious diminishing return in agriculture has been modified. At the same time, the gap between industrial productivity in an advanced and a less advanced country has been narrowing. The less advanced countries can now acquire and absorb the expert knowledge and experience which at the beginning of the century was the monopoly of advanced countries. Improvements of technique move much more rapidly from country to country. It is almost certainly true to say that the international differences of comparative advantage which create the gains from international trade are themselves narrowing. The less developed countries are industrializing, moreover, not always because in the short term they believe this to bring economic gain but because they believe that in the longer term industrialization is necessary to their progress.

For these reasons a picture of the world in which the advanced countries are imagined to be the specialist manufacturers and the less advanced countries to be the specialist providers of primary products would have very doubtful economic validity even if it were politically acceptable. And it is abundantly clear that it is politically wholly unacceptable. The less developed economies will in practice almost certainly encourage industrial development. And the advanced countries, with limited export earnings, will find economic forces pushing them in practice towards the maintenance of agricultural production.

The purpose of this paper is to examine some of the trends of food consumption and of agriculture in the more advanced countries and to see their mutual effects.

II. THE DIFFICULTIES OF FRAMING AGRICULTURAL POLICIES

The making of agricultural policy is beyond question the most difficult of all tasks that confront a government. In any not very highly developed economy, agriculture is normally the largest industry. Food consumption represents, even in a rich country, something like a quarter of all expenditure, and the element in this of crude foodstuffs represents about one-half. Thus in a richer country, the amount of crude foodstuffs consumed (and, if net imports are small, produced also) may represent some 7–10 per cent of gross national product; in one of the poorer countries in Europe the proportion is likely to be nearer to 20 per cent.

The problems are made very much greater by the fact that in all countries the income elasticity of demand for food is relatively low. Thus, as a country increases its productivity and its income per head, the demand for food products tends to grow less rapidly. This implies, if productivity in agriculture is growing almost as rapidly as in other occupations, that the percentage of population engaged in agriculture must decline and that prices and incomes need to be such as to stimulate contraction of agricultural employment. Equally it implies that agricultural structure must continuously be adapted to a declining manpower; that sizes of farms need to be adjusted to changing circumstances; that production methods need to be adjusted to higher real wages. But agriculture is one of the most difficult of industries in which to make adjustments; in industry large firms can grow up without grave impediment; in agriculture the land itself is an essential of production, its ownership is fragmented and widely distributed and the consolidation of small peasant holdings, or the small farms created at the time of the enclosures of land in the seventeenth and eigthteenth centuries in the United Kingdom, into farms large enough to give a reasonable income to a family in the conditions of today may take many years and present great difficulties. (Agriculture is, moreover, essentially an industry of small concerns, with numbers of farms and decision-making units running in most countries into several hundreds of thousands, and with immensely powerful political influence in consequence.) Thus agriculture is at once inevitably subject to almost continuous economic pressure to adjust and possessed of powerful influence to resist adjustment.

III. RECENT TRENDS IN EUROPEAN AGRICULTURE ILLUSTRATED FROM THE UNITED KINGDOM

Pressure on agriculture has come during this century from four major technical revolutions : from the improvements of ocean transport, including refrigeration, which already before the end of the century were changing the character of competition in the European markets for food products ; the application of the internal combustion engine to farming and the more general mechanization of farming which began about the time of the first world war but made most rapid progress in the 1940's and 1950's ; the widespread use of chemical fertilizers which dated from about the same period ; the improvements of seeds and pesticides, which have further increased yields of crops, and the improvements of breeds of cattle, which have increased the yields of animal and dairy husbandry.

All of these collectively have resulted in increases of productivity in agriculture which have been as great or greater than the remarkable increases in industrial output during the same period. If I may take the United Kingdom as a single example of a trend that has been common to all European countries, the changes in agriculture and its productivity are summarised in Tables I and II.

It will be seen in Table I that, over the first sixty-two years of the twentieth century, the output of British agriculture had increased two and a half times, when defined to exclude any double-counting and to exclude imported inputs into agriculture, but not so as to eliminate the inputs of other British industries, such as those making tractors, machinery or fertilizers ; the concept used here represents in a sense a gross output, free of duplication and of imports. Since the area covered by agriculture has appreciably diminished over this period, as a result of the spread of towns and the devotion of land to other purposes, the quantum of output per acre has increased very nearly three times.

This increase in quantum of output per acre has resulted partly from increased yields per acre in the more normal sense, partly from a shift of demand and output towards agricultural activities (like market gardening or egg production) which have a higher output per acre used. But in the period since 1936–38, in which most of the increase has been concentrated, there has in fact been a very considerable increase in yield per acre in the more normal sense, which can be seen in the relation of the production of corn-crops to their acreage.

In Table II are shown the implications of this in terms of employ-

TABLE I

LONG-TERM TRENDS IN AGRICULTURE IN THE UNITED KINGDOM *

(1900–02 = 100)

Years †	Output of ‡ Agriculture	Total area of Agricultural Land	Area of Arable Land	Output per Acre of all Agricultural Land	Total Output of Corn Crops	Acreage under Corn Crops	Yield per Acre under Corn
1900–02	100	100	100	100	100	100	100
1911–13	106	99	101	107	92	98	94
1922–24	114	95	100	120	86	101	85
1936–38	132	90	83	147	59	72	82
1947–49	172	88	118	195	94	111	85
1952–54	200	88	114	227	133	109	122
1956–58	213	88	111	242	158	102	155
1962–64	256	87	118	294	248	111	223

* All figures in this Table have been derived from *The British Economy: Key Statistics 1900–1964,* published for The London and Cambridge Economic Service by Times Newspapers Ltd. They have been adjusted where necessary to cover the United Kingdom as now defined by the author with the assistance of the statisticians who compiled that volume.

† Averages of three-year periods.

‡ Output free of duplication and of all imports of seeds, livestock, etc., but *not* excluding British inputs outside the agricultural industry (*e.g.* of machinery, fertilizers, etc.); measured at constant prices of 1954–57 and of earlier years, linked for this purpose.

30

ment and productivity in agriculture and other industries. From the turn of the century to 1922–24 there was a relatively modest increase of net output per worker in agriculture ; indeed a substantial increase of population attached to agriculture in the first decade of the century was not balanced by a comparable increase of output and it was only when labour moved out of agriculture in the 1914–18 war and the 1920's that output per head was restored to the level of 1900. The first beginnings of an increase of productivity in agriculture came in the depressed years of the 1930's when mechanization was more rapid and the numbers occupied in agriculture fell heavily. Since 1936–38 the numbers have continued to fall — much more rapidly during the past ten years. By 1962–64, just about two-thirds as many workers were engaged in agriculture as at the turn of the century, representing about 3·8 per cent of the working population as against 8·9 per cent at the earlier date.

Productivity per worker in agriculture, measured in output per head at constant prices after meeting the costs of inputs from outside agriculture, has increased nearly three and three-quarter times over the whole period since 1900–02. In the period since 1936–38 it has considerably more than doubled. Over the whole period productivity per head in agriculture has risen by four thirds of the increase of productivity per head in industry. Over the period since 1936–38 output per worker in agriculture has risen one and a half times as fast as output per worker in industry. It does not, of course, follow that net output per worker in agriculture is, even now, higher than in industry ; the answer depends on the relative valuations attached to agricultural and industrial output. It has been estimated that in 1938, at the relative values of output in 1938, the ratio of productivity in agriculture to productivity in industry was about 62 per cent.[1] On that basis, productivity in agriculture today would be about 94 per cent of that in industry ; at the relative values of 1963, net output per head in agriculture is about 98 per cent of that in industry ; in the ten years since 1952–54, however, productivity in agriculture has been rising 3·9 per cent per year while that in industry has been rising 2·7 per cent per year.

What I have been anxious to stress in terms of the example of the United Kingdom is that agriculture in many of the more advanced countries of Europe is efficient and dynamic and making rather rapid

[1] See estimates in *Economic Survey of Europe for 1948*, Table 121, p. 225. The figure is very close to the figure that can be derived from the estimates in international units made by Mr. Colin Clark for the same date.

TABLE II

LONG-TERM TRENDS OF PRODUCTIVITY IN AGRICULTURE AND INDUSTRY IN THE UNITED KINGDOM *

(1900–02 = 100)

Years	Numbers engaged in Agriculture	Net Output† of Agriculture	Output per Worker in Agriculture	Output per Man-year worked in Industry‡	Output of Gross Domestic Product per Head of Working Population‡	Percentage of Total Working Population engaged in Agriculture
1900–02	100	100	100	100	100	8·9
1911–13	109	106	97	105	98	8·5
1922–24	98	113	115	126	108	7·5
1936–38	83	133	160	178	126	5·4
1947–49	86	195	227	185	126	5·4
1952–54	81	205	253	209	139	4·9
1956–58	76	204	268	225	149	4·5
1962–64	67	248	370	272	170	3·8
1962–64 (as % of 1936–38)	81	186	231	153	135	70

* For sources, see Table I, note *. All figures adjusted to United Kingdom as now defined.
† Net output, free of duplication and of imports, and excluding inputs from British industries other than agriculture.
‡ The numbers occupied in GDP are total numbers not excluding those unemployed; the numbers working in industry are those actually employed.

technical progress. This is, I realize, equally true of agriculture in North America and in other parts of the world. The immediate problem is the relation of the markets for food and agricultural products generally to the expanding capacity of the agricultural industries.

IV. THE DEMAND FOR FOOD

If, once again, I may illustrate the broad problems from the example of the United Kingdom, I give in Table III the relevant data relating to food expenditure and the sources from which it has been met.

It will be seen that, over the whole period since 1900–02, consumers' expenditure of all kinds, measured at constant prices, has just about doubled ; food expenditure at constant market prices has increased by almost exactly the same amount. The latter includes not only expenditure on food proper, but also on distribution and processing. The total supply of crude food, deriving partly from domestic agriculture and partly from imports, would appear to have increased very slightly more. This is probably to be explained partly by a relative decline in consumption of alcoholic drinks (which contain a considerable element of processing and distribution cost) but principally by redistribution of income favourable to those who in the early years were under-nourished.

The rate of growth of total supply of crude food has averaged about 1·2 per cent compound over the whole period ; it was about 1 per cent down to 1936–38 and about 1·4 per cent over the whole period thereafter ; but since 1952–54 the rate of growth has risen to about 2·6 per cent per year. Some of this, of course, represents increase of quality of consumption rather than of actual physical volume.

The increase has been partly the consequence of increasing population, which over the whole period has been about 0·6 per cent per annum ; partly it has represented an increase, over the whole period, of about 0·6 per cent per year in the quantum of crude foods per head. The latter rate of growth would appear to have risen in the last ten years to 1·9 per cent per year.

It seems likely that this rate of growth will slow down in future. Some of the growth of crude food consumption in the past ten years has represented the effects of the relaxation of controls imposed

TABLE III

LONG-TERM TRENDS OF FOOD EXPENDITURE AND THE SOURCES FROM WHICH IT HAS BEEN MET IN THE UNITED KINGDOM *

Years	Population	Consumers' Expenditure at constant Prices	Consumers' Expenditure on Food at constant Prices	Output of Agriculture †	Imports of Food and Feeding Stuffs at constant Prices	Total Supply of Food ‡	Consumers' Expenditure per Head	Consumers' Expenditure on Food per Head	Supply of Raw Food per Head
1900–02	100	100	100	100	100	100	100	100	100
1911–13	111	106	111	106	113	110	96	100	99
1922–24	117	108	123	114	132	123	92	105	105
1936–38	125	140	152	132	160	146	112	122	117
1947–49	131	142	160	172	124	148	108	122	113
1952–54	133	151	174	200	132	166	114	131	125
1956–58	136	169	189	213	154	184	124	135	135
1962–64	142	203	208	256	168	212	143	146	151
Annual Growth :									
1936–38 to 1962–64 (%)	0·5	1·4	1·2	2·6	0·2	1·4	0·9	0·7	1·0
1952–54 to 1962–64 (%)	0·7	3·0	1·8	2·5	1·2	2·6	2·3	1·1	1·9

* For sources, see Table I, note *. All figures adjusted to United Kingdom as now defined.
† Output as defined in Table I.
‡ Weighted equally in terms of 1900–02.

during the period of the Korean war and of limitations on imports of preferred foodstuffs. Part of the growth almost certainly represents a rate that one may expect when the distribution of income is changing but will slow down if the pattern of distribution is stabilised.

It will be seen that, in the United Kingdom, over the whole period and particularly since 1936–38, the annual growth of agricultural output (2·6 per cent) has considerably exceeded the rate of growth of consumption of crude foods (1·4 per cent). This has been possible only because, over both periods, the share of imports in total supply has declined.

It is not easy to calculate exactly the relative shares of imports and of domestic production in total supplies. As nearly as can be estimated they were equal in 1900. In 1911 they were estimated[1] to be almost equal in terms of those products both produced abroad and imported. If allowance is made for tea, coffee, cocoa and other products the figure of 52 per cent seems appropriate. On this basis, home agriculture probably produced about 45 per cent of total supplies in 1936–38, when little or no allowance is made for differences of quality. In the post-war period the share of home agriculture in total supply has increased greatly, and is now approximately 60 per cent. Of the increase in supply since 1936–38 well over 90 per cent

TABLE IV

LONG-TERM TRENDS IN THE RELATIVE SHARES OF HOME AGRICULTURE AND IMPORTS IN TOTAL UNITED KINGDOM SUPPLIES

	% from Home Agriculture	% from Imports
1900–02	50	50
1911–13	48	52
1922–24	46	54
1936–38	45	55
1947–49	58	42
1952–54	60	40
1956–58	58	42
1962–64	60	40
Of growth 1936–38 to 1962–64	94	6
Of growth 1952–54 to 1962–64	61	39

[1] By R. H. Rew, 'The Nation's Food Supply', *Journal of Royal Statistical Society*, Vol. 76, Dec. 1912.

has come from home agriculture. The share of imports in marginal expenditure over the past ten years has, however, been very little different from the share of imports in total supplies.

The process of adjustment in the United Kingdom of a rapidly modernizing and growing agriculture, with increasing productivity and increasing yield per acre, to a domestic market which is growing more slowly by the squeezing out of imports cannot continue much further. Already the pressures of conflict between agricultural producers and the government, fearful of establishing support prices which will lead to surpluses, are making themselves increasingly felt. The area of competition between home agriculture and imports is, moreover, narrowing. A large part of consumer demand for food is not for the food products of the temperate zone but for the products of tropical (or at any rate more southern) climates. The losers in the British market for food products have been not the under-developed primary producers but other temperate zone industrialized countries.

V. THE EXPERIENCE OF OTHER EUROPEAN COUNTRIES

The problems of a majority of the more highly industrialized countries of Europe would seem to have been similar in many respects to those of the United Kingdom. Since 1938, total consumption of all kinds in Europe has risen by about 85 per cent; in the ten years 1952–54 to 1962–64 it rose by about 50 per cent. Since 1938 population has increased by 25 per cent and over the last ten years by 11 per cent. Table V shows what this has implied in terms of consumption per head : total consumption per head has increased about 36 per cent in the ten-year period and food consumption per head by about 23 per cent — a little under two-thirds. This change reflects a combination of factors : income elasticity proper ; the effects of price elasticities in this field ; the effects of income redistributions from income-groups whith lower income elasticities towards income-groups with somewhat higher income elasticities.

For European OECD countries as a whole there is a better balance than in the United Kingdom between the growth of food production and of food consumption. The volume of food consumption has been growing between 1952–54 and 1962–64 by about 3·1 per cent ; the output of agriculture, free of duplication and net of imported feeding-stuffs and store cattle, has been growing by about 2·5 per cent over the same period (see Tables V and VI).

For Europe as a whole, moreover, there has been no radical change in imports of food from extra-European sources. The diminished ratio of net imports by the United Kingdom has been offset by a

TABLE V

DEMAND FOR FOOD IN EUROPEAN OECD COUNTRIES

(1938 = 100)

	Population	Total Consumption	Food Consumption	Total Consumption per Head	Food Consumption per Head
1938	100	100	100	100	100
1948–49	109	99	101	91	93
1952–54	113	123	128	109	113
1956–58	117	146	148	125	127
1962–64	125	185	174	148	139
1962–64 as % of 1952–54	+11	+50	+36	+36	+23
Annual rate of growth 1952–54 to 1962–64 (%)	1·0	4·2	3·1	3·1	2·1

TABLE VI

AGRICULTURAL PRODUCTION * IN WESTERN EUROPE †

Pre-war	100
1947–49	97
1952–54	128
1956–58	138
1962–64 ‡	163

Rates of growth :

Pre-war to 1962–64	2·5%
1952–54 to 1962–64	2·5%

* Net of imported feeding-stuffs and store cattle.
† OECD countries of Europe.
‡ Partly estimated.

larger dependence on imports in other Western European countries. The change in United Kingdom policies has been particularly great in the case of cereals. The general position can therefore be well illustrated in terms of an analysis of the sources of cereal imports into

Western Europe. Table VII shows that, for cereals of all kinds, Western Europe in 1935 produced 81 per cent of her consumption and imported 19 per cent ; in 1961–63 she produced 79 per cent and imported 21 per cent. Two countries greatly changed their patterns : the United Kingdom, as has been mentioned, and France. The latter from being a net importer of 7 per cent of her consumption became a net exporter of an amount equivalent to 15 per cent of her con-

TABLE VII

SOURCES OF SUPPLIES OF ALL CEREALS TO CERTAIN EUROPEAN COUNTRIES *

(Percentages)

	1935				1961–63			
	Total All Countries concerned	France	United Kingdom	All Others	Total All Countries concerned	France	United Kingdom	All Others
Production	81	93	34	87	79	115	54	75
Net imports (+)⎞ Net exports (−)⎠	+19	+7	+66	+13	+21	−15	+46	+25
Consumption	100	100	100	100	100	100	100	100

* Austria, Belgium and Luxemburg, Denmark, France, Italy, Netherlands, Norway, Portugal, Spain, Sweden, Switzerland, United Kingdom. For 1935 the figures include aslo the whole of Germany as then defined ; for 1961–63 they include only the present area of West Germany.

sumption. The countries other than the United Kingdom and France increased the proportion of their total consumption covered by imports from 13 per cent to 25 per cent.

VI. THE PATTERN OF WORLD PRODUCTION

If one pursues the pattern of the production of cereals onto its world scale, the change over the past decade has been small but significant, though the total output has increased by very nearly a half. The share of Western Europe has declined slightly. The share of the Asiatic countries is unchaged. That of Africa has slightly declined despite rapidly rising population. More important, the shares of the specialist producers and exporters — North and Central America, Latin America and Oceania — have in aggregate declined from 29 per cent to 25 per cent of all output ; Latin America and Oceania have held or very slightly increased their shares ; that of North and Central America has declined proportionately, though increasing absolutely by about 21 per cent.

38

TABLE VIII

DISTRIBUTION OF WORLD PRODUCTION OF CEREALS

(percentages of world output)

	1948/49 to 1952/53	1962/63
Western Europe	15·9	15·4
North and Central America	24·7	20·5
Latin America	3·6	3·6
Asia	22·2	22·2
Africa	4·7	4·5
Oceania	1·0	1·2
Others *	27·9	32·6
World total	100·0	100·0
Volume of production	100·0	145·1

* Including U.S.S.R. and Mainland China.

In terms of total food production the same trends can be seen. If one compares (see Table IX) the averages of the three years 1952–53 to 1954–55 with those of 1960–61 to 1962–63, world food production

TABLE IX

RECENT TRENDS IN WORLD PRODUCTION

	Food Production		Food Production per Head of Population *	
	1952/53–54/55	1960/61–62/63	1952/53–54/55	1960/61–62/63
W. Europe	100	122	100	115
E. Europe and U.S.S.R.	100	147	100	131
N. America	100	114	100	99
Latin America	100	124	100	100
Oceania	100	127	100	106
Far East †	100	124	100	106
Near East	100	125	100	105
Africa	100	114	100	95
All above	100	125	100	107

* It must be remembered that total output is being divided by population ; a decline of exports from an area with large exports will reduce the apparent output per head without implying any reduction of consumption per head of the population of the region.
† Excluding Mainland China.

rose by 25 per cent. The increases were largest in Eastern Europe and the U.S.S.R., that area apart most regions increased production by an amount very close to 25 per cent. The exceptions were North America — affected by demand — and Africa where for technical reasons expansion has been considerably less than is desirable.

Food production per head of population has risen by 7 per cent for the world as a whole. Again, Africa has lagged and presents a very serious problem. In Eastern Europe and the U.S.S.R. food supplies per head of population have increased greatly ; as in Europe and in other parts of the world there has been a problem of expanding rapidly enough to match income elasticities of demand for food.

Increasingly the world seems to be dividing into areas where the problems are those of highly productive agricultures, with no severe immediate limitations imposed by acreage and with problems of agricultural surplus as the most acute problems of policy, and areas of over-population, with the problems of the adequacy of food supplies as the major problem, with limitations of acreage acute and with severe restrictions imposed on their capacity to organize growth through a high income elasticity of demand for food and the constant danger of excessive inflation of food prices.

Such a pattern presupposes, firstly, that the countries of over-population, under the pressure of economic forces, will find themselves driven to encourage manufactures, to develop skills and to accept levels of incomes and prices which will enable them progressively to find markets. It presupposes, secondly, that the more advanced countries that are rich in agricultural resources will progressively find markets growing for their potential agricultural exports if the over-populated countries concerned are able to find in them markets in their turn for their manufactures. One can see this trend already, for example, in the commercial relations of Australia and Japan.

If one is right in visualizing this as one of the major long-term trends, where do the agricultural and trade policies of the present advanced countries fit into it? It would imply for European countries progressively keener competition with many of their exports. The trend of the past twenty years has been for the extra-European trade of the advanced European countries to be increasingly concentrated on durables and capital goods from the engineering industries. In the durables field, at least, greatly increased competition is to be expected. It would imply also that the terms of trade would be likely to be becoming less favourable to manufacture and more favourable to agriculture, as the over-populated countries come to secure greater access to the potential food supplies of the more advanced countries.

If, in the longer run, this is likely to be the trend, it would seem to be the right long-term policy for the advanced countries to seek to retain the capacity to achieve a high level of agricultural output and to prevent any serious deterioration of the potential fertility of the land. On the other hand, these longer-run objectives will not be best achieved by protecting the survivals of medieval agriculture, with tiny peasant holdings and with farms too fragmented as well as too small to use modern equipment. The demands on the agricultures of the advanced countries will only be capable of being met by an agricultural organization as efficient as modern industry, and which will be able to pay the levels of wages which are likely to prevail in the future in an industrialized country.

In the nearer future it would seem probable that the agricultural sectors of the advanced countries will be under increasingly severe pressure to rationalize, to release labour and to enlarge the sizes of holdings. If the experience of the United Kingdom is any guide, there is reason to think that productivity in agriculture may grow over long periods as fast or faster than in the economy as a whole. If, as the evidence of recent years would indicate, the growth of demand for food, resulting from income and price elasticities, is around 0·6 of the growth in total, such a rate of increase of productivity must mean a fairly rapid contraction of employment in agriculture. To attempt in this situation to protect agriculture too completely from economic pressures to make these adjustments may do more harm than good.

VII. AGRICULTURAL POLICY FOR ADVANCED COUNTRIES

If one approached the problems of world specialization and trade in food from the angle of physical resources, one might expect to find the countries that are most seriously short of land per head to be importers of food and exporters of manufactures to pay for food, and on the other hand to find the countries that are rich in resources of land per head of population to be specialized in agricultural production and importers of manufactures.

Paradoxically that is not the pattern of world agriculture and world trade in agricultural products. The reason is obvious. The distribution of the resources of skills is directly opposed to this. The highest levels of skill in manufactures tend now to be found in the countries which are encountering problems of agricultural surplus — North America, Western Europe, Eastern Europe, Oceania. The lowest levels of skills are for the moment to be found (with the obvious exceptions of Japan, China and Hong Kong) in Asia and Africa. For many types of manufacture, cheap labour (in terms of

TABLE X

LAND AREAS PER HEAD OF POPULATION IN DIFFERENT REGIONS

(Hectares)

	Land Area per Head of Population	Agricultural Land per Head of Population *	Agricultural Land per Male engaged in Agriculture *
Europe	1·1	0·6	7·0
U.S.S.R.	10·1	2·7	15·6
N. America	10·5	2·5	97·0
Latin America	9·2	2·2	8·5
Near East	8·2	1·9	20·4
Far East	1·2	0·4	2·6
Oceania †	51·8	30·0	700·0

* The area includes all arable land, land under permanent crops and permanent grass-land.

† The area of agricultural land in Oceania includes large areas used for occasional grazing only but which are included in agricultural holdings ; the areas more comparable to the practices of other continents would be about 6 hectares of agricultural land per head of population and perhaps 140 hectares per agricultural worker.

wage per head) cannot offset the disadvantages of lack of specialized skill and general educated intelligence.

The nineteenth-century pattern of agricultural trade was built on a deficiency of output in low-yielding agriculture in Europe and the early specialization of North America, Latin America and Oceania on food production. One may perhaps visualize a twenty-first century trade built on a near balance in Europe, a building up of capacity to manufacture exports in the densely populated countries of Asia, and a trade in food products between them and the countries of North America, Latin America and Oceania.

The central problem of agricultural policy in an advanced country is just how far one should permit these economic pressures to operate and how far one should attempt to mitigate them.

The principal argument for allowing them to exercise in full force their normal pressures is that agriculture must progressively find itself in competition with industry for manpower and other resources and thus must achieve forms of organization which will enable it to pay wages approaching those paid in industry — differing, that is, from those in industry only to the extent that workers may prefer to accept slightly lower real wages in a rural environment. If technical progress is at a fairly steady rate both in agriculture and in industry there is no obvious reason for thinking that agricultural adjustment should be slowed down or made to be less rapid than the general adjustment of the economy in which it is operating.

Some of the agricultural policies and measures that operate today were devised in the 1930's to insulate agriculture from the catastrophic cyclical and intermittent pressures of that period. They are — it may be argued — much less well suited to steer agriculture through a period of fairly steady high level employment with relatively small cyclical variations. There is greater danger of building up progressively a lack of adaptation to the current situation.

If one wishes to defend policies that are in some degree protective of agriculture it should be, I suggest, not through a desire to postpone adaptation, but to keep the adaptation orderly and to prevent the perpetual economic pressures to contract the proportion of all resources devoted to agriculture from achieving a distribution of income undesirably unfavourable to the very large numbers who are engaged in agricultural occupations. While some pressure is necessary, it is not immediately clear that the pressure needed is as severe as is likely to emerge from time to time in an activity in which the income and price elasticities are both low and short-term mobility out of the activity is also low.

It can also be argued (and United Kingdom experience would tend to support it) that technical progress in agriculture is likely to be more rapid where there is an assured market at pre-known prices for the current output, and that long-term adjustment requires that resources adequate to modernize agriculture shall be made available to those working in agriculture. In the 1930's agriculture was not only impoverished ; it was also conducted on a basis of minimum exposure to risk ; and it was not regarded as a credit-worthy occupation.

For these reasons some measure of protection to agriculture may be justified. But it is becoming increasingly clear that in a number of countries that are attempting such policies, the governments find it extremely difficult in practice to maintain a sufficiency of long-term pressure to adapt and contract. Measures of price support mean that for the impersonal forces of economics are substituted very personal and political negotiations which, on the one hand, leave farmers angry and, on the other hand, provide an insufficient stimulus to adaptation and so allow the preservation of an increasingly out-of-date agricultural structure. This, as I see it, is the fundamental dilemma of agricultural policy in many European countries today.

———

DISCUSSION OF PROFESSOR ROBINSON'S PAPER

Professor Gale Johnson said that Professor Robinson's paper includes interesting and relevant data on the agriculture of the United Kingdom, and he thought that the data reflects much of what has occurred in Western Europe, North America and Oceania. He said that Professor Robinson's concentration on the United Kingdom was useful because it is possible to present information for a period of six decades while, for many countries, such data is available only for two or three decades.

Professor Gale Johnson said that Professor Robinson had been asked the impossible question : 'What is the desirable level of agricultural output ?' He thought that Professor Robinson had quite rightly refused to answer it. However, he had hoped that Professor Robinson would have spent some time in at least trying to define the objectives, even if there were no empirical way of specifying them. The paper relied on production as a percentage of local consumption, as a relevant measure of actual or desired output. As a rough comparative measure this concept has considerable validity. However, by itself, it says little about a desirable level. Is, for example, the desirable level in the United Kingdom, 45 per cent (as was the case before the last war) or 60 per cent ? Is 80 per cent the desirable figure for all of Western Europe, or 110 per cent for the United States ? Professor Johnson agreed that any concept of desired output that has analytical significance is extremely complicated. He said that on another occasion he might have argued for a concept of desired output based on optimum use of world agricultural resources. However, neither agricultural policy makers nor most agricultural economists seem to show much interest in this notion.

Professor Johnson said that he was not clear what is involved in Professor Robinson's point about the appropriate agricultural policy for industrial countries. Professor Robinson had argued that agriculture should be required to adjust to the forces of competition. Elsewhere he had argued that some protection is required to avoid an undesirable distribution of revenue. Finally, in his last paragraphs, Professor Robinson referred to the argument that protection for agriculture reduces the

44

impact of uncertainty. Professor Johnson said that he was uncertain how various aspects of these points relate to one another.

Would Europe, in Professor Robinson's estimation, become self-sufficient in food even if the degree of protection were significantly reduced ? Many countries in Europe protect their agriculture by about 30 per cent. Only Denmark and France have significantly lower degrees of protection ; and in the case of France this lower protection will disappear as the Common Agricultural Policy comes into full force.

To what extent does the trend in developing countries of reduction in their exports of food depend upon the agricultural policies followed by Europe and North America ? A large amount could be added to the export earnings of the developing countries by free trade in sugar, both to European markets (including Eastern Europe) and to North America.

Is there any connection between the protection of agriculture in Western Europe and the difficulties of competition in the exporting of non-farm products ? If there is any connection, the current trends in agricultural protection become self-validating since the growth of non-farm exports will be held back by high food prices and/or direct taxes. Under these circumstances it would not be rational to contract agricultural imports to reduce foreign exchange requirements.

Professor Gale Johnson turned to the question of the possible trends in the distribution of world trade in agricultural and non-agricultural products. He asked whether the recent trends are consistent with Professor Robinson's projection. Western Europe has not significantly increased the percentage of its agricultural consumption from local sources during the past decade. In the case of developing countries it is true that the growth of food imports has been somewhat greater than the growth of exports, and also that the relative growth of non-farm exports has been greater than of non-farm imports. However, the absolute increase in non-farm imports is larger. In the 1950's the only important grain-exporting regions were North America and Oceania and to a lesser extent the U.S.S.R. These trend changes had occurred while agriculture had been quite systematically protected in Western Europe and the United States. Professor Johnson asked whether, in the light of these trends, Western Europe will be able to compete with Oceania, North America and parts of South America, and so to become self-sufficient. He did not think so. However, he agreed that if the present agricultural policies in Western Europe and the United States persist, Professor Robinson's projections may not miss the mark by very far.

Professor Papi said that the paper is not as easy to grasp as it seemed to be at first sight. He said that he thought the introductory passage in the paper needs to be further explored. The basic picture is of two distinct areas in the world : one with a highly productive agriculture, and serious problems of surpluses ; and the other, over-populated, with a serious shortage of agricultural resources, suffering from high income elasticity of demand for food products, and the consequent threat of inflation of food prices. Such a division leads to the supposition that the over-populated countries will be compelled to encourage their manufacturing sectors and to accept prices that enable them to find markets. The richer countries will find markets for agricultural surplus to the extent that developing countries find markets in the developed world for their manufactures. This tendency is already to be seen in the commercial relations between Japan and Australia.

It is this long-term picture which is perplexing. By what process can the developing countries achieve a level of incomes and a level of prices which will allow them to create a local market for their own products ? This achievement is usually taken to necessitate a long and difficult programme of development. Such a programme would include the reduction of risks to producers, the improvement of the infra-structure, incentives to investment, and the successful solution of the problem of income and production balance between the different sectors ; in short — the fulfilment of policies of structural change and commercialization. Is it essential for developing countries to follow this gradual sequence of development, or can they leap straight into industrialization and invade successfully the markets of the world, including even the markets of advanced countries ? By what means can such a result be brought about ? Professor Papi said he found it hard to find the answer in Professor Robinson's paper.

With regard to the developed world, in this picture of future world trade, the developed countries will have to limit their industrial activity to highly specialized fields in order to allow developing countries to enter their markets. This implies both abolition of all obstacles to imports and an explicit renunciation of certain lines of industrial activity. Such a solution, if it is at all possible, can only be envisaged over the very long run, and will have to be undertaken with great caution so as to avoid a serious disorganization of national and world markets.

Whatever the final outcome, the essential fact remains that in any country, advanced or poor, economic development has to start by a revaluation of human capacity, and by a liberation of the factors of production from subsistence agriculture so that they can be used in other sectors giving higher returns. There is no way of by-passing the stage of the industrialization of agriculture.

Professor Papi said that, typically, the part of the population employed in agriculture is constantly decreasing, and with this the incomes derived from agriculture become a smaller and smaller part of national income. Since all policies of rationalization, industrialization and commercialization of agriculture lead in this direction, how can this be reconciled with the giving up, by industrial countries, of a great part of their industrial activities on behalf of the development of poor countries ? What will then be the sources of growth of national income in the developed countries ?

Mr. Tracy said that as a representative of OECD he was interested not

46

only in Professor Robinson's paper but in the work of the whole Conference. Mr. Tracy said that in their work on agriculture the OECD has been placing increased emphasis on relationships between agriculture and the rest of the economy, and on the need for adaptation in agriculture which results from economic growth. This approach had led in particular to the preparation of a report by the OECD on 'Agriculture and Economic Growth'.

Turning to Professor Robinson's paper, Mr. Tracy said that he had been particularly interested by the conclusion to the effect that, in the highly developed countries, a higher degree of self-sufficiency in food may be justified because of future shifts in the balance between agriculture and other sectors in the world as a whole. It should be stressed, however, that this should not be an excuse for policies which retard adaptation in agriculture.

Mr. Tracy said that like Professor Johnson, he was sorry that Professor Robinson had not attempted to 'quantify' the desirable level of self-sufficiency, as he had in his previous work on the United Kingdom. Could he say whether he still thinks it possible, for a given country, to calculate what degree of self-sufficiency in food would be desirable ; and, in particular, whether he thinks that in the United Kingdom the degree should be higher or lower than it is now ?

Mr. Tracy said he queried those statements in Professor Robinson's paper which indicate that there has been a reduction in the degree of self-sufficiency in Western European countries, other than the United Kingdom. The data upon which these statements are based do not seem to be very conclusive for a number of reasons, and it is more likely that the degree of self-sufficiency is rising. This, of course, would be consistent with Professor Robinson's general thesis concerning the future pattern of world trade.

Professor Delivanis said that he thought that Professor Robinson was right when he stressed that industrialization is not only an economic but also a prestige question for developing countries. It should be added that the search for prestige is the ambition of those that make the decisions and who do not necessarily suffer from any unfavourable repercussions. However, even though industrialization may not pay in the short run, Professor Robinson was surely right to exclude the possibilities of some countries developing with industries and other countries developing without any industry at all. Industrial countries tend to become exporters of agricultural products.

Professor Delivanis said that Professor Robinson's paper brought out the interesting fact that, on the basis of statistics available, food consumption, measured at constant prices, from 1900 to 1962 in the United Kingdom had increased by nearly as much as total consumers' expenditure. Thus while the percentage of individual income spent on food falls when incomes increase, that does not happen with national income and national expenditure on food.

Professor Delivanis referred to the statement by Professor Robinson that those who had been losers in being excluded from the United Kingdom market had not been producers of tropical products but the producers in other temperate-zone countries. This requires qualification. The reduced purchasing power of the countries whose products had been excluded from the United Kingdom market led to losses of markets for tropical producers. The producers of tropical products had had to rely more heavily on the United Kingdom market as the markets in other countries shrank.

Professor Gulbrandsen said that he wished to put forward the price aspect of the problem and to suggest that in this way the desirable level of agricultural production in industrialized countries can be defined. Since the price in a completely free market for all countries of the world would be higher than the existing world market prices and lower than the existing protected levels in the industrialized countries, there is at least a range within which the desirable level of production may be determined. Calculations, based on linear programming micro-economic models, for the allocation of world resources in agriculture could provide information about the long-run desirable level of production. Professor Gulbrandsen said that such studies are at present being tried in order to make such calculations for Sweden and that interesting results are emerging.

Professor Dandekar said that one of the major conclusions of Professor Robinson's paper is the pattern of world trade in agricultural commodities which he put forward for the future. Professor Dandekar said that he could well understand the uneasy feeling of industrialized countries at the prospect of such radical change in world trade patterns and, in particular, at the prospect of reversing, in some cases, their roles as exporters of manufactured goods and importers of primary materials. He did not think, however, that these changes necessarily imply that those who produce food will suffer from a lower standard of living.

Professor Dandekar said that he thought that it would be illuminating to add to Table X in Professor Robinson's paper, a column of net agricultural product per acre of agricultural land. Such a calculation shows that North America and Oceania do not stand favourably compared with the less-developed countries of Asia. In many Asian countries, net food production per acre of agricultural land is twice as great as that in the United States. This makes clear that land resources in the United States are under-utilized. If, in the long run, the world population is to be fed, some land which is not at present being used intensively needs to be used intensively. Professor Dandekar said that this calculation supported Professor Robinson's conclusions.

Professor Ashton said that he wished to point out that the limiting factor, as far as the solution of longer range, strategic problems is concerned, will vary from country to country. Thus, Professor Robinson has attempted, on several occasions in his writings, to measure the contribution of British agriculture to the U.K. balance of payments. This has been, and still is, the crucial issue underlying agricultural policy in Britain. In other countries, the chief concern may not be so much the economic use of resources in agriculture as, for instance, the lack of alternative opportunities for farm population, or the need to inject a minimum level of income into the farming community.

Referring to the final paragraph of Professor Robinson's paper, Professor Ashton said that he thought that the conclusions could have been stated more forcefully. In a paper like this which contributes to the policy debate, it is necessary for economists to state their conclusions unequivocally in order to convey the message to those concerned in the political arena. It is especially necessary to be as positive as possible in view of the inherent tendency for policy measures to lag behind the general development of the agricultural industry. Professor Ashton gave three examples of this tendency. First, there is too little provision in economic policy measures for rapid technological advances that are achieved in the agricultural sector. Second, policy measures invariably lag behind the changes in consumer taste which repeatedly occur, against the background of rising incomes. Finally, institutional arrangements related to support systems frequently act as a deterrent or obstacle to necessary adaptation and change.

Dr. Odhner said that the central problem is the future structure of world trade. On the one hand there is the rather unsatisfying picture of two self-sufficient parts of the world losing the gains from comparative advantage and on the other hand there is the possibility, hinted at by Professor Lundberg, that it is more realistic to think that industrial countries will import simple manufactures than that developing countries will be able to develop their agriculture at a sufficient speed. Dr. Odhner said that as remedies to the wastefulness inherent in the first of these pictures, a successful full-employment policy and an increased structural mobility of labour and capital would alleviate the political pressure for protectionism.

With respect to the problem of the expansion of agriculture in developing countries, the prior conditions must include agrarian reform, extensive education and, as a consequence, a revolution in the social structure. Such revolutions had taken place in the present industrial countries. However, in these countries, these reforms had been taken as a result of the pull from the industrial sector. In the developing countries these same reforms are having to be done by direct 'push' on these problems. These results have not been achieved in democratic societies. There is a strong case that if industrialization is first developed, the 'pull' effects will again work.

Professor Robinson said that Professor Gale Johnson was correct in saying that he had failed to be sufficiently explicit about criteria for the desirable level of agriculture. Some of the criteria which must guide developed countries are those of strict Ricardian economics. However, with prices and exchange rates adjusted to provide full employment, there is often an advantage for advanced countries to produce their own food. In a world in which prices and exchange rates are not so adjusted, it is not easy to apply the principle of comparative advantage. The remaking of agricultural policies in 1947 had tried to take some account of long-term comparative advantage and the prospects for trade liberalization. Professor Robinson said that he had none the less been increasingly disturbed by the tendency towards the under-employment of North American and South American agricultural resources. This is a complex question of the economic relationship between North America and Europe.

Professor Robinson said that many questioners had picked up his somewhat light-hearted forecast of the possible pattern of agricultural trade in the twenty-first century. He said that the background to his own thinking on this question is as follows : The trade between Europe as an entity, and the rest of the world, had been originally built on an export of textiles and finished manufactures. Such trade is now virtually dead and has been replaced by trade in machinery. But few developing countries are in a position anywhere near approaching equilibrium in their international trade. For example, if Indian development is to go well, India has to develop the capacity to meet machinery needs from its own resources. This applies to many other countries. By the year 2000, most developing countries may be expected to follow the way of Japan. This would imply a decline in international trade in machinery. Even the trade between Europe and the rest of the world in automobiles does not seem likely to have a perennial life. From these arguments it would seem that the European Continent will tend to be more self-sufficient. The differences in comparative advantage would narrow.

Professor Robinson said that Professor Papi had asked how we could, at the same time, look forward to rapid growth in Europe and to the prospect of a reduction in exports of manufactured goods from Europe. Professor Robinson said that he had assumed that there would be an increase in intra-European trade and a decrease in extra-European trade.

Professor Robinson said that he did not feel sufficiently brave to attempt to quantify the degree of self-sufficiency that can be justified. His own attitude was ambivalent. He would like to see the United Kingdom importing as much agricultural produce as it could afford ; but balance of payments problems do not, for the time being, allow the United Kingdom to reduce support and increase agricultural imports. In answer to Professor Delivanis, Professor Robinson said that he had found extremely interesting the fact that the growth of food supply kept step with the increase of aggregate income. This implied, over the years, little indication of low income elasticity for total food production. Professor Robinson said that his explanations were : first, the ordinary one of the process of long-run elasticity, and second, the very marked redistribution of income from groups of high incomes and low income elasticities towards groups with low incomes and high income elasticities.

Chapter 3

THE SEARCH FOR NEW INTERNATIONAL ARRANGEMENTS TO DEAL WITH THE AGRICULTURAL PROBLEMS OF INDUSTRIALIZED COUNTRIES [1]

BY

DALE E. HATHAWAY
Michigan State University, U.S.A.

I. INTRODUCTION

To most observers agriculture comes the closest of any industry in modern society to meeting the conditions of competition. But, in almost every industrial country in the world there has developed a series of governmental programmes for agriculture which depart widely from the free market. This intervention is too widespread in countries where the agricultural population is a distinct minority to regard it as merely the exercise of rural political power. Instead, agricultural programmes must be regarded as a search for new institutions that will be more satisfactory than the unregulated market for the production and exchange of farm products.

To understand the desire for an institutional structure which will work better than free markets we must briefly review the economic conditions in the industrialized countries which, in the absence of intervention, would be the major determinants of the conditions surrounding the production and consumption of farm products. Any policy designed to improve on the market must recognize these conditions, and either be prepared to alter them or to operate in their presence.

An important mark of the agricultural situation in industrialized countries is the relative abundance of farm products. In these countries the population is, on the whole, well fed and clothed and the proportion of their income spent on these farm-produced goods is relatively low and declining. Of course, deficiencies in diet and

---

[1] The paper has benefited from suggestions of my colleague, L. W. Witt.

dress may exist for segments of the population of most industrialized countries, but these are largely due to the distribution of income among the population rather than to a general deficiency of income for the entire population.

Put in more precise economic terms, the total demand for farm products in industrialized countries is generally inelastic although it varies considerably among products. In addition, the income elasticity of demand for farm products is low and declining, having almost reached zero in the highest income countries. In this situation, if growth in agricultural output much exceeds the growth of population, output is expanding faster than demand, and the small additional increments in output severely depress product prices.

Another observable feature of the agricultural situation in industrialized countries is that the average income per worker in agriculture generally is well below that of workers in other industries. This condition exists despite the existence in every industrial country of government programmes designed to enhance the income of the farm population. At the same time, the farm population of each of these countries contains progressive efficient farmers whose incomes compare very favourably with those of the non-farm population.[1]

This seeming paradox cannot be explained merely by asserting that many of the workers in agriculture have low productivity and should be employed elsewhere, although it is clear that the agricultural work force in every industrialized country is too large to earn a satisfactory income in the market. It is likely, moreover, that even with a major reduction of the agricultural labour force in most countries that the re-structured agriculture would be plagued by chronic over-capacity and low earnings in the absence of government programmes.

The demand conditions for agricultural products in industrialized countries, by themselves, are not sufficient to account for the over-capacity found in modern agriculture. But, together with the economic characteristics of farm production products in industrialized countries they can and do produce the observed results.

[1] Recent data for the United States shows that operator families on farms selling over $20,000 of products annually had average incomes of $12,357 in 1963, which was about twice the average family income of the non-farm population. There were an estimated 384,000 such farms in the U.S. in 1963.

II. THE CHARACTERISTICS OF MODERN AGRICULTURAL PRODUCTION

Modern agricultural production has three important characteristics. They are :

(i) A structure that closely resembles the competitive structure of economic theory.

(ii) Rapid technical and economic change which generally increases the output per unit of input of labour, land and capital.

(iii) Specialized production resources and techniques which, once committed to agricultural production, are not likely to find alternative uses sufficiently rewarding to warrant a shift to other production. In other words, resources become 'fixed' in agricultural production.

(i) *The Competitive Structure of Agriculture*

Even under modern conditions, agriculture is composed of large numbers of relatively small firms. The number of firms is so large that the individual producer cannot influence the market price, and the products are sufficiently uniform so that a producer cannot differentiate his product sufficiently to enhance its price. Indeed, in modern industrial societies consumers may not even know the country of origin of their eggs, meat or vegetables, let alone the farm where they are produced.

Another feature of the agricultural structure is the relative ease of entry to the production of an individual product and to the industry. No licence is required to farm, no special training required by law. Except for those few products subject to governmental production or marketing controls, no barriers other than economic or technical ones exist to prevent the production of specific products. Even where governmental controls exist one usually can acquire the rights to production by buying the productive assets (usually only the land) of an existing producer.

With few exceptions, an existing or prospective farmer is free to use production methods, techniques of organization and management, and quantities of productive resources he believes will be most profitable. He is free to add new technology and capital that adds to the productive capacity and output of the industry to the extent of his financial capability, and in most advanced countries large public subsidies go into the production of new techniques and the dissemination of information about them.

One important feature of the competitive model of economic theory is lacking in agriculture, however. This is perfect knowledge about future prices and production techniques. And, such knowledge is especially difficult for individual farmers to acquire. In the absence of knowledge about the future, the present or recent past is often used to make decisions about long-term investments in farming.[1] In an agriculture undergoing rapid change this tends to lead to a chronic over-commitment of resources.

(ii) *Rapid Technical and Economic Change*

One of the main blessings of an industrial society also contributes heavily to the problems in agriculture. Both public institutions and private enterprises in industrial countries are heavily engaged in the production of new inputs, resources and resource combinations which will increase the output of the individual farmer. For, in a competitive structure, the only way the individual producer can increase his profit is by reducing his cost per unit of output. This new technology has increased the output of farm products per acre, per animal and per man-hour. Education of the farm population has been shown to be also a major contributor to increased output per unit of conventional inputs. The competitive nature of the industry means that it will pay the individual producer to adopt such measures as rapidly as possible, and it also means that the rate of adoption will vary widely depending upon the economic situation and managerial ability of the individual farmer.

These changes in agriculture would lead to an increase in total farm output if other resources could be withdrawn from agriculture into other pursuits at a pace sufficient to offset the new output-increasing changes. Economic and social conditions are such that resources already in farming do not move out of farming at a sufficient pace. Moreover, produce price levels which provide barely sufficient earnings for those using the older techniques may make it highly profitable to add new techniques, so that maintaining prices at old levels may lead to substantial increases in output, while lowering prices would severely depress the earnings of those not using the new methods.

---

[1] Government programmes designed to provide income guarantees almost always increase price certainty and thus remove this aspect of uncertainty.

(iii) *Resource Fixity in Farming*

If, under the conditions described above, a slight decline in earnings in agriculture led to a major out-movement of resources from farming there would be no serious problem in agriculture. But in modern agriculture this is not the case. Most of the new productivity is in the form of capital having a productive life of several production periods. This is also true for specialized agricultural education and training. Machines and equipment used in farming are increasingly specialized and once in use in agriculture are likely to continue there because of lack of profitable alternatives. And people who work in farming for any period find their skills not transferable to non-farm industries, so that even with full employment a farmer past middle age probably has few prospects outside farming. Thus, when farm income declines the rate of new entrants to farming is lower and the outmigration consists largely of the younger members of the farm population who are not yet committed to agriculture.

Land which is suitable for farming is not that which is most desired for recreational purposes in an industrial society, and despite casual observations to the contrary only a small fraction of farmland is taken by housing, roads and factories. Much of the new technology also is land-saving (yield increasing), so that this does not become a limiting factor even in countries with relatively limited land resources.

This immobility of resources out of farming is largely due to economic rather than the social factors that many economists imagine, for factor markets in industrial societies do not offer much reward for resources previously committed to farming.[1] But, as long as immobility exists it results in a withdrawal rate of resources from farming which is insufficient to offset the effects of the new technology. The result is a rise in total agricultural output at a faster rate than demand increases, bringing a downward pressure on farm prices, and average incomes in farming below those in industry even though some farmers prosper. Thus a combination of conditions contribute to the over-capacity which is a common feature of agriculture in the modern world.

[1] For a theoretical discussion of this condition see Dale E. Hathaway, *Government and Agriculture*, Macmillan Co., New York, 1963, Chapter 4, and the extensive series of professional papers that are footnoted therein.

III. THE INTERNATIONAL NATURE OF THE PROBLEM

Industrial countries vary in their balance between demand and national output of individual farm products and total farm products. Some countries are regular importers of farm products, some are regular exporters, and some vary from year to year depending upon local growing conditions. Regardless of the situation, however, almost all industrial countries have deemed it necessary to take some action to enhance the income of farmers. Nations which otherwise might choose not to use government intervention on behalf of their farmers have found that the national policies of other countries operated in a fashion that increased the pressures on farmers in countries not having programmes ; and the combination of 'natural' difficulties that face farmers in industrial countries together with outside national policies has made intervention almost universal.

There has been a general tendency to view the agricultural problem described in the previous section as being limited to a few countries, namely the United States, Canada, Australia and New Zealand, with France an intermittent member. Yet the conditions described are observable in every industrial country and not just those who export farm products. Increases in output in United Kingdom, the Netherlands or West Germany add to the capacity problem just as much as similar increases in the United States or France. Thus, in considering the present situation it is necessary to consider an international and not merely a national view.

The agricultural problems of industrialized countries take on quite a different view if one thinks not of the various nations individually but of the industrial world as a whole, linked either by trade or, as in the EEC, by economic union. Some of the differences are as follows :

(a) Viewed in an international context it becomes obvious that importing as well as exporting countries contribute to excess capacity and over-production for commercial markets. This is in contrast to the common view that 'surpluses' only exist when domestic production exceeds domestic demand.

(b) Viewed in an international context the individual or resources which are 'sub-marginal' may change sharply from those which are defined as such in a national context.

(c) In an international view total effective demand for farm products must be considered. The concepts of 'home markets for home producers', of 'traditional markets' and of 'market shares' are all national policy concepts applied to international negotiations.

(*d*) In the present world scene an international view must recognize the existence and problems of the under-developed countries as well as those of industrialized countries.

It would be unrealistic to assume that national interests do not exist and that agricultural policies of industrial countries will not reflect the political pressures of national legislative bodies. But in many areas nations have been able to pursue national policies which are compatible with both their national interests and those of other nations. This is not to say that there have not been conflicts between national policies and international interests, but when such conflicts arise both have been recognized as important considerations. The same cannot be said for agricultural policies. Generally, when conflicts arise national interests have been considered in a very narrow framework.

Various schemes have been tried or suggested as measures to deal with the 'surplus' problem of industrial countries. All of those used have been national in focus rather than international. It is now clear that any satisfactory solution must satisfy certain international requirements in addition to offering solutions to the problems of individual nations.

IV. CHANGING THE DEMAND FOR FARM PRODUCTS

To many persons there is sanctity to food production that is found for few other economic goods. Thus, the first approach to surpluses in agriculture generally is one of expanding demand rather than one of adjusting output to the demand preferences of consumers. There are three important approaches used in dealing with the demand side of the agricultural equation : (1) to change the demand facing producers of an individual nation or product without altering the basic demand of consumers, (2) to change the basic structure of demand at the national or international level, and (3) to use market discrimination programmes within and between nations to increase producers' incomes without altering the basic demand curves.

(i) *Changing the Demand Facing Producers*

One of the reasons for the difficulties caused by excess productive capacity in farming is the highly inelastic demand for farm products in developed countries. Numerous national agricultural policies have been and still are used to change this in so far as it affects the prices and income of their farmers.

Countries which have continuing imports of farm products have used import barriers in the form of quotas, tariffs, levies, etc., as a method of maintaining their internal farm price level in the face of the general over-capacity in agriculture in industrial countries. This amounts to the creation of a completely elastic demand for the domestic agricultural output up to the point where total output equals domestic demand. This has the political advantage of minimizing government budget costs for agricultural programmes by shifting their burden to consumers in the form of higher prices. It is an approach which shifts the entire burden of adjustment of over-capacity to exporting countries, thereby increasing their problems.

Another widely discussed approach, used largely by the United Kingdom, is to allow the market supply and demand conditions to exist and then to supplement farmers' incomes by deficiency payments. In its most extreme case, with generous deficiency payments on unlimited quantities of production, this amounts to underwriting additional excess capacity and can be very expensive to the government involved and to competing producers in other countries.

Exporting countries face somewhat more complex problems. First, if they guarantee an internal price for an unlimited quantity of production they are also underwriting additional excess capacity in agriculture ; and if the internal guaranteed price is above that existing in international markets the export market is lost. This leads to export subsidies of one type or another which may put additional pressures on the markets of importing countries. This is the programme followed by the United States over much of the last decade, and there is ample evidence that it leads to very high budget costs even if the internal price guarantees are relatively modest.

These approaches — import controls, price supports and export subsidies, and deficiency payments — are all programmes which make the demand for farm products appear completely elastic at the national level as far as the relationship between farm output and national farm income is concerned. Since the market demands for farm products is clearly inelastic in industrial countries, the illusion that it is otherwise can only be maintained at great cost in consumer prices, public expenditures or reduced income for farmers in other countries.

These costs appear as public expenditures and lower income to other farmers in world markets if the protection is by unlimited deficiency payments or by export subsidies. The costs appear as very high consumer costs when import barriers are used to maintain high internal prices for products which could be purchased for much less from external producers. And this cost also falls on the producers in the under-developed countries who see world markets depressed by subsidized exports and import barriers. Maintaining the illusion of unlimited demand for farm products solves only one problem — the income problem of producers receiving the protection — and creates many problems. This is because it leaves the basic supply and demand conditions unaltered and deals only with the farm income problem.

(ii) *Demand Expansion*

Why not actually change the demand for farm products rather than merely offer programmes which appear to change it as far as producers are concerned? Within industrial countries some of the population has its consumption of food limited by inadequate income, and in the under-developed countries this restriction is severe and commonplace.

The use of subsidies to expand the demand for farm products at a sufficient pace to absorb the rapid growth in agricultural output in industrial countries has been suggested.[1] Such a programme is in use in the United States via the food stamp plan which subsidizes the food consumption of many low-income families. Experience has shown that these subsidies do expand the retail demand for food, especially for higher quality products. Experience also has shown that not all of the full value of the subsidy goes to expand the demand for *farm* products. In a country where food consumption levels already are high and the income elasticity low, all indications are that a very large public subsidy would be required to absorb present surpluses. Most consumers in industrial countries probably would prefer public expenditures on other items.

If, however, the poor countries of the world enter our calculations the picture may change. In those countries the population is pressing against the food supply and the general level of the diet is low. The

[1] For a discussion of the theory of subsidizing food consumption see Herman Southworth, 'The Economics of Public Measures to Subsidize Food Consumption', *Journal of Farm Economics*, Feb. 1945, pp. 38–66. This paper is oriented towards domestic subsidies to expand food consumption but its theoretical implications hold for international programmes as well.

income elasticity for farm products is higher than in wealthy countries. It has been suggested that the situation be met by large cash grants to poor countries which could only be used to purchase farm products. It is asserted that this would increase the effective demand for farm products sufficiently to provide satisfactory prices to producers. Such a scheme would encounter serious problems in practice. First, part of the grant funds would replace normal commercial demand, so that total demand would expand by less than the grants. Second, a very large grant would be required to increase demand sufficiently to raise world market prices to the level now maintained by many industrial countries. Finally, if the money were given to governments and not to the population of poor countries the programme might have adverse effects upon local prices and production in developing countries.

All demand expansion programmes face a common problem of inducing even faster rates of increase in farm output. The higher prices generated by the programmes will induce additional resources to enter production. Subsidies might induce a significant shift in the demand for food, but over the longer period they would not alter the basic problem of the tendency towards chronic over-capacity.

(iii) *International Price Discrimination*

Programmes which ignore the basic demands for farm products or attempt to alter them by subsidy are costly and cumbersome. A third alternative is to accept the basic demand structure and enhance producers' incomes under it. One way of doing this is to apply the well-known principle of price discrimination to agricultural markets. This could be done by charging a higher price in the less-elastic markets of the industrial countries and a lower price in the under-developed countries where the demand is presumably more elastic. Such a scheme would increase the revenue of producers of farm products (in industrial countries) and would increase farmers' welfare without heavy governmental budget expenditures.

If this programme operated with a single price to producers at the national level it would be another price support and export subsidy scheme. If, however, the different prices in the different markets were reflected directly to individual producers, then the programme will avoid some of the problems of other types of programmes.

The multiple-price plan is close to the pricing plan used by public utilities in that it is a method of pricing the product so that some of the consumers' surplus under the demand curve is transferred to the producer, thereby raising his income. The 'consumer surplus' in this case would be the consumer expenditure obtained by charging higher prices in industrial countries than in under-developed countries. However, in order to function, a market discrimination programme must be able to discriminate between markets by controlling the quantity available to each market. This can be done only where markets are physically separate (as for utilities) or where product differentiation can be achieved in the minds of consumers.

This can be achieved in several ways. The United States wheat programme differentiates between the market for wheat for domestic human consumption and the market for feed and export use via the use of a processing tax on wheat going into domestic food use. The value of the tax largely goes to farmers, based upon their share of the domestic food market. Importing nations which for some reason wished to maintain a high internal price level for all farm products could achieve the same result by the use of import tariffs and producer tax on each unit of production in excess of a certain quantity of production. The producer tax should be exactly equal to the tariff *so that all producers in all countries would receive the same world price for additional farm output*. The value of this output would be its economic value to the under-developed countries, which might be further enhanced by aids, grants and general economic growth.

An essential feature of such a programme is that the returns to producers clearly provide a differential price that is obvious to producers. The use of blend pricing or pool pricing, where the producer is paid an average price amounting to the average revenue obtained from high and lower priced markets, is self-defeating. Blend pricing is self-defeating because the marginal revenue of the producer exceeds the marginal value of additional product in the secondary market. This induces expansion of uneconomic output. There has been a great deal of experience in milk, fruit and vegetable pricing in the United States with blend or pool pricing and it all has been as indicated.

Such a scheme has economic and political advantages as a method of dealing with the agricultural problem of industrial countries. It will maintain farmers' incomes above that obtained in a free market without large direct government expenditures. The income transfer will be from the consumers in industrial countries. It does not require state trading or state interference with the private farming decisions. It does not artificially remove efficient resources from production, although hopefully it might retard the entry of new resources. The entry of new and more efficient producers is not prevented, but they will only do so if it is profitable at the lower prices. If such a system were adopted by all of the industrial countries at present output levels it would equalize the competition for the growth in markets while protecting the incomes in the industrial markets they now have.

There are economic and political problems in such a programme. It is unlikely that such a scheme will be effective for products which have close substitutes in consumption, such as cotton and man-made fibres. It is not readily adaptable for intermediate products such as feed grains.[1] It requires government allocations of production bases to producers and requires tax collection somewhere in the marketing system.

Advocates of simplicity in agricultural programmes will quickly point out that this scheme is merely a form of limited deficiency payments financed by consumption taxes in wealthy countries. It is exactly that, yet many practical men of political affairs believe that in wealthy countries where farmers are in a distinct minority there are many reasons for preferring such schemes to large direct governmental outlays for farmers.

The impact of price-discrimination programmes upon underdeveloped countries is a topic that cannot be treated adequately here. A variety of concessional price programmes of various types already are in use on a bilateral basis, and there are serious political objectives arising from them. A programme which provided a universal low market price to under-developed countries would at least reduce political control on the part of either sellers or buyers. The realities of population pressures and lagging agricultural development in many countries suggests that some type of differential pricing will be used, probably on an even wider scale than at present.

V. CONTROLLING AGRICULTURAL OUTPUT

Recognition of the fact that commercial demand for farm products is limited and that it is growing less rapidly than farm output in industrial countries, has meant that increasing attention is being given to ways to adjust farm output more nearly in line with market

---

[1] This is only true for exporting nations. Importing countries could maintain high feed-grain prices via tariffs and use the tax to reflect world prices to producers for increased output.

demand at price levels which are consistent with social objectives regarding producers' incomes.

The EEC in its agricultural policies apparently assumes that the problem of modern agriculture in an industrial society can be successfully dealt with by modest changes in absolute or relative product prices. Price policy is one in which the United States has had a great deal of experience, and this experience indicates that modest downward adjustments in farm product prices in the face of rapid technical advance and economic change will not slow the rate of increase in output sufficiently to produce satisfactory incomes for farmers in a 'free' market. Therefore, it is likely that the EEC policies of guaranteed prices at present levels will encourage additional farm output and add significantly to the general problem of excess resources in the agriculture of the developed countries. The ability of the EEC to shift the costs of these additional resources to the consumers within the EEC and to producers outside the Community, by displacing lower-cost outside supplies, does not alter the economic realities of the policies.

The EEC spokesmen have proposed that fixing of the margin of support be the basis of international agricultural policies of the industrialized countries. Under this proposal the upper level of support and of the export subsidies would be fixed for the various countries by negotiation. The level of support could vary from one country to another and the method also could vary.

This policy is doomed to failure as a successful international policy for industrial countries on two counts. First, it assumes that the increases in output of agriculture of industrialized countries at rates exceeding the growth in demand will not occur in the absence of price increases. Second, it makes no provision for sharing the adjustment problems with countries outside the EEC who are affected by world market developments.

It is true that higher farm prices will bring a faster rate of expansion of production in modern agriculture. But it is probably equally true that output of modern agriculture would expand more rapidly than commercial demand even with farm prices so low as to cause great social distress among farm people. In the absence of agricultural production controls, the fixing of price supports at anything like recent national levels is likely to result in the chronic expansion of output at a faster rate than commercial demand expands.

The United States' experience indicates that even substantial reductions in farm prices are insufficient to reduce the rate of increase of farm output sufficiently to bring it into balance with demand. Price declines sufficient to reduce output would be inconsistent with social policy by increasing the income disparity between farmers and non-farmers. Therefore, there has been continued experimentation with devices other than market prices to reduce and restrain the rates of increase in farm output.

In the United States, control has been exercised largely by reducing the acreage of total crop-land used and the acreage planted to individual crops under price supports. Success has varied but certain problems have arisen. First, the importance of land as a factor of production is steadily declining in modern agriculture. Fertilizer, irrigation, drainage and other capital items all can increase the output per acre. Thus, controlling output via land is a control steadily declining in effectiveness.

Second, modern mechanized agriculture brings certain economies of scale in the use of labour and machines. Control of land inputs on individual farms retards or prevents farmers from achieving these economies and thus retards the structural adjustments to changing technology needed in all countries. Finally, control of output via land results in rises in returns to land relative to other factors of production and consequently in rising land prices. This has important effects upon resource combinations on individual farms and for agriculture in total. Both of these results are likely to be considered undesirable on several counts.

Recognizing the structural situation facing agriculture and the difficulties inherent in controlling output by controlling inputs, a scheme has been proposed for sharply controlling farm output by drastically changing the economic situation facing individual farmers. Its chief architect has been W. W. Cochrane, Professor of Agricultural Economics, University of Minnesota, and for three and a half years (1961–64) chief economic adviser to the U.S. Secretary of Agriculture.[1]

Cochrane's proposal runs as follows : Recognizing the characteristics surrounding supply and demand in modern agriculture, what is needed is a way of reducing and then maintaining agricultural production at the quantity that will be absorbed by commercial markets at 'fair' prices. Fair prices would be prices that would produce satisfactory incomes on the average for the farm population.

[1] A detailed description of the proposal can be found in Willard W. Cochrane, *Farm Prices, Myth and Reality*, University of Minnesota Press, 1958 ; and W. W. Cochrane, 'Some Further Reflections on Supply Control', *Journal of Farm Economics*, Nov. 1959, pp. 697–717.

The limitation of production is to be achieved by requiring that only production covered by a marketing certificate can be marketed without the payment of a substantial cash penalty by the seller. Marketing certificates would be issued to producers annually in advance in quantities sufficient to fill the aggregate demand at the target price levels. The marketing certificates would be issued in proportion to a quota base assigned to each farm. Initially the quotas would be distributed among producers on the basis of historical production shares, but the quota bases would be saleable between producers.

This scheme basically alters the shape of the revenue curve facing individual farmers. Under competitive conditions the demand curve for the individual producer is completely elastic, so that the individual's revenue is always increased by increasing production as long as his marginal cost does not exceed the market price. This holds for every individual farmer even though the aggregate demand curve is inelastic so that any additional output decreases total revenue. The scheme of saleable marketing quotas would result in individual farmers facing a declining revenue curve for output in excess of their quota. This would come about because the possibility of buying or selling quota must be added to or subtracted from the gross revenue available from the farm production marketed. It can be shown that the resulting revenue curve for the individual is similar to the aggregate revenue curve.[1]

This proposal has never been tested in actual operation, but it has been widely analysed and several observations can be made about it. First, if this proposal were applied to a situation where excess capacity existed and factor rewards were low, marginal returns to factors producing farm products probably would not rise and would be likely to fall to the level earned in the next more profitable use outside farming. Thus, returns to labour, land and capital might fall or remain the same as before, but total income to those selling farm products would rise with the increase going to the holders of marketing bases. Thus, total revenue to producers of farm products could be increased and maintained but the distribution of the increased revenue would depend largely on the initial distribution of marketing bases.

Second, under such a scheme the rate of advance of new technology

---

[1] A theoretical discussion of the Cochrane scheme and its implications can be found in Lyle P. Schertz and Elmer W. Learn, *Administrative Controls on Quantities Marketed in the Feed-Livestock Economy*, Technical Bulletin 241, University of Minnesota Agricultural Experiment Station, Dec. 1962.

would not necessarily be slowed ; in fact it might increase as a result of the additional price stability offered to producers. But the net rate of return to new resources would be no greater than under free markets since they would have to buy marketing certificates to retire other resources sufficient to maintain output at the desired level. Therefore, the scheme would be neutral in so far as productive efficiency in the economy is concerned.

Third, such a scheme of marketing quota certificates would not represent governmental controls over individual farmers' operations in the way that acreage allotments and other such schemes do. The individual would be free to produce with the optimum combination of resources. An individual farmer could expand output if his costs were low enough to make it profitable by merely buying the marketing rights of others. New entrants would essentially find the acquisition of quotas an additional fixed cost much as land or capital are now.

The scheme breaks down on two counts. First, its administrative feasibility is doubtful for non-storable products, especially those subject to large unplanned fluctuations in output. It would be extremely difficult to administer for livestock products and even more difficult to correlate the food grain and livestock sectors wherever there are inter-farm sales of feed grains. It would require a large measure of acceptance by farmers to be administratively feasible, and such acceptance is unlikely unless their only alternative is extremely low farm prices for an extended period. Even if acceptable to farmers, a significant administrative machinery would be required to inaugurate and carry on such a programme, especially where marketing channels are diverse and loosely structured.

Another obstacle to such an approach to the agricultural problem is its international complications. It is inconceivable that a country that is currently an importer of farm products would find it politically possible to place an effective maximum quota on its farm output at a level less than the national consumption level. It is equally inconceivable that countries with an agriculture geared to large exports could set quotas only to cover their domestic consumption, for the quantity of resources that would have to be displaced from farming would be too large. Thus, such a scheme would entail export subsidies and the same international problems as the national support programmes of the past few years.

In order to deal with some of the international complications of supply control, the concept of market sharing under international commodity agreements has been put forth. This means that international agreements would be negotiated whereby importing countries agree to import either some fixed quantity or some fixed proportion of its requirements from a given exporting country.

The idea of market sharing of markets for farm products in industrial countries is an extension of the output control mechanism to an international level. Market sharing is not likely to be effective in dealing with the general problem of excess capacity in advanced agriculture unless it includes effective production controls and government market intervention in all of the industrial countries. Importing countries cannot guarantee access and make the guarantee effective unless they restrict their domestic production to provide a place for the guaranteed quantity in their market. An alternative is to use deficiency payments as the British do, so that access agreements primarily affect the cost of deficiency payments. However, the British have found it necessary to negotiate import controls to control their programme costs.

The idea of guaranteed access is an international extension of the domestic proposal for market shares via marketing certificates. It will require some mechanism, either economic or political, to remove excess resources from agriculture as technology advances in the participating countries. The negotiated market shares would determine the distribution of the burden of adjustment between producers in the different countries. Although international commodity agreements have discussed the removal of excess resources from agriculture, no real evidence of success is found.

The basic idea of production control in agriculture is to achieve a downward adjustment in the quantity of resources producing farm products. Ideally this adjustment would be sufficient to allow the remaining resources to earn satisfactory incomes in the market without continuous large government expenditures for surplus storage or surplus disposal. In practice, however, supply control is not likely to be effective enough to more than keep the magnitude of the other programmes within bounds.

It is clear that supply control only by the exporting countries is not sufficient to deal with the total problem of the industrialized countries. To deal successfully with the present international situation, some kind of effective supply control will be required in all of the industrial countries, importers and exporters alike. The political possibilities of this occurring seem remote at the present time.

Even if effective supply control could be instituted for all of the industrial countries, they would not have solved all of the international problems. Should the supply control be such that it will raise the world price of agricultural commodities to rich and poor countries alike? If so, it will certainly encourage poor countries to expand output of commodities which might be produced more efficiently elsewhere. Some would benefit from and others suffer from serious exchange difficulties if the world market prices for agricultural commodities were increased significantly. Thus, effective supply control by the developed agricultural economies would require some provision for special programmes for some poor countries.

VI. CONCLUDING SUMMARY

Industrial countries rather universally experience agricultural problems that are similar in nature, but which may be temporarily hidden by import policies of importing countries. The supply and demand conditions in modern agriculture tend to result in a chronic excess capacity in agriculture, so that farm output exceeds the effective commercial demand at prices necessary to provide a satisfactory income for many of the producers in agriculture. The immobility of resources and people from farming is such that large average income gaps persist between farmers and non-farm workers.

All of the past agricultural policies of the industrial countries have created some kind of international difficulties. These difficulties have become increasingly apparent in recent years in terms of trade negotiations, disruption of traditional markets and the disruption of export markets upon which developing countries are heavily dependent.

Much of the difficulty lies in the failure to recognize that the agricultural problems of the developed countries are common problems incapable of individual solution on a satisfactory basis. There also is inadequate recognition that the agricultural problems of the industrialized countries and the methods used to deal with them have significance for the developing countries.

Recognition of the fact that the economic conditions facing farmers in industrial countries are widely shared is but a first step in the process of developing better policies. A second step must include recognition that individual unco-ordinated national policies cannot suffice in a situation that involves a substantial number of countries. A third step must include recognition that the political economy of an international problem is different from that of a national one in important ways. Finally, a stable solution, either politically or economically, of the agricultural problems of developed countries can only be developed in an international context taking into account the economic and political interests of both the industrial countries and the under-developed countries.

The economic and political pressures that lead to a search for institutions to supplement the market for agricultural products in advanced countries are not likely to diminish. There are indications that the multitude of unco-ordinated national programmes is increasing the desire for some kind of international programme. The World Food Program experiment is one of these indications. Hopefully, others which deal with the fundamental problems will follow at a rapid rate.

———

DISCUSSION OF PROFESSOR HATHAWAY'S PAPER

Professor Campbell said that he thought that the survey of the agricultural situation in industrial countries presented in the first part of Professor Hathaway's paper applies more particularly to the temperate-zone agricultural exporting countries. The picture which Professor Hathaway drew may be a fairly accurate representation of the agricultural situation of the United States, but it cannot be generalized. Professor Campbell said that there are some who would argue that there is no chronic over-capacity in some of these countries, either in terms of effective demand or in terms of the prospective rate of increase in world population. What is the evidence for Professor Hathaway's assertion that the income elasticity of demand is as close to zero in all these countries ? The evidence on consumption trends presented in Professor Robinson's paper do not seem to support the contention. Even if Professor Hathaway was correct in his general assertion, the outlook for some food products, such as meat, is not nearly as depressing as he suggested. The differential between agricultural and industrial earnings in some of the named countries — particularly Australia and New Zealand — is very narrow as compared with the United States. The importance of labour immobility depends very much on whether an absolute decline in the use of labour resources in agriculture is necessary over time. Professor Campbell said that, more generally, he questioned the adequacy of the economic notion of asset fixity as the sole explanation of the lack of responsiveness of agricultural output. In short, the Conference needs to give some attention to the universal relevance of Professor Hathaway's description of the state of the agricultural industries. It might also be asked whether the adjustment of the situation described by Professor Hathaway represents the sole goal of the international commodity arrangements sought by the industrialized countries.

Professor Campbell said that Professor Hathaway while expressing optimism about the international co-ordination of agricultural policies, claimed that agriculture is lagging behind as compared with the progress made in other areas of international collaboration. Professor Campbell said that he disagreed with this contention as a contemporary view, and would return to the point at the conclusion of his remarks. He thought that much of the international problem as stated by Professor Hathaway stems from lack of restraint in income-support measures by relatively few countries, in particular the United States and specific countries of Western Europe. In countries which are poor or where agriculture is a significant sector in the economy, there are definite limits to the amount of income transfers that can be achieved via agricultural support measures. In many cases the income transfers are in the opposite direction. Thus, while it is true that the surplus problem in the world wheat market would be much greater were it not for the storage and surplus disposal policies of the United States, it is too superficial a diagnosis to claim that the international state of excess capacity has been inevitable, given the inherent characteristics of the agricultural industries.

Commenting on the third type of programme for the manipulation of world demand as discussed in Professor Hathaway's paper — the one which involves the use of the principle of price discrimination — Professor Campbell said that no clue is given as to how the respective price levels are to be determined. Despite their political and economic advantages, schemes which involve price discrimination raise certain problems. For instance, it is difficult to develop effective schemes for products with close substitutes, or for intermediate products such as feed grains. It would involve substantial administrative machinery to allocate production bases to producers and to collect taxes. Professor Hathaway recognized that programmes of price discrimination might have some adverse effects on the under-developed countries, but he did not elaborate them.

Professor Campbell said that while he shared, in some measure, Professor Hathaway's apprehension about the effects of EEC price policies, he challenged his proposition regarding the ineffectiveness of price movements as a means of influencing output, particularly the output of individual commodities. Professor Campbell suggested that, whatever the U.S. experience, there is evidence of considerable shifts in agricultural resources in, for instance, the United Kingdom and Australia, as a result of movements in product prices. Moreover, the political acceptance of lower prices may be a function of the pattern of income distribution within agriculture. If the particular problem of the low income farmer is resolved, political pressures inhibiting flexible price policies may be substantially reduced.

Professor Campbell said that while he had appreciated the comprehensive and logical paper, he had been disappointed that Professor Hathaway had approached his subject in the way he had. Professor Campbell thought that it would have been much more fruitful to have discussed specifically the progress already being made in the international arena towards the co-ordination of national agricultural policies, rather than to have discussed, as Professor Hathaway had, a series of rather hypothetical (if logically patterned) schemes based on the battery of policy instruments either used or proposed in the United States. For example, there are significant pointers to be found in the changing pattern of international wheat agreements, in the discussions that stemmed from the so-called 'French Plan', in the United Nations Conference on Trade and Development, in the various negotiations concerning sugar and, perhaps most importantly, in the activities of the Cereal, Meat and Dairy Product Groups of GATT. To have discussed the nature of the proposals under current discussion, to have analysed their possible economic consequences, to have examined their political feasibility, to have talked more specifically about price levels and how they might be determined and to have suggested, if the price mechanism is to be superseded how market quotas would give proper recognition to comparative advantage and efficiency of resource allocation in a world context — these exercises would surely have been more pertinent and helpful in the contemporary situation than the approach actually followed by the paper.

Professor Campbell said that in the session of the Council which had preceded the Conference, Sir Ronald Walker had pleaded for the forging of closer links between academic economists and those working in the international agencies. In this vein, Professor Campbell suggested that more progress towards establishing such a bridge would be achieved by discussing the important developments which have been occurring, say, in the GATT Cereals Group, than by discussing Cochrane's proposals which, in Professor Hathaway's own judgement, are unlikely to be adopted even in the United States.

Professor Hathaway said that he thought that there is an irreconcilable difference between Professor Campbell and himself on what is the most fruitful topic for discussion. Professor Hathaway said that he did not think that progress has been made in international agricultural policy; there is a long way to go before countries even begin to get together on the underlying problems. Even the plans to which Professor Campbell referred at the end of his comments are little more than nebulous proposals full only of good intentions.

Professor Hathaway said that he thought that Professor Campbell's views on who has chronic over-capacity may be determined by the experience of a country that has never tried production controls, whereas his own views were certainly affected by the fact that his own country has withdrawn fifty million acres from production so that other countries could sell wheat. Moreover, the U.S.A. has been the only stockholder of agricultural commodities at times when serious downward pressure made itself felt on international commodity prices. Thus, satisfactory incomes for farmers in some countries of the world are a result of the generosity of the U.S. Treasury and the Canadian Wheat Board.

Professor Hathaway said that he agreed with Professor Campbell that income elasticities are not as low for some countries as they are for the United States. However, the elasticities to which his paper referred are those for total agricultural output and not simply those for food. It is clear that a major factor in the higher income elasticities for food is the high income elasticity for marketing services.

Referring to the question of the effectiveness of prices in the attempted control of production, Professor Hathaway said that the United Kingdom has in fact had trouble with the payments she makes and that the level of income support is quite high. However, he did not argue that relative prices are unimportant in controlling the mix of agricultural products, but rather that the effectiveness of using price policies to move resources out of agriculture is to be questioned. Price adjustments necessary to force people out of agriculture would have to be very large ; larger, indeed, than they have been in the United States. Professor Hathaway said that he was not convinced that price adjustments of these magnitudes would be politically acceptable. Until other countries tried to deflate the absolute level of farm prices, there can be no evidence that it would be effective in reducing output. The experience of the United States is at present all that can be cited, and this evidence is conclusive in the opposite direction.

Mr. Allen, referring to the last few remarks of the paper, asked if Professor Hathaway literally meant the last decade when he referred to the apparent inability of lower prices to limit output in the United States ? Mr. Allen said that 1955–64 takes us from the trough of the post Korean war slump in total farm incomes through a steady and substantial recovery to a peak around 1963 and a subsequent levelling off. He said that this was partly the consequence of the technological revolution which Professor Hathaway had mentioned, but it was also assisted by extremely favourable weather conditions in the Corn Belt in 1957–63. Mr. Allen asked if it is possible to conclude anything about the effect of price on output during this period when other factors most dramatically did not remain constant.

Professor Mundlak said that to someone unfamiliar with the problems of the international flow of commodities, the most difficult question is how to measure the success or lack of success of different policies. Is it to be measured in terms of increases in excess supply ? Or is it rather a problem of measuring the extent of the burden of the excess supply which already exists ? Even if income elasticity in many parts of the world is high, that does not solve the problem unless the income of these people can be increased. How are we to increase the income of people overnight ?

Meanwhile, factors such as the advancement of technology act to aggravate the excess supply.

Professor Hathaway had said that in the United States prices for agricultural products had gone down with little effect on production. The important question is what would have happened to production if prices had *not* gone down. If production would have remained the same even if prices had not gone down then this would support Professor Hathaway's contention ; but if production would have been higher if the prices had not gone down this would indicate that prices had not been lowered enough to bring about an absolute decrease in output. Professor Mundlak suggested that if grain prices had gone down enough, grains would have been used in meat production and with the higher price elasticity for meat much of the surplus might have been taken up.

Professor Mundlak said that there is a general tendency at the Conference to play down the role of prices, and yet at the same time to argue that people are moving out of agriculture which indicates that prices are having an effect. A similar contradiction seems to be involved in the points made about the changes in capital use. Professor Hathaway had said that the role of land is decreasing in importance, and that agriculture is becoming more capital intensive. Is this flow of capital into agriculture to be regarded as exogenous or is it in response to price ? This question is the real concern of economists. In the long run, when the farm population will have gone down, the welfare problem will lose its urgency, but the problem of how to control the flow of capital into agriculture will remain.

Professor Hathaway said that he had clearly given Professor Mundlak the wrong impression. Prices certainly do slow down the rate of capital entry, and as agriculture becomes more and more commercialized, it becomes more sensitive to price changes. However, the important point is that farming remains profitable at drastically lower prices and that to lower the prices by enough to stop the flow of new technology means that a heavy income burden is placed on those who do not make structural changes. A serious attempt should be made to separate the two functions that price is now attempting to fulfil. If prices are maintained in order to give adequate income to older technology, a great excess of production with the newer technology is inevitable. The dilemma consists in the fact that a price reduction sufficiently drastic is not politically acceptable in most countries.

Professor Gulbrandsen said that Professor Hathaway had suggested in his paper that an effective way to increase the total revenue to producers would be to institute price discrimination between rich and poor countries. Such discrimination already exists in as much as there are high food prices in rich countries and the earnings from these higher prices are distributed directly to producers in poor countries. However, the nature of the distribution is not optimal as far as the producers in rich countries are concerned. Professor Gulbrandsen asked whether Professor Hathaway had any concrete suggestions as to how the transfer of earnings from the consumers in rich countries to the consumers in the poor countries can be achieved.

Professor Hathaway said that it is clearly true that most countries operate a double price system. He said that he objected to the way this is done. There is no direct relation, for the producer, between the marginal value of their production on world markets and what they in fact receive for their output. Thus there is one price internally and then an export subsidy which is used to render the farmer competitive in the world market, so that effectively each country is competing with the treasuries of other countries. A scheme is clearly needed so that producers who can supply food at drastically lower prices do so at the real value of that food to the purchasers. On the problem of how to transfer the earnings from rich countries, Professor Hathaway said that he had no answer. However, any system which involves the farmer producing for the world market at the true value of his production to the consumer would be a marked improvement over what has obtained in the past.

Professor Bishop said that he was concerned with the tendency to take a short-run view of these problems, when the long-run view gives even greater cause for pessimism. Does it really make sense for the highly developed countries to think in terms of pulling commodities off the world market and raising prices to obtain reasonable returns? Such an approach completely disregards the commitment of these advanced countries to the world scene.

Professor Hathaway said that he agreed with Professor Bishop that it does not seem to make sense in the long run for the output of those countries which, by any standards, are the most economic producers, to be reduced. The retiring of the most economic producers in this way would then have to be offset by the use of high prices to make less efficient producers fill the gap. Modern agriculture is highly capital intensive so that the comparative advantage lay with developed countries. Moreover, countries with the most acute food shortages have high man/land ratios. This question clearly needs discussion, and the present topic has to be brought into relation with it.

Professor Haley said that Professor Hathaway's paper was a very pessimistic one. This led him to make two comments.

First, the inelasticity of supply of agricultural commodities in the long run may not be as low as thought by Professor Hathaway. After all, resources do shift, and lower support prices should encourage the shift. Professor Hathaway kept saying that a reduction in support prices to a level sufficiently low to produce such a shift would be politically unacceptable. But, Professor Haley said, no one professes to use a reduction in support prices as the only device for achieving the desired shift of resources out of agriculture. This device should be combined with other appropriate measures. In any event, we should not dismiss any proposal that is otherwise acceptable simply on the ground that it is not politically feasible. In the United States a great reduction in support prices would certainly have to be accompanied by direct payments to farmers. Even if this is not politically acceptable today, this is not necessarily going to continue indefinitely to be the case.

Secondly, Professor Haley said that Professor Hathaway stated in the second last paragraph of his paper that, 'a stable solution . . . can only be developed in an international context . . .', yet in the body of his paper Professor Hathaway discarded all proposals for international action either on the basis that they do not involve production controls or because they are not feasible politically. Surely, Professor Hathaway is too pessimistic ? We cannot expect to achieve the perfect form of international action all at once. We have to proceed small step by small step. For example, even though the EEC proposal for an international agreement limiting the height of support prices may not go to the heart of the matter, it could be a step in the right direction.

Mr. Robinson said that he felt emboldened by the final paragraph of Professor Hathaway's paper and by the comments of Professor Haley to suggest that perhaps the view the Conference has taken on the possibility of agricultural development in developing countries is too pessimistic. A great deal of the capital required in developing countries can be provided by labour. It has been estimated that nearly half the cost of labour-intensive projects can be paid for in food. Professor Hathaway had not mentioned the experience of the United States in providing food aid. The World Food Program, operating in small rural public works, is improving human resources in developing countries. The first steps were already taken in channelling resources into developing countries at lower cost than the straight transfer of cash.

Professor Dandekar said that the basic problem seemed to be that, at prices which the free market provides, agriculture fails to produce enough income for the rural population of a country. If, however, producers are paid more than the product will fetch, then more food is produced than the country needs. This seemed to be the experience of many countries. At what stage of development does this happen ? Is this related to a stage in economic growth indicated by some proportion of the population still remaining in agriculture ? Can it be argued that, even in the Indian situation, if a standard of living above what the market provides is introduced, agriculture will pick up, and produce more — even with the available technology ?

In answer to Professor Dandekar, *Professor Hathaway* said that the situation develops when agriculture achieves large shifts in the supply curve as a result of re-structuring and the introduction of improved technology. He said that he had no idea what is required to achieve this in terms of proportion of population in agriculture. It certainly is not a question of relative income levels or price levels in any absolute sense.

In reply to Professor Haley, Professor Hathaway said that he did take a pessimistic view, if only because optimists are chronically disappointed. He thought that there would be a series of international arrangements which would benefit developing countries. These arrangements would probably continue to do a bit of everything. What particular conglomeration of arrangements would ensue is very uncertain, but it would almost certainly include some programme of international price discrimination and international production controls.

Chapter 4

FREE TRADE AND PLANNING IN THE COMMON AGRICULTURAL POLICY

BY

MARIO BANDINI

I. INTRODUCTION

In order to understand the general problems of the common agricultural policy, which are the subject of this short paper, it is necessary to know a few essential facts concerning agricultural development during the past decades which are at the root of the situation today. The best that can be done in the space is to recall these facts, almost in the form of a list.

The free trade system in international trading, which had its greatest vogue in the middle of the last century, was progressively disrupted by the reappearance of protectionism in the typical form of customs duty. First of all, it was of an industrial nature and then later, with the purpose of ensuring adequate returns, it also affected the agricultural sector. In agriculture, protectionism manifested itself mainly with the duty on corn, followed by tariffs on other products, including, notably, sugar beet. Agricultural protectionism was aimed at opposing the so-called American competition, which came about with the planting of vast, new areas of land capable of producing cereals at a low production cost.

In the period between the two wars, simple protection was made stronger by a series of measures, basically nationalistic or autarkical, introduced by nearly all the big European nations. Added to the relatively simple weapon of tariffs, there then came rigid quantitative restrictions. All this was done in the forms and with the results which we must regard as being well known to everybody. The consequences of the nationalistic policy were, in our opinion, extremely serious for the economic development of Western Europe. It is not only the fact of the contraction (or the lack of expansion) of international trade which confirms us in this judgement. As far as agriculture is concerned, we note other consequences of great weight.

77

II. THE CONSEQUENCES OF NATIONALISTIC PROTECTIONISM

First of all, there was the illogical international distribution of the principal agricultural products. With economic nationalism, every country wanted to cultivate even the products which could have been better obtained abroad at considerably lower cost. This mainly applies to those mass products, of which cereals are the chief example, which, in a refined agriculture like that in Europe, ought partially to have given place to quality products (vegetables, fruit, wine, prime meat, products of agrarian industries, by-products of milk and the like) which were more convenient for European agriculture. Protectionism and economic self-sufficiency checked this healthy trend. In this way, it came about that the advantage gained by the cultivation of corn was dearly bought with the loss of outputs which were more natural for the European agrarian economy. All this, obviously, resulted in keeping agricultural costs relatively high, in a lower competitive capacity and in a failure of supply to meet the demands progressively tending towards quality products.

The nationalistic policy also had an influence on the employment of means of production ; this was, on the whole, slowed down. The development of rich or quality crops would have brought about a greater employment of fertilizers and given greater place to stock breeding, which is so beneficial in improving the land. It would also have brought about the application of more advanced techniques as regards planting and rational systems of fighting disease, not to mention a definite improvement in the commercial presentation of the products. All this was hindered or slowed down. In the Mediterranean territories the development of irrigation and quality crops was also slowed down ; instead, an excessive cultivation of corn, often poor and primitive, continued.

We turn now to a point of great importance for understanding the problems of today. The trends, which we have briefly touched upon above, have not only influenced the course of farming but also the productive structures which are practically permanent or at any rate permanent over a long period of time. The observation is important. If it had only been a question of anti-economic productive trends, a few years of cultural reconversion would have quickly rectified the situation and eliminated the influences of the errors committed. But that agrarian policy led to the formation of agricultural structures not easily alterable or, if so, only in the long term. In other words, it may be said that the effects of protectionism or nationalism cannot be eliminated or reduced today except after a long process of reshaping and adapting agriculture to different situations.

What, broadly speaking, may be defined as the defective structures of European agriculture are varied in type. For example, excessive protectionism has determined the persistence of poor, primitive and socially and economically backward latifundian structures. There was no incentive to divide up the land and no new business classes capable of modernizing agriculture. Southern Italy, Spain, Eastern Germany, many parts of Central Europe and certain parts of Southern France give clear examples of this. The nationalistic policy has also brought about the retention on the land of excessive quantities of labour which, in turn, hindered further modernization, such as by mechanization. Some examples of protectionism (sugar, for example) have given rise to the formation of important industrial structures which today one obviously cannot think of abolishing.

Turning now to the problems that arise specifically with the establishing of the Common Market, there is one aspect of the nationalistic policy which makes it particularly difficult to reverse our steps. It consists in the fact that each country having operated from its own viewpoint, which was not necessarily right in its own interests, the productive and market structures, the price levels and the advantage ratios of the individual crops are seen to be quite different from country to country. This has now rendered the process of European unification, which the Common Market is pursuing, particularly difficult. There are very many examples of this. Within the Common Market there persist economic problems and agricultural structures which are basically different. There is France with a vast extent of good land and relatively low prices (particularly for cereal products); it can produce more of these than it needs and become an exporter of cereals and meat. Then there is Germany, which produces about three-quarters of its foodstuff needs, but which desires to maintain the high level of the peasant incomes and so has prices slightly higher than those in France. Italy also has relatively high prices, particularly for cereal products but, as opposed to Germany, it is interested in the growing exports of its quality products in or outside the Common Market. Then we have Holland which also has problems for the export of the greater part of its livestock products.

To reach an agreement over these different situations in a single view of agricultural policy was not and is not an easy undertaking.

III. THE COMMON AGRICULTURAL POLICY

(i) *Introductory*

The Community agricultural policy pursues certain aims laid down in general terms by the Rome Treaty, in language which at times leaves economists dissatisfied, but which is normal in documents of this kind. It even sets out to achieve contradictory results such as, for example, parity of agricultural incomes in relation to those in other economic sectors, without compromising the interests of the consumer. But rather than examining what the Treaty says, let us have a look at the actual common agricultural policy. We must necessarily take its principles as known, remembering only that by the end of 1970 (or perhaps the middle of 1967 according to the latest views) the Common Market has to reach a situation whereby the movement of goods, men, businesses and capital is perfectly free within the ambit of the six countries.

In the Treaty, agriculture is dealt with in a special way, since it is established that the extention of the Common Market to it must be *accompanied* by the application of a common agricultural policy. Note that the Treaty does not say *preceded by* or *followed by*, but *accompanied*. This common agricultural policy which is laboriously and slowly being put into operation, was found in reality to be extremely hard to formulate, much more so than might be imagined. The objective reasons are those set out above. It is only progressively and with co-ordinated action possible to emerge from the situation of agricultural disagreement between the various countries and from the lasting effects of economic nationalism. A sudden change-over to an unshackled situation is inconceivable. In other words, it is necessary to plan the march towards free trade which the Rome Treaty envisages. Public intervention must be logically studied, in accordance with the transformation that is to come, and this goes for all the sectors covered by the intervention. After all, in the other partially successful attempts that preceded the Common Market (*e.g.* those made by OEEC), it was recognized that liberalization, even though of smaller scope than the full freedom envisaged by the Common Market, had to be accompanied by a comparison of the structural and economic policies so as to bring them into accord. Thus, the common agricultural policy can only be based on large-scale public intervention, and there is no need to argue this point. What we do ask is that this public intervention be planned in such a way as to arrive at a regime of greater freedom, and that it be a logical and feasible regime.

It does not seem that the common agricultural policy is yet clearly based on these premises. This is probably a result of its youth and of the fact that, having to move and obtain certain essential results within the established time, the politicians responsible have indulged in a practical hustle without, in our opinion, a severe and critical examination of the premises. In any case, in a sense nothing has been compromised on condition that, at a given moment, the problem of the limits of public action and the problem of the need, sooner or later, to arrive at a system of free private activity, be adequately considered and solved.

(ii) *The Policy of Markets*

Up to the present, the common policy has been mainly concentrated on the so-called policy of the markets. The second great branch of the common agricultural policy, which deals with improving the agricultural structures, has only taken its first steps. The policy of the markets aims at giving economic stability to agricultural production and its various sectors. In a world characterized by hysterical markets and, often, by situations stemming from origins of long ago, in a chaotic system of agreements and trade, dominated more by political than economic factors, and with extremely variable prices, it seemed necessary to achieve a certain amount of order in the agricultural markets of the Six. This was done by means of a technique, now well known, aimed at stabilization of agricultural prices at levels made known before sowing. Once the so-called 'indicative price' is fixed (*i.e.* the target price as a general indication), on the basis of that an intervention price is established, which is generally 5 or 10 per cent lower. In point of fact, this is the price at which the authorities promise to purchase the products that may remain unsold. By means of the market regulations, the agricultural policy tends to consolidate the indicative price by a system of sliding duties, changed daily, the measure of which is equal to the difference between the world market price and the indicative price. Actually the market machinery is very much more complicated and, for some products, the system is a little different; but, in the main, that is how it works.

The system of dues aims at providing secure prices for farmers. This, according to some views, is a good thing; but according to others, it can only give rise to profound concern. From the strictly juridical point of view, there is also something to be said: to be exact, the Rome Treaty does not provide for stabilizing prices but for stabilizing the markets, which is quite a different matter. Stabilizing markets means removing from them every reason for functional deviation but leaving their functional role intact. In other words, the swing of prices up or down cannot be suppressed, because only in this way can the markets direct production, adapting supply to demand. Fixed and stable prices, on the other hand, suppress the physiological function of the markets, and that ends in maintaining production levels and productive structures which should be eliminated or transformed by competition.

It thus seems to us that the system of stabilizing prices may only be admitted in the initial phase of setting the markets in order but not as a permanent principle of the life of the economic-agricultural community. It may also be maintained for the purpose of avoiding too rapid fluctuations, *i.e.* as a temporary expedient ; but, in the long run, it has the danger of removing the functional role from the free trade system.

Other reasons for concern are found in the practical realization of the price policy. The most difficult and serious situation undoubtedly is in the corn sector. The events that accompanied the introduction of a community cereal policy are known to all and they are expressions of extreme difficulty only overcome by compromise. The result arrived at, holding the prices of corn not far from what were the highest levels in Germany and Italy, gives a very great advantage to French production and an incentive for France to extend and intensify her production to limits which before today were inconceivable. It should not be forgotten that corn cultivation can easily be mechanized, that France already possesses a million tractors, that this crop is adapted to the vast size of French farms and that it may be carried out with very little labour. The danger of the community policy is clearly that of obtaining, within the community, a mass product, such as corn, in quantities far beyond the internal needs. This excess corn will be sold on the world markets at exceptionally low prices. The burden of paying the difference between the price paid to the farmers and the world price, which is sensibly lower, will be borne by the Community at a cost to be divided between the six countries in accordance with the proportions laid down by the Treaty. Now, in the long run, all this can bring about a false orientation of the community agricultural production which, as we said at the start, is naturally inclined towards quality products. It is also obvious that trends which leave wide scope to cereal production arouse violent reactions from the countries that used to sell these products to Europe. The greatest protest came from the United States, which brought up the problem at the Geneva conference known as the Kennedy Round. The American Minister, Freeman, does not lose any opportunities for making his displeasure felt. The problem is complicated. Europe certainly cannot become the dumping ground, at any cost, for the alarming agricultural surplus which the United States has allowed to grow with its policy favouring agriculture. But in a normalized system, without the nightmare of those surpluses (which weigh on all the world markets, not only on those of North America), it is perfectly reasonable to consider the possibility of an economic-agrarian co-ordination between Europe and the United States, the former tending more towards quality products and the latter towards the export of mass products. This would be mutually convenient. So also in this case, rather than a conflict we must reach a co-ordination between planning and free trade. In other words, public intervention and international agreements must aim at leading to freer systems and make it possible for them to work.

(iii) *Policy of Structures*

The second aspect of the community policy is the improvement of the structures. This is a distinct question because, if the efforts of the common agrarian policy tend above all to eliminate the old defective structures and establish, in every country, those which are best reconciled to the necessary international competition, it will be possible to render the community system logical, and eliminate many of the perplexities arising from an excessively controlled policy.

The policy of the structures should above all determine types of farms which are more competitive, increasing their size, making mechanization possible and developing the processing industries. This policy should determine a better and more natural system of locating the various agricultural productions in the different territories of the community, so as to obtain wider and more rational trading on the basis of the old doctrine of comparative costs.

If the improvement of the agricultural structures, linked with the reduction of costs and better localization of the productions is successful, presumably this will be a good instrument for simplifying the market policy and rendering it more valid economically with the passage of time. After all, this is one of the points on which those most responsible for the community agricultural policy all agree.

IV. CONCLUDING REMARKS

We regarded with pleasure the statements that Mr. Mansholt made to the European Parliament commenting on the resolutions passed by the Council of Ministers of the European Community on 15 December 1964. He declared that 'an important simplification of market machinery is now in sight. During the period of transition this complication was necessary, but in the long run, the flow of regulations can become fatal. In the course of a few years, it is important to establish a free market governed by the conditions of competition.'

As regards the system of protection from outside countries (which undoubtedly it is planned to maintain) and in consideration of the Kennedy Round discussions, Mansholt clearly stated that 'protection from outside countries will have to be consolidated and the dues will lose their variable character'.

One cannot but agree fully with these trends and be most satisfied with them, in the hope that the manias for controlling the markets and planning production do not get the upper hand and render vain these good and wise intentions.

We may therefore close this short report by saying that the interventions in the programme or planning stage of the community agricultural policy must be regarded as the necessary forerunner in the elimination of the heavy incrustations of the past and in the preparations for the advent of an agriculture with modern structures and a more efficacious economy. We admit that with the situation in the world today, the task cannot be left to the free play of the economic forces. We must aim at bringing about these transformations with the least possible jolt and the least possible damage.

But the goal should be a freer and less controlled economy, with a minimum of planning affecting the logic and rationality of public interventions, which must be limited to their own field. To work, therefore, according to a programme, with a view to reaching a situation which permits the reduction to a minimum of the needs for programming and planning — this would seem to be the essence of a healthy community policy.

DISCUSSION OF PROFESSOR BANDINI'S PAPER

Professor Ashton said that he had found it illuminating to contrast the historical perspective provided by Professor Bandini in the introduction to his paper with the completely different sequence in the history of the United Kingdom. The repeal of the Corn Laws in 1846 had given rise to the liberalization of trade with important consequences for British agriculture. Especially in the last quarter of the nineteenth century British agriculture was in a state of rapid change, and there had been no special provision for the welfare of the agricultural population. The necessary adjustments had been painful to those who were squeezed, but employment opportunities elsewhere had been relatively numerous. There had been a rapid migration of farm population and by 1900 only about 10 per cent of the working population remained in agriculture. Moreover, the United Kingdom had experienced 'structural reform' in the enclosure movement and this had given rise to a size structure which permitted a more extensive agriculture. The size pattern is now taken to be rather static, but it is in a relatively healthy state in comparison with other countries in Europe. While the average size of holding is between 30 to 35 hectares, the average size of effective farm businesses lies somewhere between 50 and 55 hectares. The parallel development on the marketing side is based on free trading and there is easy access to the British market for producers throughout the world. The British public benefited from this situation. They enjoy a varied diet, and the supply of agricultural produce from many parts of the world is rapidly adapted to the growth in income.

Professor Ashton said that experience in the United Kingdom tends to confirm Professor Bandini's analysis in that structural and marketing problems are not necessarily as acute as those in the rest of Europe, which largely stem from national policies of self-sufficiency.

Professor Ashton said that economists may well ask why agriculture has to be singled out for special provision. For instance, can the problem be subsumed under special social policies for low income groups ? Is this not really what it is all about ?

Professor Bandini had expressed doubts on the cereals policy in the Common Market and had argued that there are dangers in the over-extension of cereal production beyond the requirements of the community. Professor Ashton said that some economists see danger in the extension of cereals even to the extent of fulfilling community requirements. Such extension might, as Professor Robinson had already warned, imply a serious failure to utilize fully and economically the agricultural resources of North America and Oceania.

Professor Ashton said the development and resulting conditions have not been the same in Holland and Denmark. These cases make an interesting contrast to the rest of Europe. The essential difference lies in their relative lack of natural resources and the degree of their involvement in international foreign trade.

Professor Ashton said that he was concerned by the possible effects on other countries of the policies adopted by the Common Market, particularly those at earlier stages of development. British agriculture could to some extent, with its structure, adapt to the requirements of the situation. But tendencies towards economic nationalism inherent in the Common Market policies could only have a damaging effect if they gave rise to similar policies in other less developed countries. Professor Ashton cited the case of Greece where there are advocates of the use of expensive irrigated land for the production of beet sugar. This sort of emulation of North European self-sufficiency policies could only detract from economical use of resources.

Despite his large measure of agreement with Professor Bandini's arguments, Professor Ashton wondered whether Professor Bandini is expecting too much from a free market. From the point of view of the farmer, at least, there are advantages in ensuring that there is some sort of fair play in agricultural policies. This means that the farmer should not be denuded of all countervailing power in his relations with consumers and factor suppliers ; and, generally, most countries would accept this as a reasonable proposition, even though it might detract from the operation of a 'free-market' in a text-book sense.

Professor Valarché said that he had two comments to make on the historical discussion which is so important for understanding the present position. First, he had some doubt about the implied causal arguments that Professor Bandini had put forward and Professor Ashton had stressed, particularly with respect to irrigation. The argument is that irrigation developed slowly in the Mediterranean countries because the production of wheat was preferred over the production of other crops. Professor Valarché thought that the relation is probably the opposite. Wheat was preferred because irrigation developed slowly. There is a high social cost to irrigation ; for example, eighteenth-century Italy was a poor country and irrigation had perhaps been well beyond its reach.

Professor Valarché said that his second point referred to the general judgement on European protectionism. There certainly had been a time when it had had catastrophic consequences, for example, between 1919 and 1939 ; but he thought that the judgement should be expressed in flexible terms. Protectionism should not be presented just as resistance to innovation stemming from a capricious decision to keep up the production of wheat. Protectionism at the end of the nineteenth century had attenuated the short-run disruption and damage that could arise from rapid specialization. It had in general not proved an obstacle to modernization as the case of Germany illustrated. In particular commodities, foreign trade in agricultural products had increased. Professor Valarché cited the cases of French wine and Swiss dairy products as commodities in which trade continued even in the period of high protectionism. Professor Valarché stressed the important strategic role played by wheat during the two world wars. He said that Professor Bandini had been a little severe in his judgements on protectionism.

Professor Gale Johnson said that he wished to comment on — even protest about — the implications of Professor Bandini's sentence on page 83 of his paper : 'Europe certainly cannot become the dumping ground at any cost for the alarming agricultural surplus which the United States has allowed to grow with their policy favouring agriculture.' This seems clearly a case of the pot calling the kettle black. While United States policy has admittedly had some positive effect in increasing agricultural production, there is little doubt that Western European policies have had an even greater effect in this direction. The European attitude, that their efforts to increase European output have not contributed to surplus problems, is curious. The United States has at least gone through motions to restrict output but there is no regard for this in Europe whatsoever. Western Europe has contributed significantly to the size of agricultural surpluses.

Mr. Allen said that he thought Professor Gale Johnson had been too generous. He wondered why Professor Johnson did not also take issue with the fifth sentence on page 83 of Professor Bandini's paper : 'But in a normalized system, without the nightmare of those surpluses (which weigh on all the world markets, not only on those of North America), it is perfectly reasonable to consider the possibility of an economic-agrarian co-ordination between Europe and the United States, the former tending more towards quality products and the latter towards the export of mass products.' Mr. Allen said that if the United States had not supported farm prices, especially cereals, prices in world markets would have been lower. He said U.S. support policies have provided an 'umbrella' in many world markets, thereby limiting competitive pressures on other major exporters or potential exporters. Mr. Allen said that if anyone has suffered it is not the Common Market countries, but major food deficit countries such as Japan and the United Kingdom who have paid more for their food imports than would have been the case without the U.S. system of price supports.

Professor Robinson said that he hoped a broad view of what has happened in agricultural trade in the past would not be irrelevant. Before 1914, Europe had been the hub of the world trade system. As exporters of manufactures (especially textiles), as the main source of machinery and as beneficiaries of long-established connections and interests throughout the world, European countries had had no difficulty in paying for their imports. The catastrophe of two world wars had had serious effects on financial relations with the rest of the world, and, in particular, had resulted in a severe reduction in income from overseas investment. The hard task of working out a new system of world trade had had to be undertaken. In the early years of the century we had been able to import from Canada and North America, and to pay by direct export earnings or from investment income. After 1918, there had been a rearrangement of North American trade. Canada had fallen increasingly into the American orbit. For the United Kingdom this had meant a serious problem in trying to provide dollars for Canada to pay for imports from the United States.

After the second world war the dollar problem had been general throughout Europe. Europe was able to pay for only half of its imports during 1947 and 1948. In the face of these problems a new pattern of world trade had been created. But, in general, the reconstruction had not been achieved on the basis of a high level of trade between Europe and North America. It is regrettable that a low trade level solution had been established. The consequences for North American agriculture are sad. Without access to adequate markets, stocks of agricultural produce have been built up. Some of this surplus now goes to developing countries, but this is not the right pattern of world trade to continue into the future. It is unlikely, according to FAO studies, that a market for United States and Canadian wheat would emerge in developing countries on the basis of trade. Furthermore, Europe is not an area that is running balance of payment surpluses in the long run. The only chance open to Europe of expanding her imports of agricultural goods from the new world lay in finding ways of exporting a larger quantity of her manufactured products. Professor Robinson said that he was very pessimistic about the future.

Professor Robinson said that his concern was not with the merits and demerits of free trade over economic nationalism, but rather with the structural adjustments required in North America and in Western Europe to give the best use of resources. With a freer international trade policy after 1945 we might by now have been moving towards a better adjustment in world trade ; but this would have been at the cost of a series of crises and a serious slowing down of European reconstruction. There would certainly not have been the creation of a prosperous Europe between 1945 and 1952.

Professor Nicholls said that he would like to call attention to two aspects of European policy which are having very serious repercussions on the many Latin American countries which are heavily dependent upon one or a few primary tropical crops, particularly coffee. He said he was referring to :

    (*a*) the extremely high excise taxes levied internally on coffee in certain European countries, which seriously curtail the consumption of coffee ; and

    (*b*) the discriminatory import duties favouring the former European colonies (especially in Africa) over the Latin American countries. Such policies have greatly strengthened the appeal of the advocates of forced-draft industrialization under heavy tariff protection in Latin

America. They are grossly unfair to the economic development of a major and under-developed region of the world.

Professor Georgescu-Roegen said that the present discussion had brought to light some vital questions. Other papers in the Conference referred to the important wage differentials between agriculture and industry and to the fact that one preoccupation of the Common Market agricultural policy is the protection of the incomes of people working in agriculture. In the present discussion, the problem of surpluses was under debate and we had heard that they are to be found in the countries of the Common Market, in Canada and in the United States, in Australia and possibly in Latin America. However, most of these countries are well or fairly well developed. In this picture we have a glaring contrast between surpluses in prosperous economies and the intense scarcity of food in numerous under-developed countries. By what means can these surpluses be directed to where they are so urgently needed ? Professor Robinson emphasized the difficult problem of the United Kingdom in paying for her imports of agricultural products from Canada, another developed country. The same basic question applies with even greater force to the problem now under discussion : how are the non-industrialized, under-developed countries to pay for their much needed imports of food if surpluses exist mainly in industrialized, developed countries ?

Professor Bandini said that he had been impressed that almost all speakers stressed the importance of historical analysis in understanding contemporary problems. He said that he fully agreed with this approach and said that we should try to understand history rather than to judge it. The sorts of problems that faced policy-makers were contingent on the conditions of the times and so also were the solutions. However, he agreed with Professor Ashton's comments on the effects of nationalistic policies which, in the inter-war period, had led to such damaging economic rivalry between all groups.

Professor Ashton had asked why special measures are required for agriculture. The basic aim of economic policy should not be to accept individual decisions of those who operate in the economic field, but rather to try to make economic movements smoother. The two main reasons for treating agriculture with special policies flow from this. In an environment of rapid industrialization, if things are allowed to go their own way without policy regulations, there are serious risks of damaging agriculture and of creating a situation of disorientation and collapse which in ten years would have to be rebuilt. In the second place, the demand for agricultural products both for food and for industrial purposes might be more important in the future than our present short-run requirements lead us to believe. World population is on the increase and industrial demands are likely to grow. If agricultural population were kept at too low a level of purchasing power, this would lead to difficulties later on. The policies of support are not simply the consolidation of existing positions but an intelligent provision for the future. We have to act on guesses of what the future trend of development is likely to be.

Professor Ashton had also asked about the danger of excess production of cereals in Western Europe. Professor Bandini said that this is what he himself feared most in the Common Market policies. Professor Bandini thought that the future of European agriculture should be based on quality production and not on the mass food products which can be grown at lower costs in other areas. Professor Bandini agreed with Professor Ashton that the productive structure of Holland and Denmark and certain other countries is already satisfactory in these respects to solve future problems.

Professor Ashton had talked about the psychological effects of agricultural policy in other countries. Professor Bandini said that he is himself considered a strong advocate of free trade, and he is certainly sympathetic to this view ; but it must be remembered that in the different sectors and areas theories have to be adapted to the facts.

Professor Bandini said that Professor Valarché had distinguished between the protectionism between the wars which had been catastrophic and the protectionism at the end of the nineteenth century. Professor Bandini said that in a period when all European countries were facing the competition of new countries, we should not regret the lowering of the costs of transport. But we had to remember that those countries which in the past 100 years had not had intensive industrial development, had, under circumstances of decreased protection, been obliged to face a crisis because of the increase in competition. This crisis was largely of a social nature. It had been a time of massive emigration from Europe to America and to North Africa which had come largely from Poland, Ireland, East Germany, Southern Italy and the valley of the Danube, where the lack of both quality agriculture and industrial development had been most striking. By these emigrations the social repercussions were reduced. In these periods the protectionist policy was concentrated on wheat. Professor Bandini argued that in East Germany and in Southern Italy in particular, such protectionism had had disharmonizing consequences.

Professor Bandini said that there seems to be something of a contradiction in Professor Valarché's point that irrigation is expensive, but made possible production of high value. The question has to be considered from many standpoints. Irrigation would only have covered small areas and there were other potential improvements such as the development of the growing of citrus and vegetables which could pay for the high costs of irrigation. Such improvements were hampered by the persistence of structures such as latifundia. In general, there is a strong argument for the modification of protectionism.

Professor Bandini said that there were two main arguments brought up by Professor Gale Johnson's protest about agricultural surpluses. With respect to actual United States surpluses, these could, as Professor

Georgescu-Roegen had said, be gradually sent to under-developed countries even while world agricultural prices are maintained at a satisfactory level. These gifts of surplus food to under-developed countries brought about their own set of problems. As Colin Clark has argued and as the neo-Ghandian criticisms of the sending of food as gifts have declared, the problem is to get farmers to work more efficiently and not simply to give them gifts. An exaggerated policy of gifts is said to be undesirable and even dangerous if it does not give rise to a new impetus for investment in agriculture. With respect to the Common Market, the problem is somewhat different. Long-term agreement might be possible, but first of all the Common Market position has to be consolidated to ensure harmonious development. After we have put our own house in order and worked out a more soundly based economic structure, then the solution of problems of integration with other countries would become pertinent. On these arguments it is necessary to have the present Common Market policy but it should not be taken as being a permanent feature for the future.

Professor Bandini said that he agreed wholeheartedly with Professor Robinson. We are dealing basically with the question of the relationships of Europe with a great many other regions of the world. To those Professor Robinson had mentioned, he would add Eastern Europe. But at the moment the challenge is to try to arrive at the structure which would make these desirable results possible.

Professor Bandini said that the problems of taxation on coffee, chocolate and tobacco centred on their importance as sources of taxation. The international trade aspects are directly linked with fiscal policies.

Professor Bandini agreed with Professor Georgescu-Roegen that it is easy to take too short a view and that, in the long run, the present position of a surplus of agricultural stocks might utterly change and be replaced by problems of general overall deficits.

In conclusion, Professor Bandini emphasized that it is necessary to avoid an over-liberal view of development and so exclude public intervention. We have to forecast population and consumption figures and adapt policy to these long-run trends. A policy which restricts itself only to immediate trends would involve reduced production in the short run and this would produce regrettable circumstances later.

Chapter 5

THE EUROPEAN COMMON MARKET AND THE MOVE TOWARDS SELF-SUFFICIENCY IN FOOD PRODUCTION

BY

CLAUDE MOUTON

I. INTRODUCTION

THE aim of this paper is to show how the creation of middle-term forecasts (provisional projections) can be used to help define objectives in agriculture and food production and, consequently, to facilitate the drafting and implementing of political measures needed to realize these objectives. The economic importance of the Six and the changes which mean that a Common Agricultural Policy is, from the outset, essential, have naturally led me to compare the actual dimension and prospects of European agriculture with the basic principles of the Common Agricultural Policy.

II. TREATY OF ROME

A. Introductory

On 25 March 1957, after painstaking negotiations, West Germany, Belgium, France, Luxembourg, Italy and the Netherlands created the European Economic Community which established a common market and provided for the gradual standardization of economic policies of the member states. The declared aim was the promotion, throughout the community, of harmonious development of economic activities, continuous and balanced growth, increased stability, accelerated raising of the standard of living and closer relations between its member states (Article 2 of the Treaty of Rome).

The Treaty of Rome proposes, at one and the same time, an economic union and a customs union between the member states ; progress in dealing with problems of tariffs implies simultaneous progress in economic integration and vice versa. In view of this complex double objective, the signatories of the Treaty provided for a precise time-table to avoid all delays in setting up the Common Market. At the same time, this time-table would compensate for the relative weakness of the Commission of the EEC ; for the Commission, being a *non supra-national* body, can only present recommendations to the Council of Ministers, and although the Council is theoretically empowered to make decisions, it can act only on specific recommendations of the Commission without being able to amend them.

Article 8 of the Treaty lays down that the Common Market shall be effectively in operation by the end of a transition period of at least twelve years and at most fifteen, starting 1 January 1958, and ending between 31 December 1969 and 31 December 1972, this last date being the extreme limit. The transition period is divided into three stages of four years each. To each stage is allotted a group of actions which must be undertaken and pursued concurrently. At the end of each four-year period, the Council of Ministers considers what objectives within each stage have been achieved and, acting by means of a unanimous vote on a report of the Commission, decides to pass on to the next stage, to extend a stage for one year or to reduce the length of time spent on a stage. Since the four-year stages can be extended, and, since the extreme limit of the transitional period is set *ne varietur* at 31 December 1972, the Council must, if extensions exceed three years, find a way to reduce the time alloted to subsequent stages ; and, in fact, the Council acting unanimously on a proposal of the Commission, can reduce the second and third stages.

(i) *The Tariff Union.* The tariff union is to brought into being during the transition period by the following steps :

(*a*) the elimination of customs duties between member states ;
(*b*) the setting-up of a customs duty common to all Six ; and
(*c*) the elimination of quantitative restrictions ;

so that by the end of the transitional period :

(*a*) there will be no obstacles to the free movement of goods, persons and capital ;
(*b*) the external common tariff will be completely effective in all trade with third party countries ; and
(*c*) all quotas and other quantitative restrictions, in force on 1 January 1958, will be abolished.

A common commercial policy follows necessarily from these uniform provisions and the principles behind them. But, as early as 12 May

1960, at the recommendation of the Commission, the Council decided, in view of the favourable situation, to set the time-table forward and bring the Common Market into being in less than twelve years and perhaps by 1 January 1966. On 1 July 1962, it was again decided to speed up the rate at which tariffs were dismantled. Finally the Commission recommended a third acceleration so that duties between the countries would be completely eliminated as early as 1 January 1967.

(ii) *Economic Union.* Economic integration, which increases aggregate purchasing power through an increased rate of economic development of member states, ought, in my opinion, not merely to go hand in hand with the customs union, but ought even to precede it, provided that a certain threshold of tariff reduction has been achieved. Certainly economic union should not *follow* the customs union. The example of the OEEC, which never managed to pass the critical figure of 40 per cent in its attempt to free European trade, illustrates plainly the difficulties and limitations suffered by all tariff unions that are not supported by economic union.

This economic union is to be brought into being as a result of the implementation of common economic policies, and these *by the end of the transition period* will have been translated into the following provisions :

(*a*) the drawing up and progressive adoption of a common agricultural policy based on the rulings of the Council, which, from the beginning of the third stage, will have acted simply with a qualified majority, unanimity being no longer required ;
(*b*) the establishing of a common transport policy ; and
(*c*) the guarantee of complete freedom of movement for labour.

The following restrictions will have been abolished :

(*a*) restriction on the right of establishment of enterprises within the Community ;
(*b*) restriction on the free supply of services ; and
(*c*) restriction on capital movements, though only to the extent necessary to the proper functioning of the Common Market.

The regulations or directives needed for the application of the articles 85 and 86 will have been put into force in order to promote the conditions essential to healthy and fair competition. Distortions brought about by existing disparities between the legislative or administrative provisions of different nations will have been eliminated ; the fiscal systems will have been brought into line with each other. Economic policy must, then, be co-ordinated both in the short and long runs. The European Investment Bank is to facilitate the application of a common economic policy.

In conclusion, it should be noted that the initial application of the provisions of the Treaty of Rome has benefited from the favourable economic conditions in Western Europe ; these conditions have accelerated the rate of tariff dismantlement. The synchronization between the tariff union and economic union provided for the Treaty, ought therefore, to mean that there would now be an acceleration in the creation of economic union.

B. *Agriculture and the Provisions of the Treaty of Rome*

One of the dominant features of the Treaty of Rome is the attention given to agricultural problems ; the attention is so marked that some economists have argued that the success of the Common Market depends on the manner in which the problem of the integration of the agricultural sectors of the different nations into one 'community' agricultural sector will be resolved. Title II of the second part of the Treaty of Rome defines the principles of the common policy in agricultural questions in articles 38 to 47.

Article 38 states :

' 1. The Common Market shall extend to agriculture and trade in agricultural products. Agricultural products shall mean the products of the soil, of stock-breeding, and of fisheries, as well as products after the first processing stage which are directly connected with such products.

2. Save where there are provisions to the contrary in Articles 39 to 46 inclusive, the rules laid down for the establishment of the Common Market shall apply to agricultural products.

3. Products subject to the provisions of Articles 39 to 46 inclusive are listed in Annex II to this Treaty. Within a period of two years after the date of entry into force of this Treaty, the Council, acting by means of a qualified majority voting on a proposal of the Commission, shall decide as to the products to be added to that list.

4. The functioning and development of the Common Market in respect of agricultural products shall be accompanied by the establishment of a common agricultural policy among the member states.'

Article 39 specifies the aims of the common agricultural policy :

' 1. The common agricultural policy shall have as its objectives :

    (*a*) to increase agricultural productivity by developing technical progress, by ensuring the rational development of agricultural production, especially labour ;

    (*b*) to ensure thereby a fair standard of living for the agricultural population, particularly by the increasing of the individual earnings of persons engaged in agriculture ;

    (*c*) to stabilize markets ;

    (*d*) to guarantee regular supplies ; and

    (*e*) to ensure reasonable prices in supplies to consumers.

    2. In working out the common agricultural policy and the special methods which it may involve, due account should be taken of :

    (*a*) the exceptional character of agricultural activities, arising from the social structure of agriculture, and from the structural and natural disparities between the various agricultural regions ;

    (*b*) the need to make the appropriate adjustments gradually ; and

    (*c*) the fact that in Member States, agriculture constitutes a sector which is closely linked with the economy as a whole.'

The objective of the Common Agricultural Policy with respect to agricultural markets is to create a single market for different agricultural products within which exchanges will take place under conditions similar to those existing in a particular domestic market, as is the case in industrial products and services.

Taking into account the divergencies between the different national agricultural policies, and the differences in the productive structures, a preparatory stage, lasting about six years, was set up to allow the progressive alignment of national prices, the co-ordination of national market organizations, the harmonization of legislation and the defining of a common commercial policy *vis-à-vis* third party countries ; in short, the steps necessary to facilitate the adaptation of national agricultures to the new market conditions.

To attain these ends, the Commission of the European Economic Community presented to the Council of Ministers, on 30 June 1960, a series of proposals on the working out and implementation of the Common Agricultural Policy. These proposals have been brought into effect by an initial series of enabling regulations adopted after lengthy discussion on 14 January 1962. Thus, the first time that agriculture in the community received concerted attention was during 1962–63. The most significant progress achieved was in the area of the organization of agricultural markets ; in matters of structure, as well as of common commercial policy towards third party countries, the Community still appears to be trying to find the proper course.

After this condensed description of the general framework, it seems proper to examine, in the next part of the paper, the main features of the agricultural and food-producing sectors of the Six, and then to compare them with the basic principles laid down explicitly and implicitly in the drawing up of the Common Agricultural Policy.

III. THE AGRICULTURAL AND FOOD-PRODUCING SECTORS OF THE SIX

A. The Situation in 1961

First we consider, in summary form, the agricultural conditions in the Community, both on a structural level and with respect to market organization, before the first spate of activity at a community level during 1962–63.

(i) *Basic Agricultural Structures.* Table I brings together a number of facts on the different agricultural structures within the Six at approximately the period 1960–62 ; these facts are taken from the publications of the Statistical Office of the European Community to which, for more detailed analysis, the reader is referred.

The agricultural sector of the Six consisted of 6,758,000 production units of more than one hectare taking up a surface area of 77·2 million hectares, that is, giving an average 'statistical' surface area of 10·7 hectares per productive unit.[1] Fifty-one per cent of this surface area was farmed under conditions of outright ownership, 31 per cent under mixed tenure arrangements and 18 per cent under pure tenant farming. In France and Italy, owner-operated farming just predominated over other forms of tenure, while in Germany very few units were operated under conditions of unmixed leasehold. Italy was also characterized by a share-cropping system (10·9 per cent of productive units).

---

[1] France, 15·1 hectares ; Luxemburg, 13·4 hectares ; West Germany, 11·5 hectares ; Netherlands, 9·9 hectares ; Italy, 9·0 hectares ; Belgium, 8·3 hectares.

TABLE I

THE STRUCTURE OF AGRICULTURE IN THE SIX 1960–62

	Germany	Belgium	France	Luxemburg	Italy	The Netherlands	EEC
Number of productive units of over one hectare as % in EEC	19·6	2·9	31·4	0·1	42·6	3·4	100 = 6,758,000 *
Average surface area of units in hectares	11·5	8·3	15·1	13·4	9·0	9·9	10·7
Agricultural area as % in EEC	19·7	2·1	41·6	0·2	33·5	2·9	100 = 77,200,000 †
Permanent labour force as % in EEC	18·6	2·7	32·9	0·2	42·3	3·3	100 = 12,384,000 *

* In units.　　† In hectares.

These production units were typically worked by 'independent' labour, either family labour or hired labour. Generally speaking, the 'independent' labourers, both male and female, in the agricultural sector were much older (in the over-50 age group) than workers in other sectors of the economy. While the proportion of male to female in family labour of the age group 30–60 was lower in agriculture than in other sectors, the structure of hired labour in agriculture was essentially the same as that in other sectors. The permanent labour force on these units consisted of 12,384,000 people (3,845,000 women) of whom 42 per cent were the managers of units, 38 per cent were family helpers and 20 per cent hired labour.

Italy employed 42·3 per cent of the permanent labour force of the community as against 32·9 per cent by France and 18·6 per cent by Germany. Thus it was in France that the ratio of labour to land was lowest in spite of her relatively large share of the permanent labour force. It must be emphasized that the concepts of full employment and of active labour force unfortunately do not have the same meaning in agriculture as in industry. Too often the employed labour force is confused with the labour force available to the productive units. In fact, however, agriculture is still a seasonal activity subject to peak periods of intensive labour which are characteristically unpredictable in the short term, and technically difficult to circumvent under normal economic conditions.

(ii) *Market Organization*. When the agricultural provisions of the Treaty of Rome were first set forth, agricultural production in the member countries provided 87 per cent of the requirements of the community. An analysis of the situation, taken product by product, is given in Tables II and III. These show that on the average, agricultural production in the Six either almost reached, or surpassed, demand for most food products with the exception of luxury fruit, fats and oils, and secondary cereals.

With the development of technical progress, the movement towards self-sufficiency in food in the Six as compared with the pre-war period had become predominant from 1954 onwards. This date approximately marks the end of the period of shortage and the beginning of a period of relative abundance.

(*a*) For certain products, particularly for bread cereals, sugar and potatoes, production increased more rapidly than consumption.

(*b*) Consumption of secondary cereals and beef increased more rapidly than production. Self-sufficiency in pork meat had been achieved for some years. However, considerable imports of eggs and

E   99

TABLE II

AGRICULTURAL OUTPUT OF EEC AS PROPORTION OF TOTAL COMMUNITY SUPPLIES

(as percentage of total supplies)

	Period before 2nd World War	Average 1952/53–1956/57	Average 1954/55–1958/59	Average 1957/58–1959/60	'1970' *
Bread cereals (excluding rice)	86	86	91	93	107
Secondary cereals	77	80	78	78	78
Total cereals (excluding rice)	*81*	*84*	*85*	*81*	*89*
Rice	44	—	92	86	100†
Potatoes	98	102	102	102	100
Sugar	75	99	101	98	
Vegetables	102	—	103	104	
Fruit and nuts	89	—	87	83	
Meat : beef and veal	96	96	94	92	94
Meat : pork products	96	102	102	102	100†
Total meat	*96*	*98*	*97*	*96*	*97*
Eggs	101	93	91	90	100†
Cheese	105	100	100	99	100†
Butter	104	100	100	101	103
Other fats and oils	41	40	40	46	50
Milk (all uses)				103	105

* On assumption of the same surface area being devoted to the output in '1970' as in 1958.

† Self-sufficiency assumed *a priori*

Sources : Statistical office of the Economic Community, EEC Study No. 10: *The Common Market in agricultural products: prospects for 1970.* The data for '1970' with respect to France and, as a result, those for the EEC, take into account the results of recent studies undertaken by the author with C.R.E.D.O.C. and S.E.D.E.S.

TABLE III

GROWTH OF DEGREE OF SELF-SUFFICIENCY FOR CERTAIN PRODUCTS IN THE SIX

	Wheat							Other Cereals						
	1909–1913	1925–1929	1934–1938	1950–1952	1957–1959	1961–1963	1969–1971*	1909–1913	1925–1929	1934–1938	1950–1952	1957–1959	1961–1963	1969–1971*
Germany	67	61	89	58	68	74	65	85	85	91	82	79	74	70
Belgium Luxemburg	22	27	23	44	71	72	77	56	56	47	56	42	46	42
France	90	86	98	101	114	120	144	89	90	80	90	101	115	114
Italy	77	74	94	85	100	94	114	92	84	93	95	78	54	68
Netherlands	18	18	42	27	30	40	30	41	34	42	61	37	34	34
EEC			86	80	93	95	107			77	82	78	75	78
U.S.A.	...	...	112	162	149	...	161	...	...	112	110	113	...	122
United Kingdom	21	20	23	35	36	...	38	47	57	38	68	58	...	64
Sweden	13	68	103	94	90	103	...	92	86	94	95	105	103	...
Denmark	47	52	60	90	69	101	...	77	75	91	98	85	...	...

TABLE III—*continued*

	Total Cereals					Sugar						
	1934–1938	1950–1952	1957–1959	1961–1963	1969–1971*	1909–1913	1925–1929	1934–1938	1950–1952	1957–1959	1961–1963	1969–1971*
Germany	90	74	76	74	68	152	105	100	57	92	86	87
Belgium	37	51	51	54	52	...	129	92	119	104	110	103
Luxemburg												
France	91	96	108	117	127	103	81	90	98	93	108	104
Italy	94	88	91	75	90	96	85	99	65	112	74	85
Netherlands	42	50	35	36	33	...	91	76	99	99	85	90
EEC	81	81	84	83	89			75	90	98	92	93
U.S.A.	112	118	121	...	130	...	...	29	28	30	33	28
United Kingdom	30	53	49	...	...	...	8	21	31	29	29	32
Sweden	97	95	90	103	...	99	56	97	79	...	82	...
Denmark	83	97	85	86	...	110	77	102	162	125	98	128

* Assuming constant prices and constant surface area as of 1957–59.

... : Not available.

Sources : OECD, EEC, FAO, Geneva, SOEC, Brussels, Étude No. 10 EEC.

especially of poultry had emerged : the European countries being content to leave to countries which produced or had control of feed grain at low final price (rather than cost price) the sizeable value added in the production of final poultry outputs.

(*c*) In the case of dairy products, the EEC continued to have a small surplus, although butter was still a luxury item in certain countries (*e.g.* the Netherlands) ; the same was true of sugar in which the EEC was basically self-sufficient.

(iii) *Basic Types of Agriculture in the Common Market.* From an agricultural standpoint, the EEC may be grouped into three types of agriculture.[1]

Type I : 'industrial' agriculture increasingly geared to intensive animal processing on a feed system initially based on locally grown feeds (primary fodders), but more and more based on feeds (grains and processed feeds) bought on the world market at prices often abnormally low. Germany, the Netherlands, Belgium and Luxemburg are examples of this type.

Type II : 'non-industrial' agriculture whose outputs are not significally processed after initial productions. Italy and, to a lesser extent, France belong to this category.

Type III : 'mixed' agriculture which still has large potential for the production of crops (primary fodders and cereals) but which has already started to industrialize, though, more often than not, on an extensive, not an intensive basis ; France is, *par excellence*, a case of this type.

How these three types of agriculture were to evolve in the years following the setting up of the Common Agricultural Policy was the problem faced by those responsible for the development of the Policy from 1958–59 onwards.

B. *Middle-Term Projections*

(i) *Studies on Projections.* Since the end of the second world war, studies on economic prospects in Western Europe have multiplied. From 1954 onwards, the Committee on Agricultural Problems of the United Nations Economic Commission for Europe set out to estimate the long-term trends in agricultural production, in food consumption and in trade in agricultural products. This Committee considered that such research could contribute to the drafting and setting up of political measures promoting stabilization of markets in agricultural

[1] See particularly Tables I and IV.

products and increases in agricultural returns. At the time that the Commission of the EEC was beginning to set out its first recommendations on the Common Agricultural Policy, two independent groups of experts, working on similar lines to the United Nations European Commission, started work in April 1959 on studies aimed at describing and comparing the main trends in production and consumption of major foodstuffs in the member states during the period 1956–65. The studies had in common a number of basic hypotheses.[1] Along with the studies at the EEC, and in close and permanent collaboration with it, the agricultural division of the Economic Commission for Europe, the economic division of FAO, and the food and agriculture division of OEEC got together to study the trends in consumption and production of foodstuffs in other European countries.

The results of these studies, based on common principles, were published in a document entitled, 'European Agriculture in 1965'.[2]

At the end of 1960, the FAO decided to follow up on efforts already made, by expanding the scope of study both with respect to time and to the regions covered, so as to be able to draw up a table of projections in world trade in agricultural products for '1970'.

Since the relevant agricultural section of the EEC required projections for the transitional period, collaboration with the FAO was clearly called for to avoid duplication of effort. The provisional work already done at the EEC by the end of 1961 was used by the FAO. The results of the FAO were published for the United Nations Commission on Primary Products.[3]

A number of other studies based on projections to 1975 have been undertaken. There are also studies on a national basis of which the outstanding example is that undertaken in France by the Commissariat Général du Plan which is more comprehensive than the studies by international organizations in that it sets up a model of French agriculture specified in terms of sectors, sub-sectors and individual units. In the near future, this model may be regionalized. The

---

[1] *Tendances de la production et de la consommation en denrées alimentaires dans les six pays de la C.E.E.* (1956–65), CEE Étude no. 2, Série 'Agriculture', Brussels, 1960.

[2] 'L'Agriculture Européenne en 1965', Doc. no. Agri/167, Economic Commission for Europe, Committee on Agricultural Problems, United Nations, Geneva, 1960.

In this connection, the work of a number of senior international civil servants should be acknowledged. From within their several organizations they contributed to the success of the common effort : Mr. Rabot and Mr. Krohn of the EEC, Mr. Sinard from the Economic Commission for Europe at Geneva and Mr. Goreux of the FAO.

[3] 'Agricultural Products, Projections for 1970', FAO Report on Commodities, special supplement ; E/CN 13/48, C.C.P. 62/5 FAO Rome, 1962.

international organizations have directed their attention to the outlook for agricultural markets with special emphasis on the import and export requirements of the countries studied. This highly market-oriented focus means that the work done in analysing basic structure is, more often than not, too restricted to be of use in studies of an international scope.

(ii) *The EEC in the Light of the Projections.* By '1970', assuming continuous and vigorous economic growth and assuming stability in agricultural price (the so-called 'constant price hypothesis'), the trends towards greater self-sufficiency will be increasingly evident. These trends will be even stronger since technical progress will affect an increasing number of farms (the intensification of productive systems), and since the Common Agricultural Policy aims at a price alignment at a level determined by the 'upper average' of agricultural prices in the community. Thus we can expect the level of self-sufficiency to increase for the commodities shown in Tables II and III between now and 1970.

The studies at the EEC, as well as those carried out in some of the member states, in particular the work done in France stemming from studies by C.R.E.D.O.C. and S.E.D.E.S., indicate that in '1970', on the 'constant price hypothesis' and assuming the same total area in use as in 1958, the Six will show the following characteristics :

(*a*) a surplus of milk which may remain moderate if the number of milk-producing cows does not increase ;

(*b*) increasing surpluses of wheat, especially if consumption by animals stays at the same level ;

(*c*) self-sufficiency in sugar, pig products, poultry and eggs fully realized, and a trend towards self-sufficiency emerging in secondary cereals ;

(*d*) small increases in the overall requirements for fats and oils (excluding butter) ; and

(*e*) increased demand for beef, which will not be met by the agriculture of the Six as long as beef remains predominately a by-product of milk-production.

With respect to (*a*) and (*c*), it should be noted that by virtue of the high elasticities of demand with respect to final prices for milk-products, milk surpluses may increase in considerable proportions if it is assumed that there will be an upward trend in these prices.

To the regret of some, it would seem that an explosive increase in French agricultural outputs must follow from the situation given in this section. The active agricultural population in France is still too large in relation to the present state of technology.[1] The reduction in active population has meant, and will mean, an increasing mobilization of a vast natural potential; a potential for so long so under-exploited, that public opinion abroad and even the politicians in France's partner countries remained quite unaware of it. This vast potential will be realized as the use of increasing doses of non-agricultural inputs becomes viable with the reduction in active population. The intensification of agriculture in Europe, particularly in France, may also be regarded as inevitable in that it is part of a world-wide trend, a trend that is already affecting more recently settled countries.

(iii) *World Outlook.* The distinction between 'new' countries and 'old' countries, when dealing with agriculture, is based on traditional, rather than practical, considerations. As Professor Michel Cépède argues, such a distinction involves us in using old and largely outdated slogans.[2] In fact, the contrast becomes weak when dealing with the differences between the EEC and the newly settled countries, in particular the U.S.A. On the level of the structure of farming, similar regional factors — geographical and economic — affect family farming on both sides of the Atlantic; the U.S.A. has its critical regions just as the EEC does. With respect to productive methods, all countries are having to achieve an effectively higher level of fertility by substituting capital for labour. Extractive agriculture has to be done away with. Industrialization of agriculture and food production in countries of primary crop production (cereals, etc.) is making the sorts of agriculture which produce crops for further processing obsolete, at least over the intermediate run. This sort of agriculture is to be found in certain Western European countries.

The classic problems of international trade and the distinctions upon which they depend are also found to be illusory in face of the evolution of modern economic life. Differences in levels of productivity become smaller and smaller and with this comes a standardization in national production costs. This standardization will be accelerated by intensification of productive methods in the new countries, and by costs associated with soil conservation policies

---

[1] In the short run, regional problems of re-establishing the correct labour/land ratio will arise with the exodus of labour; but in the longer term, they can hardly be expected to affect the general trend.

[2] I wish to thank Mr. Michel Cépède for allowing me to reproduce in this section the conclusions drawn in his article, 'L'Agriculture dans les relations Europe–États-Unis', published in *Progrès et agriculture*, Cahiers de CISEA, Nov. 1964.

TABLE IV

COMPARATIVE DATA FROM NATIONAL ACCOUNTS (1959–61 AVERAGE): RELATIVE IMPORTANCE (I) OF VARIOUS ITEMS IN RELATION TO TOTAL EEC FINAL OUTPUT; AND COMPOSITION (C) OF SUPPLIES WITHIN EACH COUNTRY.

(estimated at current prices: in percentages)

(percentages)

	Germany		Belgium		France		Italy		Luxemburg		Netherlands		EEC
	I	C	I	C	I	C	I	C	I	C	I	C	
Total crop output	7·4	26·7	1·7	34·5	12·7	38·0	16·6	62·9	0·0	17·0	2·5	34·3	40·9
Total animal output	19·9	72·0	3·1	65·5	20·2	60·1	9·6	36·5	0·2	83·0	4·9	65·7	57·9
Total agricultural output	*27·6*	*100·0*	*4·8*	*100·0*	*33·6*	*100·0*	*26·4*	*100·0*	*0·2*	*100·0*	*7·4*	*100·0*	*100·0*
Total intermediate consumption	9·6	34·9	1·6	32·8	7·8	23·3	4·1	15·3	0·1	34·9	3·2	42·9	26·4
Gross product at market prices	18·0	65·1	3·2	67·2	25·8	76·7	22·3	84·7	0·1	65·1	4·2	57·1	73·6
Gross product at factor cost	17·5	63·3	3·3	68·0	25·3	75·3	22·3	84·7	0·1	63·2	4·5	61·7	73·0
Net product at factor cost	15·5	55·9	3·0	62·4	23·4	69·8	20·4	77·4	0·1	50·4	4·2	57·3	66·0

Sources: SOEC; *Agricultural Statistics No. 3*, 1964.

which will have to be added into final cost prices. (See Table VI.)

The crucial question would seem to be whether the traditional exporting countries should be allowed to retain their old advantages as a permanent right, at a time when the structural and dynamic properties of different national economies reacting to the impact of technological and sociological change seem to call this right in question.

(iv) *Conclusion.* This survey of the outlook for EEC agriculture in a world context, suggests that all countries are going to 'move towards a type of agriculture both technically progressive and more intensive with highly productive use of both land and labour'.[1] For geographic units of the size of the U.S.A. and the EEC, such a trend towards agriculture of Type III — mixed but intensive — will entail real self-sufficiency in foodstuffs ; at least with respect to those outputs which are 'natural' to the respective geographical units. There is, however, a great risk that strictly commercial considerations,

TABLE V

SHARE OF AGRICULTURAL OUTPUT * IN NET DOMESTIC PRODUCT †

(at constant prices, as %)

	Germany ‡	Belgium	France	Italy	Luxembourg	Netherlands
1953	8·9	7·8	11·8	24·1	—	11·9
1954	8·5	7·4	11·7	22·6	—	11·7
1960	6·2	6·7	9·5	16·1	—	10·1
1961	5·7	6·1	8·7	16·6	—	9·4
1962	5·3	6·4	9·0	16·0	—	8·6

* Including for France both agriculture and forestry products (series broken after 1959).
† For France Domestic Product measured at market prices ; for other countries measured at factor costs.
‡ Including the Saar and West Berlin from 1960 onwards.

Source : SOEC, *Agricultural Statistics*, 1964, No. 3.

often of a purely private nature and pursuing very short-term objectives, will delay the adoption of political measures that work in harmony with the inevitable world-wide reduction in the share of agriculture in Gross Domestic Product (see Table V). It is to be expected that this reduction in agriculture's share will produce the well-known, and often self-defeating, psycho-sociological reactions shown by all minorities. These are likely to entail pressure-group

[1] Michel Cépède, *op. cit.*

activity, both political lobbying and direct anti-economic obstructionist action, directed at overcoming the problems of the adaptation to requirements of economic development.

IV. THE COMMON AGRICULTURAL POLICY

The principles on which the Common Agricultural Policy at present rests have been laid down step by step from 1958 onwards by a team within the *Direction Générale de l'Agriculture* in Brussels, directed by Mr. Mansholt, former minister of agriculture in the Netherlands. For some years, a number of studies in agricultural policy has been pursued, either within the framework of the OECD and the *pool vert* or as part of the preparatory work at Val Duchesse. Also, at the Stresa conference, the representatives of the Six had been able to inform the offices of the EEC about their views on the future orientation of the Common Agricultural Policy. However, at that time, no forecasting studies dealing with the six countries were available. Not until the end of 1960 did the first forecasting results, obtained by two independent groups of experts at the request of EEC and dealing with the period 1956–65, become available. That is to say, these results only became available at a time when the Commission had already published its recommendations relating to the setting up of the Common Agriculture Policy (30 June 1960) ; and these recommendations are still, on the whole, the basis of regulations currently in force or still being discussed within the community.

A. *The Principles of the Common Agricultural Policy*

It must be re-emphasized that only in the area of agricultural markets is agricultural policy properly defined. The policy of structural improvement is often mentioned as being the basis for improvement in returns to producers. But certain commentators have come to wonder whether this is not just a verbal gesture : the abundance, which is out of all proportion, of texts dealing with the organization of markets, contrasts strangely with the paucity of texts dealing with the improvement of structures.[1]

[1] Several studies have, in fact, been made, but it is not at all clear to the informed reader what the logical connection between them is. There are, however, two notable exceptions : the preparatory work for the setting up of an input-output table for the Community, and the preparatory work for standardizing the models.

With respect to agricultural markets, the aim of the Common Agricultural Policy is to create, for different agricultural products, a single market within which exchanges will take place under conditions similar to those existing in an internal market. This policy necessarily leads to a system of uniform prices ; in fact, all regulations are based on the same premise, namely, that prices are the determining factor in Common Agricultural Policy. The strategic role allowed to prices goes hand in hand with a strong reluctance to consider measures of direct intervention. The inauguration of a single market must lead to production reallocations at the national level, as well as in Europe as a whole, as a result of the behaviour of a single price system and of 'healthy' and 'fair' competition. However, priority is, in fact, given to the wishes of consumers, since all techniques for establishing agricultural prices are based on the level of price in the marketing centre of the area with the larger deficit ; so that by allowing for the variations in transport, handling and storage costs,[1] it is possible to determine, on this basis, the commercial prices for agricultural products for each productive region. Furthermore, pride of place is given to the organization of the cereal market, since the organization of the animal processing markets (pig products, poultry and eggs) depends heavily on the measures taken with respect to cereals.[2]

A number of further points stated or implied by the Common Agricultural Policy need to be set down :

(*a*) When the single market stage is reached, the consumer centres (deficit areas) upon which pricing will be based will be set in the Rhineland ; from this it follows that the outlying areas, such as the south of France and the south of Italy, will on average receive, *ceteris paribus*, the lowest prices when they become more highly commercialized in production.

(*b*) It is through the cereal market organization alone that producers are provided with revenue security. The organizations responsible for market interventions at a national level (or at a community level at the common market stage) are committed to take, at a fixed price (called the intervention price), all that is delivered to them by producers. It should be added that the intervention prices are not prices guaranteed to producers, but are prices guaranteed to wholesale outlets.

[1] Hence the importance of a Community transport policy.
[2] It should be emphasized that the shift to a belief in the dominant virtues of the competitive economy, albeit slightly managed, represents a radical departure from the organization of markets set up after the years of economic crisis between the two world wars.

(*c*) The full application of the body of regulations will mean free movement of commodities, with the only obstacle being the setting of the sluice-gate price for pork, and the threshold price for cereals — the difference between the threshold price and the price on the world market establishing the size of the external levy.

(*d*) The setting up of a common market implies the abolition of all financial aid granted by a member state to any line of production (Article 92, Paragraph 1). The regional redistribution of crops must take place normally through the free play of the competitive price mechanism. Because of this, and in view of the rules and present provisions of the Treaty of Rome, it is not possible to predetermine the allocation of particular crops to particular regions.

(*e*) The financial implications of the Common Agricultural Policy can then no longer be assumed by each state, but must give way to Community responsibility to be exercised through the European Agricultural Guidance and Gurantee Fund. Moreover, the operation of the preference system of the Community implies that the yield from the levies collected at the external frontier will become a fund which the Community can use to offset the fact that it will still be in the interests of certain member states to go on importing from third party countries items which are, in fact, available from within the Community. These refunds, which are like negative levies, are paid automatically to a member state as an exporter of the commodity imported from third party countries.

We are now in a position to see that the provisions and regulations of the Common Agricultural Policy turn on two central ideas :

(*a*) That the fundamental role be accorded to price as the one and only determining variable for achieving a balance between supply and demand ; and this in an 'industry' where the conditions of supply are technically complex and the regulation of supply therefore burdensome.

(*b*) That the Six is inevitably to depend upon net imports to balance consumption and production.

The working basis which these two assumptions provide deserves serious critical analysis. While the implications of the theory of perfect competition justify the central place given to prices in the Common Agricultural Policy, analysis of the conditions of production specific to agriculture, and, in particular, of the nature of the demand for food products, makes it quite clear that price is but one of many variables determining equilibrium at a given level — a level which may well, and indeed should, vary over time. Furthermore, studies made in a number of countries on the price elasticity of supply, above all those based on time-series data, are, to put it mildly, misleading. To place any reliance on these in trying to establish the true price relationship, is surely to forget that in agricultural societies of the Western world, the farmer has typically played a number of roles and hence has given evidence of many different sorts of behaviour. The farmer is, at one and the same time : a capitalist, either as full landowner or at least as the owner of circulating capital ; an 'entrepreneur' in F. Perroux's sense of the word ; a supplier of manual labour ; the head of a more or less extended family system which may be more or less closed to the outside world ; and even, on some occasions, a figure in professional organizations or head of an integrated and articulate interest group. The motives underlying the farmer's behaviour cannot, and will not, be identifiable through the examination of one variable alone, even though that variable be the 'queen' in classical economic theory. Moreover, it is surely to be expected that the natural conditions of uncertainty which characterize his productive efforts will induce the farmer to seek security by concentrating on getting the average return for his line of production rather than to seek a particular price or fluctuating price. On the other hand, the search for security may also lead him to work for particular solutions within his system of production — solutions which will be dominated by extra-economic considerations. Further, past experiences have shown that agricultural policies have not been able to limit themselves merely to variations of price to achieve certain objectives ; this is especially so when price is used as part of a social policy of redistribution or equalization of incomes, whether between several types of producers or between the different social classes of the nation.

The above criticisms of the central role of price in the Common Agricultural Policy would not be so serious if the second basic assumption, that of the inevitability of imports, could be substantiated. This assumption implies a number of things :

(*a*) that there be a permanent supply for the Six from world market sources. The difference between the low world price and the threshold price would then permit, through the system of levies, an automatic financing of the European Agricultural Guidance and Guarantee Fund, without there being any necessity for direct contributions from member states (see Table VI for the relevant

TABLE VI

PRODUCTION PRICES IN 1959–60

(In DM per 100 kg.)

		Germany	Belgium	France	Italy	Netherlands	EEC	U.S.A. I 59	U.S.A. I 60	U.S.A. II 59	U.S.A. II 60	European ports CIF 59	European ports CIF 60
Wheat	Price	42·40	39·40	31·32	43·32	33·58	37·48	27·30	26·88	30·66	29·82	28·14	28·56
	Index	113	105	84	116	89·6	100	73	72	82	80	76	77
Rye	Price	38·50	30·24	24·32	33·23	33·43	36·35	16·38	14·28	20·58	18·00	—	—
	Index	106	84	69	92	92	100	45	40	57	50	—	—
Barley	Price	42·10	31·75	26·70	33·00	32·87	32·25	16·38	15·96	21·00	20·58	24·78	23·52
	Index	131	99	83	102	102	100	51	50	65	64	77	73
Maize	Price	—	—	32·70	28·12	—	29·62	17·22	16·38	19·32	18·00	24·36	25·2
	Index	—	—	110	95	—	100	58	55	65	61	82	85
Sugar Beet	Price	7·33	7·44	5·94	6·07	5·96	6·43	6·00	6·12	—	—	—	—
	Index	114	116	92	94	92·7	100	93	105	—	—	—	—
Milk	Price	32·50	31·08	32·20	32·25	29·84	32·01	39·22	39·77	—	—	—	—
	Index	102	97	101	101	93·2	100	123	125	—	—	—	—

Notes: EEC prices obtained by taking the prices of member states weighted by the quantities they produce. U.S.A.—I: average production prices; II: average wholesale prices.

Sources: SOEC and FAO.

output prices) ; thus an improvement in the sub-standard agricultural structures of the Six would be assured with little difficulty ; and

(*b*) that the refunds from the Guarantee Fund be negligible in relation to the levies, with the member states trying, in the first place, to supply themselves from within the Community. This presupposes that the net import needs of the Six remain large even after allowing for the dictates of commercial policies of certain member states to third party countries.

Part III of this paper showed that a middle-term trend towards self-sufficiency has to be accepted as long as the assumption of constant prices to producers is maintained. In view of this, and in view of the general principles of the Common Agricultural Policy explained so far in this part, it is necessary to turn to the practical aspects of imposing the Policy.

B. *Contradictions*

An analysis of rules governing the application of the Common Agricultural Policy reveals serious contradictions between the general principles and reality. First, there is the fact that those responsible for the Community have always said that they wanted to see the EEC with markets wide-open to world trade : the adoption of a common external tariff at a very moderate level would, in itself, be manifest proof of this. To adopt this sort of policy constitutes a fundamental political choice which is not at issue here. But the political measures taken by the Council of Ministers, on the recommendation of the Commission, must at least be seen to be consistent with the choice made.

(i) *Contradictions in the Common Agricultural Policy.* Since there is a definite trend towards collective self-sufficiency at constant prices in the case of a number of foodstuffs and temperate agricultural products, and since price is, according to the way the Common Agricultural Policy is set up, the equilibrating variable, the EEC must, for its own sake, clearly adopt a careful price policy, so as to restrain the increase in production within the Community and so as to cut back on production of those outputs which are already excessive. Now, the decisions taken on the common prices for cereals for

Table VII

Base Indicative Prices and Import Prices for Cereals

(in U.S. dollars, per ton, standard EEC quality)

	Soft Wheat	Hard Wheat	Rye	Barley	Maize
I. Base indicative prices from July 1964					
Germany	118·87	—	108·13	103·00	—
Belgium	104·60	—	83·60	89·00	—
France	100·22	117·26	81·79	83·00	98·32
Italy	113·60	143·20	—	72·22	68·42
Luxembourg	117·00	—	108·00	89·00	—
The Netherlands	104·83	—	74·59	82·32	—
II. Indicative prices from July 1967	106·25	125·00	93·75	91·25 *	90·63 †
Percentage deviation of 1967 indicative prices from French indicative price in 1964	+6·0	+6·6	+14·6	+9·9	−7·8
III. *Most favourable import prices in 1964* Rotterdam	100·86	75·15	60·78	58·21	60·01

* Proposition Mansholt 92·50. † Proposition Mansholt 93·75.

Sources : EEC ; FAO.

the common market stage, beginning 1 July 1967, show (see Table VII) :

(*a*) an alignment of prices with the 'upper' average bringing with it a rise in French prices ; that is, a rise in prices for the country which has by far the greatest potential for expansion in cereal production ;

(*b*) a substantial rise in the price of wheat and barley even though, as Tables III and VIII show, a trend towards self-sufficiency is clearly apparent ; and

TABLE VIII

DEGREE OF SELF-SUFFICIENCY IN CEREALS IN THE SIX

(as % of total supplies)

	Wheat	All other Cereals apart from Wheat	Total Cereals	Rye	Barley	Oats	Maize	Other Cereals
1955/56	94	80	86	93	68	94	70	28
1956/57	79	87	84	98	90	96	71	37
1957/58	94	77	85	100	71	88	63	24
1958/59	91	78	84	99	74	92	66	20
1959/60	94	78	85	99	84	90	61	15
1960/61	89	82	85	95	99	92	63	18
1961/62	86	71	78	74	82	88	56	17
1962/63	108	75	89	86	94	94	45	19
1963/64	91	79	84	96	106	94	52	28

(*c*) with respect to the price ratio between less-common cereals, the price of barley will be slightly higher than that of maize even though maize has a higher nutritive value. However, with an increase in the consumption of maize, the degree of self-sufficiency for that crop will diminish, while that for barley will approach full self-sufficiency. This seems to suggest that there is a wish to encourage production of barley and wheat for animal consumption, while holding back on providing attractive prices for maize!

Under these circumstances the economist cannot but ask the following questions :

(*a*) Do the decisions taken by the policy-makers mean that price is, in the final analysis, to play a social role without regard for the effect on production ? The weakness of studies based on price-output analysis alone has been stressed already, but we know that the producer, though he hardly reacts to a drop in price (within reasonable limits), always reacts to a rise in price. Since the price guarantee will mean an increase in price, producer reaction is to be expected. Thus, France will have record wheat crops. Moreover, many maize producers seem to have given up maize production in favour of wheat, thus aggravating the imbalance between the different cereals.

(*b*) Is it really the intention to favour maize-producting countries outside the Community?

(*c*) Is it the case that the foreseeable glut on the markets for wheat and barley will lead to an inevitable series of readjustments in target prices, albeit at a time when electoral considerations are pressing?[1]

(*d*) Has not the Commission, by the proposed new provisions of 30 March 1965, made it clear that it is worried that the trend towards a reduction in net imports will lead to a fall in income to the European Guarantee Fund ; especially if the setting up of world agreements by products brings with it an increase in world prices? Such an increase in world prices would decrease still more the Fund's income and would increase its expenses.

The text of the Commission states : 'The refunds relating to exporting being carried out within the framework of international agreements can be granted only if these agreements have a "community character".' Thus, the refunds would no longer be granted automatically to exporting countries. There can be no doubt that this expresses a desire for a built-in discouragement of production, but by a provision which, to be effective, would have to abolish the price guarantee on the entire crop.

Does it not seem that there is the threat of policy dominated by quantity restrictions in the Six, with serious ramifications for the economic liberation so strongly favoured by the Common Agricultural Policy?

(ii) *Contradictions in the Treaty of Rome.* That, for a real common market to be set-up, it is absolutely imperative that there be a common transport policy, has been argued above. But no clear priority has been given to such a policy. Under these circumstances, how can a Common Agricultural Policy in which a hierarchy of prices is established as a function of distance, develop healthily?

At a more general level, the effective implementation of both a common transport policy and a common agricultural policy means,

---

[1] A very serious omission is to be found in the provisions of the Treaty of Rome. Provision for simultaneous elections should have been made ; this would certainly have helped the work of the Council and Commission of the EEC.

ipso facto, the creation of a common economic policy — in the broadest sense of the word 'common'. But, in fact, only *co-ordination* has been provided for. But if a surrender of sovereignty is essential, it must be total ; indeed, as things now stand, the problem is precisely to reconcile, within the framework of a single 'national' economy, the co-existence of sectors of the community run under a common policy and sectors merely co-ordinated. Surely sovereignty must be completely given up or not at all ?

This fundamental problem, which the Six have yet to solve, results from the fact, as the history of United Europe since 1946 clearly shows, that the Treaty of Rome is only a compromise between the supporters of a Federal Europe and the supporters of a *Confederate* Europe. This is the block which those who took part in the marathon discussions in Brussels are well aware of, but which is not, unfortunately, to be found explicitly stated in the documents on the founding of the Common Market.

Must this section on the Common Agricultural Policy end on such a pessimistic note ? Perhaps not, if the contradictions that have been pointed out can be resolved. Such a solution only seems possible if principles are made to give way to reality, and if definite objectives are assigned to agriculture in the Six and, more generally, to the Community's economy.

As was stressed by a French expert in 1963, 'the Brussels provisions only laid down the mechanisms — the means of carrying out a Common Agricultural Policy, but it laid down neither the foundations of this policy nor the long-term objectives which are to be assigned to it'.[1]

That is why the policy-makers in the Community must set out clearly the main options facing the EEC in its relations with third party countries which export agricultural products.

V. CONCLUDING REMARKS

This paper suggests the following conclusions.

(1) The usefulness of middle-term forecasts in defining political measures or in amending them is undeniable. It is certainly to be stressed that such forecasts will be approximate, but this holds true in any attempt to estimate *a priori* the effectiveness of a given economic policy.

---

[1] Y. Malgrain, *Société française d'économie rurale* (22–23 janvier 1963).

(2) In the agricultural sector, the economic allocative role of prices is weakened or even made to disappear as its use for social objectives increases. In view of this, is it not simply evidence of wishful thinking to hold that there can be 'true' prices and that the price reflects a true equilibrium between supply and demand ? It is, then, essential to give up the simplistic and dangerous expedient of trying to subsidise incomes with a single price system which is being used in providing overall direction to the agricultural sector. For this reason, it is regrettable that a policy such as that of deficiency payments was rejected by the policy-makers in Brussels. Such a policy in effect allows the whole body of the 'Nation' to know properly the costs of public intervention on behalf of agriculture, and allows freedom to market forces.

(3) The setting up of forecasts and targets presents the problem, which is in my view fundamental, of the relations between research services and the centres of political decision. The value of the forecasts and studies depends, *ceteris paribus*, largely on the objectivity of those in charge of research. This raises the problem of the independence of researchers *vis-à-vis* the centres of decision ; for these centres most often have full control of the researchers' professional careers. It is essential to provide them both with the independence enjoyed by university researchers and with a working background adequate to the job in hand which means close links with the relevant administrations.

There are a number of possible solutions to conflicts of this sort.

(1) Each administration may set up its own research centre. For technical reasons, this arrangement would lead to a rapid encroachment of one research centre into the domain of other administrative centres, thus creating a technocratic interministerial feudalism ; any democrat would condemn such a solution.

(2) The work of the research services might be kept rigorously secret with the centre of political decision only making generally available the answers which justify its decisions. Moral considerations condemn — or rather should be accepted as condemning — such practices in a society founded on the concept of liberty. The danger of this sort of practice is aggravated if the research services centre upon a single ministerial body.

(3) The research centre may be linked to a parliamentary body where all political and professional groups are in the normal course of affairs represented. The research service can then benefit from a definite independence of thought and adequate working conditions.

Agriculture in the World Economy

Under these conditions, objective information for all interested parties is assured and each political group can select, from among the diverse possible answers to a particular problem, that answer or those answers which is (are) consistent with its particular ideology. Within the framework of the Six, it would be possible to realize a solution of this sort with intersectoral research services dealing with middle-term and long-term problems while being attached either to the European Parliament or to the Social and Economic Council of the EEC ; the latter would seem to be the better choice of institution to which to attach the research services.

DISCUSSION OF PROFESSOR MOUTON'S PAPER

Professor Ashton said that Professor Mouton's paper was directly involved with the possible conflicts between what is economically desirable and what is politically possible. This issue confronts economists both in government service and in international agencies including such bodies as the FAO and the Commission of the EEC. Governments are much more conscious today of the value of economists in public service, but few politicians, national or international, are likely to abdicate their prerogatives entirely to their economic advisers.

Professor Ashton said that the declared purpose of the paper by Professor Mouton was to show how middle-term projections can assist in the formulation of a coherent food and agricultural policy in the EEC. Professor Ashton said that the paper does many things but he suggested that it does not really fulfil this purpose. Much could be said on the subject of projections when discussing both their methods and their uses. Of the methods used, Professor Ashton asked whether it is justified to keep prices and areas constant, dealing only in effect with changes in yields. In addition, Professor Ashton doubted if a single projection is much use. He thought there is a need to vary the assumptions since one is attempting to determine limits of possible responses in the examination of alternative policies.

Professor Mouton's views concerning farmers' reaction to price might require some clarification. At one point Professor Mouton disputed the central role of price in relation to agriculture. He argued that prices are only one of a number of variables affecting the equilibrium of supply and demand. He stated that the studies of the price elasticity of supply with respect to price, based on time series, showed disappointing results. Professor Mouton asked whether we had forgotten the many functions of the farmer as a capitalist, entrepreneur, manual worker, head of a family and a bearer of professional responsibilities arising from group activity. Professor Ashton asked whether this was Professor Mouton's way of saying that the farmer is not just an 'economic man' but an all-round good fellow. Professor Mouton continued by saying that price changes had not been a successful instrument of agricultural policy except to achieve some degree of income redistribution. However, in contrast to this argument, Professor Mouton also pointed out that farmers appear to have reacted to the decline in the price of maize and increase in the price of barley by switching from maize to barley. Producers seem to have reacted to these price changes more according to the text-books than Professor Mouton previously argued.

Referring to that part of Professor Mouton's paper in which he offered alternative solutions to the problems confronting European agriculture, Professor Ashton said that he thought it was extremely difficult to write general prescriptions of this nature since circumstances vary from country to country, from organization to organization, as well as from one time to another. He did not think Professor Mouton's paper would have suffered if this section had been excluded. But he would accept that deficiency payments offer a useful solution in achieving a transference of incomes while allowing the market forces to operate. Unfortunately when deficiency payments become too high, other political considerations, such as the excessive burden on the taxpayer, give rise to a search for alternative policies.

Professor Ashton said that Professor Mouton pointed out in his paper that the tendency towards agricultural self-sufficiency in the Community had started to show itself as early as 1954. This was long before the question of a Common Agricultural Policy had arisen. Professor Ashton asked how far this tendency, which is similar in some respects to what had occurred in the United Kingdom, is purely a phenomenon of development in Europe, brought about by technical innovation, and regardless of a Common Agricultural Policy ?

Professor Ashton said that he thought that 'the tendency towards self-sufficiency' in Europe is perhaps only an elaborate way of saying that 'output had increased'. The important issue, presumably, is whether there is a production surplus or whether production is less than total requirements. Professor Ashton said that Professor Mouton's argument seems to be justifying agricultural self-sufficiency in Europe on the grounds of the narrowing of the gap in comparative advantages, formerly enjoyed by the traditional food exporting countries, as a result of new technology. Professor Ashton said that this is not necessarily a conclusive explanation. He said that he shared the view expressed by Professor Robinson in an earlier discussion that it is likely that the agricultural resources of North America and Oceania are under-utilized in an economic sense in relation to those in Europe, as a result of restraints on trade. The basis of this view, which differs from that of Professor Mouton, is that technology is not the only determinant of comparative advantage. Other considerations could not be ignored, in particular the relationship between the agricultural and non-agricultural sectors in the different regions.

In this connection Professor Ashton said that it should be made clear (as M. Bergmann had explained to him) that the categories called 'appelées', 'permis' or 'interdit' refer in essence to biological concepts and are not necessarily a matter of economics. Professor Ashton went on to say that he had found it difficult to accept Professor Mouton's classification of the agriculture of the member countries of the EEC into 'industrial', 'non-industrial', and 'mixed' categories ; he thought it is influenced more by agronomic than by economic considerations. Professor Ashton suggested that the aim of this classification was merely to convey a measure of intensity of farming systems while a direct economic classification would have been more useful.

In conclusion, Professor Ashton said that satisfactory progress would depend on the liberalization of the market and the removal of elaborate and complex regulations. Present interventions could be interpreted as a necessary prelude to correct the existing defects and establish an appropriate structure for the industry to perform in a free market economy. This could not be done overnight. The ultimate goal should be to reduce public interventions to a minimum and so achieve a healthy Community policy.

Dr. Krohn said that his comments were on the major questions raised by Professor Mouton. First ; how should middle-range projections be used in the formulation of any agricultural policy ? Second ; what role is played by price policy in the Common Agricultural Policy ? Third ; where does the EEC at present stand with regard to trends towards self-sufficiency in agricultural production ?

Dr. Krohn said that he did not agree with Professor Mouton that middle-term projections should be used to specify the aims of agricultural policy. This quantitative approach suggests that demand projections should be used to give hard-and-fast objectives. Dr. Krohn said that the approach of those responsible for the realization of the Common Agricultural Policy is rather to see the projections as indicating economic trends which would be used to guide and illuminate the decisions that would have to be taken. The projections are not treated as prophecies of what is bound to happen, but as indicating trends which are implied by the present socio-economic conditions. Thus, on the demand side, expected population growth, income increases, and on the supply side, technical progress in agriculture and other expected changes, are made explicit in the projections so that it could clearly be seen what is likely to happen if nothing is done to alter these trends. The projections are to be seen as part of the preparation for decision-making, and are not to be taken as being an integral part of the decisions themselves. To use the purely quantitative approach, with the projections used as targets, is not acceptable to the Community, because it implies a policy of specific production objectives in the various states and regions, which is clearly not the way to move towards greater specialization and structural improvement. Political bargaining on specific quotas is a far less efficient or desirable way of moving towards specialization than free competition. Moreover a 'target' use of the projections would automatically bring about an import quota system with respect to third countries.

On the question of the fundamental role of price as the main equilibrating device, Dr. Krohn said that he thought it had been rightly emphasized that its use is an 'innovation' after the autarkic national policies of the inter-war period. The Treaty of Rome had been a happy occasion to return to the use of price as the regulator in the agricultural sphere as well as in other spheres. The view embodied in the Common Agricultural Policy is that 'agriculture is an economic activity following economic laws, and not simply a way of living'. To deny this would be tantamount to accepting that agriculture should permanently receive charity. Professor Mouton's emphasis on the special sociological nature of agriculture is misplaced ; price must be given pride of place in the analysis. The idea of using the English deficiency payments system is not applicable to the Common Agricultural Policy. It is one thing to apply the system in a country where agricultural population is 3–4 per cent of the total, but quite another to apply it where the percentage is 30–40 per cent of the total. Furthermore, while England imports some 45 per cent of its food requirements, the same figure for the Common Market countries is 13 per cent. In the European situation, the deficiency payment method would involve a serious rupture between consumption and production so that production would not adapt itself ; agricultural structures, requiring dynamic change, would be frozen if the system were adopted. In general, the problem of the right degree of intervention has to be related to questions of the elasticities of supply and demand for the different products. Where the elasticity is high, market forces would normally achieve sufficient stability, while in some markets intervention is justifiable on arguments of stability alone.

Dr. Krohn pointed out that Regulation No. 25 of the EEC does not say that the European Fund is fed from taxes on imports from third countries. The Fund is part of the budget of the Community. The supply of monies for the fund comes from a number of sources, including agricultural taxes, customs duties and margarine tax, and, if these are not adequate, from contributions drawn from the member countries. The contention by Professor Mouton, that continued imports from third countries help the fund, is not supported.

The third major question raised by Professor Mouton's paper, was the suggestion that there is an inevitable trend to self-sufficiency within the agriculture of the Six. Dr. Krohn said that there is no such inevitable trend in the present policies. While Professor Mouton thought that quantitative production targets are required, those responsible for policy are convinced that they can achieve the same result by price policies.

Professor Dams said that he intended to comment on the link between market (price) policy and the structural policy in overall policy determination in the EEC. Professor Papi had earlier in the Conference remarked that the policy-makers had failed to define structural policy very closely and had seemed to ignore structural problems. Now, Professor Mouton criticized policy-makers for employing the term 'structural policy' when it seemed that the use of the term was little more than a verbal gesture. In fact, in accordance with Article 39 of the Treaty of Rome, account had been taken of structural problems in agriculture, and since 1960 the EEC had held to the basic ideas of co-ordinating structural policies in the member countries and encouraging the financing of structural improvements in agriculture within the Six.

Professor Dams said that there are two ways of trying to deal with structural problems and so of putting the policies referred to into effect. Either it can be attacked by dealing directly with structures themselves, and attempting, by a step-by-step integration of these structures, to form a workable market in agricultural products : or, alternatively, the problem can be approached indirectly by trying to organize the markets for agricultural produce in the Community and then see what repercussions arose and take appropriate measures to improve structures in the light of these repercussions. There is little doubt that the second of these alternatives should be followed. Structural policy cannot be made in a vacuum : it must be intimately linked to market policy and to the regional development of the economy as a whole.

Since 1962 the Commission had adopted a number of decisions on the co-ordination of agricultural structural policies. However, the weighting of these measures within the whole complex of measures that are to be implemented depends on the degree of development of markets, and had to be held subordinate to this development. In the setting up of the agricultural guarantee fund — the European Fund — provision was made for some of the proceeds to be set aside for the financing of projects of interest to the Community as a whole, and for the improvement of agricultural structures. As the money in the fund increases the Commission can decide how it can best be used.

Professor Dams said that Dr. Mansholt had already stressed the importance of structural policy, and had, in 1959, stated that some 8 million people would have to be absorbed into other sectors if agriculture was to compete with these other sectors successfully. It is true that structural policy is lagging behind other developments ; but this is necessary and even desirable. The first modest steps had been taken.

Professor Hathaway said that he was concerned that the fundamental issue is being overlooked. Whatever projections of output proved to be

correct — whether those of Professor Mouton or those of Dr. Krohn —
the basic question is whether Western Europe can bear the cost of a high
food policy. The United States and Canada would be able to bear the
costs of adjusting their agriculture to reduced exports as made necessary
by the European policies. But Western Europe depends in the long run
on the successful exporting of manufactured goods, and with food costs
remaining at a high level, she would not be able to compete in the world
market for industrial products.

Mr. Tracy said that we could be grateful both to Professor Mouton for
having stimulated this discussion and to Dr. Krohn for his clear and
interesting statement. As Professor Mouton had prepared his paper in a
hurry, he had left himself vulnerable on a number of points. Nevertheless, he had raised some very important questions, and Mr. Tracy said
that he was not sure that Dr. Krohn had replied to them fully.

Professor Mouton had made it clear that the EEC policy so far is based
on the determination of prices. He had also pointed out that the flexibility of price policy tends to be limited by political factors. If prices
serve the social purpose of supporting farm incomes, it is very difficult to
have them serve also the economic function of guiding production. Mr.
Tracy said that Professor Mouton had also shown that problems of imbalance in markets are likely to arise and that there would be a need to
adjust production. He said that the questions which then arise are :
(1) Does the EEC expect to be able to change prices to the extent which
might be necessary to guide production ? (2) If not, do they intend to
adjust production through other means ? (3) Why has the EEC been so
reluctant to make use of some kind of 'compensatory payment' (not
necessarily the British deficiency payment) ? Some such system makes it
possible to support farm incomes at the desired level while leaving the
necessary degree of flexibility to price policy. Mr. Tracy said that he
could understand the reasons given by Dr. Krohn as to why the EEC does
not adopt deficiency payments as the basis of their policy, but he asked why
they were unwilling to admit such arrangements in any circumstances ?

We were entitled to ask how the EEC is going to face up to the problems
which seem likely to arise, although it might be difficult to get the answers
at this session.

Professor Mouton said that he did not think that there was the basic
clash between himself and the *porte-parole* of the EEC which Dr. Krohn
had implied there was. Professor Mouton said that he was emphasizing
the spirit of the Treaty of Rome rather than the letter of the Treaty to
which Dr. Krohn adhered. This is an issue on which the economist has
to take account of the political issues involved and of the social background
with which he is dealing. Since political issues are inevitably involved at
every stage in the formation of a Common Agricultural Policy, it is surely
desirable that those who study the economic aspects of the problem and
who present recommendations on the basis of these studies, should take

account of the political realities of the situation, and, within reason, reflect the democratic forces which lay behind them. The automatic nature of the passage from stage to stage as spelled out in the Treaty of Rome could not be taken to mean that the development of the Common Market is to be a purely mechanical operation. The solutions which are achieved would have to be lived with in the future, and it is essential that they should embody substantive agreement among the countries who are obliged to live together.

Professor Mouton turned to the question of the adequacy of the European Fund to deal with structural problems by subsidy. Fifteen per cent of the funds — which, at present, is all that is earmarked to deal with these problems — seems to be quite inadequate for the purpose. At least three times as much as the amount now expected to be available would be needed. Professor Mouton said that there were a number of similar points about which he disagreed with Dr. Krohn and Dr. Dams though time did not permit him to deal with them more fully than he had already done in his paper.

In conclusion, Professor Mouton said that the EEC policy should have been one of *low* prices for agricultural produce, and of rationalization in resource use. In this matter he agreed with the view expressed by Professor Hathaway in the discussion on the papers concerning the family farm.

(*The Conference returned to the problems raised by Professor Mouton's paper at the end of a later session. This subsequent Discussion is for convenience recorded here.*)

Professor Robinson said that the discussion after Professor Mouton's paper had not had time to get round to the question of the repercussions of the Common Market Agricultural Policy. It had largely been about administrative problems. It is perhaps of more general interest to try to find out what the economic philosophy of the policy makers of the Common Market is, and to discuss strategic rather than purely tactical questions. The distinction between the two could be illustrated with the case of German agricultural policy. It seems quite plain that German agriculture is structurally out of date. There is a large proportion of smallholdings of doubtful economic viability set in a country of vast and successful manufacturing areas. Yet, as Professor Delivanis had pointed out, the increase of requirements of industrial manpower in Germany is being obtained not from migrants from agriculture, but from immigrants from Greece and Italy. Professor Robinson said that he was not arguing that this is necessarily wrong, but if it were continued as a long-term policy it could lead to an imbalance inside the country between very rich manufacturing centres and poorer rural outlying areas. Professor Robinson asked whether the importing of labour is to be a permanent policy or whether it would, to some degree, be modified by industry-location policy.

Another basic problem of strategy is the relationship with developing countries. For example, it had originally appeared from the Monet plan that Europe might become a willing importer of food. Has that policy been abandoned ?

Dr. Krohn said that, in the discussion immediately after Professor Mouton's paper, on the agricultural policy of the Common Market, he had said that price regulation is not taken to be the most important aspect of their work. Structural reform is regarded as essential. The efficient farm has to be safeguarded from extreme pressures by price policy, but there are major parts of European agriculture where structural reform would be necessary. In the view of the policy-makers of the Common Agricultural Policy, structural improvement entails not only improvement in the size and viability of holdings but also the improvement and development of regions. But from the start it had been stressed that manpower in agriculture should be reduced.

On the specific example of Germany, Dr. Krohn said that migration from agriculture has been going on for some time. Between 1950 and 1960 the agricultural population was reduced by 30 per cent (some million units). This process is continuing, but it is inevitable from the nature of the problem that the outmigration would not be as rapid or easy as it had been in the past. During the period from 1950 to 1960 the migration consisted mostly of wage-earners followed by the sons and daughters of the farmers. But, today, it is only by the reduction in the number of small farmers that further migration would be possible. This is a far slower process. Moreover, the regional pattern also affects the ease of migration. Most of the early migration had taken place in areas where there were plenty of industrial opportunities near by. Now it has become necessary to concentrate on regions where the opportunities are slight. Thus, the problem of migration is part of the complex problem of regional development. The aim is not to attract labour into the present large agglomerations of industry, but rather to improve the infrastructures of regions so that industries could be brought there.

In short, the process of outmigration from agriculture continues in spite of the international migrations, and regional policy underlies the aims of the Common Agricultural Policy in attempts to re-structure the agricultural framework.

Professor James said that the case of France confirmed much of what Dr. Krohn had said. The growth of the French economy since 1945 is largely attributable to the movement of population from agriculture into industry. This had been in part a spontaneous movement and in part the result of incentives emanating from policy. The planning which had abetted this movement had been 'decorative' and vague in its application, rather than fundamental to the framework of the economy. However, the movement out of agriculture is recognized as part of the strategy of development. Over the last fifteen years the decline in agricultural population was not far short of 50 per cent of the original figure.

The policy to aid this migration from agriculture flows partly from the fact that France has joined the EEC, and has had to re-structure its agriculture in preparation for joining the full agricultural policy. Under this stimulus, France moved along the path of development so that the relative share of agriculture in national income declined and that of the secondary and tertiary sectors increased.

Professor James added that, as had been noted by Dr. Krohn in referring to other countries of Europe, there is, in France, a regional problem created by the transfer of labour from agriculture into industry. There is the threat that the country may be cut into two parts, with the half of the country lying south-west of a line from Cherbourg to Marseilles being condemned to a stagnant or even declining future, while the north-eastern part of the country developed into a vast industrial complex. Regional policy of industrial location is the best remedy to this threat, just as, on the plane of the full Common Market itself, policy has to be used to prevent the division of Western Europe into areas of immense industrial development and areas of rural stagnation.

Professor Bićanić said that Western Europe has made vast strides in the re-structuring of agriculture. Even a cursory examination of such areas as the Loire region — where large-scale wheat growing has replaced mixed farming — indicates the extent of the change. However, the main question under discussion is the effects on the labour force of this re-structuring, of the outmigration that goes with it, and of the importing of foreign labour. How far is it intended that agricultural populations should decline ? What difficulties might there be for the foreign immigrant in comparison to the internal migrant ? Are there indications that they would settle ? What would be their fate in case of an economic recession ?

Professor Bićanić said that specific policies are prone to treat individuals as ciphers — or worse. What estimate could be made, for all types of migrant, of the social costs that might accrue from policies which encouraged migration ? Social effects, such as the rise in juvenile delinquency, etc., should receive careful treatment.

Mr. Tracy said that in the context of Professor Robinson's remarks, it is useful to refer to some calculations in the OECD report on 'Agriculture and Economic Growth'. These calculations concern the contribution made to economic growth by the movement of manpower from agriculture to other sectors, and reflect both the extent of this movement and the differences in labour productivity between the sectors.

Germany, with Austria, appears to be the country where the movement of labour from agriculture has made the greatest contribution ; the figure which indicates the extent to which national income in 1960 would have been lower if there had been no shift of labour from agriculture since 1960 is between about 6·5 and 11 per cent, probably nearer the upper figure. For France, the figure is between 3 and 5 per cent ; for the United

Kingdom it is probably not more than 1 per cent. Mr. Tracy said that one could draw various conclusions from these figures. The contribution so far has been important in countries like Germany, but could perhaps have been still greater. In the future, the extent of the contribution would tend to decline as the absolute movement of labour became smaller and other countries approached the position of the United Kingdom.

Professor Fauvel said that he would like to know the answer to Professor Robinson's question about the apparent policy of high food prices in the Common Market and its repercussions on the developing world. This high price policy seems to be partly the result of German high prices, which have committed the policy-makers to a *politique du pire*.

Professor Gulbrandsen said that Dr. Krohn had seemed to be arguing in an earlier discussion that prices in the Common Market should be set to achieve self-sufficiency for nearly every product, irrespective of the comparative advantages between the EEC and the rest of the world. In the present discussion Dr. Krohn had argued that a reduction of the agricultural population is intended. Professor Gulbrandsen said that the price policy which has been described implies that the population in agriculture would decrease more slowly and the productivity of labour in agriculture increase more slowly than it would if prices are established which reflect the comparative advantage of Europe as compared with the rest of the world. Furthermore, there would be severe repercussions for the exports from developing countries.

Dr. Krohn said that what he had said in the earlier discussion had been misunderstood. He had not said that the price policy is aimed at self-sufficiency for the EEC. He stressed that one of the aims of the policy is to maintain part of the EEC market for the products of countries of the third world.

Dr. Krohn said that it is fully appreciated by the policy-makers of the Common Agricultural Policy that it would be wrong to try to achieve fair wages for the whole agricultural population by means of price policy alone. Structural changes would be necessary.

Dr. Krohn said that, in a memorandum of the Commission, it had been indicated that a preference is to be given to direct aid rather than to reductions in tariffs and taxes in order to help the third world. This does not mean that the policy-makers aim at self-sufficiency, but that price and structural policies are mainly directed towards the problems internal to Europe.

Dr. Krohn said that price policy has to take account of political and social issues. It has been argued that the price decisions for cereals, which would be applied on 1 July 1967, is a 'social price' and is too high. Dr. Krohn said that social and political factors have certainly played a part in determining the price level. However, this level is only an increase of 6 per cent, and this in an inflationary world cannot really be taken as too high.

Chapter 6

PROBLEMS ARISING FOR THE AGRICULTURE OF A DEVELOPING COUNTRY BY VIRTUE OF ITS ASSOCIATION WITH THE EUROPEAN ECONOMIC COMMUNITY [1]

BY

D. J. DELIVANIS
University of Thessaloniki, Greece

I. INTRODUCTION

IN this paper I mean to do three things : (*a*) to outline the reasons which may lead a developing country to seek association with the European Economic Community ; (*b*) to analyse briefly the problems which such a country will have to resolve if it is to derive any economic benefits from its association with the European Economic Community ; (*c*) to examine the particular nature of the effort which an associated developing country will have to make in its economy to solve the afore-mentioned problems with respect to its primary sector.

It is with these three points that the three, rather unequal, parts of the paper will be concerned.

II. REASONS FOR SEEKING ASSOCIATION

A developing country may be led to consider association with the European Economic Community (EEC) for a number of reasons, as follows.

(1) The structure of the developing country's economy and its degree of development in relation to its development targets may seem to preclude its simply joining the Community, notwithstanding all the reservations and escape clauses provided for in the Rome Treaty of

---

[1] Translated by Elizabeth Henderson.

25 March 1957. When this treaty was negotiated, one of the countries concerned was still far from any sort of satisfactory external balance, and the other five had all had to battle since the end of the last war with problems of the achievement and maintenance of external equilibrium. In these circumstances none of them could have signed, let alone ratified, a treaty which would restrict their freedom of action to deal with any threat to their internal or external equilibrium, without safeguarding their right to act as they saw fit in a crisis. When, therefore, a developing country decides to seek association with the Community rather than outright entry, it must be assumed that the authorities concerned feel that the problems that they foresee will not be susceptible of solution within the framework of the relevant provisions of the Rome Treaty of 25 March 1957.

(2) On the other hand a developing country may feel that it has to seek association for fear of losing a major export market, especially when it seems difficult to acquire or keep other markets. This applies particularly to countries exporting only one commodity, which have little hope of altering this state of affairs in the relatively near future for a number of reasons, including the producers' inertia and lack of initiative, the influence of technical factors which can be offset only at great cost, if at all, and finally the entrepreneurs' and exporters' scant know-how or their distrust of official intentions and wavering attitudes towards them.

(3) A developing country already suffering all the disabilities peculiar to its state of development, can ill afford to risk isolation in a world economy which today, more than ever before, tends towards integration. Indeed, the countries which stand to gain from integration, want to achieve it on the most far-reaching scale, and to this end are prepared to do something for developing countries so as to speed up the latter's economic growth. By this means the advanced countries hope to acquire not only new markets as quickly as possible, but also new sources of supply from which they can expect priority treatment in case of shortages of imported agricultural commodities, raw materials, semi-manufactured goods and manufactured goods. A developing country whose association with the Community is under consideration or negotiation may well be able to furnish such goods. The help offered may be in the form of grants, but most often is limited to very long-term loans at rates of interest distinctly lower than those which the developing country would have to pay elsewhere abroad — assuming, which is by no means always the case, that its credit-worthiness enabled it to raise loans abroad at all. Association

F

may also be connected with trading facilities, such as tariff reductions for the developing country with respect either to all its exports to countries of the European Economic Community or to a certain volume of exports laid down in the treaty of association or by the agencies responsible for its implementation.

(4) Lastly, a developing country may be swayed by political or social motives in seeking association with the European Economic Community. The government may hope that at least some of the country's military expenditure will be taken care of by the Community after association, or that would-be aggressors may think twice in the future, since the associated country could then count not only on the moral support of the Community but also on its financial support, even though this is not mentioned in the treaty of association.

III. PROBLEMS OF ASSOCIATION

In section A of this part of the paper I shall make an attempt to classify the problems to be resolved by the economy of a developing country by virtue of its association with the European Economic Community. In the three subsequent sections I shall analyse these problems, with special reference to agriculture.

SECTION A

There are many problems which a developing country has to face when it enters into association with the European Economic Community. These have been discussed over and over again, and I myself had occasion to deal with some of them at a Congress, held in the autumn of 1964 at Thessaloniki, on European integration and its implications for Greece. At that time I distinguished two categories of problems, those connected with the associated country's freedom of action in determining and applying its economic and monetary policy, and those connected with the inadequacy of the associated country's economic infrastructures and with the consequences of this inadequacy for employment and external balance. On further reflection, it now seems to me that another classification might be rather more meaningful. We may distinguish three categories of problems concerning, respectively :

(1) government ;
(2) producers and exporters, together with all those who work with

them, without any chance on their part to do anything about these problems ;

(3) the same groups as in (2), but with an opportunity to take remedial action.

The problems of association may be grouped into these three categories as follows :

(1) One of the chief problems concerning government is the danger that the economy will be dominated by the EEC, and it is one which varies greatly in intensity in different cases. Other problems concerning government are that it may not be able to make its own country's point of view prevail in the Community, and that, with its restricted freedom of action, it may find it more difficult to keep the balance of payments in equilibrium.

(2) The difficulties which producers and exporters, and their helpers, have to endure passively are of three kinds. First, there is the disappearance of tariff protection, which producers previously enjoyed at least within the frontiers of their own country ; then there is the difficulty of competing, when the output of the country is produced under conditions of poor infrastructure, against producers in the European Economic Community, where enormous external economies greatly reduce cost ; and finally there is the wage rise consequent upon massive emigration to the countries of the Community at times of full employment.

(3) Lastly, there is a set of problems upon which producers and exporters, and their helpers, in the associated country can exercise some active influence. They will have to adopt every possible technical innovation, lest failure to do so give their competitors in the EEC additional elements of monopoly. They will, likewise, have to try to do something about their excessive production cost even at modest wages and salaries, about their small volume of output where decreasing costs are practically out of the question and about their own lack of knowledge and experience when it comes to gaining access to large foreign markets or indeed to defending their own domestic market.

Section B

In this second section I propose to analyse the problems of the first category (*i.e.* (1) above), in the light of what was said above and with special reference to agriculture.

(*a*) So far as agriculture is concerned, the problems connected with economic domination by the EEC find expression above all in the government's restricted capacity to introduce changes in farming methods, even if such changes may be necessary with a view to raising farm incomes, diversifying agricultural production in the interests of internal and external balance alike, increasing public revenue or diminishing underemployment and unemployment. There may indeed be plenty of other reasons which in the government's view make changes in farming methods necessary. Furthermore, it goes without saying that domination of the associated country's economy by the EEC means that industrial products for use in agriculture or in any other sector may have to be purchased within the EEC, even if they could be bought elsewhere on better terms. There is, of course, no danger that this will happen at present, when there is full employment everywhere in the Community except in Italy; but it may well happen if there is a depression, which, after all, is not to be excluded if deflationary policies are applied, especially to avoid devaluation. Any devaluation would create very great difficulties for the Community, especially now when, since the end of December 1964, wheat prices are fixed no longer in the currencies of member states but in common units.

(*b*) So much for the disadvantages which the agriculture of an associated country may suffer through economic domination by the EEC. These are reinforced by the inability of that country's government to insist effectively that the treaty of association be strictly applied whenever this would be to the country's advantage and designed to mitigate or indeed avoid the afore-mentioned consequences. An associated country may be persuaded not to avail itself of the opportunities open to it under the treaty of association, in the case where the Community is afraid that this would create an awkward precedent and believes it right to avoid it. There is every reason to think that such a state of affairs would have unfavourable repercussions on agriculture in the associated country, for it is generally agriculture which most needs assistance and it is also agriculture which is still the most important sector of the economy in all the countries now associated with the European Economic Community.

(*c*) Developing countries which enter into association with the EEC invariably have balance of payments difficulties. These difficulties are enhanced by the development process itself, because of the growing propensity to consume foreign goods. At first sight it would seem that this would not affect agriculture, but that the sole sufferers would be craftsmen and industry. But there is competition from tinned foods, or attractively packaged foods, from the countries of the European Economic Community. These countries can in such cases step up their exports to the new associate on a large scale, if the treaty of association obliges the latter, as it does Greece, to reduce tariffs *vis-à-vis* EEC countries. Two difficulties must then be expected. First, the agricultural products concerned will be harder to sell at home, and there is no hope of increasing exports of them to the EEC, because the products are marketed in less attractive form. Secondly, the associated country's balance of payments will suffer at the very moment when the treaty of association makes it impossible to take direct counter-measures. It is true that such measures are applied only too often by developing countries, and it may be argued that it is all to the good to tie their hands in this respect. But if the government of an associated country is not free to do what needs to be done to maintain or re-establish external balance, it may have to invoke much sooner the right to apply the special clauses which the Treaty of Rome provides for protective measures in case of extra-ordinary difficulties connected with external balance.

The unfavourable repercussions of this state of affairs on the agriculture of an associated country are further aggravated by the danger of losing other foreign markets, if the treaty of association wholly or partly rules out compensation agreements or bilateral trade agreements. Even countries which have long since completed their development process sometimes find that they cannot be sure of selling their agricultural output without such agreements.

Section C

We now come to the problems of association with respect to which the producers and exporters of agricultural commodities, and their helpers, can do nothing.

(*a*) First, there is the disappearance of the tariff protection which agricultural producers previously enjoyed. The abolition of customs duties is, of course, slower in the agricultural sector than in the industrial one. But first of all it is known for certain that, barring amendment or repeal of the treaty of association, the customs duties must eventually disappear. Quite apart from that, it may well happen that in the negotiations for the treaty of association the government concerned has to sacrifice the producers of some farm products in order to secure other concessions. This was the case of ham and cheese when Greece became associated with the European

Economic Community. Finally, association may make it more difficult to subsidize agricultural exports to the Community, quite apart from other difficulties of a budgetary and institutional order which habitually impede export subsidies.

(*b*) The disappearance of tariff protection involves additional hardship for farmers in an associated country where social overhead capital is well below needs — which is certainly not the case of the great majority of the territory of the European Economic Community. Deficiences of infrastructures in the associated country raise farming costs, especially as a result of the unsatisfactory condition of transport and storage facilities. Bad transport also impairs the quality of agricultural produce. It follows that the farmers of the associated country suffer a deterioration of their competitive position *vis-à-vis* the farmers of the European Economic Community. It is only natural that the propensity to consume foreign goods, which in any case is always strong in developing countries, should be further reinforced in such circumstances. The consumers of the associated country will no doubt be anxious to buy products of better quality, but this must be subject to the explicit condition of sufficient international liquidity in the associated country. There can be no certainty in this respect, and the special escape clauses may then have to be invoked.

It does not necessarily follow that production costs in the agriculture of the associated country are always lower than in the EEC at large just because farm wages are lower. There are other points to consider in the comparison : in the EEC, production costs are reduced thanks to the use of machines and to low interest rates, and interest rates in any event have but scant influence on costs, especially in agriculture ; in the associated country on the other hand the wages actually paid out to workers generally, and to farm workers in particular, tend to rise in a situation of growing employment and emigration, such as obtains today at a time of full employment in the European Economic Community.

(*c*) This last consideration leads us to the unfavourable repercussions which massive emigration from an associated country to the European Economic Community has on the former's agriculture. First of all, this causes domestic demand for farm products to drop in the associated country. It is, of course, not impossible that the consumption, and hence the demand, of other population groups may rise, if demand was not previously saturated, if emigrants send money home to their families and if association raises investment in the associated country. It is also true that farmers feel the impact of rising wages only to the extent that they employ non-family labour. It might be argued that nowadays large farm estates tend to disappear and that the relative importance of farm wages therefore tends to diminish. But even small farms do, at certain periods, need more labour than the family can furnish, unless they can use machines. Whenever purchase of machines is impossible, production of certain farm products diminishes or may be discontinued altogether as a result of wage increases. Emigration to the EEC certainly is a contributing factor in such wage increases, and the result has been seen in Greek cotton production since 1963.

Section D

It remains now to discuss the problems of association with respect to which producers and exporters, and their helpers, can take some remedial action.

(*a*) First, there is the problem of the intensity and rate of adaptation to technical progress abroad, and especially in the European Economic Community. Before the treaty of association came into force, producers in the associated country could postpone adaptation until such time as their equipment, machines and implements became totally useless from the point of view of costs, production time, quality and marketing. This meant that equipment, machines and implements which had long been fully amortized, continued to be used and to yield their owner a profit long after they had gone completely out of use in the countries of the Community. The producers nevertheless had their safe market at home, thanks to tariffs and import restrictions.

But once a country enters into association with the European Economic Community, producers are led to apply technical progress as fast as possible, even if they are not certain that it is indispensable, even without waiting to be guided by experience abroad, and even if it means rising production costs. They do it because they want to forestall as far as possible any loss of markets both at home and abroad. They feel that if they postpone the necessary investment they may not be able to catch up again later, and they want to avoid such a situation at all costs. It seems to me that we are thus in the presence of a new concept of marginal returns. The producers of the associated country need to learn to appreciate the distinction between what is really indispensable and what is not. As it is, they suffer losses of capital or at least of profit whenever their decisions prove wrong.

(*b*) Another problem demanding action from producers and exporters, and their helpers, in an associated country is their excessive production cost, even though wages are low in relation to the workers' needs, though not in relation to their productivity in quantitative and qualitative terms. There can be no question of reducing wages in these circumstances, especially in the light of the growing tendency towards emigration, which was discussed in section C above. The solution, at least on the micro-economic level, seems to lie in a continuous and systematic effort at rationalization. Human labour can be replaced by machines which can be operated without much skill and can be bought without much capital and hence, in most cases, without much foreign exchange ; this can be done without detriment to the profit rate. Division of labour can be increased in the context of a growing volume of output. There is no doubt at all that all this is within the reach of producers in an associated country, if only they realize their opportunities, examine them carefully and find not only the technical, but also the financial means of introducing the necessary changes. It would obviously be best if finance funds could be found in accumulated profits or raised by an increase in the firm's capital ; but if for one reason or another neither of these solutions is feasible, recourse to medium- or long-term credit is indispensable.

(*c*) In considering the problem of excessive costs for the producers of an associated country (among which we must mention in particular the high costs which inevitably follow from the defective and slow working of public administration), we must not forget the cost-increasing effect of a small volume of production, as is to be found in most cases. Overheads have to be spread over only a small number of units, and there is no point in installing plant capable of a large volume of output at decreasing cost. In agriculture, which is all that we are concerned with in this paper, cost reductions can be achieved in a number of ways : the use of seeds giving higher and better-quality yields, as well as quicker yields, so that producers have the benefit of the higher prices paid for early produce ; the use of machinery so as to reduce labour requirements ; the use of speedy methods of harvesting, whereby the risk of rain and of expected or unexpected changes in temperature can be diminished ; improvements in the storage and handling of crops ; and finally the transporting of crops to market at a time favourable from the point of view not not only of weather and temperature, but also of domestic and foreign demand. It happens only too often in developing countries that producers — and not only in agriculture — are used to their customers' accepting whatever quality and appearance of goods may be offered to them. While this might be acceptable in a country surrounded by high tariff barriers made even more effective by import restrictions, it will certainly not be acceptable when the latter are abolished and the tariff barriers lowered. And this applies with even greater force to buyers abroad.

(*d*) This leads me to the last problem which I want to stress here ; namely, the lack of knowledge and experience on the part of producers, exporters and their helpers in the matter of defending the domestic market when tariffs and import restrictions are abolished, or of gaining access to foreign markets under these conditions. All these people need training and instruction to enable them to take effective action in this area. They must learn how to adapt themselves not only to rational requirements but also to the preferences of such customers as they have or hope to acquire, if these preferences make themselves felt on a large scale. Adaptation may be necessary with respect to quality, presentation, delivery dates, strict fulfilment of obligations, prices, terms of payment, publicity, stock-holding at the point of sale, and, finally, to the policies of other suppliers abroad. It will be a continuing struggle for producers and exporters, and their helpers, and they will need always to be alert to any opportunity to defend and enlarge their domestic and foreign markets. Success depends not only on being able to back this struggle with substantial funds, but also on firms knowing how to react, how to move ahead and how to put imagination, know-how and constructive ideas at the service of customers who are for ever in search of cheaper and better supplies and easier terms. Unfortunately, these conditions are by no means always present in a developing country associated with the European Economic Community.

IV. EFFORTS TO SOLVE THE PROBLEMS

In this third and last part of the paper, I want to examine the particular nature of the efforts which an associated developing country will have to make in its economy to solve the problems of the primary sector, such as they were analysed above. Apart from the developing country's own efforts, there are others which are incumbent upon the Community ; I shall discuss the two categories separately in the two sections of this part of the paper.

SECTION A

As regards the efforts to be made by the associated country, a distinction has to be made between (*a*) those pertaining to government, and (*b*) those pertaining to producers, exporters and their helpers, with the assistance of government and administration.

(*a*) The government of an associated country should do everything in its power to minimize the consequences of economic domination by the EEC and to retain freedom of action in taking such measures as may be necessary to modernize, develop and improve domestic agriculture. There are several things the government of an associated country can do to this end.

   (i) It should keep within the narrowest possible limits its requests for loans and credits from the institutions of the Community and other EEC governments, in so far as these funds are to be used otherwise than for an improvement of infrastructures.

   (ii) It should try by all means at its disposal to obtain new outlets for the country's agricultural produce, without giving up subsidies altogether — unless, of course, their use entails reprisals on the part either of the Community or of third countries. There is no doubt that tax reliefs and infrastructural improvements in the broadest sense are much more effective in this context, and also much more in keeping with the established principles which guide Community policies. These efforts can, of course, be successful only on condition that the country's agricultural output increases.

   (iii) The government should do what it can to persuade member states and, even more so, the Community's own institutions to fulfil to the letter all the obligations they assumed in the treaty of association, and to this end can appeal not only to economic, but also to political and moral considerations. There is a case for every 'agreed' violation of the treaty of association by the Community to be compensated by some concession on its part, provided the latter can be put into effect without recourse to the Parliaments of the Six, which would be rather time-consuming. From this point of view it may be better to be content with lesser advantages, if these, only, can become operative at once. I have in mind, among other possibilities, an increase in agricultural quotas, especially when stocks are unduly high.

   (iv) The government should do everything which helps to

strengthen external equilibrium and refrain from doing anything which might weaken it, even if this means that the interests of farmers may be hurt at first. Thus the refusal of subsidies will certainly not raise the income of farmers, but it will be an incentive towards rationalization, which subsequently will raise their exports and thereby their income.

(v) In the fifth place, the government should do its utmost to improve infrastructures. Better infrastructures can do more than anything else to raise returns from farming and farm incomes, yet there is little that farmers themselves can contribute to this end, even through co-operative organization. The only exceptions are drainage, irrigation and flood control projects on the strictly local scale.

(*b*) It is up to the producers, exporters and their helpers, on the other hand, to make every effort to reduce costs and increase output, not forgetting such matters as quality and presentation of the product, delivery and payment terms, and price. To be successful in these efforts, farmers again need government help ; in particular, the government should

(i) steadily enlarge and improve infrastructures ;
(ii) give farmers appropriate inducements to do what needs to be done to meet the preferences of foreign and domestic consumers, provided that these measures can be applied without creating worse problems than they are meant to solve ;
(iii) teach farmers by what principles they ought to be guided in deciding whether or not to adopt some technical innovation.

It is an unfortunate truth that the public administration of a developing country is not always perfect and cannot always take care of everything it ought to. Consequently, there is wide scope for the activities of agricultural producer co-operatives, of the bank, or banks, dealing exclusively with agricultural credit, and also of exporters, especially when these latter dispose of large capital funds or can count on the support of the authorities.

It seems to me that demand for agricultural products, except cereals, is still far from saturation point in all the countries of the European Economic Community, and that consequently there is still plenty of room for an increase in production and in exports to these countries, even if their own output inevitably grows. But it must be added that any widening of markets in the Community presupposes prices, terms of payment and delivery, quality and presentation such as to be acceptable to consumers in member states. It will, furthermore, be easier to sell agricultural products in the Community if they undergo a certain amount of processing prior to export, for this saves the consumers time and labour. Once they engage in this sort of activity, farmers in the associated country may eventually succeed in exporting the products of agricultural processing industry, provided they can satisfy consumers in the Community. Nor are other industrial exports to the Community precluded, subject to appropriate quality, prices and delivery and payment terms being offered. The example of Japan in the nineteenth century, and of Hong Kong in the twentieth, may serve to remind us that this is more than just a pious hope — that it is an effort which can succeed if only it is pursued systematically, and, above all, without expecting results in a few months time.

Section B

The Community, in its turn, can make an effort towards the desired end by avoiding the creation of institutional or other obstacles to agricultural imports from the associated country — without prejudice, of course, to agriculture in member states themselves. In addition, the Community could help by facilitating the rational execution of infrastructural work in the associated developing country, thus contributing to an organic reduction of its agricultural production costs. Even if it might seem in certain cases and in the short run as if loans to this end were more in the nature of outright grants than anything else, they may, if used rationally, turn into highly profitable investments in a few years' time, for the Community itself, under every aspect, no less than for the associated country concerned.

————

DISCUSSIONS OF PROFESOR DELIVANIS'S PAPER

Mr. Allen said that Professor Delivanis had stressed the difficulties faced by Greek farmers and distributors in defending domestic markets and gaining access to foreign markets. In the general context, associated status is designed to minimize loss, or maintain the status quo, *vis-à-vis* competitors. However, if the benefits of the international division of labour are to be reaped, it is essential that authorities in the countries seeking association recognize the need for developing marketing organizations. Mr. Allen asked what social and political restraints might be working to prevent this development, and so be deterring the few progressive entrepreneurs from going ahead. He cited the examples of California and Arizona, as 'associated' regions of the United States, where there is no restraint on movement of capital into the states and no restriction on competition. Similarly, in the north-west of Mexico, the production and export of vegetables has flourished, since the Mexican government has not imposed the restrictions, which apply for other parts of Mexico, on land development and on capital imports. Mr. Allen asked how much the difficulties of countries associating with the EEC, for example, Greece, are due to the fact that they were not ready to accept the logic of developing the competitive system that was required.

Professor Robinson said that, in view of the conflicting objectives, it was difficult to decide what is gain and what is loss for any particular country. Marginal farming is becoming more difficult to conduct profitably, and this seems to be true for all fast developing economies. Low productivity and low income farming present development problems in places as diverse as the United Kingdom, Central Africa and Southern Italy. Should one not welcome the difficulties of farmers in marginal areas, as evidence of progress ? Can the Greek economy, as suggested by Professor Allen, adapt itself to obtain a high level of economic activity, or will it be forced to restrict the progressive forces because of balance of payments difficulties ? These payments problems at least show that part of the difficulties of transition are short term.

Professor Martinov said that it is recognized that association with the EEC might facilitate the development of agriculture in a particular country by encouraging rationalization of exports in the light of new opportunities. However, since over-production of agricultural goods is becoming increasingly acute within the EEC, what measures should be applied to protect the agriculture of an associated developing country from the danger of over-production and the resultant drop in the prices of agricultural goods from that country ?

Professor Bićanić said that the paper is of great interest to many other countries in Europe. Sooner or later there would have to be active relations with the EEC, and he thought that the sooner they are initiated the better. He agreed that twenty-two years is too long a period of transition ; the psychological pressure for action exerted on governments by a short transition period is important.

Professor Bićanić asked to what extent remittances from emigrant workers from Greece have been a source of capital investment in agriculture, and so off-set any loss of manpower, which in any case might ease rural over-population. What effect has the development of tourism (especially from EEC countries) had on changes in Greek agriculture ?

Finally, what evidence is there of increased regional specialization as a result of association ?

Professor Ashton said that, in the context of the development of the Greek economy, the lack of alternative employment opportunities for farm labour is a striking feature. Surplus labour has to emigrate to find employment opportunities. This tendency should be viewed against substantial developments in the last decade in the non-agricultural sector. Much new industry and other economic activity has been introduced. But it has generally been capital- and not labour-intensive and has provided very little employment opportunity in relation to the available labour surplus in agriculture. At the same time, preoccupation with the need for continuing industrial development should not be allowed to lead to neglect of the development of the agricultural sector. This could present special difficulties. For instance, there is a wide discrepancy between modern technological possibilities and the existing structure and organization of Greek agriculture. It is difficult to make the best use of a modern irrigation scheme, for example, with small and highly fragmented holdings. As recommended by Professor Delivanis, there is a pressing need for rapid progress to be made with structural reform in Greek agriculture.

Turning to the competitive position of Greek agriculture in the light of its alignment with the EEC, Professor Ashton said that one would expect, in the short run, that comparative advantage, in soil, climate or technology would largely determine the future pattern. Thus, livestock and dairy products could be expected to enter Greece from those member states with more highly developed livestock industries. But there are distinct possibilities that Greece could benefit from the association in the marketing of fruit, vegetables, cotton and tobacco.

Professor Ashton said that he would like to take up Professor Bićanić's point about emigration. Some of the emigration from Greece is from urban areas. But the rural population, which more than reproduces itself, can withstand emigration without the abandonment of land. In fact, some grazing land at present in use is certainly sub-marginal and there is a loss of social capital as a result of erosion and the spoliation of forests.

Finally, Professor Ashton wondered whether Professor Delivanis, following his remarks on page 138, would include wheat among the commodities where output should be increased.

Professor Delivanis said that it is true that as yet no marginal land remains uncultivated, and that the degree of fragmentation is uneconomic. The latter cannot be separated from the question of the Greek farmers' individualism. This individualism also explains why tractors are bought, not only when needed in the fields, but also in order to go to town, and even for simple reasons of prestige. It is also true that there is a serious lack of alternative employment opportunities in industry. For this reason he favoured tourism and migration. Also in reply to Professor Ashton, Professor Delivanis said that the price paid by the Treasury to wheat-growers is excessive, and is attributable to political desire for the good-will of the producers.

In reply to Mr. Allen and others, Professor Delivanis said that the importance of monopolistic elements in preventing satisfactory development should not be over-emphasized. In Greece, quotas were abolished in 1953. Foreign entrepreneurs are free to settle and operate. Foreign capital, as far as service and repatriation are concerned, enjoys protection which is linked with the Greek constitution so that the law cannot be amended. Certain professions are to a certain extent closed to outsiders, but this is true even of the United States.

In reply to Professor Bićanić, Professor Delivanis said that regional specialization in agriculture is not favoured. The fear of mono-culture led to the promotion of diversification.

Professor Delivanis said that the investment of savings by emigrants tends to be into city flats or small handicraft firms, rather than into agriculture.

Professor Papi said that even without association with the EEC, a developing country would encounter most of the difficulties listed in Professor Delivanis's paper. However, he thought that Professor Delivanis was quite right to stress the need to improve the infrastructure in this instance.

The weak point in the planning of the EEC policy, particularly with respect to agriculture, is that it has not respected the logical sequence of the developmental process. Thus the Mansholt Commission has, from the beginning, submitted proposals for the stabilization of markets without due attention to the complexities of joint markets in different products. Professor Papi said that the present Common Market policy is on the wrong path. The Commission proposes an arbitrarily fixed price system with threshold maximum and minimum prices, etc. But these are not the result of any real developmental tensions. These prices prepare the basis for customs rates within the community and outside. The indicative price is the main factor in calculating the amount to be paid at the frontier. But this does not allow for development, so that the agriculture of the Six is simply to be crystallized in its present form, while indicative prices and frontier taxes block the way to further evolution. Under these circumstances, the likelihood of third countries enjoying the advantages of regional integration are reduced.

Moreover, this system of artificial price creation completely neglects structural problems, the solution of which would have led to the lowering of prices to the advantage of the Six themselves, and of countries outside. Indeed, the committee to consider structure in agriculture was set up eight years later than the initiation of the policy on markets. Price policy alone will not lead to the advantages which European agriculture might desire, and, at both the national and international level, these advantages can only be obtained if greater advance is made in structural policy.

Professor Nussbaumer said that the problems of integration and association depend not only on the degree of development of the country in question, but also on the types of agricultural products involved and their possible future development. The comparison between Greece and Austria with respect to the EEC demonstrated this.

The fear that existing markets might be lost because of subsidies paid to producers within the integrated area is not the only worry. It is often the case that elasticities of demand for home-produced agricultural products seem to be more favourable on foreign markets than on the home market. This meant that increased sales at existing prices often seem more likely abroad than at home. It is thus important to know whether an importer into a large integrated area will be more likely to have to bear the burden of protective tariffs than the consumer within the integrated area. Clearly, the elasticities of supply and demand would have to be studied in detail.

Professor Nussbaumer said that in the case of Austria there are export restrictions on some products of agriculture and forestry. These restrictions were set up to ensure cheap supplies to domestic industry and consumers. This is at the expense of farmers, who would presumably profit from the higher prices of the integrated area if these controls were removed. However, reduction of internal sales of home products might result from increased competition of canned and stored food, and of agricultural staple commodities competing in home markets. The depressing effect of this on real agricultural disposable income might be outweighed, however, by the reduction of prices of industrial factors and of manufactured consumer goods, as a result of greater international competition.

Finally, the attitude taken towards integration would depend largely on the relative accessibility of factors of production. Since Austrian farms (in terms of improved agricultural land) are, on the average, larger than, for instance, German farms, and since consumption trends for some typical Austrian farm products like beef are positive, Austrian farmers generally take a favourable view of integration with the EEC.

Professor Ruttan asked for explicit data on the question of wage rates paid to hired labour as compared with the income on small farms.

In reply to Professor Ruttan, *Professor Delivanis* said that unfortunately the statistical sources are thin. In Greek agriculture, the family is the main source of labour ; only at harvest time is hired labour used, and this comes from young people working in the city, and from the unemployed. The problem of labour shortage is not serious for wheat, but affects cotton and tobacco production. The difficulty of obtaining hired labour, and the heavy increase in wages, have contributed to an absolute decline in cotton production. However, the high prices paid to wheat-growers are said to have played a part in the decline of cotton production.

Professor Robinson said that underlying Professor Delivanis's paper seemed to be the general idea that if a developing agricultural country were associated with a more advanced area, then, with tariff reductions, there would be a tendency for that country to increase imports by more than would be offset by increased exports. Is there any theoretical reason for thinking this would happen ? Are the transport costs low for imported industrial goods, and high for exported agricultural products ? Is it a reasonable presupposition that a peripheral developing agricultural country is likely to be damaged by a reduction in tariffs over the short period, and so will require greater internal adjustment ?

Professor Eastman referred to Professor Delivanis's statement that there is a danger that the associated country would be obliged to purchase industrial products from EEC countries even if they could have been bought elsewhere on better terms. He said that he believed the purpose of a customs union is, in part at least, to ensure that this will happen. The real cost (in terms of exports) of industrial products will thus rise for the associated country, though the price will fall for the individual consumer.

Professor Eastman said that the increase in foreign exchange necessary to purchase a constant volume of industrial products will tend to turn the terms of trade of the associated country against itself. The effect of a particular association on the associated country will depend on its competitive position on the export market and the elasticity of demand for its goods, and also on the real costs of EEC industrial products as compared to those of third countries. These factors are likely to vary between particular cases.

Professor Haley said he wanted to add that even though a country to be associated with the EEC might have to anticipate a worsening of its terms of trade as a result of association, it might nevertheless choose association on the ground that to refrain from it might mean such a serious loss of exports to the EEC, that it would be even worse off.

Professor Haley asked what was meant in the paper by 'domination by the EEC'. Surely this did not mean political domination ; it presumably did mean economic domination ? Was Professor Delivanis perhaps simply referring to the fact that the associated country's manufactures would now be subjected to the serious competition of goods imported from the EEC ?

Professor Neumark asked a question along the same lines. He said there seemed to be two standards of reasoning. The paper expressed anxiety about certain disadvantages of association, but drew attention to indisputable advantages. Did this contradiction stem from the fact that there was no clear distinction between the short- and the long-run situations ? Professor Neumark admitted that in the short run, and perhaps for 5–10 years to come, there would be certain difficulties for an associating country, but these were a result of the fact that in the past the country concerned did not have the economic policy that it should have had. For example, it had often been heavily protected in its agriculture.

Professor Delivanis said that he agreed with Professor Papi and others that, even without association, the associating developing country may well encounter more or less the same difficulties. Thus along with development come problems of balance of payments, increased need for foreign products and increased consumption of home-made products. However, without association, the freedom of action of the Greek government would have been less restricted.

In answer to the questions about the advantages and disadvantages for Greece as a particular case, Professor Delivanis said that the nation had had to consider the possibility of being excluded, while important competitors such as Turkey had associated. The case is complicated by the fact that the negotiated transition period has been made so long that nobody in the public or private sectors has felt obliged to make decisions and carry them out. The attitude is rather that of '*après moi le déluge*'. Thus the Greek manufacturing sector has taken only slight advantage of the 70 per cent cut in duties on goods imported into the EEC. Lack of experience, high initial costs and lack of competitive drive are the contributory causes. In the agricultural sector, the high prices to be gained for tobacco have not led to the rise in output that was expected. However, Greek manufacturing industry has not yet suffered from increased export of raw materials it uses, although this is a possible threat for the future. As a further disadvantage, Professor Delivanis referred to the high propensity to import in Greece, which is due to the preference for Western European presentation and quality in products which can now be compared with the home produced goods. These disadvantages are largely attributable to past protectionist policies. These are aggravated by the unrealistic prices fixed by the EEC, referred to by Professor Papi.

In comparison with Austria, Greece suffers from a poorer infrastructure, high freight charges over long distances and the very slow transport system south of Belgrade.

On the question of domination, Professor Delivanis said that he did not mean it in a political, but rather an economic, sense. Thus Greek importers naturally make their choice in favour of suppliers whose products (provided that the quality was not different) cost less after payment of import duties. So goods from the EEC, which are originally more expensive than the same products from other parts of the world, would be effectively cheaper because of the lower import duties. This weakens the Greek balance of payments position, which in any case is not strong. Fortunately the treaty of association allows Greece to import certain quantities of commodities from the socialist countries, so that she does not lose her markets there.

On the question of the elasticities of demand for Greek exports, Professor Delivanis said that these tend to diminish as the incomes of EEC countries continue to increase.

PART II

REPERCUSSIONS ON DEVELOPING COUNTRIES OF
PRESENT POLICIES IN INDUSTRIAL COUNTRIES

Chapter 7

THE CHANGING PATTERNS OF INTERNATIONAL TRADE PROBLEMS OF UNDER-DEVELOPED AREAS

BY

J. H. RICHTER

International Federation of Agricultural Producers, Washington, U.S.A.

I. INTRODUCTORY

THE general theme of this Conference is economic problems of agriculture in industrial societies and repercussions in developing countries. My specific assignment is to deal with international trade problems of under-developed areas. I should, therefore, try to throw some light on the question whether or not, or to what extent, there are influences upon the trade of less developed countries that emanate from the economic problems of agriculture in industrial societies — or from the circumstances which create these problems — and from policies which these societies, all of them or some of them, apply as a result.

To anticipate some of my conclusions, and as I have stated on previous occasions, whatever the changes in the patterns of the trade problems of the under-developed world, these changes, patterns or problems have not in significant degree been caused by the problems of agriculture in industrial countries, or by resultant policies of agricultural protection and support. In my view, the trade problem of the under-developed world must be seen in a different setting. I shall touch upon this subject in the concluding part of the present paper. The limits set for it will not permit more than a cursory reference, but I expect to have an opportunity for more extensive discussion elsewhere.

II. THE DEVELOPING WORLD AND PROTECTIONISM BY DEVELOPED COUNTRIES

The argument that agricultural protectionism in the developed industrial countries is largely responsible for the plight and trade problems of under-developed areas has held a strange fascination for some economists in international secretariats and industrial countries. Trade of the under-developed countries — it was contended — has lagged in relation to total world trade and world production ; trade in foodstuffs has lagged compared with total world trade ; there has been a relative lag in the trade of foodstuffs from under-developed to developed countries ; and, in all this, agricultural protectionism in the industrial societies has been a significant factor and, hence, has also been a significant impediment to economic development in the under-developed world.[1]

Although I am assured to the contrary, I doubt if many economists in the under-developed countries themselves truly subscribe to this proposition. And indeed even a preliminary analytical look gives reason for considerable scepticism. The industrial areas are areas of the temperate zone. The agricultural products that countries there protect are products of the temperate zone. Conversely, the non-industrialized countries are, for the most part, located in the tropical, sub-tropical and monsoon climates ; and their agricultural exports are thus largely of a non-competing type, or have great competitive advantage, or are admitted duty free and unrestricted. The great agricultural earners, actual or potential, of export income of the under-developed countries are coffee, cocoa, tea, rubber, tropical fibres and fruits, oil-bearing materials and spices — not as a rule agricultural products that compete with those which the temperate zone supplies and protects. There are, of course, exceptions which I shall mention later.

There is also another aspect of this question that needs to be considered. The pattern of trade with its share of food aid which has resulted, and will result, from the growth of agricultural production in the developed countries has in some respect been or will be in support of, and partly in response to, the development needs of low-income areas.

It is strange that it should seem necessary to emphasize these points. Current literature is full of warnings that the under-developed world needs to make a desperate effort at raising its food output to satisfy urgent domestic requirements and cannot possibly hope to have any food to spare for export to the industrial areas. And, of course, it is well known that for some time past the flow of food has actually been

---

[1] For a sober look at the contention that trade in general operates to the disadvantage of the less developed and with a bias in favour of the richer countries, see Gottfried Haberler's Presidential Address of December 1963, *American Economic Review*, March 1964, pp. 14–21.

from the industrial countries *towards* the under-developed regions, rather than the other way around. There are few who would doubt that, for some time to come, things will continue to be that way ; in fact, for a prolonged period of transition to more balanced growth the under-developed countries may require food aid from the developed countries in even much larger amounts.[1]

Of course, if we take the under-developed areas not as a whole but look, as we should, at individual countries as well, we do perceive exceptions, as I said, both with respect to products and with respect to individual countries. We are indebted to the Bureau of General Economic Research and Policies of the UN and to FAO for having provided, in analytical fashion, a great amount of data and studies of broad scope which help to isolate cases and to gain a balanced view of this whole issue.[2]

Although the pertinent findings of the UN Bureau of General Economic Research and Policies are difficult to condense into a single judgement, they do not seem to contend that more than a small gain in the earnings of under-developed countries could result from reductions in *agricultural* protection in industrial areas. In view of an often higher elasticity of supply and low elasticity of demand in the industrial countries as a group, such liberalization would permit only a very modest expansion of trade by the developing countries.[3]

[1] For a recent review of the food/population prospect see Egbert de Vries, 'Prospects for World Agricultural Development', in *World Agriculture*, Vol. xiii, No. 2 (April 1965). While most experts assert the need for continued food aid from developed to developing countries, it is clear to them that the real solution of the food/population problem in the areas of great population pressure must be sought in an expansion of domestic production. (Besides, such expansion is essential to 'balanced growth' in most of those countries individually and all of them taken together. I shall return to this point.) Also see Sir John Crawford's paper on 'The World Agricultural Situation' in *World Food Forum Proceedings*, Washington, D.C., 1963 ; Fritz Baade, *Der Wettlauf zum Jahre 2000*, Oldenburg 1964 (6th ed.) ; and Gunnar Myrdal, *Priorities in the Development Efforts of Under-developed Countries and their Trade and Financial Relations with Rich Countries*, paper presented before the Società Italiana per l'Organizzazione Internazionale, Rome, Italy, March 1964. The need for a balanced view of present policy desiderata, intermediate-term prospects and long-term requirements, with references also to studies by Sukhatme and Cochrane, is discussed in J. H. Richter, 'A Long View of the Short Run', in *World Agriculture*, Vol. xii, No. 2 (April 1964).

[2] I should like to make particular reference to the UN's *World Economic Survey 1962*, Part I : 'The Developing Countries in World Trade' and *World Economic Survey 1963*, Part I ; as well as to FAO's *Special Studies No. 3* on 'World Agricultural Commodity Trade : Prospects, Problems and Policies', and *Trade in Agricultural Commodities in the UN Development Decade (1964)*, all submitted for purposes of the UN Conference on Trade and Development of 1964. These excellent studies indicate a profound knowledge of the situation in many commodities.

[3] See the first-named UN study, especially pp. 38–40, and 'International Commodity Problems' (para. 49) in Volume III of the *Proceedings of the UN Conference on Trade and Development* (1964).

So far as the FAO studies go, and even though the consistency of the argument is not maintained throughout, it appears to me that in them FAO has likewise dismissed the idea that agricultural support and protectionism in the industrial countries is a significant factor in the export problems of the under-developed areas so far as their trade in temperate-zone foodstuffs or in agricultural raw materials is concerned.[1] That *tropical* products are not affected by *agricultural* protection in the industrial countries is, of course, also obvious. Only in the case of so-called temperate/tropical agricultural products do we find an assertion by FAO that 'the principal obstacle to an expansion of exports from [the under-developed areas] lies in the heavy support and protection given to producers in the [developed temperate zone regions]'. (Para. 93.)[2] However, even with respect to this group of commodities it appears that *sugar* is the only example where that argument applies with any degree of importance to under-developed countries' export and production interests.

As regards *fats and oils*, direct competition from high-income industrial countries is mainly in terms of United States output and exports. But the large amounts of soybeans, soybean oil and cottonseed oil which the United States put on the commercial market have been essential to meet world needs. Several large pre-war tropical exporters have expanded their oilseed production, but their internal

[1] Thus in the first-named FAO study, section on 'Main Commodity Problems and Obstacles to Trade', paras. 87 to 116, and in paras. 22 to 31 on export problems, problems of commercial trade in temperate-zone foodstuffs are properly treated as export problems of *developed* countries; and the agricultural trade from less-developed to developed countries is properly indicated as consisting primarily of trade in tropical products. With respect to agricultural raw materials, the study's argument (paras. 111 to 116) is reduced to pointing up the problems arising not from agricultural protection but from competition by synthetics (admitted to be largely an inevitable, and mostly desirable, result of the general process of technological evolution) and from the discrimination of import policies in favour of imports in raw rather than processed form (which is industrial, not agricultural, protection). Similarly, the other FAO study contends, with regard to temperate-zone foodstuffs, that 'owing to the small contribution these commodities make to the export earnings of developing countries as a group, these countries would benefit far more through an increase in commodity aid than through an increase in export earnings'. (Para. 146, p. I–54.) And 'as a group these countries stand to lose more than to gain by any rise in world grain prices'. (Para. 31, p. II–9.) With respect to agricultural raw materials the study correctly maintains that 'the main problem . . . is that of man-made substitutes' and also mentions the need for better supply responses in developing countries. (Paras. 309–312, p. II–76/77.)

[2] Temperate/tropical agricultural products are products of similar or closely competing nature produced in the temperate zone and/or in the tropical, semi-tropical and monsoon regions where most now under-developed countries are located. These products include sugar, citrus fruit, some vegetable fats and oils, rice and tobacco.

consumption has far outpaced such increases. Some of these countries are now net importers.[1]

Another indication that the United States has not been depressing or limiting the oils market for tropical countries lies in the effects of its large export of soap fats (tallow). As I am reliably informed, the increasing availability of this material, shipped to developed and less-developed countries alike, made it possible to upgrade for edible use coconut oil and the other palm-type oils from tropical areas as well as marine oils — thereby increasing their price level.

The United States has also shipped edible oil under its food aid programme (known as P.L. 480) to needy areas. Most of these shipments are likely to have filled nutritional requirements which would not have been supported by effective demand. They have not in significant measure displaced U.S. or other countries' commercial supplies. In fact, many consuming interests assert that P.L. 480 operations have tended to increase prices of oil or, at any rate, prevented larger declines. The United States has been carrying sizeable stocks of soybeans and oil, variously interpreted as a price-depressing factor and as a factor of price support. More balanced views have emphasized that at most times from 1954 onwards (when the United Kingdom liquidated its Food Ministry stocks) the American stocks were the only world reserve of any size, not really adequate to cover a serious world emergency or even a partial failure of the country's own soybean crop in any one year.

As for the other products of this temperate/tropical category, *rice* is a commodity which is mainly grown in, and traded between, under-developed countries. With a sizeable flow of rice from developed to under-developed areas on concessional terms and in response to urgent needs, there has been no serious suggestion that industrial countries' policies have hampered under-developed countries' export interests.[2] Nor does rice in any way compete with other cereals in

---

[1] These judgements are corroborated, and even more generally applied, by the first-named UN study (*loc. cit.*, pp. 39/40) where it is stated that 'in many cases developments in the industrial countries reflect responses to supply difficulties in the primary exporting countries : the growth of North American production of oilseeds . . . for example, was largely related to the absence of growth in supplies from the developing countries'. The lags in supplies from developing countries 'lie partly in more general problems of economic development', among them 'the competition for supplies created internally by population growth . . ., by rising incomes . . ., and by the expansion of domestic industries which have absorbed increasing proportions of local supplies of various raw materials'. And similar findings are mentioned in the second FAO study (paras. 132–137 and 147–156, pp. II–36–41).

[2] Some impediments to less-developed countries' (small) rice exports to Europe may, in the future, result from the European Economic Community's new agricultural policy for rice.

the consumer's demand schedule in high-income areas. Also, where in low-income areas wheat from developed countries has been substituted for rice, this too was a food aid operation in response to needs and not a displacement of commercial trade for which there was no effective demand.[1]

With respect to *citrus fruit* it appears that export availabilities from under-developed countries who market a quality product have found favourable outlets both in West European countries, where import policies thus far have been liberal, and in the centrally planned economies where incomes also have increased and policy restraints on consumption have been relaxed. The growth of citrus processing, too, has been a dynamic factor in demand. This relatively favourable picture might deteriorate in view of protectionist import policies of the EEC, although, on the other hand, it is quite conceivable that the trade of some of the less-developed export areas will in the future profit from their possible association with EEC. Also, there might be a continued formidable increase in import demand from the centrally planned economies.

With respect to *tobacco*, import demand by industrial countries for the product of developing areas has, on the whole, been favourable and steadily increasing, despite protection and fiscal impediments in the former. Some of this trade moves through bilateral channels and on preferential terms. In general, United States price support policies have facilitated the expansion of production and exports of developing countries.[2]

Among the *agricultural raw materials* of export interest to the under-developed countries, *cotton* is one in whose case *agricultural* protection in the industrial world could have had an appreciable effect upon those countries' export possibilities. And, again, the United States is the only developed country where cotton is an object of agricultural protection. Subject to a minor qualification which I shall mention presently, United States cotton supports, coupled with acreage controls and stocking policies, if anything have supported rather than depressed the world market and have promoted the expansion of production in other countries. The United States has been in the role of residual supplier and there has been no sizeable accumulation of surpluses in other producing areas. The cotton export trade of other countries has increased and 'market opportunities would have existed for larger quantities of cotton had supply

---

[1] Also see second-named FAO study (para. 110, p. II–31).
[2] See also second FAO study (para. 218, p. II–54).

obstacles been overcome', as yet another FAO study has pertinently pointed out.[1] Abandonment of cotton supports in the United States would not reduce total U.S. supplies, would depress prices, would benefit the large growers at the expense of the small ones and of those in other countries, though it would probably also tend to raise the share of cotton in total fibre consumption.

The exception to which I referred is some long-staple cotton in whose production under-developed countries like Egypt, the Sudan and Peru have a decided comparative advantage. The United States maintains quantitative import restrictions on these varieties, though it should not really produce them. The quantities involved are small (100,000 bales) but could be an important additional export for those countries. There have also been United States concessional disposals of raw cotton under P.L. 480. They may well on occasion have unfavourably affected commercial export interests of some under-developed areas. Other under-developed areas, the recipients of such aid, have of course profited from these operations.

Sugar, then, and *cotton* are the only agricultural products for which a relaxation or abandonment of domestic agricultural support policies in individual countries could produce a significant addition to the export income of under-developed areas. Sugar is far and away the more important item. If the further increments of demand in the industrial countries could be reserved for imports from the tropics, this would be of sizeable assistance to under-developed areas and would be good economics for developed countries. And, indeed, it is perhaps not unrealistic to expect this much of those who are always so vociferous in acknowledging the need for helping the under-developed countries' export trade. More immediate help could, of course, be derived from a policy in developed countries gradually to reduce their sugar output to the extent of its expansion since the early fifties — in concert with a concurrent increase in output for export in the tropical areas. Some 500 million dollars' worth of import buying power might result from such an operation for the beneficiaries. There is no likelihood, however, that such a development will occur. Sugar beets have become an integral part of the intensive crop rotation and livestock economy, contributing much to their efficiency and productivity at least in Western Europe; it would be unrealistic to expect any deliberate policy to upset this balance.

[1] Cf. *Commodity Bulletin Series 38*, 'Synthetics and Their Effects on Agricultural Trade', FAO, 1964. See also the second-named FAO study, especially para. 317, p. II–79.

III. THE EFFECTS OF PROTECTIONISM BY
DEVELOPED COUNTRIES

It is in the light of the facts and probabilities set out above that I think I am justified in reaffirming the view that agricultural protection has not been a *significant* factor in the international trade problems of under-developed areas, or in changing their patterns, or in impeding economic development in those countries. It would even appear that the food surpluses which have resulted from those policies in industrial countries have, on balance, been of assistance to the recipients of so-called food aid and have helped economic development. In any case, they have added to the total amount of resources that flowed into the under-developed countries on an unrequited basis.[1]

The countries of recent settlement — Oceania, North America, Argentina — actually have been significantly affected by protectionist agricultural policies in the importing industrial countries. But they no longer are 'the developing regions', which they were in the nineteenth century. This historical association of agricultural exports with a developing part of the world may, to some extent, have tinged the views still held with respect to our problem. And a similar influence may have resulted from much confusing but often inevitable terminology and statistical classification : under-developed countries interchangeably described as primary producers or non-industrial areas ; non-industrial areas defined so as to include such high-income countries as Australia and New Zealand ; industrial areas including such large agricultural producers and exporters as Canada and the United States ; primary products including food-stuffs ; agricultural products or foodstuffs including tropical products and beverage crops (of which much is true that is not true of temperate-zone foods) ; a small curtailment of imports by importing countries and a large expansion of exports by North America, possibly in response to import needs by developing countries, shown as a big curtailment of *net* imports by 'the industrial areas' ; and so forth.

While denying that agricultural protectionism in the developed countries has greatly intensified the international trade problem of

---

[1] I am aware of the argument that such aid might weaken efforts at domestic agricultural expansion in the needy areas. I also have a high regard for Professor P. T. Bauer's reasoning and the context in which he has warned against 'the easy ways' in both developed and under-developed countries : see P. T. Bauer and J. B. Wood, 'Foreign Aid — the Soft Option' in *Quarterly Review*, Banca Nazionale del Lavoro, Rome, Dec. 1961. But I doubt that interrelationships between food aid and agricultural effort are quite as direct ; and it should be possible to prevent adverse repercussions by appropriate policies.

the under-developed world, I have myself pointed out *some* unfavourable effects. I believe it is impossible to measure them or to say more than that they have probably been small. In any case, I think we cannot gain a correct impression of their magnitude by contemplating statistics of the kind that show what percentage of the primary products exported from the low-income countries consisted of agricultural commodities identical with or substitutable for those produced and supported in some or many high-income countries.[1] If this percentage is high, it does not of itself mean that exports of such products from the developing countries *could* have been increased, or sold at better prices, had protection or support in the high-income countries been less. No adequate judgement of this issue is possible without a consideration of the elasticities of demand and supply and comparative cost advantages and output potentials in low-income as against high-income areas, including those other *developed* countries that are the main suppliers of such products and might also be the main beneficiaries of reductions in agricultural protection ; reduced protection might even result in a formidable increase in producers' efficiency in the previously sheltered areas themselves.[2] Other factors to be considered would be the great need and desire for improved nutrition of a rapidly increasing population and increasing demand for food in the developing countries ; the increasing demand for raw materials by developing industry ; and given policies and policy priorities in these areas.

Conversely, if the under-developed countries have not in the past derived a large share of their export income from agricultural commodities sharply affected by competition from those which the industrial countries produce and protect, this fact does not itself mean that that share *could* not have been larger, had protection in the industrial countries been less. There is plenty of reason for watchfulness, certainly with respect to possible future policies and developments.

As for the rest, I would like to repeat that policies of agricultural

---

[1] This is probably the only important point on which I have some reservations with respect to Dr. Gerda Blau's approach in her otherwise admirable article on 'Commodity Export Earnings and Economic Growth' in *New Directions for World Trade*, a Chatham House Report, Oxford, 1964, later published by FAO and the basis of much of FAO's work for the UN Conference on Trade and Development 1964.

[2] A good example is Denmark which, unlike other countries, did not meet the European agricultural crisis in the 1880's by protectionist measures, but rather by rationalization of its agriculture and agriculture's output mix. As a consequence it became a most efficient, if not the most efficient, agricultural producer. A similar example is the Netherlands.

protection and support in high-income areas have no doubt been a major detriment to the export interests of other *developed* areas and to efficient resource utilization throughout the world. Perhaps it is mostly in this roundabout context, rather than directly, that the interests of under-developed countries have also been unfavourably affected. The problem is a complicated one. It is intimately tied up with policies necessarily or desirably based on broader concepts of social welfare and balance in the developed countries. In any case, it is an issue which the developed countries must settle among themselves. Quite logically in the GATT negotiations on agricultural protection in the 'Kennedy Round' the main contestants are the *developed* countries on both sides. In healthy attempts at reducing protection we should keep in mind, incidentally, that the world's food/population problem for the next 10 to 20 years is a big question-mark. Therefore, whatever adjustments are made in *current agricultural production* in industrial areas, they must not reduce *future producing capacity* and possibilities for rapid expansion in case of need.

IV. CONCLUDING REMARKS

I now come to the concluding remarks which are to touch upon what I really consider to be the essential change in the patterns of international trade problems of under-developed areas. I believe that the answer to this question was given by the late Ragnar Nurkse; and although his ideas have not remained unchallenged, I think that the essence of his thought still stands. He has correctly identified the content of that change as a reduced intensity of the dynamic influence which trade exerts upon economic development in under-developed areas compared with what he found in the circumstances of nineteenth-century trade.[1]

To be sure, trade between the high-income and the low-income countries still expands, will continue to expand, and will for long remain of overriding importance. But the interdependence of the growth performance of the two areas, which constituted a driving force for that trade, has greatly weakened. The present developing countries are not of the type and structure for which trade with the industrial centres proved such an 'engine of growth', in Sir Dennis

[1] See especially his Wicksell Lectures (1959) in *Equilibrium and Growth in the World Economy, Economic Essays by Ragnar Nurkse,* edited by G. Haberler and Robert M. Stern, Harvard University Press, 1961.

Robertson's phrase. And the developed industrial centres no longer show the characteristics of population growth, elasticities of demand and raw material dependence which made for such rapidity in the expansion of their import requirements for primary products.

If thus the dynamic impact of trade upon economic development has declined and continues to decline, this is not altogether an alarming process. It is, in part, a natural consequence of the world's economic and social progress. There are indications that another driving force is in the making — as a result of objective circumstances, but also supported by deliberate national and international policies.[1] That force is 'balanced growth' or balanced agricultural-industrial expansion in the under-developed world taken as one whole, also meaning growth of trade *within* that area. The balance meant is one between agriculture and industry and between various types of industries and services, with diversity of output itself developing the demand for its various products, as conceived in Say's 'law of markets'.

There is a good chance for such balanced growth within the under-developed world to become a dynamic force in economic development because it includes countries at various stages of growth and of various economic structures — a diversity that might develop pressures of interdependence similar to those that prevailed in the seventy years prior to World War I.[2] As partners in such a trade circuit for balanced growth, the centrally planned economies also offer a dynamic influence since they are at stages of development intermediate between the low-income and the high-income countries.

It may take some time for that 'new' dynamic force to gain real ascendancy. In the interim, an increasing flow of special technical

[1] Political competition for commercial ties with the under-developed countries belongs in this chapter ; and so do the various movements towards political co-operation and economic regionalism in those areas.

[2] As a part-illustration of this increasing diversity of stages of growth and structures may serve one of the many instructive comparisons which P. Lamartine Yates has drawn in his *Forty Years of Foreign Trade*, New York, 1959. While in many under-developed countries consumption goods imports in the fifties remained a very high percentage of total imports, in other such countries their share had declined drastically in favour of raw material and manufactured capital goods, for example, in India, Argentina, Mexico, Brazil, Chile, Venezuela, Egypt and Morocco, to name a few (pp. 196/197). Such changes in the composition of imports must, of course, precede any possible changes in the composition of exports for which Yates had yet found only a 'small amount of change in forty years' (p. 182). Although the idea is not specifically developed in T. W. Schultz's article 'Economic Prospects of Primary Products' in *Economic Development for Latin America*, edited by Ellis and Wallich, London, 1961, I consider his conclusions also to point in the direction of increasing possibilities for dynamic development of trade *as between* the under-developed countries themselves.

assistance and of loan and aid capital (and, hence, unrequited exports) and other income transfers from the developed countries — contrary to the slogan 'trade, not aid' — is likely to fill the dynamic void. Notwithstanding a possible temporary setback in international aid, the prospect for such a course of events seems to me to be quite good, even if few of the resolutions of the UN Conference on Trade and Development were to be carried out.

Chapter 8

THE CHANGING CHARACTER OF INTERNATIONAL TRADE AND THE PROBLEMS OF UNDER-DEVELOPED REGIONS

BY

V. MARTINOV

Association of Soviet Economic Scientific Institutions, Moscow, U.S.S.R.

I. INTRODUCTORY

The subject of this paper, on which I was invited to speak, has been amply covered in special investigations of the UN Secretariat, FAO and other international and regional bodies, in the works of many economists, and in numerous papers.[1] Finally, it has been comprehensively discussed at the UN Conference on Trade and Development the proceedings of which offer a wide field for further investigations into the present-day trends in the development of trade. For this reason my task, as I see it, does not consist in attempting to elucidate all the aspects of the problem in hand. It would appear far more expedient to limit my report to brief comments on some problems associated with the trends in the development of world agricultural production.

II. THE CHANGES IN THE RELATION BETWEEN THE DEVELOPED AND DEVELOPING PARTS OF THE CAPITALIST WORLD

The determining feature of the world capitalist economy is its division into two unequal parts: (1) the economically underdeveloped countries which account for 66·6 per cent of the global

[1] Among others, we could refer to such investigations as UN *World Economic Survey 1962*, Part 1 : *Developing Countries in World Trade* ; FAO : *Agricultural Commodities-Projections for 1970* ; FAO : *The State of Food and Agriculture 1958–1964* ; R. Nurkse, 'Patterns of Trade and Development', *The Wicksell Lectures* ; Stockholm, 1959 ; R. Prebisch, *Towards a Dynamic Development Policy for Latin*

population and only 15·6 per cent of the total of national products, and (2) the industrially developed capitalist countries with only 33·4 per cent of the global population and 84·4 per cent of the total of national products. Until recently, in the world's mechanism of capitalist reproduction and accumulation, the under-developed countries only played the part of agricultural and raw material appendages and sources of additional surplus product for the imperialist countries. Suffice it to say that in 1959–61 the export of agricultural products provided 58 per cent of their total income from export. Moreover, if we exclude petroleum (whose export is of great importance for only a few developing countries), the share of agricultural exports in the total exports of the developing countries increases to 86 per cent. This structure of the world economy has clashed with the new distribution and correlation of world forces which is connected primarily with the downfall of the colonial system and the development of new sovereign states.

Further, the necessity of replacing the old colonial system of commercial and other ties by a new system based on equality and the economic rise of the liberated countries, is dictated also by the fact that, under the conditions of the current technical revolution which has embraced the industry and agriculture of the developed capitalist countries, the importance and role of the agricultural and raw material branches in the global mechanism of reproduction is materially decreasing. According to the calculations made by a Soviet economist, Yu Olsevich, on the basis of value added, the relation between agricultural production, primary industry and secondary industry in the capitalist world has changed in the manner shown in Table I. Moreover, there is a trend towards a relative growth of the utility branches (gas, electric power, water supply, etc.) as well as the various services.[1]

America, 1963 ; B. Belassa, *Trade Prospects for Developing Countries* ; Homewood, 1963.

Among the Russian sources mention should be made of the fundamental work by V. Rymalov and V. Tyagunenko, *Underdeveloped Countries in the World Capitalist Economy*, Moscow, 1961.

[1] The indicated shifts are due to a number of tendencies in the sphere of production and consumption, including : (*a*) the increasing complexity of finished articles (particularly in mechanical engineering) ; (*b*) the rise in the percentage of relatively less material-consuming types of goods in the finished product (the growth of the share of mechanical engineering, chemistry, etc.) ; (*c*) savings in raw materials and fuel ; (*d*) taking over by the secondary industry of a number of functions which were previously performed by other branches ; (*e*) the change in the structure of effective consumer demand : that is, the increase in the consumption of durable goods and the reduction in the share of food, etc. ; (*f*) the development of militarism ; (*g*) some acceleration of economic growth. See Yu Olsevich, *On the Crisis in the Structure of the World Capitalist Economy*.

TABLE I

CHANGES IN THE RELATION BETWEEN AGRICULTURE AND INDUSTRY
IN THE CAPITALIST WORLD

| | | | Secondary Industry | | | |
| | | | | including | | |
Year	Agriculture	Primary Industry	Whole	Heavy	Light	Total
1938	44·8	7·0	48·2	21·9	26·3	100
1948	35·9	7·1	57·0	30·0	27·0	100
1958	32·6	6·7	60·6	36·1	24·5	100

At the basis of the shifts enumerated above lie technical, social and economic factors. In particular, the drastic curtailment of the share of agricultural production against the background of hundreds of millions of starving people is clearly associated with the abnormal concentration of effective consumer demand in the hands of a small group of developed capitalist powers.

In the world market these shifts are manifested, for instance, in the fact that trade is developing mainly to the benefit of the products of the secondary industry. As regards international trade in agricultural goods, its principal development trends within the last decade are (as was noted in the FAO's investigations) : (*a*) a decrease in the share of agricultural products in total world trade ; (*b*) a lag in agricultural exports from the developing countries as compared with the industrially developed countries ; (*c*) a sharp drop in prices for agricultural products in the world market in general, and especially in prices for agricultural products exported from developing countries; (*d*) a deterioration of the terms of trade in agricultural goods.[1] The drop in prices for agricultural goods with a simultaneous rise in prices for manufactured articles has led to a 32 per cent drop in the purchasing power derived from the exports of the developing countries between 1950 and 1962. The slight increase in the prices for individual agricultural goods in the world market in 1963 and 1964 has generally brought about no substantial change in these trends and, according to experts, is essentially of a temporary nature.

The shifts in the world trade in agricultural goods which are extremely unfavourable for the liberated countries are due to many causes of economic origin, although in some cases political motives

[1] FAO, *World Agricultural Commodity Trade-Prospects, Problems and Policies.* UN Conference on Trade and Development E/Conf. 46/42 pp. 9–10.

played a more important part (the Suez crisis and the drop in prices for Egyptian cotton, etc.). Among the economic causes the following

TABLE II

TRENDS IN THE WORLD TRADE IN AGRICULTURAL GOODS

| | \multicolumn{4}{c}{1959–61, % of 1952–53} | | | |
	Volume	Value	Unit Cost	Real Unit Cost *
World trade	150	150	100	100
World agricultural trade	136	114	84	78
Agricultural export of developing countries	128	107	83	75
Agricultural export of developed countries	146	124	85	83

* Unit cost of agricultural export related to the unit cost of export of the secondary industry. See FAO, *The State of Food and Agriculture*, 1964.

are usually emphasized : (*a*) the technical revolution taking place in the agriculture of the developed capitalist countries ; (*b*) the expansion in the production of synthetic raw materials displacing natural raw materials, and the introduction of more economical utilization of raw materials per unit product (the trend towards reduction in material consumption) ; (*c*) the relatively low demand on the part of Western Europe and other developed capitalist countries, which is connected with the relatively slow growth of the population and the decrease in the percentage of the population's expenditures on foodstuffs.

The principal aspects of the technical revolution and the consequent economic and social problems are the subject of special discussions in other panels of the Conference. I will, therefore, only note that the general law in the developed industrial countries today is a more rapid growth of agricultural production as compared with the population (both total and urban). Moreover, the attained growth of agricultural production has, as a whole, surpassed the relatively slowly expanding demand for foodstuffs and agricultural raw materials in those countries. The over-production of agricultural goods has become an important feature in the economic position of agriculture in many capitalist countries. The over-production of agricultural goods, the main centre of which is the United States of America, has become the basic reason for the drop in prices for the products of the temperate zone in the world market. And since the industrially developed countries are economically in a position to carry on the policy of maintaining agricultural prices in the home market, the price drop in the world market has hit, in the first place, those producers who are the weakest economically.

III. THE EFFECTS OF THE TECHNICAL REVOLUTION ON THE DEVELOPING COUNTRIES

In accordance with the agenda of the Conference one should first ask what is the effect of the technical revolution and related processes in the agriculture of industrially developed countries on the economy of the developing countries. In answering this question it is essential to distinguish, firstly, the general (theoretical) statement on the mechanism of the relationship, in the world process of reproduction, between the developed and developing countries, and, secondly, investigation of the direct influence upon the export of agricultural goods from the developing countries.

(i) *The Historical Background*

The theoretical aspects of the general world process of reproduction, no matter how interesting they may be, are beyond the framework of our discussion. Therefore I will limit myself to brief remarks concerning this problem. In its essential features, the present-day structure of the international division of labour was formed in the epoch of colonialism in the nineteenth and the beginning of the twentieth centuries. Leaving aside all the political and economic forces which have led to the economic and political enslavement of the vast majority of the peoples of Asia, Africa and Latin America by a handful of powers, I will only point out that one of the most important causes for the amazingly uneven development of the individual parts of the capitalist world was the lag of agriculture behind industry. In fact, up to the thirties and forties of our century agriculture was chiefly based on manual labour, whereas industry began to employ factory machinery as far back as the end of the eighteenth and the beginning of the nineteenth centuries.

In Western Europe, the birthplace of machine development, capitalism mastered agriculture extremely slowly, although the fact that the peasantry was partly dispossessed of their land was a precondition for the development of capitalism in agriculture. Having extricated agriculture from its medieval and patriarchal clutches, and having set it moving within the flow of commerce, capitalism, on the one hand, undermined the bulwark of the old feudal society — the 'peasant' — and speeded up the development of agriculture, and, on the other, proved incapable of converting agriculture to an adequate industrial basis during an entire historical period, and so began to conserve the backward peasant method of production.

The lag of agriculture behind industry, on the one hand, resulted in high costs of agricultural goods and a rise in land rent in industrial countries and, on the other, fostered the colonization of new lands and the drawing of the backward countries into the orbit of the world capitalist market. As a result of the industrial revolution the nineteenth century witnessed the creation of a new international division of labour, corresponding to the location of the main centres of machine production, and this turned part of the globe into an area of predominantly agricultural production. Destroying the traditional crafts in under-developed countries by the cheapness of its goods, big machine industry transformed these countries into areas of production of the corresponding raw materials. Causing emigration of 'redundant' workers, it promoted the colonization of other countries (the United States, Canada, Australia, etc.), also turning them into suppliers of mineral and agricultural raw materials and foodstuffs. In view of the lag of agriculture behind industry and the rapid growth of the urban population, the demand for food and agricultural raw materials in European countries was increasingly covered by imports. England is a classical example, but this trend prevailed to a certain extent in other countries as well. Roughly speaking, the world mechanism of reproduction was based on the exchange of European industrial goods for agricultural products of the other countries. The effect of this mechanism, however, was 'lopsided' : the rate of development of production in agricultural and raw material countries was determined by the development of the industrial countries.

Since then the world pattern has changed. The United States has long since become the leading industrial power in the capitalist world. Canada, Australia and others have turned into industrial-agricultural countries.[1] What is more important, in most of the industrially developed countries agriculture, as has been stated before, is developing faster than the growth of the population. Thus the impetus given by the industrially developed countries in the past

---

[1] The history of economic development, which, of course, would be helpful in elucidating present-day trends, is, however, beyond the framework of the report.

in the development of developing countries has been considerably undermined. Up to the present time there has been no question of measuring the extent to which the import of foodstuffs and agricultural raw materials is replaced by home production in these countries. However, it is important that the demand on the part of industrially developed countries is no longer the determining force responsible for the expansion of agricultural production in the developing countries, although it is still true that the entire system of relationships is such that trade turnover between the industrially developed countries and those of the 'third world' still determines the development of many developing countries.

The structural crisis of the world economy primarily manifests itself precisely in the fact that the trends in the world economic development undermine the very foundation of the existence of the under-developed countries as suppliers of agricultural raw materials and foodstuffs, although the whole system of imperialism is, as before, directed to a considerable extent at keeping these countries as agricultural and raw-material appendages. Of course, the crisis of the structure of world agricultural production should not be understood as totally denying the importance of this agricultural production for industrially developed countries. I only wish to emphasize that the developing countries cannot any longer base their development exclusively on agricultural production.

(ii) *The Effect on Developing Countries*

There follow brief remarks on the question: How and to what extent does the growth of agricultural production attained on the basis of the structural reorganization of agriculture in the developed capitalist countries affect the positions of the developing countries in the world market?

I find it a difficult and complicated task to answer this question, and not only because economic processes in real life are greatly influenced by political and social movements. These processes themselves are contradictory in view of their capitalist nature. Further, in estimating the effect of the technical revolution and its implications in the agriculture of the developed capitalist countries one should not ignore such questions as the following: How long will the centuries-old backwardness of agriculture in the developing countries persist? In what ways and by what methods will it be overcome: by radical agrarian reforms beneficial for the peasantry, or by partial reforms which are in the long run only in the interests of the reactionary forces? What are the possibilities for drawing the tremendous agricultural over-population into material production? etc. It is a well-known fact that the great diversity in the political and economic development of individual developing countries also explains the difference in the positions of these countries in the international system of division of labour and in the world market. Obviously we need not consider these and other related matters. Then the problem facing us can be limited mainly to the question how and to what extent the growth of agricultural production in the developed capitalist countries influences the import of agricultural products by these countries from the developing countries.

Evidently, analysis should not be restricted to the statement that agricultural production in industrially developed countries, which are chiefly located in the temperate zone, does not compete, in the case of most goods, with the tropical and subtropical goods of the developing countries. Firstly, many products of the temperate-zone countries can be successfully grown in a number of developing countries and, besides, many foodstuffs are interchangeable. Secondly, the increase in agricultural production in the industrially developed countries has already inflicted some damage upon the export of the developing countries. This was discussed at the UN Conference on Trade and Development. Thus, for example, some temperate-zone foodstuffs (grain, dairy and meat products, eggs) are produced by certain countries of Latin America and North Africa.[1]

Although internal political and social causes seem to have played the decisive role in the reduction of the share of these countries in total world production and export of these goods, the growth of production of the indicated goods in the industrially developed countries taken as a whole has been quite an important factor in this process.

If we take the group of goods produced both in the temperate, the subtropical and tropical zones (rice, fats and oils, citrus fruits, tobacco), many developing countries are already competing with the developed countries in the trade in these goods and as a result have lost part of their markets in the developed countries in recent years.[2] For instance, the export of rice from the United States, to say nothing of the changes within the internal market of the United States itself,

[1] In 1959–61 the developing countries accounted for 10 per cent of world exports in these goods.
[2] In 1959–61 the share of developing countries in the world export of these goods was 44 per cent.

has led to the ousting of the traditional suppliers from the market in Canada and in a number of countries of the Far East. Fats and oils produced in the industrially developed countries show a trend towards displacing vegetable oils arriving from the developing countries. Even fruits from the temperate and subtropical zones replace those imported from the tropical zone, etc.

It is actually only the export of tropical foodstuffs (coffee, cocoa, tea, bananas, spices) from the developing countries that does not meet with direct competition in the markets of the developed countries.[1] However, on the one hand, these products account for only less than one-third of the total value of exports from the developing countries and, on the other, the consumption of these goods in the developed capitalist countries is already at a rather high level, so that it cannot be expected to increase to any considerable extent.

As regards agricultural raw materials (cotton, wool, jute, hard fibres and rubber), the export of these goods from the developing countries suffers mainly from the competition of synthetic substitutes.

IV. THE PROSPECTS FOR DEVELOPING COUNTRIES

The evaluations of the prospects for agricultural exports from the developing countries to the industrially developed ones, which have been made by FAO, the UN Secretariat and individual economists on the basis of the current development trends, are generally more or less unanimous. They all agree that one cannot expect any serious expansion of agricultural exports to these countries or a considerable improvement in the terms of trade between the products of secondary industry exported by the industrially developed countries, and agricultural products and raw materials exported by the developing countries.

Thus, specialization in the production of raw materials and, in particular, agricultural products for export cannot remain the economic basis for the development of the liberated countries. At the same time, in order to reorganize the structure of their economy they will have to expand their imports of capital goods and so they must increase their currency receipts to cover these imports. According to a widely accepted estimate, the developing countries should

---

[1] The share of the developing countries in world export in these products was 99 per cent in 1959–61.

expand their imports by 6 per cent annually so as to achieve a 5 per cent rate of growth (UN decade target). Consequently, the position of the liberated countries is paradoxically contradictory. On the one hand, they have to look for ways of expanding exports of their traditional goods and so have to orientate themselves in some measure towards the development of the raw material sectors. This results in a weakness of the positions of the liberated countries relative to the imperialist powers. On the other hand, they must, in general, within a fairly short period, put an end to the emphasis on agricultural products and raw materials in their economy. Of great importance in achieving this aim is the unity of the liberated countries. This was demonstrated, in particular, by the UN Conference on Trade and Development.

In general, it should be noted that most of the recommendations adopted by the Conference, such as the suggestions aimed at eliminating or at least narrowing the gap between the prices for industrial and agricultural goods (commodity agreements, etc.), and measures for facilitating access to the markets of the industrially developed countries (curtailing protectionism in the industrially developed countries, granting general preferences to the liberated countries, reducing tariffs not only on raw materials, but also on processed products from these countries, increasing the export of finished articles from these countries, etc.) are of great importance for the developing countries.

The Role of the Soviet Union

The exchange of opinions and the recommendations made at the Conference are being studied by Soviet economists. Trade between the Soviet Union and the developing countries began to develop comparatively recently; however, marked successes have already been achieved, although the volume of this trade is not very great for self-evident reasons. It is well-known that the U.S.S.R. grants considerable long-term and low-interest credits to more than twenty-five developing countries, which are covered by deliveries of traditional goods, as a rule. The economic and technical contribution of the Soviet Union to the economic development of the liberated countries is an important factor promoting and extending international trade and economic ties.

At present there is every reason to expect a further considerable expansion of trade between these countries and the U.S.S.R. It is envisaged in the twenty-year programme for the economic development of the Soviet Union that the total national product will increase about five-fold by 1980. As a consequence of its rapid economic development, the Soviet Union will be able to enlarge its export resources, and, naturally, its import requirements will grow. From the point of view of future development we may well expect that with a favourable international situation, the trade of the U.S.S.R. with the developing countries, as was stated by the Soviet delegation at the Geneva Conference, will be greater than the 1963 level by approximately eight times and will reach eleven thousand million dollars. Soviet experts are quite optimistic about the possibilities for a considerable expansion of trade by means of long-term bilateral agreements and contracts. They believe that these agreements will also be an important factor in the stabilization of market prices for goods exported by the liberated countries, as in the case of the sugar trade agreement between Cuba and the Soviet Union.

During this twenty-year period the Soviet Union will be able to expand their exports to the developing countries of such goods essential for their development as machines and equipment, rolled steel, raw materials, chemical products, etc.

The Soviet Union will expand its imports from the developing countries both of traditional goods and of new articles produced by the young industry of these countries. The tentative estimates of Soviet economists, as was stated at the Conference, indicate that the Soviet Union will be able to increase the import of cocoa beans from 54 thousand tons in 1963 to 120 thousand tons in 1970 and 350 thousand tons in 1980, the import of coffee from 29·1 thousand tons to 60 thousand tons and 120 thousand tons, that of citrus fruits from 60·2 thousand tons to 180 thousand tons and 750 thousand tons, respectively.

However, in spite of the great opportunities for expanding exports from the liberated countries to the Soviet Union and other socialist countries, it appears that the industrially developed capitalist countries will remain for a long time to come the principal importers, in particular, of the agricultural products of the developing countries.

V. CONCLUSION

The economic upsurge in the liberated countries greatly depends on the solution of the market problem. The ways to solve this problem cannot be the same for the individual liberated countries. It seems, however, that the diverse particular solutions of this problem in the individual countries do not exclude the possibility of a common approach to the liberated countries as a whole.

At present few economists or political leaders in the West (at least openly) adhere to the conception that the developing countries can attain an economic upsurge without breaking up the old structure, that is by doing no more than expanding their agricultural and raw material exports and raising its efficiency. This, however, by no means signifies that this colonial philosophy is extinct. Many gigantic monopolies, often with the direct support of their governments, continue to pursue their old colonialist policy of economic enslavement of a number of countries (for instance, United Fruit in the countries of Central America). However, old colonialism is coming to an end on our planet.

Some Western economists, advocating the necessity of maintaining the Western orientation of the production of the developing countries, shift the centre of gravity from the export of raw materials to the export of semi-finished and finished industrial articles. From the standpoint of the general world shifts noted in the structure of demand this approach may be considered as justified. It should be borne in mind, however, that international trade expanded at an increased rate mainly through trade in the products of highly equipped and technically complex industries (mechanical engineering, instrument making, electronics, the chemical industry) with which the developing countries will hardly be able to compete in the near future. Moreover, the production of semi-finished goods for export yields only a limited effect, as a rule. The trends of the last decade have shown that specialization in the production of certain articles of light industry may in the long run become a source of foreign currency for the developing countries, but they also show that such specialization cannot become the basis of economic development. With regard to this conception it is also important to note that with the orientation of the production of the developing countries towards the export of semi-finished goods and products of the light industry, the rates of large-scale reproduction in those countries will be determined, as before, by the tendencies and policy of the industrially developed countries. Although under conditions of the growing unity of the developing countries with support on the part of world socialism, the march of individual liberated countries towards a rapid economic upsurge may begin with the stage of building up a light industry, for the liberated countries as a whole this path could mean only partial industrialization carried on under the actual control of the Western powers.

In my opinion, an economic upsurge in the developing countries can be achieved by creating a high-capacity home market within the framework of one or more countries. In this case the expansion of exports of raw materials and manufactured articles from the developing countries will be an important, but subordinate, means for building up a new economic structure in these countries.

JOINT DISCUSSION OF DR. RICHTER'S AND PROFESSOR MARTINOV'S PAPERS

Professor Robinson said that the initial development of most developing countries was largely based on the export of primary products, and the growth of enclaves within which modern commerce and industry grew out of the handling and processing of these primary products. The problem, common to most of these countries, is that in recent years the volume of this activity has not been growing rapidly. This has been aggravated by adverse terms of trade against these countries, so that the rate of growth of foreign earnings has been even slower than the growth in physical output. Professor Robinson said that he agreed with Dr. Martinov that there is a need for radical change in the structure of trade undertaken by these countries. This implies basic changes in the structures of production. At present there are structural maladjustments in the developing world, and no serious solution of the problems can be achieved without a readiness on the part of many countries to change their productive structures so as to afford a new break-down of exports. This is true of Europe and America as well as of the developing countries. Since the beginning of industrialization of the United Kingdom the emphasis of production has had to be changed from decade to decade in order to keep abreast of the requirements of export markets ; such continuous flexibility should reasonably be required of the developing countries, and aid policies should be directed towards this end. Considered in these terms, the agricultural policies of the advanced countries do not play a very important part of the whole.

Professor Robinson said that although he did not agree with Dr. Martinov's political interpretations of past and present colonialist forces, the analysis of the economic problem discussed in the paper would meet with wide acceptance. The first table in Dr. Martinov's paper presented the changes in composition of the output of the non-socialist world. The explanatory footnote to that table might be usefully paraphrased in the following way : there is a relative decline in the raw material content in world output as a result of : (*a*) a shift in demand in favour of products with a high content of manufactured output ; (*b*) a shift in demand towards higher quality goods (this also implies a higher manufacturing input content) ; and (*c*) savings in fuel consumption ; there are low income elasticities for food and higher income elasticities for finished products with lower raw-material content ; there are increases in expenditure on defence goods which use materials that do not originate in primary producing countries ; and finally there is an increase in the use of synthetic substitutes which, from having been regarded as inferior to the natural materials they replaced, are now taken to be superior. Professor Robinson said that this gave a list of some of the difficulties under which developing countries labour, but Dr. Martinov had not sufficiently stressed the high import content of development in developing countries. In India the import content of capital investment is approximately 50 per cent. There are thus two trends against developing countries : that implied by Dr. Martinov's footnote — the declining ratio of exports to national income ; and that implied by the high import content of development — a rising ratio of imports to national income. Many of these problems are remote from the policies of agriculture in developing countries. Whatever these policies had been, the developing countries would have found it necessary to reshape their economic structures and, as in the United Kingdom, to make possible a growth rate of the economy which was faster than the growth of exports. These needs would be served by the development of internal manufacturing resources and import substitution by the production of capital goods.

Even if many of the problems of developing countries would have arisen in spite of the agricultural policies of advanced countries, it is necessary to ask how much the problems have been aggravated by these policies. Dr. Richter rightly argued that this issue cannot be discussed in general terms, and that the gross figures have to be disaggregated into particular products. Dr. Richter's conclusion was that the effect of the agriculture of developed countries on the developing world has been slight because of the differences in climate and output composition of the two groups of countries. There was a strong difference in emphasis on this point between Dr. Richter and Dr. Martinov. Dr. Martinov showed that the case is complicated by the fact that, in Latin America, temperate-zone agricultural produce is in direct competition with the agriculture of more advanced countries. Furthermore, there are some products that are common to the developed and developing worlds : citrus fruits, tobacco, rice, fats and oils, and sugar. Professor Robinson said that he thought that Dr. Richter treated the case of sugar rather lightly, and that the cases of Latin America and North Africa mean that his climatic distinction between the developed and developing worlds is oversimplified. It is not only the agricultural policies of developed countries which affect the agriculture of developing countries. The taxation policies of the developed world are also significant for a number of products of which the most important is coffee.

Professor Robinson said that the end of Dr. Martinov's paper made an interesting contribution to the discussion of what should be done to share out the world's requirements for structural change and change in trade patterns. The opening of the Soviet market for the products of developing countries will help in the period of transition from the dependence of these countries on primary product exports. It would have been encouraging to find in Dr. Martinov's list of those commodities which might be increasingly traded, a reference to sugar, and — even more desirable — a reference to manufactured goods.

Professor Robinson said that he did not want to end by exonerating the agricultural policies of advanced countries from all blame for the difficulties of the developing world. In this connection he hoped that further discussion would be forthcoming on the impact of the Common Market agricultural policy on world trade. In that way discussion would revert to the fundamental dilemma of a world with food surpluses in some countries and acute food shortages in others, and the difficulties of matching the two sets of countries by flows of exports. If advanced countries are to maintain their exports of capital goods and yet are determined not to import food and primary commodities, trade between countries of very different degrees of development will only remain possible if the advanced countries become prepared to import finished manufactures.

Professor Fauvel said that on the specific issue of sugar, the problem is that Cuba is the only country in the world which produces sugar in huge quantities, and which depends on sugar as its main source of foreign exchange, and that there are only two potential acceptors of that quantity of sugar — the United States and the Soviet Union.

Dr. Martinov said that his paper had necessarily been rather general in attempting to indicate the world mechanism of trade between the three major areas — socialist, non-socialist and developing. It has to be emphasized that the trade between the Soviet Union and the developing world is small and young. This trade will expand and, as reported at the United Nations Conference on Trade and Development, will increase by eight times by 1983. But the opportunities for trade with the Soviet Union will be restricted, and trade with the Western industrial countries will be of paramount importance.

The international trade patterns and the international division of labour which had been set up in the nineteenth century and the early part of the twentieth century had been the determinants of the situation in which most of the developing countries now find themselves. These trade and division of labour patterns resulted from an amazingly uneven progress of development in capitalist countries. Agricultural development had lagged behind industrial manufacturing development. Mechanization of agriculture on a wide scale only dated from the 1930's and 1940's. The result had been that for much of the nineteenth and twentieth centuries the price of agricultural goods had remained high and the extension of agricultural activities into other parts of the world had been essential. The United Kingdom had been the classic case. The colonial countries had been simple appendages of the advanced countries, so that their development could only progress at the pace of the increase of demand for raw materials in the industrial countries. When independence was won by the colonial territories, there seemed to be a golden opportunity for them to capitalize on the role they played with respect to the advanced countries. This had not come about, but instead the trade of these countries had declined; that is, there had been a decrease in the agricultural component in world trade and the prices of primary products had declined. Dr. Martinov said that in his own view this was chiefly due to the technological changes in advanced industrial countries. These changes had been so great that the relationship between agriculture and industry, which used to be that of agriculture lagging behind the industrial sector, was reversed. Thus the industrial advance of developed countries is no longer available as a drive behind the agricultural advance of developing countries. Furthermore, the growth potential of those industries in developing countries which depend on trade in agricultural products is heavily reduced. The developing countries will therefore have to supplement the old patterns of trade, and develop production for high capacity home markets. However, the exporting of primary products will remain important in earning the foreign exchange needed in development.

Dr. Martinov said that he agreed with the ideas put forward in Professor Dandekar's paper. The agricultural policies of advanced countries do have serious effects on the prospects of the traditional trade of developing countries. He did not think that Dr. Richter was right in discounting these effects on the immediate needs of developing countries.

Professor Georgescu-Roegen said that while he agreed with much of the analysis in Dr. Martinov's report he found an underlying inconsistency in the ideological framework of his paper. Dr. Martinov criticized the capitalist system for the basis on which the capitalist countries developed their early relations with the under-developed countries. However, from Dr. Martinov's discussion of the present situation, we gather that there is no difference between the attitude adopted by the socialist countries and the old capitalist countries towards the developing world. Indeed, Dr. Martinov tells us that the socialist countries plan to help the developing countries by increasing their trade in agricultural products. But is not increased trade in agricultural products and in raw materials precisely the practice of advanced capitalist countries?

Professor Georgescu-Roegen said that the issue is not just that some countries are under-developed, but that their economic activities are such that they have a low income. If they are encouraged to stick only to these activities, their labour productivity is bound to remain low. A true plan for the development of the under-developed economies must include more than foreign trade expansion : it must include direct aid.

Professor Dandekar said that at the end of his paper, Dr. Martinov argued that a solution for the problems of developing countries lies in 'creating a high capacity home market within the framework of one or more countries'. Dr. Richter argued along the same lines. Professor Dandekar said that he agreed that trade between developing countries is important, and that it has not been sufficiently explored by the governments of these countries, or by the international agencies. At present such trade is usually conducted on a bilateral basis, and the move towards multilateral trade between developing countries is restricted by the fact that the individual currencies of these countries are not wanted for other than bilateral trade. A special monetary agency is needed to promote multilateral trade among the developing nations.

Dr. Odhner said that in discussion there is a tendency to treat the expansion of agriculture in developed countries and the expansion in developing countries as alternatives to each other. It is argued that either international specialization of labour and trade in agricultural produce should be achieved, or countries should work towards self-sufficiency. But in view of the rapid expansion of world population, is it not essential that agricultural expansion should be striven for in all countries of the world ? Temporary arrangements would have to be made to deal with the surplus problems of advanced countries and the acute shortages of developing countries, but the aim should be the intensive development of agriculture throughout the world.

Professor Nussbaumer said that the competitive positions of the developed and developing countries have to be compared, since we are dealing with trade not aid. The problems of the immediate terms of trade have already been discussed ; but there is another crucial factor determining the long-run competitive position of these countries. If the long-run trend of unit costs in under-developed countries is rising faster than that in developed countries the situation is bound to get worse. The costs involved here are not only the individual plant costs of production, but also the potentially heavy social overhead costs. At present the relative change in long-run unit costs between the two worlds seems to be unfavourable to the developing countries. The costs of mechanization are justified in most cases, but to these costs have to be added social costs of breaking up large families so that such families are denuded of support, and heavy social costs are incurred in replacing the traditional breadwinners.

Professor Ruttan said that he found it confusing to discuss as part of the same problem countries whose *per capita* income varies by as much as do those of the so-called developing world. The *per capita* income in India is about $100 while that in the Argentine is over $500. Any typology which includes a range of this magnitude is too broad for analytical purposes.

Professor Ruttan said that the second question that worried him in the discussion was the constant analysis of the problems of both areas from the supply side only. If, in developed countries, there were a 3 per cent annual growth in demand there would be no surplus. Similarly, if in the less-developed countries demand were expanding by only 1·5–2·0 per cent per year, as in most developed countries, the problem of shortages would largely disappear. Policies to modify rates of growth in demand are at least as important as policies to modify rates of growth in supply.

Dr. Gerda Blau said she wished to comment on Dr. Richter's paper, where he refers to her own paper (pages 158–9). Dr. Blau said that in using the estimates which were in her paper her purpose had been to give a general review of the categories of commodities and sort out which are not affected by serious competition. She had started with the intention of warning the developing countries that they should not be optimistic about the advantages of developing certain commodities because of the competition that stands against them from developed countries. She was able to point out that there are a few commodities over which industrial countries do not have command, but she disagreed with Dr. Richter that there were many of them. In fact, the number of commodities which are non-competing is small and most of them are unimportant. Competition with synthetics is heavy. In order to get a true picture of the situation it is necessary to go below the surface and conduct market analyses to find out what are the effective areas of competition.

Dr. Richter's negative conclusion concerning sugar that 'there is no likelihood, that such a development [the reduction of sugar output in developed countries] will occur', lay at the heart of the problem. Dr. Blau said that though she accepted the view of Nurkse that there is a reduction in the dynamic influence of trade on the process of development, it is clear that for some time to come — until balanced growth can take over — the requirements of foreign currency will have to be provided by the export of primary commodities. The purpose of the development decade is to find ways of achieving such development in order to get to the stage of balanced growth. It is important that the case of the competing products should be treated seriously and imaginatively if this is to be achieved. For example, the possibilities of exploring complementarities in the processing of primary products should not be disregarded.

Professor Campbell asked Dr. Martinov whether the substantial increase of imports to the Soviet Union from developing countries, foreseen in his paper, would be on a purely bilateral basis, or whether the Russian payments would be freely convertible. If they would not be freely convertible, would not this trade have much in common with the trade relationship which existed between metropolitan countries and developing countries in the so-called colonial period, and which is frequently the subject of criticism?

Dr. Stipetić said that Professor Nussbaumer had brought up the trends in long-run costs which lie behind the changes in terms of trade. These need to be supplemented by data on the long-run elasticities of demand for food. The developed capitalist countries show high long-run elasticity of supply and low long-run income elasticity of demand. In socialist countries where population and GNP are rising very fast, the income elasticity of demand is such that supply fails to cover the increases in demand at the planned levels of resource allocation. In the third world the demographic explosion is combined with high income elasticity of food demand.

Dr. Stipetić said that at present the labour productivity in the agriculture of socialist and advanced capitalist countries is rising at a high rate (in some cases reaching 7 per cent increases per year), and much faster than the increases in most developing countries (about 2 per cent per year). How is this to be altered? The opportunities for introducing the techniques and discoveries of temperate zones into developing tropical countries are not great. Such techniques cannot be transplanted without expensive and careful adaptation to the requirements of the different climatic zone. There is a danger of discussion remaining at a superficial level if we discuss the possibility of shifts in the structure of world trade without examining the possibilities of radical technical change in the agriculture of developing countries.

Professor Zemborain said that tropical products typically use a very high level of labour input and there are great technical problems in mechanization. In the case of coffee and cotton, hand labour has proved impossible, or prohibitively expensive, to introduce in most places.

Professor Zemborain said that the Argentine is more highly developed than most of the countries of Latin America. Most of these other countries are heavier producers of tropical crops. The general level of wages in the Argentine is double that of the average level in Latin American countries, and the social services are more highly developed than they are in the other countries. Any Argentinian producers of crops who compete against these other countries are at a disadvantage. In the event of the opening up of larger volumes of trade between Latin American countries these Argentinian producers would not be able to continue.

Dr. Martinov said that he had not been trying to criticize the behaviour of the capitalist countries. He had wished only to analyse the world production process, which, he found, does not favour the developing world. There are important historical factors which explain much of the unfavourable position.

Professor Martinov said that the Soviet Union has examined the problem of bilateral versus multilateral aid, and has come to the conclusion that there are opportunities for multilateral trade within the ambit of the socialist common market.

Chapter 9

REPERCUSSIONS OF FOOD SURPLUSES IN INDUSTRIALIZED COUNTRIES ON ECONOMIC GROWTH IN DEVELOPING COUNTRIES

BY

V. M. DANDEKAR

Gokhale Institute of Politics and Economics, India

I. INTRODUCTORY

Food surpluses in industrialized countries owe their existence mainly to two phenomena, namely, spectacular improvements in agricultural science and technology and effective price and income support policies for farm products and farm producers. The former have made it technologically possible and the latter have made it economically inevitable to grow more food than is needed by the populations concerned. This is what some of the industrialized countries, notably the United States, have succeeded in doing. At the same time, nearly all developing countries, accounting for two-thirds of the world's population, subsist on national average diets which are nutritionally inadequate. The reason is twofold. In the first instance, their food production is less than adequate to meet their needs. Secondly, they lack external purchasing power to buy their food requirements from the industrialized countries who have the food in surplus. The food surpluses in the industrialized countries thus are not only physical surpluses over their own requirements but are also commercial surpluses which they cannot sell.

Ordinarily, this would lead to a fall in the international price of these commodities and make them available to the developing countries who needed them at a low enough price at which they could buy. This was prevented by mutual agreement between the developed countries and the surpluses were stockpiled rather than released on the international market. The food surpluses if they were thus merely stockpiled in the producing countries could not have many

182

repercussions on the economic growth of the developing countries except for the influence of the potential they represented. However, the burden of the stockpiles soon became unbearable and means had to be found to reduce them. Two standard methods would be either to gift the surpluses away or to give them on loan that is for deferred payment. In fact, a certain amount of the surpluses was disposed of in these manners. However, over the last decade, a third mode of disposal, seemingly more normal and which in a sense represents a *via media* between grant and loan, has been developed. It is to sell the surpluses for local currencies of the countries concerned. Clearly, this particular mode of disposal is especially appropriate for disposing of the surpluses in the market of the developing countries which lack foreign exchange to buy their food needs normally. From the point of view of the developed countries, the disposal programme, besides reducing the burden of the surpluses, seems to serve certain objectives of the domestic and foreign policies of these countries. Consequently, the disposal programme has expanded rapidly in recent years and has now become an important element in the international food market. It is obvious that such massive disposal of surplus food in the markets of the developing countries will have several repercussions on the economic development of these countries. The purpose of the present paper is to invite attention to a few possible consequences.

II. REPERCUSSIONS ON DEVELOPING COUNTRIES

(i) *On Food-exporting Countries*

The developing countries as a class are net importers of food and are therefore direct beneficiaries of any concessional disposal of the food surpluses of the developed countries. However, there are a few developing countries which are net exporters of food and for most of them the market for their food exports is in the developing countries. These are mainly a group of rice-exporting countries of Asia and the Far East. In the general approval that the disposal programmes of the developed countries have received, the interests of the rice-exporting developing countries have been largely brushed aside on the superficial ground that food surpluses of the developed countries were mainly in wheat and coarse grains and that therefore there was little danger from them to the trade interests of the rice-exporting developing countries. There is, however, an important circumstance which makes it necessary to examine the situation more closely. As mentioned above, the main export market for the rice exports of the developing countries are other developing countries and these same developing countries also provide the main ground for the disposal of the food surpluses of the developed countries. The trade interests of the rice-exporting developing countries therefore deserve greater attention than they have received so far. In Tables I and II are shown the trade statistics of rice exports and imports by regions.

TABLE I

INDIGENOUS EXPORTS OF RICE REGIONS

(thousand metric tons milled equivalent)

	1957–59 average	1960	1961	1962	1963 provisional
Non-communist Far East *	3,442	3,738	3,732	3,450	4,038
Communist Far East †	1,256	1,338	442	598	657
Other developing countries ‡	438	466	604	510	695
Developed countries §	1,018	1,270	1,214	1,346	1,447
World total	6,154	6,812	5,992	5,904	6,837

* Burma, Cambodia, China (Taiwan), Korea Rep., Pakistan, Thailand, Viet Nam Rep.
† China (Mainland), Viet Nam (North), Others (principally North Korea).
‡ Africa, Near East, Western Hemisphere other than U.S.A.
§ U.S.A., Europe, Oceania.

Source : *FAO Rice Report, 1964*, Table 6.

TABLE II

RETAINED IMPORTS OF RICE BY REGIONS

(thousand metric tons milled equivalent)

	1957–59 average	1960	1961	1962	1963 provisional
Far East (mainly non-communist)	3,654	3,859	3,728	3,383	4,082
Africa, Near East and Western Hemisphere (mainly developing countries)	1,101	1,241	1,303	1,300	1,338
Europe and Oceania	1,199	1,528	728	1,073	983
World total	5,954	6,628	5,759	5,756	6,403

Source : *FAO Rice Report 1964*, Table 8.

Thus nearly 80 per cent of the world import market for rice lies in the developing countries. In view of the possibility of obtaining

184

sufficient wheat on concessional terms, many of the developing countries have promoted self-sufficiency policies in rice, either through expanding production or curtailing demand. Thus the entire increase in Latin American demand has been met from local production. Self-sufficiency policies are also becoming widespread in West Asia and the Near East. In many Asian countries, self-sufficiency in rice is difficult to achieve. But they have replaced part of their commercial rice imports by wheat and other surplus grains available on special terms from the developed countries. This has retarded the expansion of trade in rice and made rice-exporting countries cautious about increasing production.

Thus the rice-exporting developing countries are clearly affected by the surplus disposals of wheat and coarse grains by the developed countries. Furthermore, they are likely to suffer more directly from the growing rice surplus of the United States. In 1962, the United States exports of rice exceeded 1 million tons for the first time in any calendar year and the government permitted a 10 per cent expansion in the rice acreage. In 1963, the rice exports rose by 15 per cent to 1·2 million tons. Much of this rice has moved under special disposal programmes. In 1954–55, the special programmes accounted for only 5 per cent of the United States rice exports. In 1960–61, they accounted for over two-thirds of the total exports. Since then, practically the entire increase in rice exports is in special sales for local currency. The main beneficiaries of this expanded programme have been India and Indonesia, both of them principal consumers of the rice exports of other Asian countries. In spite of the ample imports of U.S.A. wheat under special programmes, India's requirements of rice still remain substantial and since the government hopes to build its central rice reserves to 2 million tons by 1967, it has requested that the quota of U.S. wheat be halved and the rice allocation doubled, namely, to 2 million tons under the next four-year P.L. 480 agreement. This is more than double the quantity the government of India has contracted to purchase from Burma during the same period. Similar demands may be expected from other rice-eating Asian countries at present content with imports of U.S.A. wheat under special programmes. The growing rice surplus in the United States offered on special concessional terms may therefore soon become a real rival to the rice market of the rice-exporting Asian countries.

In defence it may be pointed out that the rice-exporting countries of Asia have failed to expand their exports not because of any threat from the food surpluses of the developed countries, but because of a failure of their own production to meet the growing demand. Superficially this is so and there are many reasons for the same. The relevant point here is to ask whether the unfair competition from the food surpluses of the developing countries is not an important contributory factor. This needs more careful examination and sympathetic consideration before it is too late and irreparable harm is done to these countries.

The situation in fats and oils is quite similar. Fats and oils, mainly of plant origin, have long been of importance in the trade of developing countries. In 1934–38, the developing countries provided over 60 per cent of the world volume of exports of oils and fats. However, by 1959–61, their share had dropped well below half. The share of the developed countries increased correspondingly. Again, the performance of the United States has been the most spectacular. From a pre-war net importer, the U.S.A. has now become the world's largest exporter, accounting for 30 per cent of the world exports. The main product is soyabeans of which the production has greatly expanded during the last decade largely because of technological improvements but also because of domestic price support policies buttressed by concessional export disposals in developing countries. The soyabeans surplus in the United States is likely to grow further and will cost many developing countries their traditional export revenues.

Because most developing countries have similar types of economy, there is no obvious scope for trade between them. Nevertheless, some of the developing countries have exportable food surpluses and potential for expanding them. Others have a growing food deficit and some of them do not possess comparative advantage in the production of food-grains, either now or in the long run. On the other hand, some of them have either a comparative advantage or at least a comparative lead in the production of manufactured goods, including durable consumer goods, light capital goods and chemical products. Therefore a deliberate development of trade relations and economic co-operation between such developing countries is likely to be beneficial to both in the transitional stages of their development and possibly also over a longer period. Therefore injection of food surpluses from the developed countries into these countries might prove positively detrimental to the development of economic co-operation between these countries.

(ii) *Food-importing Developing Countries*

Let us now examine the repercussions of the food surpluses on the food-importing countries. A majority of the developing countries belong to this category. Their own food production is short of their nutritional requirements and they lack external purchasing power to purchase the balance of their requirements in the international market. Therefore, if the food surpluses of the developed countries are made available to them on concessional terms, it is obvious that these countries should be greatly benefited. The extent of benefit to be derived will of course depend upon the quantum and the terms and conditions on which the food surpluses are offered. It will also depend upon the particular conditions obtaining in the country concerned and the manner in which the food surpluses are used. A great deal of experience has been gathered over the last decade and a large body of literature is now available on the subject. Here, we shall be content with raising only a few relevant issues for reconsideration.

III. CONTRIBUTION OF FOOD SURPLUSES TOWARDS FINANCING ADDITIONAL ECONOMIC DEVELOPMENT

The main concern of the country having a food surplus is naturally how to dispose of the surplus and in the process try to expand the market for such commodities and also help promote certain domestic and foreign policies of the government concerned. There is only one international obligation to be recognized, namely not to harm the trade interests of other food exporting countries. Promotion of domestic and foreign policies through every means is, of course, natural to all governments. Seeking to expand the market for the surplus commodities is also natural. It is also complementary to the main international obligation, namely not to harm the trade interests of other exporting countries. This was how the United States Congress viewed the problem when in 1954 it approved the U.S. Agricultural Trade Development and Assistance Act commonly known as the P.L. 480 providing, among other things, for the sale of U.S. food surpluses for local currencies of the buying countries. On the other hand, an international organization such as the Food and Agricultural Organization is concerned with the interests of the food-receiving countries as well and, in particular, with the possibilities of using such food surpluses to promote economic development of the developing countries. Accordingly, by means of a pilot study in India in 1953–54, the FAO worked out a detailed economic rationale for such use of the food surpluses.[1]

It led to the conclusion that where there was idle manpower in the recipient country, and proper precautions were observed, surpluses could be used to offset part of the increased demand for food, which would result from putting part of the idle manpower at work on new capital-formation projects. Men do not live by food alone, and surpluses cannot cover the whole range of food requirements. The part of the increased consumption resulting from the increased employment not satisfied by surpluses would need to be met either by increased output from domestic industries previously operating below capacity, such as textile mills, or by increased imports financed by additional financial assistance made available to the recipient country. . . . In India incomes are low and elasticity of demand with respect to income is quite high, even for grains. The study concluded that under those favourable conditions, and with a wide 'market-basket' of many surplus foods including some preserved meat and dairy products, only about half the capital cost of the additional development projects could be offset by surpluses.[2]

In the following, we propose to re-examine the case and hope to demonstrate that the FAO study has overstated the contribution that the food surpluses might be supposed to make towards financing such additional development.

The conclusion that 'about half the capital cost of the additional development projects could be offset by surpluses' was based on a detailed analysis of the multiplier effect of additional development expenditure. The following assumptions were made : (*a*) of the total project costs, 10 per cent go into imported equipment and supplies, 40 per cent into domestic goods and services such as cement, steel, transportation, etc., and the remaining 50 per cent into wages paid to direct labour employed on the project ; (*b*) of the wages paid to the direct labour, 34 per cent are spent on surplus foods (at wholesale) and the remaining 66 per cent on other consumer goods and services ; (*c*) of the other derived incomes, 9 per cent are spent on taxes, 9 per cent go into savings and 8 per cent are spent on imported consumer goods ; of the remainder of this income, 26 per

[1] *Uses of Agricultural Surpluses to finance economic development in Under-developed Countries: A Pilot Study in India* ; Commodity Policy Studies No. 6, FAO, Rome, June 1955.
[2] Mordecai Ezekiel, 'Impact and Implications of Foreign Surplus Disposal on Developed Economies and Foreign Competitors : The International Perspective', *Journal of Farm Economics*, Vol. 42, 1960, pp. 1065–1066.

cent is spent on surplus foods (at wholesale) and the rest on other consumer goods and services. On these assumptions, a capital expenditure of 100 units on additional development projects may be seen to lead to the situation shown in Table III.

TABLE III

FINAL DISPOSAL OF PROJECT EXPENDITURE AND
INCOMES GENERATED BY THE SAME

| | Project Expenditure | | | |
Item of Expenditure	Direct Labour	Other than Labour	Derived Incomes	Total
1. *Capital goods and services*				
Imported		10		10
Domestic		40		40
2. *Consumer goods and services*				
Imported				
Non-surplus			13	13
Surplus foods	17		31	48
Domestic	33		88	121
3. *Taxes and savings*				
Taxes			14·5	14·5
Savings			14·5	14·5
4. Total	50	50	161	261

Thus the capital expenditure of 100 units on additional development projects is expected to create 48 units' worth of additional demand for surplus kinds of foods. If such surplus foods in adequate quantities are placed in the hands of the governments of the countries concerned and if the governments are allowed to sell these on the domestic market and treat the sales proceeds as their revenue receipts, then we may say that out of the 100 units of capital expenditure, 48 units would return to the governments. It is in this sense that 'about half the capital cost of the additional projects could be offset by surpluses'.

It must be emphasized that the above formulation is correct only from a strictly limited governmental point of view. It amounts to suggesting that the governments of the developing countries may sell on the home market foreign food surpluses made available to them in aid and use the sales proceeds to finance about half the capital costs of additional development projects to be concurrently initiated and that if they agree to do so, they will need to finance only the remaining half of the capital costs from other fiscal resources.

Let us now examine the same operation from the point of view of the total economy. From this standpoint, the total cost of the additional development project is not 100 units but 232 units as shown in Table IV.

TABLE IV

REAL COSTS OF ADDITIONAL DEVELOPMENT PROJECT WORTH
100 UNITS OF DIRECT CAPITAL COSTS

		Units
1. *Imported components*		
	Capital goods	10
	Consumer goods other than surplus foods	13
	Surplus foods	48
		71
2. *Domestic components*		
	Capital goods	40
	Consumer goods	121
		161
3. *Total*		232

Thus from the standpoint of the total economy, the surplus foods contribute only 48 out of the 232 units of real costs and hence the contribution of the surplus foods is only about 20 per cent and not 45 per cent. The error in the FAO formulation lies in taking into account the total absorption of the surplus foods in the economy but relating it to only the initial capital costs. A correct assessment of the contribution of the surplus foods must be based on one or the other basis. Either we might confine attention to only the capital costs of the project in which case the contribution of the surplus foods would be only that amount which could legitimately be given as wages to labour directly employed on the project ; this is only worth 17 units and hence the contribution of the surplus foods to the capital costs is only 17 per cent of the total. Or, we might consider the entire additional demand for the surplus foods generated as a consequence of the capital expenditure on the project in which case we should relate it to the total additional demand for all goods generated similarly as a consequence of the capital expenditure on the project ; on this basis the contribution of the surplus foods is 48 units in a total of 232 units which is only about 20 per cent. Clearly, the FAO formulation overstates the contribution of the surplus foods to additional development.[1]

Financing of the Additional Development Projects

Let us now return to a consideration of the operation from the limited point of view of the government of the developing country. From this limited point of view, as we have seen, the surplus foods help finance about half the capital costs of the projects and hence the government need finance only the remaining half from other fiscal resources. On a closer examination, it seems that the total financial position of the government *vis-à-vis* the additional projects will be even easier than this. This is because proceeds of sale of the surplus foods is not the only channel through which a part of the expenditure incurred on the project returns to the government. In the first instance, there are additional tax receipts worth 14·5 units which will return to the government. Secondly, there are additional savings worth 14·5 units. These will of course be in private hands and will not automatically reach the government. However, the government can acquire them through additional borrowing. In order to simplify the argument, let us suppose that the government is in fact able to do this. Thus through the sale of the surplus foods, through additional tax receipts and through additional borrowing, 48 plus 14·5 plus 14·5, that is, in all, 77 units' worth of expenditure initially incurred on the projects will in due course return to the government. The remaining 23 units do not return because of leakages through imports. In the first instance, 100 units' worth of project costs involve 10 units' worth of imported capital goods. If the project is to be additional, it is obvious that these imported capital needs must be provided by additional foreign aid. Let us suppose that this is done. The other leakage of 13 units is because out of the 161 units of derived incomes, 13 units are spent on imported consumer goods other than the surplus foods. These therefore do not return to the government and it is only this part of the capital expenditure which ultimately remains uncovered or deficit-financed. Thus the total financing

[1] The former of the two bases was what was suggested in a memorandum prepared by myself working as a Consultant to the FAO Pilot Study in India. See V. M. Dandekar, *Use of Food Surpluses for Economic Development*, Gokhale Institute of Politics and Economics, Publication No. 33, pp. 127–128.

of the project would appear as shown in Table V.

TABLE V

FINANCING OF ADDITIONAL DEVELOPMENT PROJECTS

	Units
Food aid	48
Additional foreign aid	10
Additional tax receipts	14·5
Additional savings	14·5
Uncovered expenditure	13
Total	100

Thus only 13 out of 100 units of project costs will have to be financed through other fiscal sources or through deficit financing. We might consider three alternatives.

(1) As we have seen, the 13 units of uncovered expenditure represents additional consumer expenditure on imported goods. Therefore an appropriate source of finance would be additional foreign aid. Thus the total foreign aid necessary to complement food aid worth 48 units would be 23 units — 10 units' worth of imported capital goods and 13 units' worth of imported consumer goods. This means that in order to enable the governments of the developing countries to make effective use of food surpluses for financing additional development, food aid must be accompanied by complementary foreign aid almost half the size of the food aid. Let us suppose that this is available. The entire capital costs of the project would then be fully financed.

(2) Alternatively, and more realistically, we may suppose that only 10 units of foreign aid to cover the capital costs of the project is available and that therefore 13 units' worth of capital expenditure remains uncovered. In whatever manner this is financed, it can be no substitute for foreign aid and hence no additional imported consumer goods can appear on the domestic market. To prevent limited foreign exchange resources leaking into additional consumer expenditure on imported goods will require stricter exchange control and import restrictions. However, if these are effectively enforced and no part of the derived incomes is allowed to be spent on additional imported consumer goods, it can be shown that the size of the derived incomes will be larger and that the expenditure on surplus foods, on additional tax payments and on additional savings together will be larger exactly by 13 units so that the entire capital

costs of the projects will again be entirely financed but now with a much smaller complementary foreign aid.

(3) Finally, let us suppose that no complementary foreign aid whatever is available, which is of course closer to the facts of the case. In that case if additional development projects are to be undertaken without disturbing the planned use of the available foreign exchange resources, the additional projects will have to be so selected that they do not require any foreign capital goods. This might put a severe restriction on the choice of the projects, but is not impossible. Let us, therefore, suppose that the additional projects do not require any foreign capital goods and that therefore 50 per cent of the capital costs is spent on domestically produced capital goods and services and that 50 per cent is spent on payment of wages to labour directly employed on the projects. It is obvious that under these conditions the derived incomes will be even larger than before and that the surplus foods together with additional tax receipts and savings will fully finance the capital costs of the additional projects. In the following Table VI, we show the size of the derived incomes and their disposal under the three assumptions mentioned above.

TABLE VI

DISPOSAL OF DERIVED INCOMES UNDER VARYING ASSUMPTIONS
REGARDING THE SIZE OF COMPLEMENTARY FOREIGN ASSISTANCE

(in units)

	Complementary Foreign Assistance as Per Cent of Total Capital Costs		
	23%	10%	nil
Items of disposal of derived incomes			
Surplus foods	31	39·6	45
Imported non-surplus goods	13	—	—
Domestic consumer goods	88	112·6	128
Additional tax payment	14·5	16·7	19
Additional savings	14·5	16·7	19
Total derived incomes	161·0	185·6	211

In order to obtain the full expenditure on surplus foods and domestic consumer goods, we must, of course, add, in each case above the expenditure of the wages paid to the direct labour, namely, 17 units on surplus foods and 33 units on domestic consumer goods. The full financing of capital costs of additional development worth 100 units under the three assumptions regarding complementary foreign assistance then appears as shown in Table VII.

TABLE VII

FINANCING OF ADDITIONAL DEVELOPMENT PROJECTS WORTH 100 UNITS
UNDER VARYING ASSUMPTIONS REGARDING COMPLEMENTARY FOREIGN
ASSISTANCE

(in units)

	Complementary Foreign Assistance as Per Cent of Total Capital Costs		
	23%	10%	nil
Financial resource			
Sale of surplus foods	48	56·6	62
Additional foreign aid	23	10	—
Additional tax receipts	14·5	16·7	19
Additional savings	14·5	16·7	19
Total	100·0	100·0	100

Thus, in the first instance, provided the government is able to mobilize the entire additional savings in private hands, the capital costs of the additional projects will be fully financed and no part will remain uncovered. Secondly, though smaller complementary foreign aid makes the choice of additional projects more difficult and requires greater degree of exchange control, it does not leave any part of the capital costs of the additional projects uncovered. Finally, the smaller is the complementary foreign aid, the greater is the use of surplus foods and the proportion of the capital costs which might be offset against them grows from 48 per cent to 62 per cent.

The principal and, from the standpoint of the government of a developing country, the most attractive result of the above analysis is that the capital budget of a development project founded upon the use of surplus foods is always balanced provided the capital account of the project is credited with (1) sales proceeds of the surplus foods sold on the domestic market to meet additional demand for food generated by the project expenditure ; (2) additional tax receipts following additional incomes generated by the project expenditure ; and (3) additional savings created by the additional incomes generated by the project expenditure. The first two are direct receipts of the government and the third can be secured by additional borrowing.

The capital budget of the project is thus always balanced. It is, therefore, implicitly suggested that in developing countries where there is enough idle manpower, the governments should avail themselves of the food surpluses, sell the surplus foods on the domestic market and use the funds to finance additional development projects. Such sales proceeds may finance anything between 50 and 60 per cent of the capital costs of the projects and, what is more, the remaining part of the government expenditure will in due course return to the government in the form of additional tax receipts and public borrowing. Complementary foreign aid makes the operation easy. But smaller complementary assistance does not alter the basic proposition, which is that the entire operation is self-financing, that it does not cost the government, and presumably the country, anything and in the process there is so much additional economic development.

IV. FALLACY OF THE BALANCED BUDGET

The formality of a balanced capital budget and the self-financing process of development which the food surpluses apparently initiate and support have proved to be their best selling points. The underlying fallacy may be seen from the fact that the budget remains balanced even if the government fails to mobilize the additional savings. For in that case, the private savings go into private investment, generate additional incomes and ultimately return to government as either sales proceeds of additional sale of surplus foods or as additional tax collections. Hence the government need not bother about mobilizing the additional savings and many of the governments do not. In fact, the government need not even bother to collect the additional tax receipts. If it does not, the additional tax receivables will go into either private investment or private consumption, will generate further incomes until the additional tax receivables flow to the government in the form of sales proceeds of additional sale of surplus foods. The capital budget is again fully balanced and now the additional development is entirely financed by the surplus foods. To demonstrate the point, let us take up the last case discussed above, namely one of no complementary foreign aid and pursue the consequences of (1) no additional savings accruing to government and (2) no additional savings or taxation accruing to government. In Table VIII we give the size and disposal of the derived incomes under these assumptions.

TABLE VIII

DISPOSAL OF DERIVED INCOMES UNDER CONDITIONS OF NO COMPLE-
MENTARY FOREIGN AID AND UNDER VARYING ASSUMPTIONS REGARDING
ADDITIONAL ACCRUALS OF TAXES AND SAVINGS TO GOVERNMENT

(in units)

	Additional Tax and Savings Accruals		
	9% in Tax and 9% in Savings	9% in Tax Only	Nil
Items of disposal			
Surplus foods	45	60·1	83
Domestic consumer goods	128	171·2	236·2
Additional tax payments	19	22·9	—
Additional savings	19	—	—
Total derived incomes	211	254·2	319·2

Here again, as before, in order to obtain the full expenditure on surplus foods and domestic consumer goods, we must add, in each case above, the expenditure of the wages paid to the direct labour, namely, 17 units on surplus foods and 33 units on domestic consumer goods. The full financing of the capital costs of additional development worth 100 units then appears as in Table IX.

TABLE IX

FINANCING OF ADDITIONAL DEVELOPMENT PROJECTS WITH NO COMPLE-
MENTARY FOREIGN AID AND UNDER VARYING ASSUMPTIONS REGARDING
ADDITIONAL ACCRUALS OF TAXES AND SAVINGS TO GOVERNMENT

	Additional Tax and Savings Accruals as Per Cent of Derived Incomes		
	9% in Tax and 9% in Savings	9% in Tax Only	Nil
Financial resources			
Sale of surplus foods	62	77·1	100
Additional tax receipts	19	22·9	—
Additional savings	19	—	—
Total	100	100·0	100

Thus, in all cases, the capital budget of the additional projects is fully balanced, no matter whether the government makes any effort to collect additional tax dues and to mobilize additional savings in private hands. In fact, the smaller the tax and savings effort, the greater will be the use of the surplus foods for financing additional development. With no additional tax and savings effort whatever the entire additional development may be financed by the use of surplus foods.

This is the economic rationale of using surplus foods for financing additional development. Developing countries generally have not been slow in extending its application to financing normal development. It is then only one step further to use food surpluses to finance normal or abnormal government expenditure. This too is not unknown. For instance :

> In some countries, local governments lack sufficient strength and stability adequately to finance their expenditures by taxes or sound borrowing. In such cases the sale of US aid-commodities for local currency can provide the local government with funds needed to run its domestic affairs. In Viet-Nam, for example, the sale of US aid-commodities for Vietnamese currency provides the local government with roughly two-thirds of its revenue receipts. The situation in Laos and Cambodia is quite similar.[1]

V. CONCLUSION : PROSPECTS

Such are the logical consequences of a smart piece of international salesmanship. These are not inevitable. Nor do we need rule out the possibility of an economic and purposeful use of the food surpluses of the developed countries for promoting development in the developing countries. The Food and Agriculture Organization has emphasized some of the precautions necessary but they have gone mostly unheeded. For instance, it is recognized that food alone does not provide all the capital needed for employing idle manpower on development projects and that some of the other capital requirements must be supplied by complementary foreign aid. This has generally not been forthcoming. It is also recognized that 'men do not live by food alone' and that 'part of the increased consumption resulting from the increased employment not satisfied by surpluses would need to be met either by increased output from domestic industries . . . or by increased imports financed by additional financial assistance'. For instance, if we may refer to our last illustration where the

[1] 'The Problem of Excess Accumulation of U.S.-owned Local Currencies', Findings and Recommendations submitted to the Under Secretary of State by the Consultants on International Finance and Economic Problems, April 1960.

surplus foods are shown to provide hundred per cent finance for additional development, we may see that the real costs of the projects involve, besides 100 units of surplus foods, 50 units of capital goods and 269·2 units of consumer goods. It is conveniently believed that these will be supplied by expansion of domestic production in response to additional demand. Presumably, the only thing that will not respond to anything is food production, particularly those items in which the developed countries have a surplus.

Unfortunately, most of the developing countries are not able to put forth additional resources in capital and consumer goods necessary to make productive use of the food surpluses. In most developing countries with ambitious development plans, domestic supply of necessary capital goods and services such as cement, steel, transportation, engineering and technical skills, and administrative services for planning and supervision is extremely limited and incapable of quick expansion. Here is a classic example. The *FAO Rice Report 1964* makes the following observation regarding wheat imports into India : 'Any large-scale increase in the imports of wheat, which are already heavy, is handicapped by the inadequacy of port and storage facilities' (p. 26). Thus the domestic resources needed even to import the surplus foods are wanting. There is lack of foreign exchange needed to pay the shipping expenses to be paid in dollars. The port facilities are inadequate. There is shortage of storage and internal rail and road transport is overburdened. Thus there is shortage of resources required merely to get the surplus food on the spot — let alone the several things needed to use the food for additional development projects. So few additional projects are undertaken and fewer of them utilize the idle manpower available. Generally, a make-believe book accounting is resorted to and certain projects already in the plan if not already implemented are shown as having been financed by the food surpluses. Consequently, the net result of the use of surplus foods is relaxation in tax and savings effort necessary even for the planned development.

In the developing countries there is a general shortage of all consumer goods. In the absence of sufficient tax and savings effort, the pressure of demand on the limited supplies pushes up the prices of these commodities and the prices of domestic equivalents of the surplus foods follow the lead. This creates an ideal situation for the expansion of the market for the surplus foods which, from the standpoint of the developing countries, is one of the primary objectives of the programmes of surplus disposals. Consequently, demands or requests for these commodities from the developing countries increase. Increased supplies of surplus foods help these governments to provide a minimum quantum of food to the low-income classes who cannot afford to buy it in the domestic market. Further, they bring in additional revenues to governments of the developing countries so that they may further relax in their tax and savings effort. Thus grows the dependence of the developing countries on the food supplies from the developed countries which ultimately helps promote certain foreign policies of the developed countries which they recognize as another primary objective of their surplus disposal programmes.

The reason why the developing countries have generally failed to make good use of the food surpluses to promote their own economic development through putting to employment idle manpower available to them, which was the basis of the FAO proposals, is the neglect of one crucial factor needed for the purpose. As the FAO proposals emphasize, transitional employment of idle manpower on direct capital-creating works requires food, other consumer goods and certain capital goods and services in order to make the employment efficient. There is one more and the most crucial requirement. It is organization — organization for discovering and planning useful works and organization for physical handling of huge unskilled manpower for employment and deployment from one work to another. Most developing countries, at any rate outside the centrally planned economies, lack such organizational ability. There is also evident a lack of effort to develop and build the necessary organizations. In the final analysis, there is reluctance to accept the national responsibility of guaranteeing to every individual a national minimum living through productive employment. Unfortunately, promotion of such organizations and of national attitudes necessary to sustain them is regarded as contrary to certain foreign policy objectives of the developed countries who have the food surpluses. Not only the importance of such organizations and attitudes is neglected but it is de-emphasized by presenting economic growth on a plate alongside the surplus foods. Consequences to the economic growth of the developing countries concerned are too plain to need further comment.

DISCUSSION OF PROFESSOR DANDEKAR'S PAPER

Professor Gerhardsen said that it had been impressed upon him by Professor Dandekar's paper that in the matter of studying bilateral aid arrangements it is necessary to use political science as well as economics. However, he did not fully understand what was meant in the final paragraph of the paper by the statement that 'the promotion of such organizations [for the discovery and planning of useful works and for the handling of huge unskilled manpower] and national attitudes necessary to sustain them is regarded as contrary to certain foreign policy objectives of the developed countries who have the food surpluses.' However, Professor Gerhardsen said that if food grants which are made anywhere in the world were 'anonymous' and given without strings to be disbursed by the FAO, any such problems would be avoided.

Professor Gerhardsen said that what is now required is some testing of the ideas in the FAO model and the ideas in Professor Dandekar's criticisms of it. The clashes on the effect of P.L. 480 on normal trade are now explicit and the need is to face them with empirical evidence.

In the discussion on Professor Hathaway's paper, Professor Bishop had asked whether the efficient agriculture of the developed world should be curtailed and the inefficient agriculture of the developing world be encouraged. This question is pertinent in the present discussion of the effects of the agriculture of developed countries on the agriculture of the underdeveloped world. Professor Gerhardsen said that his own answer to this question is, yes. There should be a conscious effort to spread the benefits of technology by this means. In the world of today no-one can be regarded as having an incontestable right to the benefits of their superior technology. There have been extensive attempts to improve efficiency in grain production in India, and the report of a Ford Foundation team has indicated good results. Efficiency can be improved in various ways and it is important that the short-run solutions of the advanced world do not reduce the chances of successful improvements in the long run in under-developed countries.

Professor Hathaway said that during recent years the export of edible fats and oils from the United States (in particular soya bean oil) has grown very fast, but these by and large have been in the form of normal convertible currency sales, indicating that the increase in exports is a demand phenomena rather than a surplus situation. One of the largest farm organizations, which includes most producers of soya bean oil in the United States, has in fact argued for the removal of support prices. In view of this Professor Hathaway asked what, in Professor Dandekar's view, constitutes a surplus in world trade.

Professor Hathaway said that he appreciated the difficulty of this question, but he asked Professor Dandekar if he thought that India would have been better off than she now is had the United States curtailed its production of wheat either by direct controls on production or by maintaining internally the agreed world wheat prices.

A training in economics makes it hard to dodge the question of whether we can really afford to ignore economic efficiency in world production by failing to work for the optimum output from limited world resources.

Mr. Tracy said that he wanted to comment on the part of Professor Dandekar's paper where he dealt with the amount of other aid that ought to accompany food aid. Mr. Tracy said that he was uncertain as to what Professor Dandekar's conclusions are. He said that Professor Dandekar seemed to demonstrate that it is not necessary to have any other aid at all and stated that, even in the absence of other aid, the budget of a food aid project can be balanced. But, he said, at other points, Professor Dandekar seemed to be referring to this as a fallacy. Mr. Tracy said that he thought this is indeed a fallacy because this approach seems to be based on purely financial considerations. He said the economic implications seem to be different. In particular : the derived demand arising from a food aid project is bound to involve some demand for imported produce other than food ; the derived demand for home-produced produce, other food will raise their prices unless there are under-employed productive resources ; and the derived demand will also occur for home-produced food, which again is likely to raise prices. Consequently, Mr. Tracy said, unless there is an increase in imports and in output of various goods, both of which are likely to necessitate extra foreign aid, the effects of the food aid project are likely to be inflationary. Professor Dandekar had stated that if there is no complementary aid, derived incomes would be even larger. Does he agree that this increase is purely monetary and that *real* incomes would not be any higher ?

Professor Dandekar said that Mr. Tracy was quite correct in his interpretation of the argument in the paper. Professor Dandekar said that he was trying to make the same distinction which Mr. Tracy had emphasized, namely, the distinction between the case viewed from the strictly limited viewpoint of governmental finance and the case viewed from the standpoint of the total economy. The FAO model viewed the matter from the limited standpoint of governmental finance. It is only when it is so viewed that it can be argued that surplus food contributes about half the capital costs of development projects. On this argument it seems that there is no need for any complementary foreign aid or greater tax and savings effort by the developing countries. In fact it can be shown that the smaller the complementary foreign aid and the smaller the tax and savings effort by the developing countries, the greater is the contribution of surplus foods to economic development. Here lies the fallacy of the FAO argument. Professor Dandekar thought that the contribution of surplus foods to economic development can be correctly assessed only when it is viewed from the viewpoint of the total economy. It is only then that the importance of complementary foreign aid and of increased tax and savings effort by the developing countries on the one hand and the positive but limited contribution that food aid can make to their economic development would become clear.

Professor Robinson said that aid under The United States Public Law 480 is in exchange for local currency. It is important to notice that the analysis under discussion is based on the assumption that the local currency once in the hands of the United States is not in fact used by her.

Professor Gulbrandsen said that he would like to make some remarks on the effects of a surplus disposal programme on world prices. If the surplus is given away, it means that it is sold at a zero price, and this from the giver's side is a form of price differentiation. Is this method of differentiation the most effective ? Let us consider the effects on the demand side. Suppose that the income elasticity is unity, that the surplus is given to the government of the receiving country and that the government sells it through ordinary channels at market prices. In this case the change in demand for food depends on the increase in disposable income, which is created by a redistribution of government earnings. The net percentage increase in demand for food will be equal to the percentage increase in disposable income, that is, the percentage that the redistributed government earnings are of initial disposable income. For example, if the surplus represents 10 per cent of total food consumption value, if this value is 60 per cent of total disposable income, and if 70 per cent of the government's earnings are redistributed as disposable income, then the net percentage increase in food demand will become 4·2 per cent, and 58 per cent of the surplus will substitute ordinary food supply. This might usefully be stated another way : If the donor country gives the government of the developing country 100 bushels of wheat, if the government then sells the wheat internally for $1 per bushel, and if the government distributes 70 per cent of the proceeds of the sale, then the increase in disposable income will be $70. If the population spends 60 per cent of all disposable income on wheat, the net increase in demand for wheat within the country will be 60 per cent of $70, that is $42 worth or 42 bushels. Thus the remainder of the 100 bushels of wheat, after allowing for this demand for 42 bushels of wheat which is generated by the way the government dispenses the gift, would displace commercial imports purchased on the world market. So the displacement of ordinary food supply will be 58 bushels or 58 per cent of the original gift. Consequently the demand on the ordinary world market will decrease and the pressure of low prices on producers will become heavier. If the food is given directly to the consumer, the substitution effect would be less, since nothing of the value of the gift would leak to the government. On the same assumptions as the previous example, there would be a substitution effect of only 40 per cent of the ordinary commercial food supply. Only if the food is given as help in conditions of extreme starvation would no substitution effect appear.

However, even in this case, in the next period certain repercussions would arise. As a number of people would be saved from starvation, the demand for food would be expected to increase compared to the situation without the help of the food gift. This implies a faster growing population, which in turn implies for the developing country a slower increase in *per capita* income, or even a decrease. If this occurs, then the demand for food would increase more slowly or even decrease, and this would adversely affect, yet again, the market demand and so the world price for food. Professor Gulbrandsen said that it is certainly very cynical to discuss the economic effects of starvation on world prices for food, but this analysis does suggest that more effective ways of solving price and starvation problems should be sought.

Professor Gulbrandsen said that the effect of low prices for food on the world market is, of course, favourable for poor countries in as far as they are demanders. But since these countries often specialize in food production, the low prices mean that they suffer from low national incomes. Depending on whether the contribution of agricultural output to national product is larger or smaller than that of expenditure from national income on food, the net effect of low prices would be adverse or favourable for the country as a demander of food. The effect would be favourable if higher world market prices increase both food supply and demand ; and it would be adverse if higher international prices increase food supply but decrease food demand. In the adverse case, consumer subsidies should be combined with the higher prices.

In general, to get an optimal allocation of food production in the world the price to producers in all countries should be the same. However, to get maximum revenues from food, the price in different countries should be differentiated, with low prices in poor countries and high prices in rich countries. In practice, this differentiation in prices already exists, but the earnings are redistributed in a sub-optimal way, that is, most of the earnings go to the producers in the rich countries in the world. Since, in spite of the support to these producers in rich countries, the agricultural sector has a very low productivity in comparison with other sectors in these rich countries, the most effective way to solve both the world food prices problem and the starvation problem, is to reduce the support and move the active agricultural population to other sectors in the rich countries.

Professor Gulbrandsen said the price system could be developed in the following way : Price supports to agriculture could be removed with international agreement, and low price systems could be introduced into all rich countries over a period of two decades. At the same time taxes which were raised on food consumption, and other taxes, could be pooled in an international food fund. This fund could then be used to buy food on the world market and give it to these poor countries. The governments of these countries could use the earnings from reselling the food either to maintain a lower domestic price at the same time as supporting domestic producers, or to subsidize food to certain population groups. The amount of food the international fund would give a particular country would be calculated on the basis of *per capita* income and the size of population. An alternative method of using the fund's money would be to give it to the poor country to be distributed to the consumers in accordance with some special agreement, negotiated by the members of the fund.

Professor Gulbrandsen said that, in short, his suggestions for solving the surplus food and starvation problems were : to even out producer price differentials throughout the world in order to achieve an optimal allocation of resources ; and to differentiate consumer prices between rich and poor countries by means of taxes and subsidies with the earnings and expenses pooled through an international food fund.

Professor Leduc said that he agreed with Professor Dandekar that not all food aid need necessarily be an aid to development. Development requires a number of complementary factors of which food is only one. If supplemental aid in other goods or loans is not available, the food aid can do little more for the government than help finance normal expenditure. A number of instances show that food aid alone is insufficient. The complementary factors have to be obtained either from within the country by diversion from other uses, or have to be obtained as further aid or loans from abroad. The examples of Tunisia and Morocco illustrate the point. In both countries systematic efforts were made under the aegis of French experts to use food aid from the United States to finance development projects, and to take up surplus manpower. In both cases the programmes were in part failures. Food aid is clearly no panacea. It clearly cannot be ignored as a major assistance in development, but high hopes should not be pinned on it.

Mr. Robinson said that Professor Dandekar's paper ought to be corrected where he states that, 'There was only one international obligation to be recognized, namely not to harm the trade interests of other food exporting countries'. This is not the only obligation as laid down under the FAO surplus disposal programme. It is also stated clearly that any disposal of surplus should not be allowed to damage the economy of the recipient country. Mr. Robinson said that in both the budgetary and economic effects of food aid there lie dangers to the recipient country.

In general, the problems of under-developed countries are clearly not simply that of nutritional needs, and any aid programme that acts as though they are would not help in development. The weakness of infrastructures and the dearth of organizing ability — right down to such problems as the administration of a school feeding programme — are the major difficulties under which many of these countries labour.

Professor Dandekar said that he entirely agreed with all those speakers who had emphasized the need for complementary aid if successful developments were to be undertaken with part of the capital costs covered by food aid. It seems that such complementary aid in terms of both capital and consumer goods should amount to at least half the food aid. Professor Dandekar said that he would accept Mr. Robinson's criticism and admit that the FAO principles on surplus disposal programme clearly laid down that the surplus disposal should not be allowed to damage the economy of the recipient country. Nevertheless, it is the FAO which is primarily responsible for selling the idea to the developing countries that no harm whatever but only good would come to them by accepting food aid in unlimited quantities. Professor Dandekar emphasized that a country which accepts food aid is usually not in a position to refuse so that the responsibility of seeing that the aid is put to proper use lies partly with the donor country or agency which can, and not infrequently does, attach conditions to the aid. Unfortunately, such conditions are often governed by political rather than economic considerations. Professor Dandekar agreed with Mr. Robinson in emphasizing the importance of organizing ability and skills and suggested that most of the failure in Morocco and Tunisia, which Professor Leduc had cited, is to be attributed to lack of organization. Recognizing this, it is sometimes suggested that along with food aid should also be given assistance in organizing skills such as for instance technical and administrative personnel for running a school-feeding programme. Professor Dandekar did not agree and insisted that to be eligible to receive food aid, the receiving country must contribute something and that the necessary organization should be one of its exclusive contributions. He argued that use of food aid for mobilizing the unused manpower resources of a country for economic development required forms of organization which are not ordinarily supplied by the market in the developing countries. Extra-market motivation and effort are therefore needed. It seems that such extra-market motivation and effort are regarded contrary to their foreign policy objectives by certain donor countries and therefore are not favoured by them. This is what he intended to refer to in the final paragraph of his paper to which Professor Gerhardsen had invited attention.

PART III

CONTEMPORARY PROBLEMS IN DEVELOPED AGRICULTURE: THE FAMILY FARM AND PROBLEMS OF FACTOR MOBILITY

Chapter 10

TENDENCIES TOWARDS CONCENTRATION AND SPECIALIZATION IN AGRICULTURE

BY

ULF RENBORG

Agricultural College, Uppsala, Sweden

I. INTRODUCTION

THE account of concentration and rationalization trends in agriculture which I propose to give in this paper will be confined to certain countries in Western Europe and to the U.S.A. I shall not deal here with the socialist countries of Eastern Europe — partly because I have wished to limit the scope of my study, and partly because the course of concentration and specialization in those countries differs, both essentially and as to influences deciding it, from that in the Western European countries and the U.S.A.

As regards the form of presentation of this paper, I shall first of all refer to the present structure of agriculture and attempt to describe the average type of agricultural enterprise of today in terms of size, organization and economic results. At the same time I shall draw certain comparisons with industry. A second section will deal with present trends towards concentration, conversion to part-time farming, specialization, division of types of production and vertical and horizontal integration, together with the relocation (geographical regrouping) of production. An attempt will then be made to analyse the processes of present conversion, using, as terms of reference, both neo-classic economic theory and a model used by the Swedish economist Erik Dahmén in his study of Swedish industrial activity during the period 1919 to 1939.[1]

In a final section, I shall attempt, with the aid of a recently completed Swedish investigation, to give some idea of how agricultural enterprises may come to look in the future, on the presumption

[1] E. Dahmén, *Svensk Industriell Företagarverksamhet*, Lund, 1950.

209

The Family Farm and Problems of Factor Mobility

of the continued acceptance of private farm enterprise and against the background of the present structure of farming enterprises in Western Europe.

Fig. 1a

Number of farms of over 1 ha in Western Europe divided according to size groups

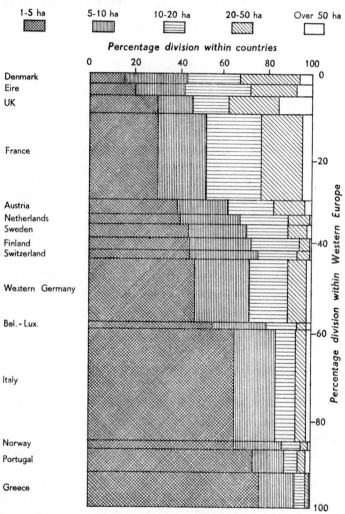

Sources for Figures 1a and b – Hjelm *et al.*: *Det svenska lantbrukets effektivioseringsvägar. SOU 1963-66*, Stockholm, 1963; and Yates, P. L.: *Food, Land and Manpower in Western Europe*, London, 1960.

II. PRESENT AGRICULTURAL STRUCTURE

A very general picture of the structure of agriculture in Western
Europe is presented in Figs. 1*a* and 1*b*, which show the total number

FIG. 1*b*

Arable areas in Western Europe divided according to types
of farm size

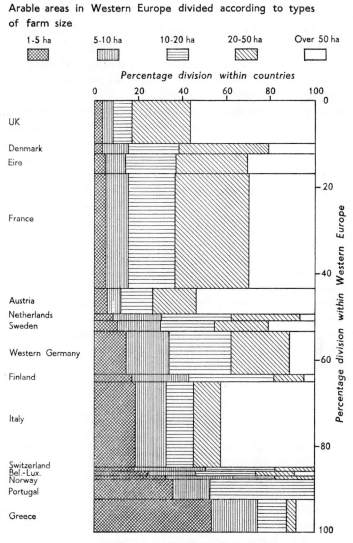

Sources : See Figure I*a*.

of farms and the total arable area of Western Europe, together with the distribution of holdings of more than 1 hectare according to size and according to country. It will be seen that numerically the small and very small type of enterprise dominate, while the greater part of land under cultivation consists of enterprises of above 10 hectares. At the beginning of the sixties an average in Western Europe would have been about 5–10 hectares of arable land per enterprise ; the comparable figure for the U.S.A. would have been 40 hectares for all farms, and for U.S.S.R. about 3,000 hectares per kolkhoz and about 9,000 hectares per sovkhoz. It should be noted however that units of 1–5 hectares in Western Europe consist to a very large extent of part-time farms, on which families earn substantial parts of their livings from occupations other than agricultural.

Figs. 1*a* and 1*b* also shed interesting light on processes in the present transformation of agriculture. A country like Italy, for instance, has a large number of small farms : but, at the same time, a large part of its arable area consists of farm properties of over 20 hectares. Even though many of the latter consist of several separate farms, the rationalization of agriculture must first of all depend upon the possibility of finding alternative openings for the employment of the large small-farming population. For the rationalization of food production, on the other hand, the arable areas on the small farms are of minor importance. In such countries as Western Germany and the Netherlands, where production is based largely upon small holdings, scope for further rationalization is restricted. In such countries, rationalization of food production would often mean the abandoning of a number of small farms in order to create enterprises of large enough size to permit the fullest exploitation of modern agricultural machinery and techniques.

The amount of arable land available to it gives very little real indication of the size of an enterprise. This applies particularly to comparisons between areas or regions geographically as dissimilar as those with which we are dealing, and to comparisons between enterprises from different sectors of the farming economy. For instance, 1 hectare of vineyard in Italy or Southern France would be quite a different proposition for an enterprise from 1 hectare of pasture in northernmost Scandinavia. The best picture of the structure of an enterprise is obtained if other factors apart from soil are taken into consideration — if possible all the factors of production. Other important factors in production can be brought into the picture of the enterprise structure by measurements according to numbers of employees and amount of invested capital. The total contribution of all means of production can be measured, for example, by registered annual sales (turnover).

Unfortunately it is not customary to use such yardsticks in agriculture. It is possible nevertheless to produce rough figures for certain countries[1] by means of completing the picture of the size of enterprises given in available statistics. In the U.S.A. it has been customary for a long time to assess the size of an enterprise according both to acreage and to sales. Farm classifications of size, according to these two measurements, are shown in Table 1. This table shows that

TABLE I

NUMBER OF FARMS IN THE U.S.A.

A. Classified by Acreage

	Number of Farms				
	1900	1925	1950	1954	1959 *
Size of Farm	Thousands				
Total	5,737	6,372	5,381	4,782	3,703
Under 10 acres	267	379	485	484	241
10– 49 acres	1,664	2,039	1,478	1,213	811
50– 99 acres	1,366	1,421	1,048	864	658
100–259 acres	1,912	1,887	1,589	1,417	1,186
260–999 acres	481	583	660	674	671
1,000 acres and over	47	63	121	130	136

B. Classified by Value of Sales

	Number of Farms			
Value of Farm Products Sold	1939	1950	1954	1959 *
(Constant 1954 Prices)	Thousands			
Total	6,097	5,382	4,782	3,704
Under $2,500	4,185	3,295	2,680	1,638
$2,500 and over	1,912	2,087	2,102	2,066
$2,500 to 4,999	1,015	882	812	618
$5,000 to 9,999	585	721	707	654
$10,000 and over	312	484	583	794

* Census data, not strictly comparable with earlier years. A change in definition of farms excluded 232,000 units which could have been classed as farms in 1954 and 1950.

Source : *Statistical Abstract of the United States,* 1961.

[1] Registration according to sales in the U.S.A. and in Canada, and according to standard labour requirements in the U.K.

not even in the U.S.A., where by European standards there are so many large farming enterprises, are the sales values of such enterprises particularly large compared with those customary in other sectors of the economy. In terms of numbers of employees, the great majority of farms both in Western Europe and in the U.S.A. are of sizes to provide 1–2 men with full employment : this also is a very low figure compared with enterprises in other sectors. The few facts I have just given should probably suffice to illustrate the well-known fact that agriculture both in Western Europe and in the U.S.A. consists predominantly of small farming. By whatever methods of comparison or measurement, the average farming enterprise is a small enterprise.

I think I can make bold to state — although I can only speak with certainty as regards my own country — that probably in the great majority of agricultural enterprises at present in existence, the techniques used both in crop and in animal production are well in arrears of present technical potentialities. Investigations in Sweden have shown that at the beginning of the sixties only some 25 per cent of farms which had taken part in official schemes of farm accounting had reached a state of technical development corresponding to what experts considered both possible and economically the most advantageous.

A broad review of the organization of existing farm enterprises of average size is made possible by the collation of calculations of production functions of the Cobb–Douglas type made for total agriculture production in different parts of the world. Such a collation appears in Table II. This table brings out the frequent occurrence of imbalance in the combination of factors of production. Low marginal products for labour and high marginal products for land and capital are very common — indicating need for increased inputs both of land and of capital per employed person.

To summarize, agriculture in Western Europe and the U.S.A. can be characterized as consisting in the main of pronouncedly small enterprises. The medium-sized farm is thus small. There is room for considerable improvement in the technical development of farm enterprises, together with a recombination of factors of production to include increased capital and land investment per employed person, as steps towards considerable improvements in the profitability of farming.

TABLE II

MARGINAL RETURN TO OPPORTUNITY COST RATIOS OF LAND, LABOUR AND CAPITAL SERVICES AS ESTIMATED FROM SELECTED CROSS-SECTIONAL PRODUCTION FUNCTION STUDIES OF FARM-FIRMS *

Location of Sample	Function for	Marginal Return to Opportunity Cost Ratio of Land Services	Marginal Return to Opportunity Cost Ratio of Labour Services	Marginal Return to Opportunity Cost Ratio of Capital Services
Australia, New South Wales	dairy	1·23	0·49	1·40
South Australia	dairy	2·12	0·61	1·53
Austria	mixed farms	0·92	0·54	1·50
Canada, Alberta	wheat, beef	2·58	1·21	1·01
„	cattle, ranches	0·88	1·58	1·38
New Zealand, Canterbury Plain	sheep	2·48	0·85	2·47
Norway, south-east	cereals	23·66	0·29	0·67
„	fodder	11·93	1·36	1·66
„	dairy	—	0·73	0·76
Sweden	mixed farms	1·78	0·31	0·64
„	dairy	—	<0·00	1·41
United Kingdom, England	dairy	—	1·06	1·21
United States, Alabama	crops	4·01	0·38	1·01
„	livestock	—	0·67	0·94
„ Iowa, Illinois	livestock-share	1·93	1·36	1·16
„ Iowa, Illinois	crop-share	1·78	1·23	1·01
„ Iowa, Illinois	full owners	0·98	0·84	1·35
„ northern Iowa	corn	3·08	0·34	0·59
„ southern Iowa	corn	3·27	0·22	1·21
„ northern Iowa	hogs, cattle	—	0·62	0·97
„ southern Iowa	hogs, cattle	—	0·62	1·13
„ Montana	wheat	2·81	0·25	2·00
„	cattle	—	0·32	1·14

* Source : E. O. Heady and J. L. Dillon, *Agricultural Production Functions*, Iowa 1961.

215

The Family Farm and Problems of Factor Mobility

III. GENERAL TRENDS IN STRUCTURAL DEVELOPMENT

(a) Concentration

Farm structure within agriculture has moved from a comparatively stable stage during the thirties and the period of World War II to a very unstable stage during the post-war period. The main indication of this instability is the rapidly diminishing number of the smallest farms over practically the whole of Western Europe and in the U.S.A. In Western Europe, except the U.K., farms of under 10 hectares are diminishing in number, whereas in the U.K. this tendency covers farms up to 120 hectares and in the U.S.A. up to 200 hectares.[1] The rate of agricultural cessation appears to be increasing. It became particularly high during the later part of the fifties, and does not look like slowing down during the present decade. Broadly speaking, however, the number of farms containing 10–20 hectares and upward of arable land appears not to have been affected : if anything, the

[1] *Low Incomes in Agriculture, Problems and Policies*, OECD, 1964.

TABLE III

CHANGES IN NUMBER OF FARMS AND IN AVERAGE FARM ACREAGE IN VARIOUS COUNTRIES

	Changes in Number of Farms, % per annum		Average Farm Acreages			
	Period	%	Year	Hectare	Year	Hectare
Denmark, 1 hectare	1937–1960	– 0·2				
and above	1950–1960	– 0·4	1950	15	1960	16
U.S.A., all farms	1939–1959	– 1·8	1939	24	1959	40 *
	1949–1959	– 2·7				
Netherlands, 1 hectare	1938–1960	– 0·1				
and above	1950–1960	– 0·5	1950	10	1960	10
U.K., 5 acres and	1939–1959					
above	1950–1959	– 0·9	1950	29	1960	32
Sweden, 2 hectares	1937–1961	– 1·2	1937	12		
and above	1951–1961	– 1·9	1951	13	1961	15
Western Germany, 1	1939–1959	– 1·0	1939	8		
hectare and above	1949–1959	– 1·2	1949	8	1959	9

* Crop land, as an average of all farms. Commercial farms are on the average considerably larger and most probably also show a larger relative acreage increase than here shown for all farms.

Source : L. Hjelm *et al.*, *Det svenska lantbrukets effektiviseringsvägar*, SOU 1963–66, Stockholm, 1963, and accompanying basic material.

number of such farms is increasing. The reduction in the total number of enterprises is shown in Table III, from which also the increase in the rate of cessation is evident.

A consequence of the closing down of the smallest farms is an increase in the average farm acreage. This increase, as it affects certain countries, is shown also in Table III. Of particular interest is the exceptionally high rate during the post-war period in the U.S.A. (and to some extent in the U.K.), as compared with other countries where the increase was, in absolute terms, inconsiderable.

As I have already pointed out, average acreage is an unsatisfactory, or in any case incomplete, gauge of the size of an enterprise and of its development. For the purpose of measuring changes in the inputs of

TABLE IV

INCREASES IN SALES VALUE PER FARM

(in percentages per annum, roughly 1939–60, at Constant Prices)

Denmark	2·7
France	1·9
U.S.A.	3·9
Netherlands	2·3
U.K.	3·4
Sweden	1·7
Western Germany	2·4

Source : Material from L. Hjelm *et al.*, *op. cit.*

all means of production, an attempt has been made in Table IV to calculate, for some countries, increases in sales per production unit in terms of percentages per annum. These figures are the result of adding to percentage increases in total production in different countries percentage decreases in the total number of production units. By means of this approximate calculation it is shown that average sales value per production unit increased by 2 per cent to 4 per cent annually, thus by between 50 per cent and 100 per cent totally, during the twenty-five year period between the outbreak of World War II and the present day. Even though increases in farm sizes reckoned in terms of acreage of arable land were only very slight during that period, most countries show large increases in the size of agricultural enterprises in terms of sales.

It must be observed that these changes have taken place alongside considerable decreases in numbers of employed workers. In all countries studied, decreases in number of employees have been substantially greater (1 per cent to 2 per cent greater per annum) than decreases in numbers of farms. Within the industrial sector, concentration tendencies are accompanied as a rule not only by reductions in the number of enterprises but also by increases in numbers of employees per enterprise. Where agriculture is concerned, in the countries within the scope of this study, we discover only the former of these two characteristics — reduction in number of enterprises — the drift away from agriculture on the labour side being so great as to reduce the number of employees per enterprise. The material we have also enables us to conclude that the tendency for farms to increase in size, in terms of increased sales per enterprise, is considerably greater than the tendency towards concentration, measured as decrease in total number of enterprises.

TABLE V

DECREASE IN NEED FOR LABOUR IN VARIOUS BRANCHES OF
AGRICULTURAL PRODUCTION DURING THE PERIOD 1935–65 *

(Man-hours per annum)

Branch of production	Unit	1935	1945	1955	1965
Spring grain	mh/ha	120	70	45	25
Sugar beets	mh/ha	800	650	375	150
Pasture	mh/ha	60	60	50	40
Milk cows	mh/cow	220	140	130	65

* in Sweden.

Source : Compiled from results of Swedish labour studies for the period 1930–64 (U. Renborg).

As we shall presently see, the possibility of running farms on a part-time basis alongside other profitable work has the effect of slowing down the pace at which concentration of agricultural production could take place. It is therefore of primary interest that we should recognize the importance of very small farms of this type in the general agricultural structure, the position being more or less identical in all the countries under present consideration. In 1960, in the U.S.A. about 25 per cent of farms were part-time, and in Sweden about 17 per cent.[1] A very large proportion of farmers, over and above these percentages, also draw substantial incomes from sources outside the farm.

[1] In the U.S.A. this type of farm is defined as one in which the farmer works 100 days or more per annum outside the farm and the family income derived from other sources is larger than the sales value of the farm products.

(b) Specialization

Unfortunately it is very difficult to describe the trend towards specialization of production in agriculture in terms of figures. As a rule the routine collection of statistics does not make possible the direct study of tendencies towards specialization. Furthermore, and quite apart from this difficulty, specialization in many countries does not appear yet to have reached such an extent as to make possible a clear discernment of its particular trends, even if the routine collection of statistics were specially organized for measuring the rate of specialization. I am obliged therefore in this section to confine myself to more or less general observations, with, however, the support here and there of such figures as have, more or less accidentally, come my way.

Agriculture is, as we know, traditionally a diversified enterprise. The main reasons for this diversification are twofold : first, the different kinds of attention needed by different crops and different animals according to season and the biological rhythm of reproduction ; and, secondly, vestiges in human nature of the entirely self-sufficient farmer of old, who had to raise all the varied requirements of his family from his own soil.

The greater part of present-day farms in the countries we have under consideration are still unspecialized, if by specialized we mean the producing of one product or very few different products. On the other hand there is to be observed in most countries an appreciable tendency towards the simplification of administrative organization in individual enterprises, with the effect of a gradual reduction of the number of the branches of production. Business-economic forces also exert powerful pressure upon individual enterprises to engage more and more in specialization. This is connected with the small farm size, which forces them to specialize in order to exploit the advantages of large-scale production, at least with respect to one or some few branches of production. The greater part of these advantages of scale can in fact be exploited if the production of one particular line can be increased to such a level as to provide in itself full employment for one man.

As I have pointed out, there unfortunately exists very little official quantitative indication of specialization trends. We have to be content here with some very rudimentary particulars from incidental studies of limited scope. From a study [1] on tendencies towards

---

[1] L. Hjelm *et al.*, *Det svenska lantbrukets effektiviseringsvägar*, SOU 1963–66, Stockholm, 1963.

specialization in the countries we are considering, conducted in the autumn of 1962 at the Department of Agricultural Economics at the Agricultural College of Sweden, Uppsala, the following general conclusions can be drawn : in those regions of the temperate zone where farming traditionally has been clearly diversified both in crops and in livestock, the post-war period shows obvious incipient tendencies towards reduced diversification on individual farms. These trends are gradually becoming more and more distinctly discernible as time goes on, and, even though during the middle of the sixties the general picture remains still dominated by diversified farming, more and more specialized enterprises are coming into existence, especially those engaging in pure grain production to the complete exclusion of cattle and those engaging exclusively in bacon, egg, broiler or milk production based upon pure hay and silage production. There are also distinct tendencies towards regional specialization.

There have been a few studies on this subject in various countries.[1] As they all show the same general picture I will take examples from my own country : In Sweden the number of purely non-livestock farms has increased during the post-war period, and accounted in 1960 for 13 per cent of total farms. Particularly high is the proportion of farms without milk cows, which has increased from about 5 per cent just after the war to a present figure of somewhat over 20 per cent. Pure pig farming has increased only at the same rate as the reduction in the number of holdings ; nevertheless the number of farms producing large numbers of pigs per year increased greatly during the end of the fifties and the beginning of the sixties. Thus in 1963, 24 per cent of bacon pigs came from producers with an output of over 500 bacon pigs per annum, whereas five years earlier the share of such producers had been only 12 per cent. As to egg production the number of poultry farms decreased during the fifties more than twice as fast as the number of holdings. This change was combined with an increase in the number of layers per farm. Another fact is that enterprises specializing in broiler production are now rapidly on the increase throughout the country.

Similar tendencies are visible in other countries. Farming in the United States has always been more specialized than in Europe, and specific production areas have often been developed ; but, now, even

---

[1] Cf., for example, Britton and Ingersent, 'Trends in Concentration in British Agriculture', *Journal of Agricultural Economics*, June 1964, and a forthcoming OECD report on the interrelationship between income and supply problems in agriculture.

in regions where specialization has been less, agriculture is moving in that direction. In meat production, specialization often means reduction in the variety of stock on individual farms. In crop production also specialization has now gone far. In California, fruit farmers now often confine themselves to one particular fruit, and in the Middle West many farms are changing over to the cultivation of corn alone. In the United Kingdom, the number of entirely non-livestock farms engaged mainly in grain production has increased during the post-war period, and, furthermore, concentration upon one or a few branches of livestock is now often found. A local investigation, conducted in the eastern part of England and concerning a selection of farms in 1958, showed a distinct increase in the number of specialized farms and at the same time a decrease in the number of mixed farms with both crop cultivation and livestock. In the Netherlands there are distinct tendencies towards specialization, with many farms engaged purely in bacon production, purely in poultry, and purely in milk production combined with the production of hay and silage. Tendencies towards specialization are also to be observed, although in a less marked degree, in Germany, France and Denmark. Examples of increases in the number of specialized farms can be shown in those countries, but they do not yet seem to feature noticeably in the official statistics.

The kind of transformation which is going on among ordinary family farms, from diversified to specialized production, can be exemplified by a characteristic study of developments in livestock farming in the Corn Belt and in the Lake States of the U.S.A. In other countries and regions developments are taking place along similar lines to this American case. The development pattern is as follows : [1]

Family farm — comparatively small acreage — diversified production

Family farm — increased acreage — 2–3 kinds of livestock and development of stock

Family farm — larger acreages — 1 kind of livestock

Family farm — larger acreages — part of production of 1 kind of livestock

Alongside this kind of development one can find examples of newly established enterprises specializing in one particular kind of livestock (usually pigs, broilers or layers) and based on very small acreages,

---

[1] L. Hjelm *et al.*, *op. cit.*, pp. 18–19.

normally in fact no more than the grounds of the house. Such enterprises can be either of the family farm type or larger with one or more farmhands. Thus in the U.S.A. there are examples of large enterprises specializing in the production of consumer milk in the close neighbourhood of large consumer centres. There are also examples in the U.S.A. of large meat-producing enterprises of limited acreage situated in the areas of traditionally diversified agriculture. Such large specialized enterprises with very small acreages of their own have expanded rapidly in the U.S.A. during the last ten years. They are to be found to a smaller extent in milk production (500–1,000 cows) and to a somewhat larger extent in egg and poultry production, but mainly in bacon and other meat production. We can take a few examples to illustrate this kind of development. In California the fattening of 90 per cent of feeder cattle has now been concentrated in 275 'feed lots' with an average fattening of 14,000 cattle per year. In Minnesota, in 1955 only 1 per cent of all turkey-breeding was concentrated in enterprises with a production of 50,000 turkeys and more per year. This percentage had risen to 10 per cent in 1960. These specialized enterprises do not yet however occupy any predominant position in the structure of American agriculture.

(c) Dispersion of Production Processes

The final step towards complete specialization is, as I have just stated, reached when only part of the production of a particular type of livestock is completed in each individual enterprise. This tendency towards the division of production branches into phases is only one step in the general trend, common also to industry as a whole, to split production into series of separate processes. Among the effects of this kind of development during the last 50 or 100 years has been the increasing conversion of the self-sufficient family farm into a pure producer of raw materials, with processing functions being gradually taken over by the industrial sector. The system is in fact the same as that in the industrial sector itself in which the manufacture of intermediate products intended for sale is becoming more and more common.

There are many examples of this splitting up of production branches in agriculture, and it is quite evident that this kind of development will go further, thereby increasing the amount of specialization. In most countries covered in the present study, the division of egg production between poultry-breeding establishments, hatcheries and commercial egg producers is already manifest. There are similar developments in many countries for the separation of suckling and bacon pig production from each other through an organized trade in sucklings in conjunction with health control. Well known is the division of meat production in the U.S.A. with the feeding of young cattle in the ranches to the immediate east of the Rocky Mountains and the sale of these living 'semi-manufactures' to the Middle West for final feeding on corn and other fodder. In my own country there is an increasing tendency to divide meat production into three phases : (1) milk producers, who rear calves up to about one month when they sell them to (2) producers, who by virtue of good access to labour and the aid of concentrated foods, either bought or self-produced, are able to continue their rearing up to the age of 6–8 months, and finally (3) sale for feeding up to the age of 18–24 months on farms with good access to natural pasture. In crop cultivation there are corresponding types of specialization. One could take as an example potatoes : seed potatoes are now being produced in districts as free as possible from infection-spreading viruses, and consumer potatoes are being produced in districts with specially suitable soil and close to consumer centres. The tendency towards specialization is clear enough and can be testified by example after example.

(d) Integration

The forces leading to specialization within agriculture are precisely the same as those which make possible the vertical and horizontal integration of the more specialized enterprises. The reduction in the number of production branches per enterprise, in the same way as the enlargement of the enterprises, make it more sensitive than it has been to variations in prices and yields. Well-organized integration can prove a substitute for the distribution of risks inherent in diversified production. Production costs can be saved through common exploitation of production resources, which can be brought about by integration. Interest in integration is of the same sort at other points of connection in the total production chain. Sellers of fodder are in need of guarantees for the certain marketing of their goods ; buyers of agricultural products require even quality and regular delivery of their purchases as well as control over the quality and amounts of deliveries over time. Agricultural entrepreneurs also have interests in common regarding the exploitation of more costly factors of production, such as machinery, management and advisory service etc. In agriculture today definite trends can be seen towards integration both at production and at processing levels. A few illustrations can be given, taken mainly from the integration links at the production level. It is appropriate to start with examples from the U.S.A., where vertical integration first began. The main concern in the U.S.A. has been with various sorts of specialization in small markets dealing with a number of perishable products which require further processing. It is estimated that vertical integration has taken place to the extent of 90–100 per cent in the following markets : sugar, vegetables for deep-freezing, seeds, tobacco, cotton and certain fruits. It appears that 95 per cent of the broiler production is either on a contract basis or owner-integrated. Vertical integration is also great in turkey production (at least 50 per cent). On the other hand there exists very little vertical integration in milk, egg, meat and bacon production.

In the United Kingdom, vertical integration, other than the integration which exists between agricultural co-operation and raw material producers, is not yet very general with respect to the principal staple products. In beet sugar and in other lines of production there is production by contract, as is to be found in many other countries. This is also the practice in egg production and in broiler production, which have become considerably more extensive during recent years. In France in recent years there has been progress in vertical integration, particularly in meat production and more especially in poultry. In 1960 about one-half of broiler production was localised in Britanny, where great processing plants have been built ; about 50 per cent of production is on an organized contract basis with local small producers. In Western Germany integration trends are particularly noticeable in broiler production, egg production and the production of certain crops. In the Scandinavian countries, vertical integration of the American type is not very usual, except in the case of broiler production, but there is close contact between farmers and their co-operatives, which in these countries are particularly strong. Here, as in the Netherlands, there has long been direct contract production of sugar beet, factory potatoes, seeds, certain vegetables and broilers.

Since the war there have also been visible tendencies towards *horizontal integration* in agriculture. In various countries joint use of machinery has been adopted with varying success. Farm accountancy and farm advisory work and other managerial services are often undertaken by larger or smaller farmers' organizations. A few attempts have been made, although without particular success, to organize milk producing co-operatives in Northern Europe. An interesting tendency towards an apparently successful horizontal integration can be reported from Sweden, where most farms contain a certain amount of afforested area, the exploitation of which is more and more being conducted jointly. Such arrangements bring farmers the advantages of mass production in sylviculture, placing them in a useful competitive position *vis-à-vis* other forest owners, private and state.

(e) Relocation of Production

Differences between regions — in terms of economic growth as a whole, the growth of population and the demand for foodstuffs, the responses of different soils to technological changes, and variations in the inputs of different means of production, etc. — give rise in most countries to gradual changes in the patterns of regional production. Examples of tendencies towards the relocation of agricultural production can be found in many countries, and I do not propose to say very much about them here. It should be sufficient for our purposes to say that these gradual changes combined with increasing competition in the food sector and the extension of vertical integration will probably, in the countries here under survey, have a stimulating effect upon the trend towards regional specialization, as well as upon the continued gradual pushing of these regions towards those products in which they now happen to possess the greatest comparative advantage.

IV. ANALYSIS OF THE CONVERSION PROCESS

I propose now to analyse the trends which I have described towards concentration and specialization. I shall proceed first in terms of generally known economic theory. The models from economic theory which concern us for present purposes are described in Figure 2. It requires very little thought to see that the processes of the conversion now going on fit very well into this model.

Changes in relative price between labour and capital, tending towards progressively relatively cheaper capital services, necessitate, in individual enterprises, the replacement of labour by capital (from position 1 to position 3 in Figure 2(a)) ; yet, at the same time the full employment of available labour calls for an increase in size to position 4 (unless the 'part-time' alternative at position 2 is preferred). Further increases in size (along the expansion path) are furthermore

FIG. 2

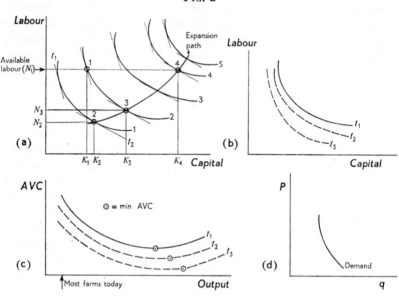

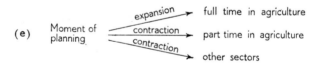

ECONOMIC MODELS FOR DESCRIPTION OF INCREASES IN FARM SIZE AND SPECIALIZATION

often profitable if the present average farm is taken as a basis. The inevitable movement towards increased size in terms of output is abetted by technical developments which in the course of time shift the isoquants of figure 2(a) leftward (t_1 to t_3 in Figure 2(b)).

The statistical picture of the development of an average enterprise is a combination of a sharp increase in the size of certain full-time enterprises, a standstill for stagnant enterprises (often run by elderly farmers) and unchanged or even reduced size of part-time farms. This illustrates the inadequacy of an analysis based only on changes in the size of average enterprises.

Trends towards specialization are engendered by reductions in average costs of particular products at times of increasing production (Figure 2(c)). This relationship is connected with the smallness of the scale on which most production branches are conducted in the greater part of present-day diversified agricultural production. The advantages of larger size in particular production branches, with the compulsion it brings upon small enterprises to increase their production, increase as time goes on and average variable costs shift down (t_1 to t_3 in (c)).

Tendencies towards expansion in particular enterprises induce an upward pressure upon total agricultural output. As we know, there is a strict limit to the possible economic advantages of increased supply, since agricultural products have a very low price elasticity of demand (Figure 2(d)). In the case of industrial countries with the highest living standards, the income elasticity of demand is also very low. This means that farmers in such countries have for their own part a choice of three alternatives — full-time farming ; part-time farming ; or going over to other kinds of work (Figure 2(e)). The flow of labour through these three channels must maintain an equilibrium within limits more or less narrowly prescribed by prevailing marketing conditions. Such are the limits within which structural changes, concentration and specialization take place.

We can proceed a step further in the analysis of the present conversion process in agriculture by applying to it one of Professor Dahmén's analytical schemes for the study of Swedish industrial development during the period between the two world wars.[1] Dahmén bases his analysis upon the presumption that the conversion process has two components, the one positive, the other negative. The positive aspect consists of the rate of development and the manner in which innovations are introduced and utilized in an industry. The negative consists of the rate at which old 'combinations of production factors' are resolved. Each of the two components can in turn be broken down into two parts — production and distribution methods on the one side and goods and services on the other.

The positive side of the conversion process in agriculture of the last twenty years has had impetus from a great deal of activity in the

[1] E. Dahmén, *op. cit.*

introduction of innovations. This period has seen a more or less general conversion from horse to motor traction. Expansion has been very rapid in the area of machinery, where innovations have succeeded one another at increasing speed. In grain cultivation during the last thirty years there has been conversion from scythe to reaper and binder to combine-harvester. In the cultivation of grass-land, the last ten years have seen mowing and the rack-drying of hay together with pitchfork loading replaced by forage harvesters with direct mechanical loading of dry or half dried hay for mechanical insertion into ensilage silos or for mechanical placement upon artificial driers. Sugar-beet cultivation has become largely mechanized. In dairy-farming there has been conversion from hand-milking to machine-milking, with the latest advance being that the milker no longer goes to the cow but the cows go to the central milking parlour. Barn-cleaning and feeding have been mechanized both in dairying and, more especially, in bacon and broiler production. All these innovations have naturally meant great demand for increased investments in mechanical equipment, at the same time however contributing to a great decrease in the amount of labour engaged per production unit. Examples of reduction in the use of labour are shown in Table V, in which norms regarding labour requirements in various important agricultural branches are compared over a period between the middle of the thirties and the present day. The material is from Sweden, but the figures should apply by and large to other countries included in this present study.

In other areas than that of basic production technique, the rate of innovation has also increased sharply during the post-war period. Vitamins and antibiotics have been introduced into livestock rearing. Veterinary practices have improved. The chemical suppression of weeds, insect pests, fungal diseases, etc., has made possible larger harvests at comparatively low costs.

All the innovations mentioned so far have been mainly in the sphere of production method. In agriculture, this has been the main sphere for innovation. Innovations in the form of new goods or types of service having played nothing like so important a role. The only really important exception here is the rapid expansion which has taken place in broiler production : even this, however, has been made possible through the intensity of innovation which I have described on the side of production methods.

Dahmén states that the expansion of Swedish industry during the period between the two world wars was supported in a particularly high degree by innovations in the form of new goods and services, innovations in the sphere of method having played only a minor role. The expansion of agriculture during the last thirty years has on the contrary been due mainly to innovations in production method.

The powerful stream of innovation introduced in this way into agriculture has increased scope for the raising of productivity per employed person. Thus, innovation has greatly decreased the need for labour, and even capital, for the production of one given unit of final product — with the effect that it has become possible to place increased quantities of agricultural products on the market at low price levels. It has thus been possible to shift the supply curve successively to the right by virtue of the introduction of these innovations. In view of the general picture of demand as applied to agriculture, which I have already described, it has not been possible to retain for agriculture the benefits from rationalization thus gained. These benefits have in fact been rapidly passed on to the consumer.

The great increase in innovations in the sphere of mechanization which has taken place has been to the advantage above all of large-scale production. The average cost curve for a given period as described in Figure 2(c) becomes successively lower during the course of that period, its lowest point moving more and more in the direction of increased production. These trends have increased the advantages of increased expansion, thereby changing the profitable proportion between labour and capital so rapidly that only very few enterprises have been able to keep pace with developments.

The negative side of the processes of conversion has worked in the opposite direction. The resolution of old 'combinations of production factors' within agriculture has obviously been taking much time. It is well known that as long as marginal costs decide how an enterprise reacts to changes in circumstances, fixed costs — which quantitatively are very important in agricultural enterprises — play a very small role in the kind of adaptations performed by individual enterprises. In many cases, especially in those of enterprises situated in areas where agricultural production undergoes contraction, salvage values both of land and of buildings fall below their marginal value product in existing production, and the marginal value product falls in turn below acquisition costs of buildings and of land. In the case of older farmers the compensation to be obtained in the form of other employment can often prove lower than the marginal value product of the labour factor. In such situations it is neither unusual nor irrational to exploit the old 'combinations of production factors' as long as they remain. The possibility of the greater part of incomes being earned from sources outside agriculture and the rest from agriculture itself — part-time farming — is also among factors conspiring to preserve the older production structure. All these factors have undoubtedly caused a good deal of sluggishness in the rejection of old 'combinations of production factors' in agriculture.

There are other obstacles to the speeding up of conversion ; for example, the inclination of many entrepreneurs to try to obtain 'utilities' other than those achievable by maximizing profits. Age of farmers in this connection plays a very strong role. Many farm entrepreneurs also have limited knowledge of technical and organizational possibilities and of the very character of the adjustment problem. Also institutional factors of various kinds introduce rigidity into farm structure. In areas where agriculture is expanding — in 'good' areas — efforts of several farmers to expand at the same time can sometimes cause a shortage of land *i.e.* expansion by some farmers curtails the opportunities for expansion by others. The fact also that it is possible for small enterprises to 'keep going' on the basis of what must prove in the longer run uneconomical combinations of factors, can in such areas help to maintain the existing farm structure.

One might, taking perhaps a somewhat drastic view, summarize the negative aspect of the agricultural conversion process by describing it as a powerful inhibiting force, ascribable to the fact that it is difficult to drive farmers all of a sudden out of business.

V. THE FARM OF THE FUTURE

In spite of the conclusion of the previous section, the analysis has shown that the positive factors in the agricultural conversion process are driving developments within agriculture forward towards expansion, concentration and increased specialization. Even though in most countries this kind of development has just begun, it has already become possible to discern certain broad lines in the structure of agriculture of the future. I propose now to outline briefly what can be expected of present developments, in the light of a recently concluded investigation in Sweden.[1] Although this investigation actually concerns the position in Sweden alone, I think that by and large what has been found in Sweden must apply also to all the countries concerned.

[1] L. Hjelm *et al.*, *op. cit.*

The introduction to the Swedish investigation states that the present average farm enterprise is a small enterprise with low profitability, which could be improved both technically and in organization, to economic advantage. The investigation then attempts to give an idea of what farms should come to look like in the future, by describing first the optimum organization of farms according to the techniques of 1960, and then the optimum organization of farms according to technical prospects for 1975.

The investigation was conducted in two stages. The first stage consisted of finding out how far increases in profitability could be obtained by improvements in technique and organization within the frame of present resources. It was found that farm incomes could reach levels equivalent to normal industrial wages only in exceptional circumstances, as long as improvements in the combination of field, forest, capital and labour were restricted to those customary to enterprises with field acreages of between 5 and 50 hectares. This result was obtained with reference to 1960 techniques. The very rough calculations which were made with reference to 1975 techniques showed that a gradual switch towards such techniques could result in a 1 per cent per annum increase in earnings only, on condition of no increase in the size of the farm. This figure is low compared with the roughly 4 per cent per annum increase in the gross national product which is at present estimated.

The aim of the second stage in the investigation was to assess possibilities of improving profitabilities by increasing the size of farms. For this purpose, average available labour force was used as a constant in the determination of amounts of other resources — field, forest, capital — which would have to be combined with it to yield optimum results. Also determined was optimum profitability. Calculations with reference to 1960 technique showed in every case that earnings level with industrial wages for equal labour input could be attained in 'optimum' farms. This, however, requires larger farms than today's average. This means that, as a rule, in order to obtain an income level up to the standard of an industrial worker, a farmer would require at least twice or three times the amount of capital as the present average capital investment in a Swedish farm. Furthermore, a farm would require sales at least three or four times as great as present sales of a medium-sized farm. The rough investigations conducted concerning possibilities with the use of 1975 technique presumes farms having become larger in the meanwhile. On the presumption of technical changes between 1960 and 1975 taking place successively, hourly earnings should increase at the same rate as *per capita* increases in the gross national product.

Conditions for the vitality of a farm are, first, that it should expand to a size well above the average present farm size, and secondly, that it should then continue to expand by successive stages. These results are not astonishing, and they tally well with the theoretical models previously reported.

I shall conclude this paper with an attempt to describe in concrete terms the anticipated future pattern of farming in my own country, on the basis of the present report.

Present-day family farming is developing — as I have pointed out — in two different directions. The first of these directions is towards *part-time farming*. This may prove in some cases successful where there are cars, where roads have improved sufficiently, and where there are employment facilities in other trades or industries within reasonable reach from home. There are many indications that part-time employment is of a transitional nature — a half-way stop on the way from farming to other occupations.

The second of these directions leads towards *the family farm of the future*, in which the farmer and his family together with one or two employees have full-time employment. Among the many plans for future farms considered in the Swedish study,[1] five main types appear to crystallize out. Described as being for family exploitation with labour input of 3,000–5,000 man-hours per annum, they are as follows :

Type 1 : Plainland farms with about 100 hectares of fields with an extensive crop cultivation, combined with pig or other meat production. Annual production of about 400 pigs or 40–50 young fat stock would seem suitable by standards of 1960 technique and price levels.

Type 2 : Plainland farms with a smaller field acreage but with milk production from 30–50 high-yield cows. It is possible that in many cases in Sweden specialized milk production based purely on forage cultivation would be included under this type.

Type 3 : In certain regions, combination of farming and forestry, with such farms consisting, possibly, of 50 hectares of farm land and 100 hectares of forest. Farming would consist of extensive grain cultivation, often combined with an annual production of 30–40 young fat stock, by standards of 1960 technique and prices. With more intensive livestock exploitation, *i.e.* bacon or milk production,

---

[1] L. Hjelm *et al.*, *op. cit.*

such combination of agriculture and sylviculture could be based on smaller acreages. Even in such cases milk production could be based on forage cultivation alone.

Type 4: In forest country, pure family sylviculture based on 200–400 hectares of forest, at 1960 techniques and price levels, either worked by the family alone or horizontally integrated, as earlier described.

Type 5: Farms with very small acreage, generally no more than a farmyard, engaging in specialized livestock production. Such farms, by 1960 technical standards, with 3,000 man-hours annually, could yield for example 4,000–5,000 pigs, 80,000–100,000 broilers, or eggs from 5,000 layers per annum. Such types of specialized farms should very largely become links in a system of vertical integration.

Alongside these part-time and whole-time farms of a family character, there appear also the patterns of two types of *large farm*. One of these is of the more traditional character, but highly mechanized. The other is a large farm, specializing in livestock production.

We in fact already have, in most parts of my own country, and in many other countries, examples of all these types of farm. They vary somewhat in size, and their particular sizes are related of course only to particular techniques and particular times. In the course of time their size, in the sense either of yield or of acreage, should come gradually to increase. My own belief, which I think is shared by others working in this area, is that the tendencies towards concentration and specialization at present to be observed in the structure of agriculture will lead us towards and gradually beyond the development of farms of the types and sizes here described.

Chapter 11

THE MODERN FAMILY FARM AND ITS PROBLEMS: WITH PARTICULAR REFERENCE TO THE UNITED STATES OF AMERICA

BY

GLENN L. JOHNSON

Michigan State University, U.S.A.

I. INTRODUCTION

IF one were to define what is meant by 'modernization of the family farm', consolidation and/or farm enlargement would be rather high on one's list of necessary requirements for modernization. So, too, would technological advance, especially the land and labour-saving varieties. Land-saving technology would be required to keep total agricultural production expanding to feed growing populations, while labour-saving technology would be required to increase *per capita* incomes in agriculture. These two, in turn, would require a substantial amount of off-farm migration as part of the modernization process. Another requirement for modernization would be an increasing level of education, both general and vocational, in order that farmers may handle the larger farms and the more advanced technologies which come with modernization. Still another requirement would be the existence of an institutional environment which would permit a modernized family farm to finance itself and prosper.

II. MODERNIZATION OF U.S. FARMS, 1946 TO 1965

Imagine an agricultural economist in the United States in 1946 and blessed with full knowledge of the oncoming very substantial rates of farm consolidation, publicly supported technological advance, off-farm migration, improvements in both public vocational and general education and aware of the expansions and changes to take place in

U.S. agricultural credit, price support and subsidy programmes from 1946 to date.

Such an economist, in 1946, would have expected the next twenty years to strengthen the family farm in the U.S. to such an extent as to virtually eliminate the agricultural problem. Actually, modernization since 1946 has been rapid but there are still agricultural problems. There has been a major re-structuring of the industry involving consolidation of farms, elimination of small farms and expansion in numbers of large farms. From 1944 to 1959 over 1·4 out of 5·4 million farms have been eliminated. In the same period, farms producing over $10,000 (1954 dollars) increased from 449 to 828 thousand while those producing less than $2,500 decreased from 3·2 to 1·9 million.

TABLE I

NUMBER OF FARMS BY VALUE OF FARM PRODUCTS SOLD, CO-TERMINOUS UNITED STATES, SELECTED YEARS, 1944–64 *

Value of Farm Products Sold	Number of Farms				
	1944	1949	1954	1959	1964 †
(dollars)	(thousands)	(thousands)	(thousands	(thousands)	(thousands)
10,000 or more	449	497	598	828	955
5,000–9,999	741	739	725	693	535
2,500–4,999	1,044	944	869	654	460
Less than 2,500	3,264	3,067	2,606	1,922	1,522
Total	5,498	5,247	4,798	4,097	3,472

* Valued at 1959 price received by farmers.
† Preliminary estimate. Figures for 1964 adjusted by Dale E. Hathaway using ERS data.

Source : 'The Expanding and Contracting Sectors of American Agriculture' by R. Nikolitch. A Forthcoming U.S. Department of Agriculture, Agricultural Economic Report.

Technological advance has been rapid. From 1945 to 1963, the number of grain combines increased from 375,000 to 1,020,000 while corn-pickers increased from 168,000 to 820,000 and tractor numbers approximately doubled. Fertilizer use increased from over 14 to over 25 million tons from 1945 to 1961.

Man-hour requirements fell from 53 to 12 hours per hundred bushels of corn between 1945 and 1960. For a bale of cotton the decrease was from 146 to 52 hours. Less spectacular decreases occurred for a hundredweight of milk (2·6 to 1·3 hours), for a hundredweight of beef (4·0 to 2·9 hours) and for pork (2·7 to 2·2 hours). Extensive off-farm migration took place. Percentage changes in the farm population for five-year periods were –0·9 for 1945 to 1949, –17·5 for 1950 to 1954 and –13·0 for 1955 to 1959. Net out-migration percentages for the same five-year periods were 13·9, 24·2 and 23·9 . The farm labour force percentage decreases were – 0·4, – 12·8 and – 12·4 for the same three five-year periods.

Educational levels also advanced quite rapidly. From 1950 to 1960, the median level of school years completed by the rural popula-tion increased 2 per cent. In 1962 almost 63 per cent of the farm operators under 35 had eight or more years of education while only 53·5 of those between 35 and 44 years had over eight years of educa-tion which indicates more accurately the changes in levels of education for those operators recently of school age.

In the 1945–65 period, the U.S. family farm maintained itself despite fears that contract farming and vertical integration schemes would eliminate the family farm. The general evidence to date is that while U.S. farms are getting bigger, as measured by output per acre and per animal units, they remain largely family operated. From 1944 to 1959 the proportion of U.S. farms hiring more than the average amount (1·5 man years) of family labour used per farm actually decreased from 5·5 to 4·3 per cent. The percentage of total farm sales produced by farms employing less than 1·5 man years of work also increased, in this instance from 66·5 to 70·1 per cent.[1]

The ratio of food prices to wage rates in the U.S. is one of the lowest on record. Output more than meets domestic needs and the inter-national commitments of the United States. By world and historical standards, the level of living of the American farmer is one of the highest attained and is matched or exceeded only by levels of living found in Australia, New Zealand and Denmark. Still further, the political and social environment in which the American farmer operates is one permitting him much in the way of freedom to move away from farming, to educate his children for governmental, professional and business work, to operate his business, to participate in political decisions, and to develop the institutional structure in which he exists.

If this Conference were a platform for Western propaganda, this paper could be closed by concluding that the U.S. and other modern Western agricultural economies are highly modernized (as demonstra-

[1] Orville L. Freeman, Secretary, U.S. Department of Agriculture, Statement before the Subcommittee on Family Farms, House Committee on Agriculture, 11 July 1963.

ted with data above) and are producing efficiently a large volume of farm products at low cost.

However, this is not a platform for propaganda. Our task, as social scientists, is to examine the problems of the family farm in a modernized agriculture. In order to do this we should look still more

TABLE II

NET INCOME ORIGINATING IN AGRICULTURE, CAPITAL GAINS AND LOSSES, GOVERNMENT PAYMENTS AND TOTAL, CURRENT DOLLARS, U.S. 1946–60

(in millions of dollars)

Year	Net Income Originating in Agriculture	Capital Gains and Losses	Government Payments	Total
1946	19,416	11,729	683	31,828
1947	20,034	12,342	277	32,653
1948	22,425	1,655	227	24,307
1949	17,141	– 4,158	162	13,145
1950	18,175	16,190	249	34,614
1951	20,793	13,197	250	34,240
1952	19,879	– 5,628	239	14,490
1953	17,632	– 5,829	186	11,989
1954	16,939	2,380	224	19,543
1955	15,965	2,235	200	18,400
1956	15,905	8,543	485	24,933
1957	16,081	11,014	891	27,986
1958	18,563	14,611	990	34,164
1959	16,336	407	622	17,365

Source : *Farm Income Situation*, July 1960. Capital gains and losses as computed by G. L. Johnson, modified by Dale Hathaway and brought up to date by Willard Sparks, Michigan State University.

closely at what modernization has done to and for agriculture in the United States. When we do this, we note that there still remain in American agriculture a hard core of subsistence, poverty-stricken farms. For the non-farm population of the United States, 17 per cent of the families have less than $3,000 gross income. In farming the figure is over 43 per cent. The median income among farm families is $3,400 while the median income among non-farm families is $6,400. In the commercial sector of the U.S. agricultural economy there are also serious problems. United States commercial agriculture is characterized by more output than can be sold at prices which will simultaneously return marginal value products for capital, labour and investments in land equal to marginal rates of return earned in the non-farm economy (see Table V and associated text). Furthermore, many of the years since 1946 have been characterized

TABLE III

CHANGES IN REAL WEALTH DUE TO PRICE CHANGES, MEASURED IN
CONSTANT 1959 DOLLARS, 1940–60 *

(in billions of dollars)

Year	Farm Real Estate	Live-Stock	Mach. and Motor Vehicles	Crop Inventories	Money Assets and Money Liabil.	Total
1946		2·4	−1·7	0·1	−1·5	−0·7
1947	−8·7	−0·2	0	1·7	−0·5	−7·7
1948	−2·5	0·7	0·6	−3·8	−0·1	−5·1
1949	0·1	−2·1	0·6	−0·6	0·1	−1·9
1950	11·8	4·4	0·3	0·8	−0·1	17·2
1951	0·6	0·2	−0·2	0·5	0	1·1
1952	−1·3	−5·6	0·1	−0·5	0	−7·3
1953	−2·5	−2·9	0·1	−0·7	0	−6·0
1954	2·4	−0·7	−0·1	−0·3	0	1·3
1955	3·7	−0·5	−0·5	−1·0	0	2·7
1956	4·7	0·4	0·6	0·1	0·2	6·0
1957	3·6	3·0	0·3	−1·6	0·1	5·4
1958	7·6	3·3	0·9	0·3	0	12·1
1959	2·6	−2·7	0·6	−1·3	0·1	−0·7

* These data are computed from a table furnished by David Boyne, University of Chicago.

by substantial capital gains and losses, whether those losses are figured on a real or a current dollar basis. Still further, the capital losses imposed upon American agriculture are considerably less than they would have been had large-scale, government-financed assistance for American agriculture not been in operation. If the income flow to agriculture from government payments amounts to as much as $3 billion a year and if the interest rate is 5 per cent in agriculture, then as much as $60 billion worth of property value may depend upon the flow of government payments. The $60 billion figure compares with the value of the entire U.S. farm production plant of something over $200 billion.

Political pressures for farm price supports and subsidies have some

Table IV

BUDGETED EXPENDITURES OF U.S. GOVERNMENT ON AGRICULTURE
AND AGRICULTURAL RESOURCES, 1961–64

Programme or Agency	Budget Expenditures			
	1961	1962	1963	1964
	(million dollars)			
Farm income support and production adjustment				
Commodity Credit Corporation :				
Price support, supply and purchases	1,331	2,051	2,853	1,808
P.L. 480 (Titles I and IV)	1,455	1,484	1,460	1,570
National Wool Act	61	65	63	106
International Wheat Agreement	76	90	78	72
Transfer of bartered commodities to supplemental stockpile	201	193	86	62
Removal of surplus agricultural commodities (excl. Food Stamp Programme)	203 *	201	174	173
Food Stamp Plan				
Present programme		14	51	27
Proposed legislation				24
Sugar Act	72	80	82	84
Other	39	67	96	120
Conservation reserve and land use adjustment	363	344	314	321
Subtotal	3,801	4,591	5,256	4,367
Conservation of agricultural land and water resources :				
Agricultural conservation programme, CCC loan for ACP, and emergency conservation	251	260	223	209
Soil Conservation Service (incl. watershed and Great Plains programmes)	146	157	183	199
Other		9	8	5
Total	397	426	414	413
Financing rural electrification and telephone	301	303	340	421
Proposed legislation				151
Financing farming and rural housing				
FHA				
Present programme	349	240	322	174
Proposed legislation				55
Research and other agricultural services	324	341	401	415
Farm Credit Administration		-6	-2	2
Subtotal administrative budget	5,173	5,895	6,731	5,696

* Incl. Food Stamp Plan.

TABLE V

RETURN PER HOUR, OPERATOR AND FAMILY LABOUR, AND RETURN PER $100 INVESTED, TYPICAL COMMERCIAL FAMILY-OPERATED FARM, SELECTED PERIODS

Type of Farm and Location	Returns per Hour of Family and Operator's Labour						Return per $100 Invested 1961
	1946–1950	1951–1952	1953–1955	1956–1959	1960	1961 *	
Dairy farms :							
Central Northeast	0·72	0·82	0·64	0·78	0·72	0·80	3·94
Eastern Wisconsin :							
Grade A	NA	1·06	0·60	0·80	0·86	1·10	4·77
Grade B	0·40	0·54	0·22	0·29	0·33	0·50	− 0·20
Western Wisconsin, Grade B	0·46	0·74	0·40	0·57	0·61	0·77	2·24
Dairy-hog farms, South-eastern Minnesota	0·65	0·87	0·64	0·61	0·41	0·65	1·54
Corn Belt farms :							
Hog-dairy	1·04	1·08	0·82	0·88	0·62	1·00	3·39
Hog-fattening-beef raising	0·88	0·92	0·41	0·58	0·32	0·53	1·11
Hog-beef fattening	2·22	1·77	0·96	1·24	0·61	1·20	4·38
Cash grain	2·21	2·15	1·54	1·10	1·04	1·55	5·47
Poultry farms :							
New Jersey, egg-producing	0·95	0·82	0·45	0·10	0·78	0·65	1·33
Delmarva, broilers	NA	1·66	1·35	1·54	2·25	2·11	10·83
Cotton farms :							
Southern Piedmont	0·38	0·60	0·46	0·46	0·34	0·60	5·24
Mississippi Delta :							
Small	0·60	0·52	0·61	0·48	0·47	0·57	6·58
Large-scale	—†	—†	—†	—†	—†	—†	11·98
Texas :							
Black Prairie	0·82	0·68	0·67	0·39	0·37	0·59	3·93
High Plains (non-irrigated)	2·17	1·48	0·15	1·85	2·94	4·91	19·42
High Plains (irrigated)	3·40	4·70	2·47	3·79	3·51	6·64	17·39
San Joaquin Valley, Calif. (irrigated) :							
Cotton-specialty crop	NA	—†	—†	—†	—†	—†	3·89
Cotton-general crop (medium-sized)	NA	—†	—†	—†	—†	—†	7·53
Cotton-general crop (large)	NA	—†	—†	—†	—†	—†	7·70
Peanut-cotton farms, Southern Coastal Plains	0·56	0·63	0·66	0·75	0·82	1·06	15·14
Tobacco farms :							
North Carolina Coastal Plain :							
Tobacco-cotton (medium-sized)	0·86	0·94	0·79	0·78	0·93	1·09	8·22
Tobacco-cotton (large)	0·93	1·04	0·76	0·81	1·07	1·29	7·71
Tobacco (small)	0·66	0·70	0·66	0·68	0·84	0·93	10·14
Kentucky Bluegrass :							
Tobacco-livestock, inner area	1·02	1·24	1·11	1·31	1·14	1·14	4·72
Tobacco-dairy, intermediate area	0·42	0·51	0·39	0·50	0·46	0·58	0·22
Tobacco-dairy, outer area	0·57	0·70	0·57	0·78	0·71	0·81	3·91

* Preliminaries.
† Not appropriate as return to labour is a relatively small proportion of residual returns.

Source : USDA Reports on Typical Family-operated Farms.

TABLE V (*continued*)

Type of Farm and Location	Returns per Hour of Family and Operator's Labour						Returns per $100 Invested 1961
	1946– 1950	1951– 1952	1953– 1955	1956– 1959	1960	1961 *	
Spring wheat farms :							
Northern Plains :							
Wheat-small grain-livestock	2·04	1·46	0·86	1·26	1·21	− 1·15	− 2·78
Wheat-corn-livestock	1·34	0·98	0·46	0·78	1·03	0·95	4·64
Wheat-roughage-livestock	1·46	0.88	0·64	0·55	1·18	− 0·26	− 2·66
Winter wheat farms :							
Southern Plains :							
Wheat	3·07	3·23	1·36	1·75	2·51	2·52	8·09
Wheat-grain-sorghum	1·74	0·93	− 0·58	1·44	2·95	3·05	10·54
Pacific Northwest :							
Wheat-pea	4·22	3·92	3·62	2·61	1·33	1·47	4·25
Wheat-fallow	4·42	4·02	2·45	3·14	2·07	1·87	4·61
Cattle ranches :							
Northern Plains	0·98	1·38	0·21	0·45	0·50	0·84	3·72
Intermountain Region	1·40	2·32	0·51	1·78	1·31	1·62	7·14
Southwest	0·76	− 0·28	− 1·85	− 0·09	0·35	0·67	4·05
Sheep ranches :							
Northern Plains	1·12	2·04	0·31	1·50	1·01	0·71	3·21
Southwest	0·34	− 0·64	− 2·23	− 0·05	0·61	0·09	3·35

* Preliminaries.

† Not appropriate as return to labour is a relatively small proportion of residual returns.

Source : USDA Reports on Typical Family-operated Farms.

of their origins in low rates of return in agriculture to labour, capital investments, operating expenditures and land investments. As noted earlier, such returns are below levels in the non-farm economy in the U.S., despite modernization. Similar economic causes appear to be behind the political pressures now being exerted by European farmers for price protection and subsidies. Thus, there is little empirical or historical basis for the generalization that modernization of the family farm is capable of solving the problems characteristic of Western agriculture. To the contrary, as will be argued later, there are even reasons for questioning the viability of the family farm in the rapidly modernizing, changing economies of the Western world. Thus, countries inclined to depend upon modernization of the family farm as a solution to their agricultural problems should examine, carefully, both (1) the conditions which produced the vast modernization of the family farm from 1946 to 1965 in the U.S. and in other Western agricultural systems and (2) the results which have been obtained by that modernization, only part of which are satisfactory. Such examination will not lead to conclusive answers as to how to proceed ; it will lead, however, to what I regard as healthy questions which, if answered, may lead to organizations of agriculture superior even to those existing in Western Europe, Australia, New Zealand and North America.

III. TOOTEL'S CURRENT APPRAISAL OF THE POSITION OF THE FAMILY FARM IN THE U.S.

With the family farm surviving as well as it has, current U.S. thought about the family farm often has to do with continued efforts to meet threats to its survival. Questions are asked less often about its functions and alternative production units for performing these functions. In this section, a recent paper will be reviewed briefly as a preliminary to the following section on family farm functions and alternative production units for performing these functions.

Robert Tootel, Governor of the U.S. Farm Credit system, has considered five threats to the family farm and four prerequisites for their survival.[1]

The *threats* considered are : technological advance, the availability of migratory workers to non-family type farms, contract farming, integrated operations and the problems of entering farming (including intergeneration transfer of ownership). He feels that the family farmer can handle technical advances and that the decreasing availability of low-cost migrant labour removes the threat of large-scale farms based on special access to cheap sources of labour. Contract farming remains, in Tootel's view, a somewhat vague threat, the seriousness of which depends on who does the contracting. While the threat of integrated operations is being studied by the newly created National Commission on Food Marketing, Tootel is not conclusive concerning the importance of this threat. As to the problems of entering agriculture and of intergeneration transfer of family farms, Tootel doubts that the family farm corporation will solve the problem or that non-family farm corporations will replace the non-incorporated family farm. He looks instead to rental arrangements, family assistance to beginning farmers and estate planning as ways of meeting the threat of these problems.

Tootel's list of *prerequisites* for survival of the family farm include : education and research, farmer co-operatives, adequate financing and public policies designed to preserve the family farm.

[1] R. D. Tootel, Conference of Production Credit Association Managers and Directors, Louisville, Kentucky, 22 March 1965.

Publicly supported research is required, according to Tootel, to expand modern technology for family farms while education is required to ensure its distribution, use and management. Both, he thinks, are being and can be provided.

Tootel looks to co-operatives (marketing and purchasing), it seems to me, as a means of meeting the threats of contract farming and vertical integration. While financing needs are handled in part by co-operatives, Tootel discusses the threat of inadequate financing separately from co-operatives. He feels that this threat can be handled by the agency he heads — that agency working, of course, alongside a substantial number of independent farm credit institutions.

Tootel's statements on needed farm policy to maintain the family farm are more than a little vague and are quoted here.

The family farm has dominated in this country since colonial times in no small part because it was desired as a matter of public policy. By the same token, certain industries or economic sectors of the economy have prospered and developed because of public policies favoring them. The family farm will survive in America only if we, the people, believe that it should survive. Our farm population now makes up only 7 per cent of the total. It may be less than 5 per cent within ten years. What effect this may have on public policy as it relates to agriculture (especially in view of voting reapportionment) is a big question. The welfare of our agriculture in the future, whether it be predominately family farm agriculture or otherwise, will be in great measure dependent upon how well the mass of consumers understand our agriculture, its problems, its needs, and its significance to them.

This vagueness, it seems to me, grows out of uncertainty concerning the basic nature of the problem faced by the modern family farm in the extremely dynamic environments of the modern countries of the Western world. Tootel's uncertainty is shared, I feel, by most of us. What are the functions performed by the family farm as a basic structural producing unit of the modern Western agricultural economy? Why do such units persist? Are there better basic producing units? Can other basic producing units be used in developing agricultural economies free of the over-investment, surplus production and substandard returns and resultant political pressures, so characteristic of the agricultural economies of the modern Western countries, without falling into problems, so characteristic of the controlled agricultural economies of non-Western world countries, either modern or under-developed? Though I go to bed thankful that the Western world has problems of excessive agricultural production rather than of shortages, I cannot rest content with our agricultural problems. And further, when I work with the agricultural economies of the under-developed world I wonder why I cannot envisage organizations of agricultural economies superior to those of the U.S., Canada, Western Europe and Oceania? Such questioning inevitably involves the problems of the family farm in modern economies and my thinking about these problems gets deeper and harder than that reflected in Tootel's presentation which was discussed here because it is rather characteristic of the thinking of many economists concerning the family farm.

IV. THE NATURE AND FUNCTION OF THE FAMILY FARM AND ALTERNATIVES TO IT

(i) *Nature of the Family Farm and of its Environment*

Modern Western agricultural economies based on family farms, as the fundamental unit of production, operate in environments characterized by :

(a) rapid change with respect to technology, population, international demand, domestic institutions, domestic demand and domestic and world supplies of farm products (often produced by large numbers of competing producers) ;

(b) imperfect knowledge on the part of family farm managers of the changes noted in (a) ;

(c) substantial divergencies between acquisition costs and salvage values of durable inputs including operator's labour, these divergencies being due to transportation and other costs (non-monetary as well as monetary) of transferring purchased inputs from seller to buyer ;

(d) excesses of rural births over rates required to maintain the farm labour force ;

(e) essentially constant returns to scale when farm operators can, in fact, have control of all inputs so that a fixed input does not trigger diminishing marginal returns.[1]

[1] This generalization is borne out by numerous Cobb–Douglas studies of individual farm data and by many budgeting studies which show (after enough size is obtained to handle large initial pieces of equipment) essentially constant returns.

(ii) *Some Theoretical Consequences of the Environment in which Western Family Farms Operate*

Imperfectly informed managers of producing units should be expected, under conditions (*a*) to (*e*) listed above, to : [1]

(*a*) make mistakes of both over and under-investment in durable factors of production ;

(*b*) be able to correct certain mistakes of under-investment ;

(*c*) be unable to correct certain mistakes of over-investment because the durables simply become fixed investments which earn more in their 'mistaken' location than they can earn 'net' in the next best location after paying moving costs, even though their earnings do not cover what was originally paid for them ;

(*d*) be able to minimize losses (relative to original cost) on certain mistaken over-investments only by further investments in other durables and expenditures which further increases production ;

(*e*) be unable in the case of simultaneous over- and under-investment in two or more durables to minimize losses on the joint over-investment without producing more than it would have been profitable to produce had the over-commitment of one durable not taken place.

Examination of the five adjustments to mistakes listed above will indicate that mistakes, even if randomly committed, lead eventually to over-commitment of durables to production. The word 'over-commitment' means in such quantity that neither the over-committed resource nor its product can be sold at prices which permit recovery of the original cost of the resource. Under these conditions, over-production and sub-standard marginal returns exist.

To recapitulate, under conditions of change, imperfect knowledge and acquisition costs in excess of salvage values, mistakes are made

[1] The theory to support these generalizations will be found in various stages of development in the following references.
Glenn L. Johnson and Lowell S. Hardin, 'Economics of Forage Evaluation', *Purdue Agr. Exp. Sta. Bul. 623*, April 1955; Glenn L. Johnson, 'Supply Function — Some Facts and Motives', *Agricultural Adjustment Problems in a Growing Economy*, eds. E. O. Heady *et al.*, Iowa State College Press, Ames, Iowa, 1958; Clark Edwards, 'Resource Fixity and Farm Organization', *Journal of Farm Economics*, 41 : 747, Nov. 1959; Glenn L. Johnson, 'The State of Agricultural Supply Analysis', *Journal of Farm Economics*, 42 : 435, May 1960; Glenn L. Johnson, 'Some Basic Problems for Economists and Statisticians Arising from U.S. Agricultural Policies', *Manchester Statistical Society*, read 11 Nov. 1959; *Economics and Management in Agriculture*, Warren Vincent, ed., Prentice-Hall, New York, 1962, Chapters 6 and 7.

which lead to fixed investments which are associated, in turn, with capital losses and sub-standard marginal returns in the sense that production is in excess of that which can be sold at prices which equate marginal value products with marginal factor costs in the short run. If changes occur continuously, mistakes are made repeatedly and the series of mistakes maintains the short-run conditions indefinitely until, to use Lord Keynes' phrase, the long run, when we are all dead.

(iii) *Speculation Concerning the Role of the Family Farm*

If, at this point, we want to become speculative we can ask about the role the family farm plays in modern Western economies. One thing is clear. Family farm managers do make the investments which account for the high levels of output typically not saleable in the aggregate at prices which will cover acquisition costs.[1] Included among the resources over-committed by family farm managers are, typically, the operators of family farms themselves, along with their livestock, machinery, buildings, wells, fences, irrigation systems, etc.

It appears that these managers make mistakes of over-commitment *so rapidly* that production is maintained at levels so *high* that prices remain *so low* that larger than family farms have not been able to develop beyond the point where they have to compete in the open market simultaneously for both labour and capital. This, in turn, raises the question as to whether managerial capacity really limits the size of business which can be established in agriculture or whether the limitation arises out of price pressures which restrict the firm's command over capital and non-family labour. With the passage of time and added observations in Western Europe and in under-developed countries, I am inclined to conclude that it is price pressure more than high managerial requirements per unit of output which limits the size of farms to that which one family can handle.

With an excess of births over replacement needs in agriculture, it is easy for imperfectly informed young men to commit themselves and their families unwisely to agriculture. Even if they commit to agriculture at age 25 at returns comparable to those in industry, a 3 to 4 per cent per annum increase in industrial wage rates soon outstrips increases in returns to labour in agriculture. It is simply hard to

---

[1] These are investments in actual factors (such as tractors, antibiotics, seeds, etc.) of production, not in something mythical such as new technology, or improvements in the human agent.

increase returns to labour in agriculture with low income demand elasticities for food, new labour-saving technologies which decrease the marginal value productivity of given quantities of labour and new land-saving and/or feed-saving technologies which make it easy to expand agricultural production to keep pressure on farm product prices despite outmigration of labour. Studies by my students and myself do indicate that young men do enter and leave U.S. agriculture on the basis of the ratio between agricultural and manufacturing wage rates. However, the farmer who discovers, on reaching middle age, that his friends who left agriculture now have higher returns to their labour than he to his, cannot just leave agriculture and obtain the wages they obtain. After all, they have seniority, union status and valuable experience to increase their real productivity. The migrating middle-aged farmer has to be content with wages paid unskilled workers in retailing, the service industries and local government. Studies by my students and myself and by others show that older farmers leave agriculture according to the ratio of farm wages to wage rates in laundries, retailing, etc. In other words, farmers who over-commit themselves to farming continue to produce products to depress prices until the MVP of their labour falls to the level of the lower wage earners in the economy. Consequently, massive out-migrations of U.S. farm labour as noted above fail to equate overall earnings of labour between the farm and non-farm sectors. How can larger than family farms which have to recruit labour in the open market compete with the output of 'trapped' family farm operators?

When capital commitments are examined a similar picture emerges. Acquisition costs exceed salvage values for many reasons including transportation and installation costs not to mention distinct suspicion of second-hand machinery on the part of buyers. Repeated errors of over-commitment keep tractors, orchards, breeding stock, drainage ditches, wells, buildings, improved pastures, etc., producing products at prices too low to permit large-scale farmers to enter the major capital markets for the massive quantities of capital required to set up modernized businesses to employ, say, 30 men.

A cynic might even assert that the family farm is an institution which functions to entice farm families to supply batches of labour and capital at substandard rates of return in order to supply the general economy with agricultural products at bargain prices.

(iv) *Speculation Concerning the Viability of the Family Farm*

If the above thoughts concerning the family farm are correct, the viability of the family farm depends on the conditions which create the problem of low factor returns in modernized agriculture. If this is so, we should find the family farm disappearing or not developing where the problems of simultaneous low returns to capital and labour have been solved or where special access to large quantities of one or both at substandard rates is available. Rough checks of this nature are easy to make :

(a) Most large-scale plantation agriculture has been based on special access to labour in colonies, in the pre-civil war south of the U.S. and in the post-war south of the U.S.

(b) Large-scale dairy operations in California have access to special price advantages not open to farmers in states where family farm dairying predominates.

(c) State-owned oil palm plantations appear advantageous in Eastern Nigeria where the state collects the marketing board revenues ; however, privately owned oil palm plantations are not developing or are being planted to rubber while oil palm production remains mainly in the hands of heavily taxed smallholders (primitive family farmers).

(d) Certain U.S. beef raising and feeding operations have involved more capital than could be handled by families of ordinary wealth. The same is true of certain modern beef feeding arrangements. Such operations cannot be started by family farmers and, hence, do not face competition from products produced by over-committed family labourers and capital.

One of the troublesome questions is that of whether it is managerial capacity *or* over-production and consequent low prices which limits the size of farm. This must be left as an unanswered question in a brief question raising paper such as this ; however, it is difficult to isolate managerial inefficiencies on larger farms which are as detrimental to farm expansion as the necessity of paying really competitive wage rates for large numbers of really competent labourers — men who will work as hard and take as much responsibility as either a factory foreman or a family farm operator.

One can now speculate concerning the impacts of policies and programmes which would solve the problem of substandard returns to labour and capital (including investments in land and acreage control allotments). Would not such problems make it possible for larger than family farms to replace family farms? Would standard returns to land and capital destroy the family farms' reason for being?

We can also ask if it would be such a bad thing to have an agricultural economy in which labour and capital earned rates comparable to what they would earn elsewhere? Farm products' prices in such an economy, I suspect, would be higher and food would be scarcer than in the modern Western economies since the mid-fifties. However, food prices would probably not be as high or food as scarce as in the present day non-Western economies which are characterized by under-investment in agriculture, high food prices, food scarcities and low returns to agricultural labour, as well. The difference might be that this hypothetical economy we are discussing would be free of the problems faced by both the modern Western agricultural economies and by the non-Western agricultural economies.

V. CONCLUDING REMARKS

I am troubled by thought of the possibility of giving up the family farm as the basic unit of production in a modernized agriculture. Yet, I had similar misgivings about the corner grocery store which has almost entirely disappeared in the U.S. since my undergraduate days. It, too, was a family enterprise in which Mom and Dad worked long hours and in which they invested their capital at low returns. I thought it would be a great loss to U.S. society to have these independent business-men replaced by chain stores with their hired employees. However, now, twenty-odd years later, I am not that sure that the development has hurt our society. The overworked and underpaid people who staffed the grocery stores had little time or money to support churches, schools and community projects. By contrast, the manager of the local supermarket has a good income. Further, the supermarket employees are paid competitive salaries or wages. Equally important, they have fringe benefits and working hours which permit them to be 'good citizens'. While I am not advocating the demise of the family farm, I am suggesting that we look at both it and alternatives to it unblinded by stereotypes.

After all, proud as I am of the productivity of the family farm, economies based on it do have some tough agricultural problems and, as I can attest from having lived on and with them, the family farm does have a few drawbacks such as low incomes, difficulty in supporting community facilities, long hours, isolation and lack of fringe benefits.

If we return briefly to Tootel's analysis of the threats to and prerequisites for survival of the family farm, it is now a little easier to see the appropriateness of his vague section on prerequisite policies. Our analysis does indicate that certain kinds of contract and vertical integration efforts might solve both the price and over-commitment problems for individual farms which might, in turn, remove the restrictions on size which probably maintain the family farm. Co-operatives, in Tootel's view, have an important function to play including, perhaps, contracting and vertical integration. Contract and vertical integration controlled by farm co-operatives would be more acceptable to many than if controlled outside of agriculture. In fact, such arrangements might lead to an agricultural economy superior to the present one. Whether the family farm is to be desired depends on our ability to devise superior alternatives. What should be done is complicated by the lurking suspicion that the existence of the family farm depends on failure to solve the price and/or over-production problems of modern Western agriculture. With neither superior basic production units yet designed and with the price and/or over-production still unconquered, it was a good thing that Tootel was vague on policy prerequisites for family farm survival.

Chapter 12

THE MODERN FAMILY FARM AND ITS PROBLEMS: WITH PARTICULAR REFERENCE TO THE FEDERAL GERMAN REPUBLIC

BY

H. PRIEBE
University of Frankfurt, Federal German Republic

I. INTRODUCTORY

THE history of Europe shows that family farms have been the predominant form of rural life of European nations. Even in the era of industrialization they have proven not only their great stability but also their adaptability to technical and economic developments, and to this day form the nucleus of European agriculture.

Moreover, for social reasons the family farm is acknowledged as being of great importance among the concepts of agrarian policy. But the question of its future problems must first be answered from an economic point of view. As a socio-economic type, the family farm will only be able to survive and to adapt itself to further development, if by its economic efficiency it can hold its own alongside other forms of management.

The subject of my paper will be to deal with the problems of the family farm in the light of a European point of view. It will thus complement the paper of Mr. Glenn Johnson. In this regard it will be important to bring clearly into view the European peculiarities while at the same time working out the common traits to be found in the basic type of the family farm — a uniform basic type which can be recognized across all continents, though in considerably different forms and phases of development.

II. SOME POINTS OF CHARACTERIZATION

The term 'family farm' combines social and economic criteria. A demarcation according to objective statistical standards is impossible.

The Family Farm and Problems of Factor Mobility

The social unit of a family constitutes the decisive criterion. Its labour potential, and its expectations of income, are the starting-point and the goal of this form of management. Both of these depend on the overall technical and economic development of the economy : the efficiency of the family depends on the means of production available to the family ; what may be expected as a return for its labour is decided by the overall standard of income. Thus both endogenous and exogenous forces are dynamic elements in the development of family farms which is consequently to be found with a number of quite different forms of combinations of productive factors.

The work capacity of the family cannot be considered a specific given quantity. Today in Europe it varies between 1·5 and 2·5 full labour units formed by several members of the family who work permanently or only part-time and with differences in composition which depend on the rhythm of the change of generations. The female members of the farm family can only be considered part of the work-force in as much as household duties leave them free ; generally they only offer a temporary supplement in times of top labour demand.

In the traditional agrarian way of life large families as a rule supply a larger work capacity. But under the influence of industrialization the small family becomes predominant even in rural areas, with descendants turning to other occupations at an early stage and with only the heir remaining on the farm. Consequently the family supplies only two male labourers at the most, the farmer with an ageing father or a growing son. More and more families can be found with only one male labourer, the son or father working in another occupation for the larger part of his time, so that the labour potential for the farm then consists only of one full labour unit.

The basic criterion, nevertheless, remains the same everywhere, *i.e.* that the labour capacity of the family represents the basis of the farm and the nucleus around which it is crystallized. Consequently, there are very many possible forms of organization and production trends. Within wide limits these are determined by natural conditions, by density of population and the consequent availability of acreage, by the way of thinking and the behaviour of the people, by the standard of development of the general economy and by agricultural technique in its widest sense.

III. THE DOCTRINAL CHALLENGE TO THE FAMILY-LABOUR FARM

For a long time the family farm was considered to be the typical form of the traditional agricultural enterprise, and its capacity to develop in a modern economy was denied. For more than a century a dogmatic controversy between the virtues of large-scale enterprises as opposed to those of small-scale farming, was a main subject of European agricultural policy debates, and the old arguments affect the discussion up to this date.

First of all there was the view that family farms were unable to keep up with larger farms in productivity and marketing efficiency. Then it was said that modern techniques definitely demand the large-scale enterprise. A hundred years ago Karl Marx adopted this thesis and thus laid the bases of economic theories for the reconstitution of agriculture on a large scale in communist countries.

Developments in Western Europe in no way confirmed the theory of the superiority of large-scale agricultural enterprise. On the contrary, family farms survived even in the era of industrial revolution ; and the natural conversion from the small to the large-scale enterprise, enforced in other branches of economy by technical and economic development, definitely did not take place in agriculture. Family farms found ways to adopt technical progress, to increase their productivity and to assert themselves successfully in free economic competition along with larger enterprises of all kinds. On the other hand, this stability in the agricultural field was misinterpreted for a long time and gave rise to ideological speculations of various types. Ideas were expressed about the value of a form of life close to nature and of the importance of rural life as the backbone of the nation. These ideas lay within a European tradition of thought beginning with Jean-Jacques Rousseau, followed by the romanticists and researchers on folk-lore of the nineteenth century, carried into the twentieth century by Oswald Spengler, and finally leading to the ideologically exaggerated evaluation of rural life during the times of National Socialism in Germany. Quite to the disadvantage of the farmers, irrealistic concepts distracted agricultural policies from the genuine problems, and created in the minds of the public a distorted picture which seemed to show that the major consideration was to defend traditional forms of life against modern development in spite of sound economic judgement.

If today we attempt to bring the discussion back to economic terms and problems, we are aided by perceptions gained from the development of the last three decades. It was not protection that kept the family farm alive ; on the contrary, in Germany, for instance, the price policy for grain and sugar beets rather favoured the larger farms. Nevertheless, within a few years the fully motorized and mechanized family farm became, without any special promotion, the most widespread type of farm in Europe. This development can be explained by reasons of farm management. It would never have been possible for the family farms to develop such dynamic response to the adoption of technical progress had not the family labour composition offered especially favourable conditions to modern agriculture.

The development of production methods results in an increasing importance of entrepeneurial behaviour, even in agriculture. Processes of production become more complicated, the importance of elementary factors and their exploitation by simple physical strength recedes, and success is decided by the ability of man to make proper allocations. Thus, the qualitative peculiarities of the family labour constitution are brought out. Even in modern agriculture, however, the prerequisites for continuity and concentration of the production phases as in large industrial enterprises, do not exist. The tasks of the labour supply remain dependent on the biological rhythm, and on soil and weather conditions, and thus are exposed to incalculable influences and to constant change. A quick ability to react and adapt is therefore of decisive importance, and economic success depends on people who are able to think and act independently, and have a personal interest, initiative, and a sense of responsibility. The family labour constitution contains these elementary prerequisites to a high degree. Therefore, the requirements in farm management in a highly developed agriculture, especially, lead to the result that the independent small enterprise of the family farm has proved well suited to development and to the achievement of economic success.

IV. GENERAL TENDENCIES OF DEVELOPMENT

Since the second world war a great structural change has been taking place in European agriculture which shows above all in changes of farm sizes and of organizational forms. In the Federal Republic of Germany, too, it has led to the breaking up of traditional forms of operation and production ; but it has led, at the same time, to an even more definite formation of the family farm.

On one side the factors and motive powers in this structural change are of an endogenous type, *i.e.* they are a result of changes in agricultural techniques ; others can be traced back to exogenous forces, especially to change in social attitudes. Biological innovations in the development of agricultural techniques, especially in breeding, feeding techniques, plant and animal hygiene, have made possible a constant increase in production. But, in addition, there is, as the most important change, the replacement of animal tractive power by motors. This has made possible — with a certain lag in response as compared with other branches of economy — the full mechanization of farm work. The result is a considerable increase in labour-productivity. Its realization has far-reaching consequences for the economic and social structure of agriculture and in a short time has led to changes reaching deeper than those observed in the first phase of the industrialization of agriculture.

The most important changes along with considerations about how they affect the family farms are briefly outlined in the following :

(1) With the recomposition of productive factors farms lose the self-sufficiency they had with animal tractive power, and are fully included in the overall division of labour. Agriculture, previously labour-intensive, now becomes capital-intensive. Thus, while in 1955 in the Federal Republic of Germany the value of the capital invested in a medium-sized family farm (with horses) of 10 to 20 hectares amounted to an average of DM30,000–50,000 per labour unit, in 1965 the assessed figure for a fully mechanized family farm would be between DM80,000 and 100,000 per labour unit.

(2) The substitution of human labour by capital goods changed the traditional relation between acreage and labour which throughout the entire first century of modernization in agriculture had remained almost unchanged. In larger farms the employees decrease in number ; in small farms where the family remains as a minimum unit, the acreage is increased or the labour converts to other occupations, keeping the farm only as a part-time enterprise.

(3) The use of modern technical means of operation comes to depend on the presence of certain productive prerequisites. Under extreme conditions of soil, climate or area, farming is given up ; a concentration of agriculture towards areas with more favourable conditions begins to take place. It becomes necessary to change the land ownership set-up and the rural infrastructure as well as to reorganize the congested villages. These changes form a central task of agricultural policy which are combined under the term

'improvement of the agricultural structure'.

(4) This development is promoted from another direction by changes in the environment. Growth of industry offers work opportunities and attractive income conditions. Thus the attraction of the other economic sectors promotes the withdrawal of labour from agriculture and accelerates the structural changes.

(5) The social integration of the rural population results in faster change in the ways of thinking and in attitudes. The traditionally conservative images lose their power and are replaced by the desire to adapt the standard of living to that of other social groups, *i.e.* the desire for parity, as it is called in agricultural policy discussion. Strong forces are thus mobilized for the economic development and the modernization of family farms.

(6) All these changes increase the importance of the social constitution of the farms. The innovations in agricultural techniques offer better economic opportunities to the people. But they also impose higher standards, not only in terms of education and knowledge of more complex data, but also in terms of responsibility, and independent thinking and action. The qualitative advantages of independent skills, as they apply to the family farm, thus gain a greater value.

Under the influence of all these changes a concentration towards the family farm, *i.e.* independent work, may clearly be observed. During the fifteen-year period since 1950 it has been stronger than ever before in so short a time.

In the Federal Republic of Germany the total size of the permanent agricultural labour force not only decreased but also changed considerably in its composition. In 1950, about 15 per cent of the 5·15 million labour force were hired hands. While the labour force declined to a total of 3·03 million, the share of hired employees has more than proportionally decreased to 8 per cent, so that today 92 per cent of the people engaged in agriculture are independent owners and their dependants.[1]

The changes in farm sizes fit this picture as expressed in a concentration towards family farms (see Table I).

The available labour on the medium-sized family farm of 10 to 20 hectares of arable land has decreased to about 2·2 labour units, on farms of 20 to 50 hectares to about 3·2, *i.e.* to just about the labour capacity of a family.[2] In 1950, true to tradition, quite a number of

---

[1] *Bericht der Bundesregierung über die Lage der Landwirtschaft 1965*, p. 37.
[2] *Ibid.*, p. 186.

TABLE I

NUMBER OF FARMS IN GERMAN AGRICULTURE

Classification by size in Hectares of Arable Land	Number of Farms		Fluctuation	
	1949 in 1,000's	1964 in 1,000's	in 1,000's	%
0·5– 10	1,562	1,055	– 507	– 32
10– 20	257	296	+ 39	+ 15
20– 50	113	130	+ 17	+ 15
50–100	12·7	14·3	+ 1·6	+ 12
over 100	3·1	2·6	– 0·4	– 16

hired people were employed on these farms. Until 1960, their number decreased by 81 per cent in the acreage group of 10 to 20 hectares, in the acreage group of 20 to 50 hectares the decrease amounted to 63 per cent, both percentages related to the original figures respectively.[1] The transformation of larger agricultural enterprises into family farms also continues even in the acreage group of 50 to 100 hectares. The labour constitution which for centuries had been typical for large farms of 20 to 100 hectares, *i.e.* the combination of family labour and single hired labour, disappeared almost completely. The majority of the farms have turned into highly mechanized family farms.

The more definite formation of family farms is the result of all the changes listed above. The farms have converted, with their sizes increased. But they have remained unchanged in their nucleus : in the basic criterion of the family labour constitution. Many things indicate that the family farm is even more a *future* form for European agriculture, since experience shows that in a highly developed economy only few people are willing to work as hired farm labourers so that, more and more, agriculture is becoming a field of independent enterprises. In view of this development the old dogmatic argument over family farm versus large-size agricultural enterprise has lost its topical value.

V. TYPES OF MODERN FAMILY FARM

Today, the modern family farm of high productivity can be found in a number of variations. While up to just a few years ago certain

[1] *Wirtschaft und Statistik*, H. 5, p. 279.

forms were typical and clearly discernible in each region, the farmer today — except in cases of extreme conditions determined by natural location, as, for instance, in mountainous areas — has open to him many possible ways of making good profits.

One important problem among the measures to improve the agricultural structure is the minimum size of the family farm which can afford an appropriate income to 1·5 to 2 full-time labour units. The acreage cannot be used as the sole yardstick but must be considered in connection with the line of production, the intensity of the farm

TABLE II

FARM SIZES OF 104 MODERN FAMILY FARMS

	Full Labour Units	Hectares Arable Land	Value of Capital Invested in 1,000 DM	Gross Return in 1,000 DM	Farm Income in 1,000 DM
All farms (average)	2·0	17·0	222·3	58·6	26·1
Farms with					
High intensive crops	2·1	11·5	169·7	44·9	24·7
Root crops	2·2	17·7	244·7	71·7	30·0
Grain crops	1·8	20·7	243·0	63·7	26·9
Forage crops	1·8	18·5	228·3	48·5	20·7

operation and the capital investment. Family farms in many forms of combinations cover a wide range of acreage groups ranging, on the one hand, from small farms with a few hectares with horticulture or special crops such as wine, vegetables, fruit, etc., to highly mechanized grain or fodder crop farms with 50 to 100 hectares or more, on the other. The safest way of defining farm size is gross return, *i.e.* the economic result of all productive factors.

Today, the minimum economic size of a full-sized family farm, managed full-time, and bringing an income comparable to that of other types of employment, may with labour-intensive lines of production be established at an approximate gross return of DM 20,000 or with a capital-intensive line of production at DM30,000. Taking farm costs (without labour costs) as 30–40 per cent and 45 to 55 per cent for these two types of farm respectively, farm income (labour and capital return) amounts to about DM12,000 to 15,000. There are, however, on the other hand, family farms which bring a gross return of more than DM100,000, thus achieving an income for the family engaged which is well above average even in comparison with the income of other sectors of economy.

In our Institute we have examined modern family farms with good income conditions in different parts of the German Federal Republic. The results of 104 selected farms give a good survey of some successful forms of organization as they have been recently developed by capable farmers. With the exception of the farms with labour intensive crops (tobacco, wine, fruits and vegetables), the majority of the farms include one or two livestock activities.

TABLE III

PRODUCTIVITY AND INCOME IN 104 MODERN FAMILY FARMS

		Labour Productivity			
	Productivity of Land : Adjusted Gross Return* in DM per Hectare	Gross Earnings in 1,000 DM per Labour Unit	Net Farm Income† in 1,000 DM per Labour Unit	Labour Income in 1,000 DM per Labour Unit	Rate of Return on Invested Capital ‡ %
All farms	2,820	30·3	12·8	10·2	8·7
Farms with					
High intensive crops	4,380	21·9	12·0	10·1	10·1
Root crops	4,200	33·2	13·6	11·1	9·7
Grain crops	3,540	38·0	13·8	10·6	7·9
Forage crops	2,100	26·7	11·2	8·4	6·1

* Expenditures for inputs descending from other agricultural enterprises has been deducted from the gross return.
† Computed with a calculatory rate of interest of 3½ per cent on the invested capital.
‡ Computed with an assumed wage of DM6,500 per labourer per year for all labour on the farm.

When compared with results on larger farms using hired labourers, the results from farms in the Institute's sample show that returns to labour and capital are just as good on the family farms as on the larger farms ; furthermore, the returns on these family farms are not below those received in other occupations. According to the computations for the sample of 104 farms, the average rate of return to the capital invested amounted to 8·7 per cent, and labour income of about DM10,000 per labour unit was typically reached : the latter figure was well above the wage of comparable employment groups which, according to the figures established in the annual report on the situation of agriculture in the Federal Republic of Germany for the same period, amounted to DM6,500.

The Family Farm and Problems of Factor Mobility

What are the necessary conditions in which these economic results become possible ; that is, what sorts of organization, what level of intensity, and what sorts of technical equipment on the farms ? The answer to this question throws light upon the next crucial question : What impedes other farms in their development and what problems must agricultural policy solve to promote these farms ?

Though the sample of 104 farms was selected on the basis of their economic results, they are, nevertheless, hardly above the general average in their production bases. Quite a number of them show structural shortcomings ; hardly any of them have a consolidated area ; one-third of them having even more than 15 plots ; some of them also suffer from long distances to the fields and congested farmsteads in traditional clustered villages ; thus their unexploited production potential is still considerable. Important reasons for the above-average economic results lie in the personal efficiency of the farmer, in the utilization of the existing production bases and in the adaptability to technical and economic possibilities. All farms are fully motorized and well mechanized; all of them have abandoned diversified production and are concentrating on one or two main production processes.

VI. MECHANIZATION AND SPECIALIZATION

For the majority of the family farms the problem is to find the most suitable form of mechanization and specialization for their enterprise. Both must be seen in a close interrelation ; both depend simultaneously on natural site-conditions, acreage, capital funds, access to credit funds, market situation and other factors. It is a rare case in European agriculture that conditions allow full specialization on just one product. A certain rotation of crops is necessary to maintain soil fertility. Furthermore, and especially for the family farm, a productive utilization of permanent labour can only be achieved in several lines of production with the times of peak labour demand in different lines being interspersed. On the other hand, in each branch of production it is important to achieve certain minimum levels so as to reach the highest technical standard without excessive costs per production unit.

So far there is no accepted standard view as to the optimal size for a productive unit. Some agricultural economists quote rather high units for profitable production, e.g. at least 40 cows, 5,000 hens, 500 pigs, and a very large acreage for the use of machinery. Since as a rule such units are today beyond the reach of most of our family farms, doubts about the family farm's competitive chances have recently been voiced again. This problem must be considered from two angles.

First, it is possible to consider the productive capacities of our average modern family farms with the addition of one full-time unit of permanent hired help working in a single line of production. This may be the correct approach for larger farms where paid specialists are to be given a certain field of full responsibility. The family farm, however, often provides part-time labour units for which full-time employment can be achieved by the addition of increased productive capacities. These additions, of course, must not be small since otherwise the most appropriate structural and technical installations cannot be realized. Under certain circumstances, however, the appropriate structure may be achieved with 20 cows, 1,000 hens or 150 fattening pigs. This offers important challenges for applied scientific research, in co-operation with agricultural advising, to find the suitable and economically profitable forms of production and the possibilities for their combination under different conditions.

Secondly, there exists ample experience of mechanization to show that joint maintenance of large special equipment enables the family farm, also, to utilize such machines economically. Several organizational and legal solutions have been developed to this end : some are groups where several farms are joined in a legally loose form, others are larger organized co-operatives ; there are in addition contracting enterprises which take over certain work with larger equipment under a contract for set fees. In any case, many examples can be used to show that full mechanization of family farms is possible without immoderate costs. Where individual machines are utilized on smaller areas the balance in costs is often reached by means of a longer life of the machine as well as low costs for maintenance and repair due to better care. In the sample of the 104 farms machine output was found to be 21 hectares for combines, 25 hectares for corn harvesters, 13 hectares for sugar beet full-harvesters, 11 hectares for field choppers. The structure of costs in these farms still remained within normal bounds ; such a high capital and labour return would otherwise not have been possible.

VII. PROBLEMS AND TASKS FOR AGRICULTURAL POLICY

The considerable difference in efficiency among family farms poses a serious problem for agricultural policy. The sample we have used may provide exemplars in this area. That many of the other farms lag behind is to be attributed to a number of different factors :

(1) The personality of the farmer always plays the decisive role. This can readily be shown by divergences in farm income between farms with equal production bases. Educational standard and subject-matter knowledge are often immediate factors of efficiency ; our survey was able to prove that farmers with a better subject-matter education achieved a labour productivity surpassing that of the average by 30 to 40 per cent. Certain supplements and aids from advisory work are of course also possible.

(2) The economic development of family farms is partially impeded by social peculiarities of the family labour composition. The close association between family life and occupation may cause human tensions within the family to hamper farm management. Conservative attitudes and all kinds of motives other than economic ones influence the economic operations and may lead to a certain immutability of conditions. In most cases overcoming these difficulties is a problem for new generations. It can be observed in Germany that the younger generation of farmers has largely overcome traditional complexes.

(3) Structural shortcomings limit output and development of many farms, even where there is the highest personal efficiency. In some regions the remnants of a medieval land arrangement still exist : widely scattered plots, poor roads, and long distances to the fields impede the application of modern production methods. At the same time, buildings and farmsteads are often too old and so congested that there is no opportunity for modernization. The individual then is powerless and has to depend on the hope that structural improvement will be carried out under a broader programme. In the German Federal Republic as well as in the EEC, structural improvement is considered an important field of agricultural policy.

(4) For many farms, extreme natural site-conditions present insurmountable limits for development. With unfavourable soil and climatic conditions, especially on sloped land, no form of management and no farm size will be able to produce profitably. In Europe this applies to many large mountain regions. Formerly, at the stage when

manual work was the norm, they could still compete, but with modern work methods, they can no longer be managed profitably. There, agriculture must in some parts be totally abandoned or limited to part-time management.

(5) Already more than two-thirds of the 1·5 million farms are managed part-time. Their share keeps increasing. In this sphere, too, it may be recognized that traditional forms are in a process of conversion. Different forms of co-operation in the utilization of machines and also in livestock production take shape, thus creating a favourable economic basis for the part-time farm. The development of decentralization of industry in the German Federal Republic offered great possibilities for the process of differentiation in the field of part-time farming.

(6) As a whole, present market development can be considered to be favourable to family farms. The elasticity of demand for the specific livestock products and intensive special plant crops are quite favourable. Within the area of EEC an increase in demand for animal products of 20 to 30 per cent may be expected for a period of about ten years ; for poultry, eggs, fruit and vegetables it may even be greater. Rationalization of the distribution of outputs is an important task of agricultural policy. It is aimed to obtain this by combining units to form larger and more uniform units, and by joining farms into co-operatives that will produce and market uniformly.

VIII. THE FAMILY FARM AND REGIONAL SPECIALIZATION

Improvement of production and market structures make it possible for family farms to be included in regional specialization over large economic regions. In the common market of the EEC there will mainly be, for the family farms, competition between each other, so regional specialization within agriculture will be intensified. Certain production centres will develop in the different countries according to their special natural conditions. Over and above that, there will remain a sensible division of work between European and American agriculture. It may become even more pronounced. With a lower density of population and with larger land resources at its disposal, North America has a natural superiority in grain production. Here the European family farm cannot compete. Its superiority lies in high-quality animal production which has its most profitable location close to the market, in order to be able to provide fresh products for the big European centres of consumption.

For the future, we may expect further considerable changes as well as increases in efficiency in European agriculture as a whole. Labour productivity in German agriculture has increased by about 300 per cent since 1950. Family farms contributed a major share. Even today, their production reserves are unforeseeable. The worries of European agricultural policy may in the near future also result from the fact that the growth rate of production exceeds that of demand. Solutions will have to be found which guarantee continuous structural adjustment to technological advance. It may therefore be necessary for the future to count on a further decrease in number of family farms, which will, however, be of better quality.

_____

JOINT DISCUSSION OF PROFESSOR RENBORG'S, PROFESSOR GLENN JOHNSON'S AND PROFESSOR PRIEBE'S PAPERS

In this discussion, arising from the three preceding papers, the order of the contributions has been altered to bring out more clearly the main features of the discussion. It may be seen to have four parts : (*a*) discussion arising from Professor Renborg's paper and centering on the measurement, identification and characterization of farms of the future (particularly with respect to U.S.A.); (*b*) after Professor Bergman's comments on both Professor Glenn Johnson's and Professor Priebe's papers, discussion of Professor Johnson's paper ; (*c*) discussion of Professor Priebe's paper ; leading to, (*d*) general discussion of the conceptual and practical problems in assessing the future of the family farm.

Dr. Stipetić, in opening the discussion, said that Professor Renborg's paper dealt with one very important aspect of the technological revolution in agriculture: the impact of that revolution upon the patterns, structure and organization of agriculture in developed countries. It faced the question : What are the economic repercussions on the individual farms themselves of the creation of capital-intensive agriculture ?

Dr. Stipetić said that while realizing the value of the paper, he would like to point out some of its limitations. The author did not deal specifically with the tendencies towards agricultural concentration and specialization in those socialist countries which are developed. The concentration of land and means of production in these countries usually came before new technology was introduced into agriculture on a mass scale. What is the author's view about such a process? Is it an advantage or a disadvantage to have such concentration before new technology is introduced?

Dr. Stipetić said that his second observation was directed at the very difficult subject of the model of a future farm given by Professor Renborg. The model is largely based on Swedish or European experiences. These, however, lag behind contemporary agriculture to a considerable extent. Farm patterns in North America, New Zealand or Australia provide a much better example of the future shape of agriculture, especially with respect to concentration and specialization. It is therefore worth examining the process of concentration in United States agriculture.

In the United States in 1959, there were some 3·7 million farms of which only 20,000, or less than half a per cent of the total, had sales of more than $100,000 each; yet they accounted for nearly one-fifth of all farm products sold in 1959 (U.S. Dept. of Commerce — Bureau of Census, Washington 1962 — Vol. 2, p. 1200).

Dr. Stipetić pointed out that the United States Census of Agriculture provides data on the division of farms by their share in sales. Professor Renborg's paper did not make full use of this measure. It would be useful to focus attention upon the economic class of farms with sales of $40,000 and more. In this so-called 'Economic Class I' there were slightly more than 100,000 farms, or 2·7 per cent of the farms in the United States. However, these farms took up a quarter of the total agricultural land, they employed 37 per cent of the total hired labour force in agriculture and they accounted for nearly 40 per cent of all expenditures on feed, seed, gasoline and oil. They shared nearly one-third of the total sales of farm products in the United States.

This is a concentration of production unprecedented in agricultural history, and clearly shows the contemporary trends of farm size in developed economies. During the 1950's, 1·6 million of the farms in the United States disappeared. More than three-quarters of those had annual sales of less than $2,500. In the main, they had been operated by full owners, and their land went to farms which had larger total sales, and of which most were in the first two economic classes.

Returning to the question of the farms in economic Class I, Dr. Stipetić said these large farms were not family farms and did not have the small scale of operation which the examples with which Professor Renborg concluded his paper had. Dr. Stipetić provided the following data on this economic Class I from the U.S. Census of Agriculture: Their average size was 2,466 acres (approximately 615 hectares); average capital per farm was $220,683 (land included); average sales per farm were $94,531. These farms provided 40 per cent of total cattle sales in U.S.A., 39 per cent of chickens and broilers, 26 per cent of eggs, 65 per cent of vegetables, and 52 per cent of fruit and nuts, etc. The greater proportion of the land was not owned by the operators of the land, but rented. Ten per cent of these farms had 10 or more regularly hired workers, a further 15 per cent had 5 to 9 regularly hired workers, not counting seasonally employed workers. Two-thirds of these farms carried cattle stock and the average size of cattle herds per farm was 275.

These huge enterprises cannot be regarded as family farms. They employ more hired labour than the family labour they possess, they do not own their own land, they operate with borrowed capital in huge quantities and they only produce for the market. This surely indicates a clear tendency away from the pattern of the traditional family farm. They have all the characteristics of capitalistic enterprises.

Returning to the model of the future farm, Dr. Stipetić said that if the United States case is anything to go by, it is hard to see how the future farm might be in the range of labour inputs of 3,000 to 5,000 man-hours per annum as Professor Renborg predicted on page 232 of his report. Moreover, if in Western European agriculture this kind of farm in 1975 might prevail, it would simply mean that agriculture on the Continent would be lagging some decades behind the United States ; nor did such farms represent the optimum size of farms even today. In the light of this it is to be hoped that Professor Renborg would be able to give some additional explanation of why he chose this particular model as a future farm target.

Dr. Stipetić said that with regard to statistics — particularly those on specialization — Professor Renborg's paper was not very fully documented. He asked whether Professor Renborg could provide the sort of statistical information to be found in the annexes of the FAO 'Whither European Agriculture?' booklet.

Finally, Dr. Stipetić wondered if the future of the *family farm* lay in huge vertically integrated capitalist enterprises, which Professor Renborg's paper described as necessary for full safety precaution against risk. The independence of the family farm, which used to be its chief characteristic, seemed unlikely to survive in such conditions. In such organizations, farmers would be gradually turned into piece-workers in huge capitalistic enterprises.

Professor Delivanis said that on page 214 of Professor Renborg's paper it was said that, 'in the great majority of agricultural enterprises, the techniques used, both in agriculture and animal production, are well in arrears of present possibilities'. If the paper was referring to the technically feasible, he agreed, but said that he was not sure that the statement was correct with respect to economic feasibility.

Professor Renborg had concluded that regional specialization is increasing. In many countries a fear of unstable monoculture leads to emphasis on the opposite tendency.

Professor Delivanis said that he was not sure to what extent the advantages of rationalization go to the consumer. The consumer has to pay taxes from which much of the costs of rationalization are financed.

It should be emphasized that the family farms in future would only be able to rely on one or two full-time employees if the living conditions they offer are comparable to those enjoyed by industrial workers.

Professor Hathaway said that it would be useful to correct some mis-information implied by some of the aggregate United States statistics on farm size. He meant, in particular, those that Dr. Stipetić had referred to for the economic Class I from the United States Census of Agriculture. The statistics refer to an untypical group of farms. These large cattle, poultry and egg lines of production are all characterized by a relatively low value added relative to sales because most of the feed is purchased. More-over, the last two categories of activity, namely vegetable production and fruit and nut production, are the major users of hired labour. Professor Hathaway said that there is much less concentration than is implied in these statistics. The 1964 Agricultural Census showed that the family farm, as measured by labour input, is now selling a higher proportion of farm products than thirty or forty years ago and that the larger than family farms has a lower share of gross sales. It is very easy to misinterpret these United States statistics.

Professor Robinson said that for a layman some of the statistics are very difficult to interpret. He said that in the study of industry, the general economist is careful to distinguish between economies of scale belonging to the size of the plant, and economies of scale belonging to the size of the business. He asked whether the data for farms is equivalent to that for plants. In the United Kingdom there is a tendency for one entrepreneur to run a number of farms and many owners of family farms collaborate and operate groups as one single unit. Professor Robinson said that he had read of the same trend in the United States and asked whether this invali-dated conclusions from an analysis which is made purely in terms of farm sizes.

Professor Stipetić answered Professor Hathaway's objection about the uniqueness of the farms represented in economic Class I. He said that he realized the importance of this objection and he hoped that it would be appreciated by the compilers of the Censuses of Agriculture.

He also realized that chicken production and cattle production are heavy users of feed inputs. However, a considerable proportion of those feeds are grown on these economic Class I farms. Thus 27·2 per cent of all field crop products in the United States (excluding vegetables and fruit and nuts) are produced by these farms.

Professor Martinov said that he completely agreed with Dr. Stipetić. Family farms are disappearing. The changes in the pattern of capitalist agriculture are not made so plain by considering the statistics of the use of hired labour as they are by noting the trend to a few larger farms. Techno-logical progress fosters this trend.

The liquidation of small family farms in Western Europe and the United

The Family Farm and Problems of Factor Mobility

States establishes a clear trend, and although in some Western European countries there are an increasing number of family farms not using hired labour, these have to increase their investment by two or three times in order to survive and it seems quite clear that the future belongs to much larger farms than even the modern family farms.

Changes in the United States pattern of agriculture show the future for capitalist developed countries. This is why studying the changes in U.S. agriculture is so important. Professor Martinov said that recent studies which are being used to imply a reduction in the effectiveness of large-scale concentration depend basically on one criterion, that of the size of the labour force. This means that many integrated farms are taken to be family farms, and this is wrong. In the beef industry, feed-lot farms are producing an ever increasing share of total production. Moreover, as Professor Glenn Johnson said of this form of production, '[these] opera-tions have involved more capital than could be handled by families of ordinary wealth'.

Professor Hathaway said that he would be the last to suggest that there is no increase in concentration, but newly available data does suggest that it is not very profitable. (See the article by Boyne, published in *J.F.E.*, December, 1965.) During the post-war period, the share in total income of the top 5 per cent of the farm income group has declined by almost 50 per cent. There is a vast decline in the concentration of farm income received by the most wealthy group, whereas there is an increase in the middle group which in the United States is classified as the modern family farm group.

Professor Johnson said that Professor Martinov might be right in assert-ing that the family farm is on the way out in the United States, but that he thought that this would be unlikely unless the pricing problem is solved.

Professor Renborg said that although he realized that the question of the development of the socialist countries is an interesting one, he had made it clear in his first paragraph that his analysis was restricted to certain particu-lar countries in Western Europe and the United States of America.

It was only for the United States that he had found official statistics measuring farm size according to total registered sales. These statistics, plus a number of smaller studies from various countries, had been used to show the trends towards specialization and concentration going on within agriculture. In his analysis of the available material he did not come to the conclusion given by Professor Stipetić, that agriculture is moving away from the family farm. The change going on is only partly a tendency towards an increase in the number of very large farms, the most important tendency being an increase in the *size* of the family farm. In this connec-tion Professor Renborg also pointed out that Professor Glenn Johnson's figures indicated that, in the United States, the proportion of production from very large farms has decreased in the last five or six years. All this suggested that Dr. Stipetić's picture of agriculture as being taken over by

Discussion of the Family Farm

huge capitalistic enterprises is, at the very least, exaggerated.

Dr. Stipetić had stated that his model of the future farm lagged behind development. It is necessary to remember that these were Swedish conditions about which he was writing. Not only must part-time farming be taken into account, but also the types of potential large-scale development. These two types of farms are also included in the model side by side with the large family farms. The picture given in the model certainly does not lag behind the development in Western Europe as a whole. Professor Renborg said that, on the available material, he did not feel justified in trying to quantify his arguments, *i.e.* indicate the proportions, at some future point in time, between the various types of farms given in the model.

He also said that his model was *not* intended as a target for 1975. There are many hindrances to rapid change and many negative forces that are coming into play. He did not expect that much would happen in ten years to alter the general pattern of the model given or to remove the close connection with the present pattern.

Professor Georgescu-Roegen said that he had already in the Conference expressed his uneasy feelings about the use of the concept of scale. The problems we have been discussing today may illustrate the difficulty. If by 'scale' we mean only the amount of output, then there is no need for a term other than 'output'. But when we talk about scale we generally think of the efficiency of qualitatively different processes of production. The problem discussed during this session turns on comparing the efficiency of a farm using a simple plough drawn by oxen and one using a multiple plough drawn by a tractor. However, a uniform measure of efficiency requires that all inputs be cardinally measurable. Directing his question to Professor Renborg, Professor Georgescu-Roegen said that if the inputs of all types of farms considered together are not cardinally measurable, they can not be represented on the same coordinate axes as in figures 2(a) and 2(b) of Professor Renborg's paper. There are no uniform units in which to measure capital and labour. Both capital and labour vary *qualitatively* with prices; hence, the current structure of an enterprise depends to a large extent on the price constellation with which the producer was confronted at the time when he bought the capital. And since prices are as much subject to evolutionary changes as technology, an answer to the problem of the optimum scale of farm could not be given in purely physical terms or by using current prices of inputs. The point is that, judgements based on current prices have a parochial, not a general, validity.

Professor Georgescu-Roegen's second question pertained to diagram (*b*) of Figure 2, by which Professor Renborg describes technological progress as a movement on a given isoquant from t_1 to t_3. By such a shift Professor Renborg represents the fact that technical progress brings about a different capital–labour structure, with capital being increasingly substituted for labour. Professor Georgescu-Roegen wished to observe that, on the very same diagram, if labour is kept constant, technical progress makes it possible to obtain the same quantity of product as before with less capital; and, if capital is kept constant, technical progress makes it possible to obtain the same quantity of product with less labour. Consequently, if the nature of technical progress were that represented by the diagram of Professor Renborg, it would follow that any technological advance consists of both labour and capital saving devices, In this case, one cannot see why technological progress always leads to a substitution of capital for labour, as Professor Renborg concluded from his diagram.

Dr. Eastman said that the evidence adduced in Professor Renborg's paper of the low elasticity of demand for farm products, as an explanation of forces inhibiting farm growth, was not convincing. It would seem, *a priori*, that the demand facing an individual farm in competition with many others is rather high and it is the elasticity of demand facing the individual firm that is relevant to its behaviour in organizing factors of production in the long run. The evidence presented by Professor Renborg is that prices fall together with productivity increases, but this is also consistent with a competitive industrial structure.

Dr. Eastman also commented on a purely expository point which related to the diagrammatic presentation of Professor Renborg's argument. He said that more important than the lowering of the average cost curve with technical progress is the changing shape of the curve, so that the minimum optimum size of a farm will be far to the right of the earlier one. Thus it is true that real costs of production of a 160-acre wheat farm in Western Canada has fallen over the past sixty years, but while 160 acres was the optimum size of farms in 1900, the present size is 6,000 acres. This would be reflected in the isoquant diagram by having the new isoquants come closer together at outputs thirty times greater, after technical progress has taken place, than before it.

Professor Renborg said in answer to Professor Georgescu-Roegen that he was aware of the fact that it was an over-simplification to present the relationship between inputs of labour and capital on the one hand and product on the other as in his Figure 2(*a*). The homogeneity of capital and labour is not high enough for this diagram to be protected from all criticism, but it did serve for him to be able to show one point. In the case of Figure 2(*b*), he had not intended to make any point about the neutrality or otherwise of technical change, but simply to illustrate the point that the amount of product that is produced, with a certain input of labour and capital, is increasing over time.

In answer to Dr. Eastman, Professor Renborg said that he did not think it very important whether the lowest point of the average cost curve is further to the right. The point he was trying to make in that diagram was simply that the modern family farm seems to be forced continually to increase its size. Another conclusion of his paper is that the major scale advantages are within the limits of the 'modern', *i.e.* large, family farm, which is, as has been pointed out by others much larger in terms of land and capital per man than the traditional average family farm.

M. Bergmann said that the subject dealt with in the papers by Professors Priebe and Glenn Johnson was vast and had been the subject of lively discussion for more than a century. The reason for the persistence of the topic is that technical developments continually brought up new problems and cast doubt on the previous provisional conclusions.

M. Bergmann said that in the main he agreed with the analysis of Professor Glenn Johnson, and that he had important points of disagreement with Professor Priebe's analysis.

With respect to questions of method, M. Bergmann said that the definition of the family farm used by Professor Priebe, is based on the labour input. If that input is provided by the family, the unit is a family farm. M. Bergmann thought that it is necessary to take account of other inputs, and of the ties of vertical integration, in order to be sure of the extent of dependence or independence of the farm. Criticism could also be made of the method used by Professor Priebe. His analysis rests on a sample of 104 farms which achieved good results. However, the economist is interested in the 'optimum' not just 'good' results. Apart from their above-average quality, these farms probably produce articles for which there is a good market which would not remain so favourable if many farms produce the same items. Furthermore, Professor Priebe used the residual imputation method to calculate returns to labour and capital furnished by the family itself. Such a method had for some time been heavily criticized. In fact the only way to find a long-run average cost curve, which would indicate the possible economies of scale, is by *ex ante* calculations.

M. Bergmann said the fact that economies of scale were found by Professors Renborg and Glenn Johnson only up to a certain size, seems quite logical. A combine harvester cannot, like a blast furnace, be a colossal machine lest it simply sink into the ground. However, there are developments in the automation of animal raising which suggest much larger units ; and these developments would change radically the tendencies which had been at work up to the present, and which were explained in Professor Priebe's paper. But in this discussion of returns to scale, it is not enough simply to deal with the internal costs of the firm. The external position of the firm, with respect to its purchase of inputs and sale of output, is also important, for it is here that external economies of scale appear. In particular, it is through such external connections that the large firm has access to sources of capital. In theory, co-operative action should make it possible for the small firms to make up for the disadvantages of their size. But the present state of the co-operative system in most Western European countries does not give grounds for optimism.

In view of these economies of scale, and of the evidence that they are becoming greater, the survival of the family farm as reflected in the statistics needs explanation, and its survival in the future is called in question.

M. Bergmann took four headings under which to examine the factors bearing on the question of family farm survival.

First : Factors which directly help to keep the family farm alive : the presence of a stream of labour entering farming for lack of other employment opportunities ; the free and efficient services of information from professional and government bodies which keep the family farmer up to date and enable him to be a 'good' entrepreneur ; progress in the 'miniaturization' of farm machinery ; a strong demand for quality (luxury) products which are not standardized, and a system of prices and marketing which pays well for such products ; and finally, the high prices for secondary cereals which encourage the raising of livestock in traditional ways on pasture, silage and hay, and discourages the automated methods which rely more or less heavily on grains.

Second : Factors which discourage the development of larger farms : low rates of return ; all those factors which render the risks of large farming serious (including ignorance of the best processes, the risk of accidents, and the erratic variations in input-output coefficients) ; rigidities in the market for land, which make it impossible for large units to obtain the large and continuous areas they require.

Third : Factors which discourage the continuance of the family farm : the technical changes already referred to and which shift the least cost point to the right ; the state of the capital market and capital requirements in modern agriculture, which demand, at each point of major technical advance, sums which are far beyond the reach of what the average man can reasonably borrow in a lifetime (Professor Bergmann said that he thought that this point was inadequately stressed in the papers) ; and finally the demand for leisure on the part of farmers who are in danger of becoming slaves to their machines and their systems of animal husbandry.

Fourth : In dealing with the factors which could further the development of larger than family farms, M. Bergmann said that he agreed with Professor Glenn Johnson's point that the prodigious 'success' of the family farm had, given the state of demand, led to the lowering of prices, the underpayment of factors of production and the reduction of profits. However, if the movement of farmers out of agriculture accelerates, this situation might well be reversed, and entry into the sector might become profitable to the large capitalist. A further encouragement to the large farm would be the presence of a cheap labour force prepared to undertake mechanical and repetitive work.

M. Bergmann said that he had not touched on the important point of the growth of integration, which would, if it occurred widely, totally transform the problem and render obsolete the concept of the family farm. Finally, M. Bergmann pointed out that even if the larger than family farm is to become the norm, the problem of the general equilibrium of markets will not necessarily be solved.

Discussion of the Family Farm

Dr. Odhner referred to Professor Glenn Johnson's thesis that, because of over-investment in family farms, prices are pushed down so that large farms cannot enter profitably. He did not think that the disadvantages of big farms are as great as it seems from the reasoning in Professor Johnson's paper. The smaller family farm is less efficient because of bad planning and lack of education in more advanced methods. Dr. Odhner thought that in his list of the consequences of environment in which Western family farms operate, Professor Glenn Johnson gave insufficient weight to the less efficient operation of small farms because of management deficiencies.

In answer to Professor Bergmann's objection to the use of residual methods of studying returns to scale with a Cobb–Douglas production function, *Professor Glenn Johnson* said that whatever method he employed he came up with the same general conclusions. Once the farm is large enough to overcome certain lumpy factors, essentially constant returns are found from then on.

Professor Johnson said that the distinction between economies of scale in plant size and those in size of business had been looked at a number of times. There are very few family farm managers who would not like to create a unit larger than the ordinary family farm and this is strong evidence that economies of plant size are not involved. Instead, their eagerness is thwarted because they cannot finance the expansion when they have to compete in regular markets for labour, capital and land.

To Dr. Odhner's question about the relative efficiency between the different sizes of farms, Professor Johnson said that the family farm is about as efficient as the larger-than-family farm (in terms of relative cost structures). However, the family farm is prone to over-commit resources so that prices are driven so low that entry for the larger unit is impossible. Many kinds of technical change, of which the manager is ignorant of the macro-implications, help bring about these mistakes. These wrong decisions confer benefits in the form of low prices on the American consumer. There is some difficulty in deciding on the right use of 'efficient' in this context as, in a sense, the family farm does deliver very cheap products to the market.

Professor Johnson indicated that he did not have time to discuss the many points in Professor Bergmann's introduction — in particular his list of factors favouring the family farm — but that the list was an interesting basis for profitable discussion.

Professor Ruttan referred to page 247 of Professor Johnson's paper where it is argued that over-production results from the over-intensive use of chemicals, etc. However, evidence from micro-economic studies in the United States indicate that the return per dollar spent on current inputs typically fell between $1·50 and $2. There is, therefore, some inconsistency between the micro-economic results and the macro-economic implications of Professor Johnson's comment that need to be explained.

Professor Johnson replied that the results he reported are typical for

273

farms in university farm account projects and for farms included in surveys. He, in turn, wondered if Professor Ruttan's data are derived from marginal analyses or from residual computations based on charges for land, durable capital items and labour. If the latter items are charged at 'off-farm' salvage values Professor Ruttan's figures might result. However, if they are charged at acquisition costs, average returns per dollar spent on current inputs would likely be less than one dollar as computed residually.

Professor Gulbrandsen said that he wanted to make a remark on the following quotation from Professor Glenn Johnson's paper : 'I am inclined to think that it is price-pressure more than high managerial requirements per unit of output which tends to limit farm size'. Professor Gulbrandsen said that he would argue that, at least in Western Europe, it is not pressure of low prices but high prices which tend to conserve the family farm. High prices capitalize into high values of land, which is the only unreproducable factor. As prices of land increase, compared with prices of other productive factors, the economic advantage of large farms using large quantities of land decreases. However, it is true that high land values also aggravate the financing problem, because frequently 70 or 80 per cent of all capital in a farm consists of estate value. This, Professor Gulbrandsen said, is probably more pronounced in Europe than in the United States of America.

Professor Johnson argued that with high, artificially supported prices, family farmers would tend to overvalue land to the point where their errors of over-investment (financially as well as physically) would reduce returns to many other factors below levels required to cover acquisition costs ; the World War I history of America's agriculture being an important case in point.

Mr Sorenson said that Professor Glenn Johnson in his paper introduced a topic that was sensitive, from the political point of view, and was also very difficult to analyse in terms of its economic implications. He said that he would like to ask Professor Johnson to comment more specifically on one point. Does he view the question of the potential viability of the family farm as primarily one of production organization, with its solution lying in adjusting farm organization, or does he view it as a broader institutional problem, the solution of which includes institutional reorganization and efforts to provide some form of increased market power for agriculture?

Professor Johnson said that he definitely regarded the question as one involving institutional reorganization though he was somewhat sceptical about 'market-structure analysis' as the best approach to the study of the problem.

Professor Bishop asked whether the theoretical consequences of the environment in which Western farms operated, as listed by Professor Johnson, are specific to family farms as such, or whether they are simply the result of technological developments giving rise to increases in size of farms. Professor Bishop said that Professor Johnson also indicated that there are substantial 'costs of wrong decisions' made by operators of family farms. Is it not possible that the 'costs of wrong decisions' might be even greater under other forms of agricultural organization?

Professor Johnson said that many of the characteristics of the family farm and its environment are characteristic of other businesses and that, in those instances, similar characteristics produce similar results. He also felt that the costs of wrong decisions in say, U.S. ffamily farms, are low relative to almost any other form of agricultural organization the world over. He felt further, however, that these costs are still so high that a major problem for agricultural economists is to devise agricultural systems which reduce those costs.

Professor Renborg said that he had been disturbed by Professor Priebe's paper. He considered that it would have the effect of giving the traditional family farmer a false sense of security. Like Professor Nussbaumer, he would like to have more details concerning the calculations about the profitability of family farms as given in the table on page 259 of Professor Priebe's paper. In note 2 to the table on page 259, Professor Priebe gave the calculatory interest rate of three and a third per cent. This could not possibly measure the opportunity cost of capital. It must therefore give a wrong answer since in all calculations of this type the opportunity costs of all resources should be taken. Professor Renborg said that his second detailed question to Professor Priebe referred to the quotation on page 260: 'Though sample of farms was selected on the basis of their economic results, they nevertheless are hardly above the general average on their productions base.' How does this statement agree with the statement on page 262 that 'there are considerable differences in efficiency among family farms'?

Professor Nussbaumer said that the results shown by Professor Priebe gave a very favourable picture of farming, especially with respect to the high returns on capital investment. He would be grateful to know how these returns were calculated.

There seems to be an interesting and serious disagreement between what he called middle-European agricultural economists and others. The two groups seem to take different views on the prospects facing family farming. He thought that the comparative optimism of the middle-European is due to the sociological importance of the family farm in their countries. This seems to be very different from the United States, in that in these parts of Europe it is thought worth while to pay for the maintenance of the conservative family farm.

Professor Nussbaumer thought that it would be illuminating if he were to list the main disadvantages faced by the family farm as seen from a 'middle European' point of view.

1. The unevenness of the working day in many specialized farms implies a low productivity of labour-hours-of-presence.

2. The labour of the family farm worker and of the other members of his family cannot be used properly the whole year round.

3. Auxiliary services might not be available in remote areas.

4. Capital is often not used for significant parts of the year which implies low average returns and, on the small units, over-capitalization and over-investment in new buildings. This over-capitalization is not so serious in the case of green grass-land farming, but in grain farming capital costs are excessive when machines are only sporadically used.

5. The educational opportunities of young members of farm families are poor ; studies had indicated that on comparable farms run by educated farmers, the net returns can be up to three times as great.

6. The low income of family farms has to be broadly supplemented or replaced by the property incentive if it is to be worth the family remaining on the farm. But this also works against the family farmer, since the younger members of his family whose labour he needs are often strongly influenced by a desire for shorter and more regular hours of work and for cash salaries.

Professor Nussbaumer said that he thought that the greatest chances for the success of the family farmer lies in their specializing in more marketable products suitable for production on a smaller scale. Generalizations are not really possible. All that can be said is that there is a strong case for intensive production on family farms. If efficient co-operative arrangements for the purchasing of the means of production, and for marketing and services, are created, the chances of survival are enhanced.

If the family farmer has access to cheap basic capital for buildings, machinery and heavy investments, then the returns on the commercial turnover capital which he employs appears to be much higher. In middle Europe the woodlands owned by the farmer are often his source of such basic cheap capital and they can be supplemented by government sources of cheap capital.

Professor Nussbaumer said that, following from what he had said, he wanted to ask two main questions of Professor Priebe. First, what assumptions had been made about access to basic capital, types of land and levels of education, in the calculations for his paper ? Secondly, what is the opportunity cost of maintaining family farms as part of the social system of middle Europe ?

Dr. Fleischauer said Professor Priebe would agree with Professor Bergmann that the concept of the family farm only applies when management decisions, and particularly the decisions with respect to capital investment, are not taken by persons or institutions outside the family farm.

Dr. Fleischauer answered Professor Bergmann's criticism of the sample taken by Professor Priebe which Professor Bergmann thought was not sufficiently random. He said that in forming the sample they had first tried to calculate models of family farms for different parts of the Federal Republic by using modern econometric methods, including linear programming. These models indicated an optimum production pattern and combination of inputs. They had worked with the restriction that the acreage of the farms should be within the range typical of German family farms. From these calculations they had found that the marginal value product of land is, as a rule, not higher than the marginal value product of the capital investment in livestock production without the use of land. As the second step, they had attempted to prove these results on existing farms. The intention was to show that the price level for agricultural products in the German Federal Republic allows profitable production with adequate technical equipment.

Dr. Fleischauer said that Professor Priebe would agree with Professor Bergman that the residual imputation method of computing returns to labour and capital is not quite satisfactory. Professor Priebe's reason for doing this was that these results would then be comparable to the annual report of the government about the situation of German agriculture, which is compiled in that way.

Professor Priebe was optimistic about the development of family farms for two main reasons. The investigations that he had conducted indicated that the point at which the economies of scale started to diminish lies within the production capacity of family farms for most of the branches of production. His second reason was that he thought that the operative family farms would be able to take the initiative and find suitable forms of co-operation for the full utilization of modern technical equipment and for meeting the requirements of marketing in a competitive modern economy.

Dr. Fleischauer said that Professor Johnson had not outlined in his paper arguments which would convince us of the superiority of other forms of management than family farms. Professor Glenn Johnson showed the Cochrane model of the agricultural treadmill with some additional variables and seemed to take the opinion that the family farm is only able and willing to exist under such conditions. In contrast, Professor Priebe was of the opinion that a number of premises in this model do not fit the situation in the Federal Republic and other European countries and he believes that right technical equipment together with suitable production methods and new ways of co-operation would allow family farms to make considerable profits. Other forms of management would arise if they were superior, but Professor Priebe believed that technical progress has not yet brought basic differences to family farms. He thought that it was more likely that organisational advances might be the cause for the development of larger commercial farms.

Dr. Fleischauer said that the problem of limited access to the capital market does not fit the situation in Western Germany since the government credit policy makes credit available to farmers at interest rates less than half the general rate of interest.

Dr. Fleischauer said that the solution to the problem of the agricultural workers' desire for leisure need not necessarily damage the prospects of the family farm. It may be another reason for seeking new ways of co-operation, but Professor Priebe did not think it was necessary to give up the concept of the family farm if farmers became more interdependent. In many respects it might increase their autonomy.

Professor Tepicht said that he hoped that he could put some sort of order into the problems under discussion. We are all tempted to treat technical progress in the investment of capital as the independent variable and we are being tempted to take as the main dependent variable some measure of agrarian structure which is seen as either helping or hindering this technical progress. But concentration, which is the main subject under discussion, takes various forms and our analysis should take account of these differences.

Professor Tepicht said that there are different types of concentration according to the different structures of production, and there are various degrees of concentration within these types. In order to deepen the analysis of the progress of concentration, we need to ask how much it depends on a particular stumbling-block; in particular, the land tenure system and relationship between the worker and the land; or in other words, the professional, occupational and class structure of the country.

Professor Tepicht said that, in this connection, Professor Bergmann, in his list of factors favouring the family farm, had mentioned the absence of the possibility of choice of type of work. But the situation is similar for the specialist. Larger units make specialization easier; but the extent of specialization — and so also the emergence and livelihood of the specialist — depends largely on the size of land available *per capita*, and the opportunities for further international division of labour. Professor Tepicht said that although constraints in Europe to further concentration might be solved by further organizational measures (progress in micro-economics), he thought that the sort of macro-economic and social conditions to which he had been alluding, should be taken into account.

Dealing with the question of the definition of the family farm, Professor Tepicht said that he thought that Professor Renborg had been right in adding the adjective 'modern'. It was easy to underestimate the differences in the economic history of different countries. One dividing line which could be used between traditional and modern family farms might be the proportion of inputs coming from outside the farm. This seems to be one of the clearest dividing lines.

Professor Tepicht said that, as we well know, the problem of the family farm is caught up with a whole series of theoretical and ideological problems. The axis about which these problems turn is the notion of the independence of the family farm. The traditional farm was a far more independent unit than its present counterpart, thanks, above all, to its entrenched isolation from the circular flow of exchanges of the rest of the national economy. Its capacity for survival, even through economic crises, was remarkable though its poverty was the concomitant of its power of survival. The modern family farm, solidly embedded in the general economic system, is necessarily far less independent. This independence is even slighter in that the family farm's bargaining power in relation to its economic environment is very small ; this is due to its fragmentation faced by the powerful monopolies determining its environment.

Clearly, the miracle of 'the expansion' of the family farm as the most numerous type of unit of production remains to be explained. One of the principal explanations is provided by a compatriot of Professor Priebe, Professor Georg Blohm, in his work *Neue Orienterung der Landwirtschaft*. He gives, as compared to the pre-war period, the following indices for agriculture in the German Federal Republic : agricultural output 215, means of production 218, labour costs 318. Under these conditions 'unsalaried' farming is a virtue by necessity. But it is, at the same time, proof that agriculture, unable to offer its labour force a return similar to that offered by other sectors, remains the weakest link in the whole capitalist system.

Professor Tepicht said that, as a Marxist, he regards it as quite essential to correct in the light of experience certain over-simplified views on the development of capitalist agriculture. The same factors which restrain the differentiation of classes within modern agriculture carry the problem of the class situation over to the area of external relations ; that is, to the relationship between farming and the monopolistic powers which determine its environment. At the same time, these factors place upon bodies external to the family farm, the main burden of realizing progress in the field of economics of scale and specialization, thus more and more removing from the family farm a series of its vital functions. Thus, that which enables it to survive, to be able to tolerate a rate of remuneration quite unacceptable in other sectors, is also the harbinger of its imminent death.

Mr. Jacobi said that some speakers had referred to the booklet, *Whither European Agriculture ?*' which had been produced in his office. In particular, he was replying to Professor Nussbaumer, who had referred to relationships between social and economic trends. These statistical observations seem to indicate that in the end economic trends tend to predominate. In Professor Nussbaumer's country, Austria, and in one or two other countries, there have been policies directed to preventing the outflow of labour from agriculture. However, the rate of outflow in these countries is the same as that in countries which favour the outflow.

With respect to the problem of maintaining the family farm, Mr. Jacobi said that Professor Bergmann had given a clear picture of the difficulties involved ; in particular, the problem of the capital requirements of a successful family farm, and the problems of vertical integration. The capital required is beyond the possibilities of most farm families. If we wish to speak of the modern family farm, as Professor Tepicht had defined it, we had to get rid of the romantic idea of independence so that the modern family farm could be taken to include a family-managed farm. With vertical integration and working on borrowed capital, the farmer is to become more and more a manager and this would certainly change his position of independence. This new idea of a family-managed farm should now enter the normal concept of a family farm. Adam Smith had argued that the manager in industry who was not also the owner would be less effective. This quite clearly had completely changed now. Mr. Jacobi argued that a similar revolution might be occurring in agriculture.

Professor Bićanić said that so far in the discussion an apocalyptic view that the family farm is dying had been put forward. We are all dying but we are not yet dead. The crucial question is how long the process would take — five, fifty or five hundred years. Perhaps, rather than dying, the family farm is changing. He was more pessimistic when he viewed the statistics and the meaning that lies behind the statistics. In only a matter of ten to twenty years' time we might see a very marked change in family farms. The senilization of the owners which has been widely noted would take effect increasingly. In countries like Yugoslavia, even in the most fertile areas, 40 per cent of the farms have no known successor, and yet a few decades ago there was over-population in agriculture. In Yugoslavia, only fifteen years ago, pressure was being put on peasants to join the kolkhoz type of co-operative but it was so strongly resisted that it was abandoned in 1953. Today, the desertion of family farms is so great that the government cannot afford to buy up all the land that is becoming available.

Professor Bićanić said that the major problem in Yugoslavia is the dearth of management ability for large farms. In general, socialist countries cannot provide a very good record of efficiency. They have, it is true, large estates and show-pieces, but these are frequently in the red, while peasant farms do not show a deficit.

In socialist, as in Western countries, family farms cannot be looked at in isolation. There are acute problems of complimentarity to industrial life which are of great concern to social policy. There is, for example, an influence keeping peasants on the land because of the costs of over-hasty urbanization. Professor Bićanić cited in this connection a letter in the Youth Journal in the Soviet Union from a young man who suggested that sovkhozs and kolkhozs be split up, in order to provide land, incentives and opportunities for workers. The question of the family farm must clearly be looked at from the point of view of interdependence of agriculture and industry.

Finally, Professor Bićanić said that, like Dr. Stipetić, he would be grateful for an answer to the question, whether there is an advantage in an increase in farm scales prior to the introduction of technical innovation.

Professor Neumark said that what worried him in Professor Priebe's and Professor Glenn Johnson's papers, as well as in many other writings, is that we seem convinced that it is possible to compare some tendency in the evolution of various types of farms without considering that such evolution in the past has been largely determined by state interventions which have been of advantage to the various types of farms in different ways.

Professor Neumark said that in trying to construct a picture of future farming it is essential to take into account the fact that the interventions of the state would continue but might continue in changed form.

Professor Le Bihan said that both Professors Renborg and Glenn Johnson stressed the increase of capital intensity of agricultural production. Furthermore, Professor Glenn Johnson, in the second part of his paper, alluded to the advantages of co-operative solutions. Professor Le Bihan asked if there is not here an internal contradiction, which might apply both to Sweden and to the United States, between an increased need of capital for any agricultural unit and the need to finance even heavier investments in co-operatives which produce the means of production. Furthermore, co-operative complexes seem to be statistically less dynamic than private concerns.

Professor Le Bihan said that a process of vertical integration acts in an extremely complex way on the specialization of farms. Professor Renborg had stressed the price guarantees enjoyed by farms that have vertically integrated. But the farmer also has to contend with the problem of the price of the resources supplied by the firms within the vertically integrated system. The most dynamic integrated organizations pay higher prices to farmers than those which they would get otherwise, and thus the farmer is relieved of some of his capital shortage problem. However, the negative and non-dynamic effect of vertical integration on the development of farms must not be lost sight of.

Professor Ashton said that he wanted to make some general comments concerning the family farm. Society seems to have a great capacity to rationalize its actions after they have been committed. The concept of the independent family farm is dear to American beliefs, but Professor Glenn Johnson's paper might lead us to ask, what other means, apart from the family farm, might there have been of developing the agriculture of the U.S.A. ? Professor Ashton suggested that the difference between Professor Glenn Johnson and Professor Priebe reflects the different stages of economic evolution of agriculture in the United States and Europe. Professor Ashton said that in the United Kingdom there is no special loyalty to the concept of the family farm, although the 'yeoman farmer' appears to have been valued in the history books and literature for his sturdy independence.

Professor Ashton said that it is interesting to note what is now regarded in the United Kingdom as the vulnerable area of the agricultural industry and in need of special assistance. Recent policy statements have dealt particularly with the extension of a farm management scheme to include farms with up to two men in full-time employment. This suggests that the concept of what could be considered to be viable commercial farming is in any case much larger in size than the family farm. Furthermore, there are proposals to encourage the amalgamation of holdings and it is clear that larger units than family farms are in prospect. In the last twenty years, when the price-environment for farmers has been highly favourable, they have seen a great deal of concentration and specialization. The risks of specialized farming have been reduced and the process of concentration seems to be following a path similar to that in the United States.

Professor Ashton mentioned the difficulty facing new entrants to farming in finding capital for these larger, specialized units and stressed the importance of education to attain the high level of management ability to run what are, in effect, large business concerns.

Professor Johnson said that he welcomed the contrast with England brought up by Professor Ashton. It seemed clear that in England institutional structure and pricing arrangements of a special kind had been set up so that the larger-than-family farm could compete.

Professor Johnson said that he had not intended to act as an advocate or an opponent of the family farm in his paper. The institution might well be worth maintaining for entirely extra-economic reasons. He intended simply that its present position should be examined with a cold and critical eye.

Chapter 13

THE MOBILITY OF RURAL MANPOWER

BY

C. E. BISHOP

North Carolina State University, U.S.A.

I. INTRODUCTION

A NET transfer of labour from farm to non-farm employment has been occurring for many decades. There have been large variations in the flow over time and among countries, but the direction has been persistent. The exodus during the last decade, however, probably exceeds that of any other previous decade.

Through employment and wage policies, immigration controls, public policies concerning education and in many other ways, government has long influenced geographic and occupational mobility of manpower. Until recently, however, the movement of manpower from agriculture was largely unstructured and was influenced primarily through indirect actions rather than through policies designed specifically to enhance or to impede a decrease in agricultural manpower. Currently, there is more support than in the past for public action to influence the rate and extent of mobility. A better understanding of the forces affecting the supply and demand for manpower in different uses and of factors causing and perpetuating disequilibria conditions is needed for effective development of manpower policy.

II. TECHNOLOGICAL INNOVATION GENERATES NEED FOR MOBILITY

Two aspects of technological improvements are particularly relevant to analyses of mobility of rural manpower. They are the resource substitution effects and the scale effects (effects on firm and industry output) generated by technological improvement.

Technological innovations in agriculture are not generally neutral with respect to the different resources. Rather they alter the relative productivities of resources and thereby create incentives to change

283

the resource-mix employed. The initial impacts of increased productivity stem from improvements in biological, chemical and mechanical technology.[1] Virtually all of these improvements in technology increase the productivity of capital and alter the technical rates of substitution of capital for manpower, reducing the amount of capital which is necessary to replace a specified amount of manpower at particular levels of output. Most innovations also alter the rates of substitution of manpower and land, making it possible to reduce the amount of manpower in relation to the land needed to produce specified levels of output. The substitution effects, therefore, are predominantly in the direction of increased use of capital and decreased use of manpower and to a lesser extent decreased use of land. Thus, the substitution effects of changes in agricultural production technology decreases the demand for farm manpower.

Technological change also generates scale effects which tend to increase the demand for farm manpower. Biological, chemical and mechanical innovations are predominantly output-increasing. Each of these types of innovations is likely to shift the marginal cost curve downward and to the right, thereby increasing the most profitable level of output for the firm. The output-increasing effects of mechanical innovations are likely to be especially significant. These innovations often involve large investments for highly specialized equipment, and substantial cost reductions are obtained only if output at the firm level is greatly increased. Since the scale effects at the firm level call for increased output, the scale effects tend to increase the demand for manpower.

Manpower adjustments are complicated by the fact that increases in investment and in scale are accompanied by greater specialization of production and by changes in organizational structure.[2] These organizational improvements may occur within the firm as a result of resource recombination and greater specialization or they may occur through vertical or horizontal integration and similar arrangements between firms. In either case organizational improvements tend to decrease the amount of manpower relative to other resources used for specified levels of output. Here again, a shift in cost functions may generate scale effects which tend to increase the demand for manpower.

[1] E. O. Heady, 'Basic Economic and Welfare Aspects of Farm Technological Advance', *Journal of Farm Economics*, Vol. xxxi, May 1949, pp. 293–316.
[2] C. E. Bishop and K. L. Bachman, 'Structure of Agriculture', Chapter xi, in *Goals and Values in Agricultural Policy*, Iowa State University Press, 1961, pp. 237–249.

The technological changes which the agricultural industry has been undergoing have provided incentives both to substitute capital for manpower and to increase the size of farm firms. At the firm level the substitution effects tend to decrease the demand for manpower while the scale effects tend to increase the demand. For most agricultural products, however, the slow rate of growth in product demand and the low price elasticity of demand mean that the marginal value productivity of manpower falls rapidly as aggregate output expands. Under these conditions the only way for the scale effects at the firm level to be fully realized is through a reduction in the number of firms. Hence, farmers attempt to increase their scale of operations to increase the quantity of other resources employed per unit of manpower by combining farms and reducing the number of farms. In the aggregate, therefore, the substitution effects between capital and manpower exceed the scale effects with a sharp reduction in the employment of manpower. Under these conditions, farm people are forced to choose between (1) occupational immobility, increased under-employment and sharply reduced returns for manpower in agriculture, or (2) large-scale transfers of farm manpower to non-farm occupations making possible a higher return to the manpower remaining in agriculture.

It should be emphasized that the processes of adjustment which have been outlined above are inherent in technological and economic progress in agriculture. They are not once and over changes. As economic development occurs, capital is accumulated and it becomes relatively low-priced in comparison with manpower. This change in relative prices causes intensification of the search for new technology which further increases the rate of substitution of capital for manpower.[1] Additional forces are introduced, therefore, which orient research and invention towards a further decrease of manpower employment in agriculture. This decrease in the demand for manpower in agriculture is a normal component of economic growth.

III. A MALFUNCTIONING MARKET FOR MANPOWER

If manpower resources were homogeneous, and if factor markets functioned perfectly, agricultural innovation would be accompanied by transfer of sufficient manpower from farm to non-farm employment to maintain equal net marginal returns for resources in both

[1] J. R. Hicks, *The Theory of Wages*, Macmillan, 1932, Chapter vi.

sectors. During the last two decades technological change in agriculture has been at a very high rate. The burden on factor markets has been very great. In Western Europe and in North America the labour market has transferred millions of persons from farm to non-farm employment. Even so, numerous studies have demonstrated that at best the market for manpower operates with a substantial lag. There are, for example, persistent differences between industries and between occupations in the returns for comparable manpower services.[1] These differences persist in spite of continued occupational and geographic transfers of manpower. If the market performed perfectly, wage differences between farm and non-farm employment would be eliminated except for costs of manpower transfers, costs of acquisition of skills and differences representing heterogeneity of manpower resources or employment and living conditions.

The persistence of low manpower returns in agriculture in spite of extensive migration is consistent with physical and economic underemployment of manpower in agriculture.[2] Although employment on a full-time basis without regard to productivity is hardly a worthwhile goal, there is substantial involuntary unemployment of rural manpower, and much manpower is employed in relatively low value productivity uses. The employment of manpower in uses where it produces a lower value product than is possible in alternative uses is economic underemployment.

Another indication of possible malfunctioning of the market for manpower is found in the large difference between gross and net transfers of manpower from farms.[3] Even if the market functioned perfectly, the ratio of net to gross migration from farms would be less than unity since some people would find it profitable to transfer from non-farm to farm employment. There is evidence, however, that a substantial part of the transfer of manpower from non-farm to farm occupations represents a return of previous migrants.[4] Much of this return migration must be presumed to represent disillusioned people who had mistaken expectations at the time of migration from farms. As such, this migration likely involved high private and social costs.

[1] J. R. Bellerby, *Agriculture and Industry, Relative Income*, Oxford University Press, 1956.

[2] See, for example, F. Dovring, 'Manpower Problems in Agriculture', OECD, Paris, Jan. 1964, or C. E. Bishop, 'Underemployment of Labor in Agriculture', *Journal of Farm Economics*, May 1954, pp. 258–272.

[3] C. E. Bishop, 'Geographic and Occupational Mobility of Rural Manpower', OECD, Paris, 1965.

[4] B. B. Perkins, 'Labor Mobility between the Farm and Nonfarm Sectors', unpublished Ph.D. dissertation, Michigan State University, East Lansing, Michigan, 1964.

Surely there must be a less expensive method of obtaining information concerning the market for manpower than through personal experience in various occupations and locations?

IV. IMPEDIMENTS TO MOBILITY

The transfer of manpower from farm to non-farm occupations usually involves geographic as well as occupational mobility. In a market context, a transfer of manpower from farm to non-farm employment is expected when the potential earnings in non-farm employment exceed the opportunity cost in farming by more than the direct costs of transfer. There are many impediments, however, to this flow of manpower.

First, it must be emphasized that people have heterogeneous characteristics. The demand for manpower is not a demand for people as such but for their services. The demand for manpower differs with the characteristics of the owners. Non-farm employers incur costs in recruiting and training new workers. They seek to obtain highly productive workers who have a potentially long tenure of future employment. Persons of young age with highest education and training are usually given priority. When relatively full employment prevails and demand is strong, selectivity with respect to population characteristics is less important in determining who migrates than when the non-farm demand for labour is expanding slowly and when rapid technological improvement is occurring in agriculture. When employment is increasing rapidly, for example, new entrants into the labour force and older workers encounter less difficulty in obtaining non-farm employment than when employment is growing slowly.

Even though technological improvements in agriculture decrease the demand for manpower services, it does not follow that these same services are freely transferable to non-farm employment. The manpower in agriculture constitutes a potential supply of manpower to non-farm employers. However, the persons owning this manpower differ with respect to age, sex, education and training and other characteristics which affect productivity and earning capacity. These characteristics of the population affect the supply of manpower to non-farm employers. Since education and training increase the non-farm earning potential, the opportunity cost to farming is increased as education is increased, and migration is more likely to take place.

Age also affects the supply of manpower which is transferred from farm to non-farm employment. Age is less of an impediment to occupational transfers which can be made without geographic transfers than for those which involve migration. Both the direct money costs and the non-pecuniary costs of mobility are greater when geographic transfers are involved. Those in the younger age groups have a longer working life after migration and, therefore, have a longer time in which to recover the costs involved. They also are likely to incur lower non-pecuniary costs than older persons who are well established in their communities.

The direct costs of transfer may be prohibitive for some families. Many people who move are in the lower income groups and have little savings. If there are significant limitations on the availability of capital or if people are reluctant to borrow to finance migration, manpower transfers may be reduced. Where capital is a limiting factor, or where it is desired to encourage people to move from an area, payments to subsidize the transfer may effectively reduce the barriers. Several countries recently have initiated a system of transfer payments to encourage selected types of manpower transfers.[1]

In the transfer of manpower from farm to non-farm occupations, it may be necessary to change the farm product-mix or to decrease farm output. In such cases the farm income forgone constitutes an opportunity cost of the manpower transfers. These costs have been increased arbitrarily by agricultural fundamentalist policies and programmes. The opportunity costs of occupational transfers are affected importantly by policies and programmes which support the prices of farm commodities, and indirectly the return for farm resources, at artificially high levels. For example, a recent study shows that in the south-eastern part of the United States an increase of $20 in *per capita* farm income is accompanied by a decrease in farm to non-farm migration equivalent to 1 per cent of the farm population.[2]

Most countries employ some form of government controls to support income levels in agriculture. Such policies increase the costs of manpower transfers, and in countries experiencing full employment may substantially decrease mobility and alter the national product-mix.

---

[1] For example, Sweden, the Netherlands, France and the United States have programmes which may be used for defraying costs of transferring from rural to urban areas. France subsidizes relocation within agriculture.

[2] W. D. Diehl, 'Farm-Nonfarm Migration in the Southeast : A Costs-Returns Analysis', unpublished Ph.D. dissertation, North Carolina State University, Raleigh, N.C., 1964.

The uncertainties associated with future employment and incomes may impede mobility. If expectations concerning the variance and range of future earnings differ among occupations, the allocation of manpower resources will be affected. In areas where people living on farms have had adverse non-farm employment experiences, future non-farm earnings may be discounted heavily.

The non-pecuniary costs of transferring manpower to non-farm employment and cultural differences among communities undoubtedly are major impediments to migration.[1] As pointed out earlier, occupational mobility out of farming frequently is accompanied by geographic transfers. Non-pecuniary factors associated with occupations, such as status, position, etc., impede mobility. Differences in the cultural milieu and in occupational and community status may generate especially significant non-pecuniary costs of occupational and geographic mobility for people in culturally isolated areas. Non-pecuniary costs are more important when both geographic and occupational mobility are involved since the migrant frequently finds it necessary to break established cultural ties and to adjust to a new environment. Other things being equal, people prefer to remain in their present location. Voluntary migration takes place only when people expect to improve their welfare thereby.

Mobility also is impeded by lack of information concerning non-farm earning potentials. Informal dissemination of information through friends and relatives is the major medium through which rural residents learn of non-farm employment opportunities.[2] People in the more remote rural areas possess relatively little information concerning employment and living conditions in industrial areas and are inclined to view demand as insufficient to warrant transfer. Furthermore, most people living in rural areas are not knowledgeable concerning existing sources of information other than friends and relatives, and they depend heavily upon friends and relatives for information concerning non-farm employment opportunities. Under such a system information obviously moves through limited channels, and the information market performs best near centres of industrial growth. It may well be that the pattern of dissemination of information is one of the major determinants of the pattern of migration.

Controlled entry into non-farm occupations also impedes occupational transfers. There is little doubt that many people employed on farms prefer non-farm employment at prevailing relative earnings.

---

[1] E. D. Smith, 'Nonfarm Employment Information for Rural People', *Journal of Farm Economics*, Vol. 38, 1956, p. 820. [2] C. E. Bishop, *op. cit.*

The specification of excessively high standards for jobs and other restrictive measures limit transfers of manpower. Such restrictive practices are likely to be more important in an economy characterized by extensive under-employment in rural areas and by involuntary unemployment in urban areas.

In some countries manpower mobility also is limited by controls placed upon the construction of housing. Many migrants encounter difficulty in obtaining adequate housing for their families. These experiences may result in a return to rural areas and in discouragement to other potential migrants. The failure to develop adequate housing and other facets of the infrastructure is a major limitation on migration to some large urban centres.

Another impediment to manpower mobility is the difficulty of altering the structure of agriculture. If agricultural technology could be developed so that the demand for manpower was uniform over seasons and resources could be recombined so as to provide full employment, mobility would be expedited. Unfortunately such is not the case. There is substantial seasonal variation in the demand for manpower. Economists are entirely too prone to consider manpower which is not used at all times to be unemployed or under-employed. The marginal productivity of manpower is very high during some seasons of the year even though it may be near zero during other seasons. Mobility decisions must give consideration to the relevant opportunity costs. If migration is involved, mobility may mean that it will be necessary to forgo the production of some commodities. In the absence of technological improvements which reduce the seasonal variations in demand for manpower, extensive out-migration may not be possible without a pronounced change in the product-mix. Furthermore, even if technological improvements are developed, they may alter the competitive position of areas and result in changes in the location of production.

The development of specific policies to encourage mobility is made difficult by the fact that some people suffer losses as a result of out-migration of others. When large-scale out-migration occurs, the productivity of non-human resources may fall and resource prices may fall accordingly. In some areas outflow of manpower may be accompanied by substantial reductions in the demand for and price of land. Many landowners are reluctant to encourage a transfer of manpower when it is necessary for them to absorb the losses. Consequently, objections to labour transfers have made it difficult to recombine land and to transfer it to more extensive uses. In recognition of this some governments have taken steps to subsidize the transfer of land where such transfers appear to be desirable.

External effects to manpower transfers also may have important economic implications for others. For example, population reduction in an area may involve decreased demand for many non-farm products and services. Public policies and programmes aimed at facilitating migration, therefore, often are opposed by local merchants and persons employed in consumer service industries as well as by agricultural service and product processing industries.

V. MOBILITY POLICY

Most countries now have developed some form of public assistance to facilitate internal migration. The kind and extent of assistance varies greatly among countries. Some countries have programmes specifically designed to encourage the transfer of manpower from farm to non-farm employment. Other countries have policies designed to slow up this transfer. Certainly no single policy will meet the needs of all countries. Neither occupational nor geographic mobility are desirable as such. Increased mobility may yield social gains in some countries while producing social losses under other conditions.

The development of mobility policy should begin with analyses of the need for mobility. Perhaps the first task is to estimate the potential of the farm sector to employ manpower services at rates of pay comparable to those paid for manpower services in alternative employment. The kinds and amounts of farm products which must be forgone, if any, as various quantities of manpower are transferred from farm to non-farm employment should be determined. Such analyses should be conducted under alternative farm organization and consolidation schemes. Changes in the amount and form of capital invested per unit of manpower as well as changes in the product-mix may be necessary to generate reasonable returns for manpower services in agriculture. In view of the necessity for co-ordinating production and marketing decisions in a market economy, such analyses should consider differences among regions and probably should be conducted on an area basis.

Individuals and families base occupational and locational decisions upon many factors. One of these factors is the payment for manpower services in alternative employment. Counselling with respect to the earnings potential in alternative employment is an important facet of any mobility policy.

Training and retraining programmes are important in developing the mobility potential of people. These programmes should include improvement of the productivity and managerial ability of those who remain on farms, training of those who will transfer to non-farm employment and the general education and guidance of community leaders.

A fourth facet of mobility policy concerns programmes designed explicitly to expedite migration. The dissemination of information concerning employment opportunities, wages, working and living conditions may affect the amount and pattern of mobility. Mobility and adjustment payments also may be used effectively to encourage mobility where desirable.

Mobility is incomplete until those who have migrated have been assimilated into the new environment. Various forms of adjustment assistance have been used effectively. The return for this assistance may be very high if it means the difference between assimilation and return migration.

In summary, although no mobility policy will apply to all countries, consideration should be given to five facets. They are (1) the employment potential of agriculture, (2) counselling with respect to income potentials in alternative employment, (3) training and retraining programmes, (4) mobility assistance programmes, and (5) programmes of assistance in assimilation in the new environment. Failure to perform effectively in any of these facets may impair the entire mobility process. Finally, as pointed out earlier in this paper, the structural changes taking place in agriculture are not once and over changes. Manpower policy can be developed more effectively, therefore, in a dynamic context giving due consideration to impending changes in production technology.

———

DISCUSSION OF PROFESSOR BISHOP'S PAPER

Professor Dupriez said that Professor Bishop's paper made it necessary to consider the problems of one of the main phases of economic development — that of active industrialization and movement of workers out of the country-side. The picture is one that is very familiar in Europe today.

The major question is how to formulate a manpower policy that is appropriate to this situation. Is it simply a question of helping those who transfer from agriculture to industry ? This is undoubtedly a difficult problem since those who transfer are commonly ill-equipped to go through with it ; agricultural migrants typically form the poorest and technically the least advanced part of the population. A more basic question brought up by the paper is whether migration should be encouraged to advance more rapidly. He said that he had doubts about this. In the past mobility has been greatest among younger groups who have left the farms more rapidly than they can be absorbed by industry. The mushroom growth of European cities in the nineteenth century was accompanied by endemic unemployment of rural migrants. This problem of ensuring that there is sufficient absorptive capacity still remains, even though in less acute form. The aim should be to attract migrants by the development of absorptive capacity.

Professor Dupriez said that the very compact first section of Professor Bishop's paper left some distinctions rather unclear. For example, there are important distinctions to be made between plant, firm and industry, and between the different sorts of technical innovation. If the different types of technical innovation are taken in the order in which they have occurred chronologically in history, the mechanical had preceded the chemical and the chemical had preceded the biological. This chronology is important in any long-run view of developmental processes, since to each type of innovation correspond different degrees of capitalization and different implications for manpower policy. For example, biological innovation does not necessarily imply greater capitalization or release of manpower. In fact, only one form of innovation — mechanical innovation — necessarily involves higher capitalization.

Professor Dupriez said that different types of land use and improvements in land use also have to be examined for their effects on the manpower situation. Improvement sometimes involves moving from one type of agriculture to another, and even in the case of the intensive agriculture of the countries of the Common Market there are important changes in the way that land is being used. In developing countries a diversity of solutions would be found, and it would not always be true that the rural population would be too large.

Professor Dupriez said that he was not clear what general arguments could safely be made about the effect of economies of scale on the manpower situation. Is it true that increases in farm scale bring about increases in manpower requirements ; and, if so, is this true of any size of farm ? Certainly it is essential to distinguish here between these scale effects at the farm level and at the national level. The national effect will depend upon whether the increases in farm size are offset by reductions in the number of units.

The low financial productivity of agricultural workers reflects the low physical productivity of labour in agriculture compared with the productivity of other factors. The labour market difficulties are not essentially different from those experienced in any industries which are in relative decline. Where sectors are declining, people invariably hang on and it becomes difficult to promote technical progress at the necessary speed to ensure viability for the remaining units. Agriculture seems to enjoy privileged treatment only because the problems of transfer usually involve so many, and involve changes in location. The scale of the agricultural problem is aggravated by the fact that in periods of economic depression, when there is not enough work for all, there is a tendency for people to try to move into, or remain in, independent occupations where it is possible to share out among the under-employed the small margins that can be made. Agricultural activities form part of the great mass of semi-independent activities which are typical of under-development and which will only disappear under the stimulus of fast and sustained development.

Professor Dupriez said that on the question of obstacles to mobility, the problems vary according to the economic and geographical context. While in the United States and many under-developed countries the costs are high, the costs of transfer in Western Europe are lower because migrants can apply first to economic centres close to their original homes and then move gradually to the larger centres of economic activity.

Professor Dupriez said that the question of the lack of adequate information depends on the point of view taken. As a welfare problem for the worker it is serious, but for the industry which requires labour, it depends on the type of vacancy they have and upon the amount of absorptive capacity they have, as to whether the spread of information is sufficient.

Professor Dupriez asked why Professor Bishop thought that area studies should be used in this problem. Might not the area approach (where an area is taken to be smaller than a nation) restrict the point of view so that national and global needs will be neglected. The problem is essentially one of transfers between productive sectors of the whole economy.

Finally Professor Dupriez referred to the problem of international migrations where the productivity of entire populations is at issue. This, he said, presents the really large structural problems, and compared with these the domestic problems of adjustment are rather a question of lags rather than of fundamental difficulties.

Professor Valarché said that Professor Bishop's analysis deals with the mobility of agricultural labour, rather than with that of rural labour. The causal factors were given as the productivity of farm labour and the nature of the demand for farm products. These factors might bring about different results for agricultural and non-agricultural rural populations. In Switzerland non-agricultural population is less mobile than agricultural population. The complementarity between these two sorts of rural populations should be included among the causes of rural mobility.

The feminization of agricultural labour is not a general phenomenon ;

both in Switzerland and in France, girls tend to leave the farms first. Thus there is at least a transitional increase in the proportion of men in the agricultural labour force. Here again, the complementary relationships between different parts of the rural work force should be examined.

Professor Valarché said that he wondered how Professor Bishop's analysis would fare when applied to centrally planned economies. The substitution process between capital and labour would not be the same since prices are not built up in the same way. Is it possible to find a common measure to express the stimuli which lie behind the rural exodus in both capitalist and socialist countries? Can man-hours per bushel of wheat be used for this purpose? There is information available to make this comparison, and he wondered whether it would bring significant results.

Professor Campbell said that the Australian situation shows characteristics that are rather different from those described in Professor Bishop's paper. First, until very recently there has been no absolute decline in the size of the agricultural work force. For most of the twentieth century, there has been a relative decline in the number of rural workers associated with a rise in absolute numbers. Second, it appears that there is no significant differential between the return to labour in agriculture and in manufacturing. The reason for this has not been fully investigated, but it seems to be related to the rate of development of manufacturing and the rate of technological development in agriculture. The results are purely fortuitous, and had not been planned.

In Australia rural–urban migration takes place over far greater distances than the 25-50 miles that was mentioned as typical in the literature. Industry is highly centralized and people commonly move three or four hundred miles to take urban jobs.

Professor Campbell thought that too much emphasis is placed on migration from agriculture to the manufacturing sector. In Australia a significant number of the people transferring from rural to urban areas are young people in the service industries. These industries, such as banking, education, railways and public utilities, are characteristically organized on a state or national basis in Australia. They are thus in a position to recruit staff in rural areas to transfer them subsequently to urban locations. In as much as rural youth are attracted to the professions and to service industries requiring superior educational qualifications, recruitment from the country-side gives rise to concern about the drain of superior intellects from agriculture at a time when the need for advanced management skills in that industry is growing.

Professor Obolenski said that socialist countries are also concerned with the spontaneous reduction in agricultural manpower which accompanies technical progress. In capitalist countries this progress is accompanied by the merging of farms and by horizontal and vertical integration. In socialist countries the reduction of manpower is achieved by central planning so that the manpower is redirected to other branches of the economy.

The Family Farm and Problems of Factor Mobility

The problem of unemployment in agriculture is to a certain extent eased by the reduction in working hours by government regulation, thus liberating the farmer for cultural and other activities. The existence of large co-operative and state enterprises makes the problem much easier to solve.

Professor Obolenski said that he would be grateful if Professor Bishop would outline what kinds of industrial enterprise can be usefully introduced in rural areas. What opportunities are there for creating agro-industrial complexes which can facilitate the use of manpower — otherwise used only seasonally — all year round ? Where manpower is not used throughout the year, there is difficulty in deciding where there is true unemployment. In the case of full-time farming families, it is perhaps misleading to call them unemployed. It is the hired labourer, who is obliged to live off his savings during periods of no work, who is truly unemployed.

Professor Dandekar said that Professor Bishop's paper approaches the problem in very general terms and gives the impression of having covered the whole field. It is important to realize that he is in fact talking about a very special case — that of technical innovation releasing manpower. This should be made very clear. There are several other situations under which agricultural labour might move. In a situation of stagnation in agriculture and a fast-growing population, agricultural labour is obliged to move even though there are no job opportunities.

Professor Nicholls said that as Latin America was not adequately represented in the Conference and as he had become 'half-Brazilian' through increasing experience in that country, he wished to make some comments about labour mobility in Brazil.

Professor Nicholls said that despite centuries of poor transport and communication, Brazilians have been a restless and relatively mobile people because of the continuing existence of an agricultural frontier. The result has been an exploitative attitude towards land with agricultural population mining the soil to near exhaustion and then moving westward to follow the same practices on new lands. Thus, even now, there continues to be a large-scale rural-to-rural migration from the over-populated and poor north-east and from the small farms of the extreme south to the rich new agricultural areas of such states as Minas Gerais, Paraná and Mato Grosso. In these new areas, labour tends to be relatively scarcer ; and, where the technology of the major crops permits it, mechanization is more advanced, properties tending to be relatively large and well managed. Here, the incomes of farm worker families and landlord–tenant relations tend to be relatively favourable. Even so, those who are fortunate enough to obtain land as share-croppers in the north-east do not have a strong income incentive to migrate, even though their marginal products are very low. None the less, their numerous children (or others unable to gain access to land as population grows) frequently find it necessary to migrate to newer agricultural areas or to the cities of the south and north-east.

In agricultural areas near to the major cities, where locational (market) advantages are bringing about rehabilitation of once abandoned lands (for example, in the Paraíba Valley between the cities of Rio de Janeiro and São Paulo), migration consists of two streams : (*a*) rural people born in these areas, who tend to shift to near-by industrial-urban jobs ; and (*b*) rural-to-rural migration which fills farm-labour needs (considerably reduced by mechanization) at relatively low cash wages by short-distance movement of people from the hillier areas where farm incomes are very low. It is interesting to note that, on the large mechanized properties of such areas, the net incomes of cash-wage worker families are actually inferior to those of share-cropper families in the north-east.

Professor Nicholls said that there is a major stream of direct rural-to-urban migration, much of it from the rural north-east to the large cities of the south, especially since the extensive construction of improved arterial highways in the last decade. Probably many of these people fail to increase their real incomes very significantly. However, they gain greater personal independence, the more diverse excitements of the city, and a better opportunity for educating their children. Meanwhile, the population pressure on provision of urban public services and infrastructure is proving to be almost insurmountable.

Professor Nicholls concluded that under these circumstances, Brazilian manpower policy certainly needs to avoid further stimulation of migration and farm mechanization until the growing non-agricultural sector gains enough capacity to absorb them satisfactorily and to support the expansion of the public services which they require. Meanwhile, an attack on the high rural illiteracy rate and a vast expansion of agricultural research and technical orientation appear to be essential.

Professor Tepicht said that he would like to give Poland as an example of special conditions which could obtain in this field. As a country at an intermediate stage of development, Poland is 'on the move' in a number of respects and increasing occupational mobility is one of them. The years of the main shock of industrialization saw the departure of $2\frac{1}{2}$ million people from the country-side; after this the rate of migration slowed down. Today occupational mobility is less often accompanied by geographical mobility. At the same time a large number of families have adopted a system of having two sources of revenue, retaining part of their work-time for agricultural activity. Out of the total of 3 million farm families, approximately 1 million engage in some part-time industrial work. This section of the population enjoy income privileges not enjoyed by the remaining part of the peasantry ; so much so that there is even a tendency to try to reduce these privileges. However, smaller farms were receiving economic support, chiefly through contracts for labour-intensive products, and this achieves greater equity. Professor Tepicht said that he thought that this part-time activity is an important factor in the overall growth of the Polish economy. In the present stage of development there is a need to increase the productivity of labour in industry, and it has to be supported by an increase in the quality, and to some extent in the numbers, of those employed in agriculture. Part-time families have effected a latent migration which continues to provide a significant share of agricultural production. At the same time, because they enjoy two sources of income, the members of these part-time families are able to provide cheaper manpower for industry at a time when it faces a difficult turning-point towards relatively rapid expansion. In the context of war damage and the need for the development of the infrastructure these people play an important role, and the skills acquired in the process of doing part-time work outside agriculture have reflected back on the agricultural activities themselves in the form of improved farm buildings and increased skill in using modern techniques. In this way a new sort of mixed rural community is coming into being ; it is a solution that might well be very attractive to many developing countries.

Professor Robinson said that there is one aspect of the mobility question which is easily forgotten. This is the immense mobility of population over short distances. The way to make use of this was learned the hard way during the war, when the diversion of labour between industries was a crucial factor in the war effort. This has important application for the present problem. If the aim is the contraction of the farm population and if, at the same time, there is a demand for labour to build nuclear power stations, the two ends can best be served by a whole series of short-distance movements. In many regions there is very high flexibility of employment within the distance of the normal journey to work. In the United Kingdom there is this very high mobility over 10-15 miles. Movement out of farms has been achieved in the first instance over these short distances with the local construction industry frequently being the first non-farm employer (just as in the past it was often the police or local service industries which played this role). After such transitional employment, these workers will often move to other occupations in widely dispersed parts of the country where their new skills enable them to find employment. By a chain of movements it therefore becomes possible to direct farm population into activities which are expanding. Professor Robinson said that in this matter there is clearly an important difference between densely settled Europe and less densely settled America, but, in most contexts, this idea of stimulating short distance mobility should guide policies of industrial location and development.

Professor Bishop said that he had intended his paper to relate specifically to conditions which exist in Western Europe and North America, so that neither centrally planned economies, nor economies at earlier stages of development, are covered by his analysis. However, it is to be expected that this analysis would apply to many of those countries in the future.

On the question of manpower policy, Professor Bishop said that while governments have shown little reluctance to develop special policies in land use, credit use, education and vocational training within agriculture, they have shown great reluctance to develop a sound manpower policy. Underlying this is the agricultural fundamentalist bias ; the idea that there is something 'super-good' about agriculture. This has meant that rural youth is at a serious disadvantage, so that their potential is not realized and they are limited in the kinds of employment that are open to them when they have completed their education and found that they have to migrate. In this respect there are two related problems : in the short run, what is to be done with the skills actually possessed by the young ; and, in the longer run, how is the skill-mix to be altered by seeing that rural youth enjoy education that is comparable to that of the rest of the population.

Professor Bishop said that whether one thought migration has been sufficient or not depends on what expectations one held. He said that he found impressive the scale of migration over the last ten years. However, OECD studies show that there is still substantial economic under-employment of labour in agriculture. At the same time it is difficult to reconcile this declared under-employment in agriculture with the tendency in a number of countries to import large numbers of people for non-farm employment.

Professor Bishop said that he had not intended to imply any special chronology in the adoption of the different kinds of agricultural innovation. In his own part of the U.S.A., the earlier innovations were largely biological and chemical, while in other parts a different order might be found. But none of these agricultural innovations are neutral, and all tend to alter the production function by increasing the productivity of capital relative to that labour. At the same time as bringing about a substitution of capital for labour, these innovations are such that firms seek to increase their scale of operations : and the only way in which the lower marginal costs associated with the increased size and the capital-intensive methods can be realized, is in the reduction of the number of firms.

Professor Bishop said that while he agreed with Professor Dupriez that under-employment is not peculiar to agriculture, it is more persistent in agriculture than in other sectors, and, because of the geographical factors involved and the complexity of the aggregate picture, far more difficult to solve.

Professor Bishop said that although his analysis is aimed at developed countries, there were important similarities in under-developed countries which can be comprehended in it. Even in terms of the U.S. scene, much of the under-employment represents lack of job opportunities, as in the case mentioned by Professor Dandekar. In many situations, including some in the United States, the supply of labour from farms to non-farm employment is to be regarded as being perfectly elastic with respect to income. Demand basically determines the amount of migration which will take place under these conditions. It is only in a full-employment economy, where labour has to be imported to operate non-farm industries at the desired level, and where the room to manœuvre is much reduced, that the dissimilarities are really great.

Professor Bishop commented, in response to Professor Obolenski's questions, that he did not feel optimistic with respect to continued expansion of part-time employment in farm and in non-farm industries. Most non-farm industries cannot afford the inefficiencies associated with part-time operations. Nevertheless, some industries which use agricultural products as raw materials or which produce products for use in agriculture may find seasonal operations in rural areas to be advantageous. Professor Bishop indicated that he prefers to use the concept of economic under-employment rather than unemployment when analysing the use of farm labour. Labour is considered as under-employed in agriculture when the economic return for the labour services can be increased by a transfer of labour to other industries. This concept places emphasis upon the contribution of labour rather than upon employment *per se*.

Professor Dams said that Professor Bishop's paper needs to be considered within the context of attempts to improve agricultural structures, and, in particular, the relationship between workers employed in agriculture and the capital invested. In a growing economy there is usually a considerable increase of cultivated area per worker, and this has to be achieved by changes in the relationship between capital and labour so that an equitable level of return is available to those who work in agriculture. In most of the countries of Europe, it is not enough simply to stimulate the exodus of agricultural populations. Direct action has to be taken to increase the size of farms. In this attempt to bring about structural change the adaptation of agriculture has to be seen from two sides : first, the mobility of man-power so that it can move into other sectors, and second, the mobility of land so that it can be joined to larger and more economically viable under-takings. These two considerations are intimately related. Thus, while mobility of manpower towards other sectors is typically realized by the young, the mobility of land can often only be realized by measures of a social character such as pensions and aided retirement. Professor Dams said that it would be interesting to compare the different schemes adopted by the major European countries to reduce the number of agricultural workers. He said that such schemes should be directed towards the rationalization he had referred to — the creation of viable units. The recent British policy statement on the development of agriculture is an example of such a scheme. In short, the simple stimulation of manpower mobility has to be accompanied by an active adjustment in rural society and the creation of viable farms.

Dr. Flek said that Professor Bishop had dealt both with the problems of professional transfer and of geographical transfer of manpower. In the Czechoslovakian experience, there had been professional migration without geographical migration. He asked whether there are any similar examples in the United States with which comparison could be made.

Dr. Flek said that he was not clear what Professor Bishop means by 'controlled entry into non-farm occupations'. Does this refer to the fact that industry is not interested in the unskilled ex-farm manpower, or does it refer to specific state regulation and control?

Professor Gulbrandsen said that the last four paragraphs of the section on the impediments to labour mobility in Professor Bishop's paper seem to him to be essentially wrong and in need of modification. The construction of housing is only a serious impediment if there is rent control. Otherwise the impediment comes from increases in the rent of apartments as the demand from those moving out of agriculture increases. Professor Gulbrandsen said that he thought this is a minor impediment, since those already using the room space will themselves adjust to the higher rents and be forced to live in smaller apartments and use a smaller number of rooms within the existing stock of houses.

Professor Gulbrandsen said that seasonal unemployment forces agricultural people to try other jobs during part of each year. This gets them acquainted with urban life, and thus promotes the permanent transfer of labour into other occupations. Thirdly, he said that if land prices fall, it becomes easier to establish large farms, not more difficult. Finally, Professor Gulbrandsen said that if external diseconomies affect other rural occupations, this will tend to accelerate migration, not impede it.

Professor Mouton said that the concept of capital used by Professor Bishop should be clarified. The question is whether one should include the underlying natural elements, and particularly the complex of energy-producing factors which derive from soil and climate. Professor Mouton said that he thought that it is necessary to include all the factors of production. To each set of soil-climate conditions corresponds a certain potential which, given some initial chemical and energy levels, and given certain techniques of agricultural production, would tend towards lower or higher entropic conditions. At the present time the soils of under-developed countries, largely because of their geographical location, are burdened with a 'congenital fragility'. This gives rise to the need, in the present stage of technology, to maintain long rotation patterns with long periods of fallow in order to avoid a permanent sterilization of the soils.

Professor Mouton said that these agronomic reasons explained why he thought that chemical and biological innovation are not neutral in the evolution of capitalist agriculture. The problem is that of assessing both on the technical level and on the economic level the value of different sorts of 'manure' in raising the productivity of particular factors. In the light of these comments Professor Bishop's discussion of the role of the factors of production is ambiguous. The concept of manpower as used by Professor Bishop is of particularly limited value. It might well be maintained that the comparison of productivity between individual workers is impossible in view of the different conditions of work in the different areas. Over and above the soil and climate factors is the seasonality of all agricultural activity. Any manpower policy has to take into account the complex problems of heavy seasonal demand for labour and the technical and economic feasibility of mechanization of labour-intensive activities. The solutions that should be adopted in the light of these considerations will not necessarily be the same from region to region.

Professor Bishop said that a number of the questions had turned upon the cost of migration, with the implication that in Europe it is not prohibitively costly. Why then has it proved necessary in a number of countries (for example, Holland and Sweden) to subsidize the movement of workers ; and why in other countries (for example, Germany, the Netherlands and France) have programmes of industrial decentralization been instituted ? Professor Bishop said that he was not persuaded that the costs of migration are inconsiderable in Europe.

Studies for the demand for labour on a regional basis are necessary, partly because studies on a national basis do not produce meaningful results because the level of aggregation is too great. Inter-regional competition within nations as well as competition between nations, render the picture too complex to be satisfactorily dealt with on an aggregate basis. Moreover, the areas taken had to be small enough to make it possible to consider altering the product-mix. In the United States there is also the need to withdraw some land from production in order to obtain a reasonable allocation of resources, and a problem such as this can only be dealt with on a regional basis.

Professor Bishop said that there is little known about the people who move from agriculture to non-agricultural employment. There is no data on occupational mobility after the initial movement away from the farms. It is known that females precede males, and that people from rural areas enter the unskilled trades, but otherwise the information is thin.

Professor Bishop said that he agreed with Dr. Dams that the relationship between the mobility of land and the mobility of labour is very important and that the re-structuring of land is, from the point of view of farming, one of the main purposes of this increased mobility. He said that he was also grateful to Professor Nicholls for emphasizing that it is not only in earnings that the agricultural workers and particularly the young in the rural population are at a disadvantage. They are discriminated against in many other ways such as in educational opportunities.

Professor Bishop said that he did not have much optimism about the future of part-time farming. In many cases the industries into which farm labour move for part-time work are undergoing similar structural changes as those which are affecting agriculture. For example, lumbering in Sweden is becoming a full year-round occupation rather than a part-time seasonal and highly labour-intensive activity.

In answer to Dr. Flek, Professor Bishop said that by 'controlled entry' he meant to refer to the host of factors which impede development of industrial employment and lend rigidity to industrial wages.

Professor Bishop said that he thought Professor Gulbrandsen and himself were not understanding one another very clearly. Professor Bishop said that arbitrary controls on the construction of housing will limit the supply of housing thereby impeding migration. While seasonal unemployment might encourage part-time farming, employment opportunities in non-farm industries are quite limited in many rural areas. Hence the choice may be one of seasonal unemployment or of migration to other areas. Thirdly, while a decrease in the price of land may be viewed with favour by potential buyers, and therefore may encourage consolidation from their standpoint, a decrease in price will likely reduce the quantity of land supplied for sale by current owners. If these owners are in older age groups and have few alternatives they may become much more reluctant to sell as land prices fall. Finally, those non-farm people who are affected adversely by migration of farm families must be expected to oppose legislation designed to expedite migration unless they are compensated for losses incurred by them. However, the main point is that part of the resistance to public policies to facilitate migration can be overcome by appropriate compensation of those who would incur losses as a result of the migration.

Chapter 14

LIFE AND INCOME OF CZECHOSLOVAK CO-OPERATIVES

BY

JOSEF FLEK

Prague, Czechoslovakia

I. INTRODUCTION: THE PROBLEM

OPENING the doors to the employment of workers (even non-qualified) in industry has removed not only all the manpower reserves from Czechoslavak agriculture, but has also led to the departure of people who would have been valuable and even necessary on the farms. In the period 1950–63, 850 thousand persons left Czechoslavak agriculture, *i.e.* 40 per cent of the initial number, and the rate of loss shows a rising trend.

Conditions for replacing this living labour by labour materialized in means of production, thus allowing a substantial increase in the productivity of labour, in the intensity of production and in the overall volume of production have, however, not been secured, or rather their formation was delayed in comparison with the rapid, uncontrolled loss of manpower. The productivity of labour (measured by the value of gross output per one stable worker) increased approximately three times up to 1963, as compared to the pre-war level (*i.e.* compared to data of the census of farms of 1930) ; but with respect to the increasing needs of the national economy and its objective possibilities this increase is still insufficient.

The uncontrolled loss of manpower was felt most sensibly in the fundamental, decisive sector of Czechoslavak agriculture — in the agricultural co-operatives, which at present work about 70 per cent of the arable land (a little more than 20 per cent being worked by state-owned farms, and not quite 10 per cent by individual farmers).[1] Since the growth of mechanization lags behind the rapid loss of manpower, a situation has come about in Czechoslovak agriculture,

[1] The average co-operative now works 600 hectares, a state-owned farm about 4,000 hectares.

304

where an optimum situation in manpower is, at the given technical level, one of the decisive conditions of the successful operation of a farm.

This of course proves that the lack of manpower in Czechoslovak agriculture is only relative, caused mainly by the fact that our agriculture did not succeed in flexibly compensating for the rapid loss of manpower by a proportional improvement in mechanization, by the introduction and utilization of new techniques and by adapting the organization of work and of the production process to the altered conditions. It is necessary to utilize to a greater degree than before the advantages of the socialistic economic system, consisting in the formation of agricultural farms with a relatively large concentration of the land (compared to capitalist farms worked under similar conditions), of livestock, of machinery, *i.e.* to utilize really large-scale forms of economy, based on a substantial increase in the professional knowledge of the working people.

The relative lack of manpower in Czechoslovak agriculture is furthermore aggravated by the *unfavourable structural composition* of the remaining manpower. Industrialization, in removing from agriculture mainly the young people, has led to a distinct deformation of the *age structure*. The nucleus of agricultural manpower is today formed by people of the higher and highest age groups. The imbalance between the youngest and oldest age categories influences not only the absolute decrease of the numbers of working people (due to a natural difference between mortality rates), but also leads to an increasing mean age of agricultural workers (which is nearly ten years higher than in industry).

The demand for manpower and the numbers available are, of course, influenced also by the *structure according to sex*. Of the people working today in Czechoslovak agriculture, 52·5 per cent are women. This high proportion is on the one hand due to the heritage of small-scale production traditions with a high proportion of manual work; and, on the other, to the complicated post-war evolution, which has caused many men to move into industrial employment. Experience from co-operative work, however, has convinced our socialist farms of the necessity of having a high proportion of highly qualified and physically capable men as a prerequisite for a growing productivity of labour and intensity of production. Therefore, they are striving to achieve a far higher proportion of boys among the numbers of young people who are newly joining agricultural farms.

II. CONDITIONS OF LIFE IN CZECHOSLOVAK
AGRICULTURE

Why does the decrease of manpower in Czechoslovak agriculture take place at an economically undesirable rate? Why furthermore does the qualitative aspect of the remaining manpower depreciate in the sense that the mean age increases, an excessive feminization takes place, professional qualification lags behind the standard needed, and differences in manpower between the social-economic sectors and between regions deepen? Why did not agriculture become an attractive profession for young people?

One of the reasons for this situation is to be found in the external conditions which affect people's lives and form their way of life. In Czechoslovakia there are no fundamental differences between the town and the country, and the evolution in post-war years has tended to confirm this situation. There are no major regions in our country devoid of all industry, which is rather spread throughout the whole country in relatively small production units. This of course has affected the character of the distribution of the population : we have few large cities ; the majority of the population lives in small towns and villages. Therefore, the Czechoslovak village cannot be automatically thought of as being identical with agriculture or the village population with farmers. The Czechoslovak village might rather be characterized as a large dormitory for industrial workers, who travel from the village to the town for their work. Over and above the natural process of historical development this has been facilitated by the rapid growth of transport. In addition to railways there is the network of motor coach services, which is one of the most dense in the world. 52·5 per cent of the whole population lives in villages in Czechoslovakia, but of these only 32·6 per cent actually work in agriculture. There are no purely agricultural villages and no purely agricultural families in our country.

This character of the Czechoslovak village has not only distinctly influenced its social composition, but also its culture, customs, consumption needs, etc. The development of housing in the near-by town has had a favourable influence on the standard of housing of the village. When small-scale production ceased to exist, together with the corresponding closeness of the places of work and residence, the farmers no longer needed to invest money in individual farms, and were able to devote substantially more money to improving their standard of housing. Statistical research into the standards of housing would, however, show that there are poorer standards for non-urban regions as a whole. In the country there are still lower standards in basic conveniences (bathrooms, plumbing, WCs) in houses and apartments. The village lags behind the town as far as communal technical services are concerned (public drainage, lighting, communications, etc.).

Differences in the level of housing are further aggravated by the fact that, owing to the low concentration of population, villages have far fewer shops and, especially, far fewer of the various facilities which help employed mothers in taking care of their children (nurseries, kindergartens, temporary children's homes, etc.). The agricultural population still has few opportunities to use joint canteens (although some progress has been made in this respect). This all results in the loss of work and leisure time for the agricultural population, and this, combined with fewer opportunities in cultural life, means that work in agriculture does not merely fail to attract but even drives people away.

However, we do not regard these factors as decisive. The majority of the village population is a non-agricultural one, living constantly in the village and travelling to towns for work : these people evidently do not mind the generally lower standard of housing, commerce, amenities and cultural facilities. We believe, therefore, that the decisive role in determining the readiness to work in agriculture is played partly by the conditions of work and the cultural level, and especially by the material incentives, expressed in the level of payment for work and in social security. Here the comparison with other branches of the national economy leads to results far less favourable for agricultural work.

Although no complex analysis has yet been made in Czechoslovakia of the difficulty of working conditions in particular sectors, there can be no doubt about their influence on manpower. Although great changes have taken place in the past twenty years, a large proportion of difficult manual work has remained in agriculture. Further disincentive aspects of the conditions of work follow from an unfavourable distribution of working hours during a day and during the year. At the same time, work in agriculture still has to be carried out under inferior conditions of hygiene and safety. Another negative factor, which is starting to take effect, is a social one — the reluctance of young people to work in agriculture at all. These negative aspects, however, are not balanced by appropriate preferential treatment in other respects, such as material incentives, etc.

III. DISTRIBUTION OF THE CO-OPERATIVE'S INCOME

Material incentive to work is, in a socialist economy, made up of, on the one hand, payment for work, on the other, the level of social facilities and security. In order to understand better the system of material incentive as operated in agricultural production co-operatives, it will be useful to discuss first the economic principles of the distribution of the co-operative's income.

The socialist principle 'each one works according to his capacities, each one receives according to his work' supposes that the wages of members of a socialist society according to work done should be *equal for equal work done* (Marx's unit of simple work), in the state-owned as well as in the co-operative sector of socialist economy. In practice, however, the concrete application of this principle is hindered by a number of problems. First of all there is the fact that an agricultural production co-operative is an independent economic unit, with a wide scope of liberty (although limited by the co-operative's statutes and lawful regulations issued by the state) to proceed according to its own considerations (for instance to decide on the proportion in which to divide the gross income). Therefore, the member's income, being that of an independent producer, is not assured by state authority and does not even take the form of wages. It is a share in the overall economic result of the co-operative's income, including wages as well as a share in the overall economic result of the co-operative's operation — gain or loss. From this point of view, the differentiation between the incomes of individual co-operatives (and thus also of their member's incomes) following from different efficiency of the labour utilized, must be regarded as justified and necessary from the stand-point of society.

The territorial differentiation of co-operatives' incomes and of the incomes of their members shows up quite wide differences, due to the fact that under varying conditions of production the same amount of labour utilized will not result in the production of the same amount of use value, and thus also in the same income. Evidently, due to some objective conditions, some co-operatives achieve an unusually high surplus product, and thus are undeservedly in a far more advantageous position than the others. How does this agree with the socialist principle of equal wages for equal work?

To answer this question, we have to study a number of problems more closely. One of these is the fact that better economic results are generally a result of a combined effect of better soil as well as a greater effort on the part of the farmers for more rational utilization of the given conditions of production. Higher income for higher labour productivity is then in full agreement with the principle of income distribution according to work done (it is of course more difficult to calculate separately the effect of these two factors).

IV. THE EFFECT OF DIFFERENTIAL INCOME

Another subject frequently debated is the question of how to use that part of the surplus product which is achieved undeservedly, without employing an increased amount of labour, by co-operatives located in more favourable conditions of production. The state can remove this by various means (*e.g.* taxes) and use this money for investment subsidies in co-operatives located in conditions of a less-than-average standard. However, this solution comes into conflict with the material interest, foremost in the interest of the whole society, of the encouragement of agricultural co-operatives in the most intensive and most efficient utilization of the soil under the best possible conditions of production. An optimal solution would probably be to remove only a part of this extraordinary surplus product following from better conditions of production ; this would support the interest of producers in further increasing their production on the best soils, and would thus be fully in accordance with the objective interest of society.

The effect of differential income in Czechoslovak agriculture is the source of numerous problems, which also affect the migration of manpower. Although our country is not large in size, its terrain is very varied, fruitful low-lands alternating with hilly and mountainous regions. According to the quality of the soil and its location, temperature, rainfall, etc., the surface area of the country is divided into five large production regions. At the same time, however, the different historical evolution of different parts of the country has also influenced the development of agriculture, so that the level of intensity of farming in Slovakia is substantially lower than in other parts of the country. Therefore, we are in fact dealing with ten different agricultural regions.

The magnitude of these differences is shown by a comparison of the levels of incomes of co-operative members from co-operative farming activities in 1963 (the overall average = 100).

This differentiation of course is not only caused by different conditions of work ; it has already been corrected by the effort of the state to compensate for the difference of economic results of the co-operatives, by a higher rate of income taxation in the case of co-operatives having better results and by means of various forms of direct or indirect aid to weaker co-operatives (various types of subsidies and investments, allowing these co-operatives to spend a relatively higher proportion of their gross income on member's

TABLE I

INCOME DIFFERENCES IN CZECHOSLOVAKIAN AGRICULTURAL REGIONS

	Czech Regions	Slovakia
Total income including :	102·0	92·3
maize-growing region	94·5	100·5
beet-growing region	108·4	87·9
potato-growing region	101·9	68·0
sub-mountainous region	94·0	62·0
mountainous	94·5	60·5

shares). This means, therefore, that the actual differentiation of the economic results of co-operatives is substantially higher than is indicated by the level of member's income.

This distinct differentiation leads to a search for means to equalize this situation. In 1960 a new system of marketing agricultural products came into being. It consisted in eliminating the previous system of two price levels, whose relationship to each other was determined in such a way that co-operatives operating under different conditions should achieve a roughly equal mean marketing price. A new unified price was introduced for the whole country, the height of which was set at roughly one quarter above the level of the previous mean marketing price.

Since the system of unified prices greatly increased the advantages of farms working under more favourable conditions, an agricultural tax was introduced with a level highly differentiated for various productive regions. The negative effect of the differential income, however, was not eliminated in this way, and the differences between the best and the worst co-operatives continue to grow.

V. ASSURED INCOMES IN AGRICULTURAL CO-OPERATIVES

It is assumed that, in future, differences will continue to be solved in a similar way, modified however in a number of cases. Besides a differentiated agricultural tax, there will be, for example, a unified price which will be modified by payments additional to the price of individual products, this being done either by a general increase of the prices of products cultivated mainly in mountainous and sub-mountainous regions, or by a direct differentiation of the additional payments for individual regions or to categories of farm according to the economic level achieved.

In solving the problem of the differentiated incomes of the co-operative members, its excessive increase can be limited relatively easily by means of a progressive income tax. It is a far more difficult problem to limit the movement of incomes which are too far below the average. This is the problem of an assured minimum reward for work. This is not consistent with the requirement that payment for equal work should be in all cases precisely equal, which is to be taken as a tendency which is felt over a longer period of time. Temporary deviations are possible, especially when putting into practice the newly introduced model of socialist national economy, which binds the payment made to workers (also in state-owned industrial enterprises) to a large degree to the economic results of the given enterprise.

To the threatened decrease of the income there is, of course, a critical limit below which normal productive activity is endangered as well as in general the reproduction of the labour force. This danger necessitates that the socialist society should assure each worker a minimum payment of a certain sum for work done. This is the case, for instance, in the newly accepted principles of the new management policy for state enterprises.

Among the means of guaranteeing a minimum payment for work in the co-operative sector, the most suitable one would be credit, which the co-operative would make use of after exhausting its own reserve fund. Since, however, the weaker co-operatives especially have only very limited funds of this nature, it will be necessary to consider the possibility of forming a centralized reserve fund, to which co-operatives would contribute a certain percentage of their gross incomes, and formation of which would also be subsidized by the state. Since in Czechoslovakia this topic is still the subject of research, I would like to mention at this point some experience gained in the Bulgarian People's Republic. There, in 1962, a joint state co-operative fund was formed, with the object of assuring, even to co-operatives working under inferior conditions, the payment of a certain minimum (in this case concretely, 1·80 Leva [1] per day of work). Co-operatives contribute 2 per cent of their gross incomes to this fund, and the state contributes to the formation of this fund by a certain fixed sum (roughly equal to the contribution of the co-operative). The first experience gained in using this new fund in 1963 shows that it was sufficient to assure payment of the planned minimum amounts in all co-operatives.

The forms of equalizing disproportionately high differences in incomes by external measures are, however, considered by Czechoslovak economists as being only temporary. The formation of equal conditions for an optimal supply of manpower in the individual productive regions, and an altered system of farming for the mountainous and sub-mountainous regions in the sense of an increased specialization in those branches of agriculture which under the given conditions are the most effective, and which allow the marketing price within the system of productive and financial relations to play a decisive role, are considered to be the most important solutions giving the best outlook for the future.

VI. COMPARISON OF THE LEVEL OF PAYMENTS FOR WORK IN AGRICULTURAL CO-OPERATIVES

Considerations of the height of the assured minimum and generally of the forms of equalizing the effect of the differential income must necessarily be based on a comparison of the level of remuneration of agricultural co-operative members with those of the employees in other sectors of the national economy. This comparison, however, is made very difficult by a number of distorting factors. One of them is the difficulty of comparing working hours. I would like to stress one more point in this connection. A *work-unit* (in Russian 'trudoden', *i.e.* work-day), up to now the prevailing measure for remuneration in the co-operatives of most socialist countries (in Czechoslovakia in about 90 per cent of the agricultural co-operatives), cannot be taken as identical with one day of work. According to our research in Czechoslovak co-operatives one work-unit on the average is accorded for 4·5 hours of work.

[1] Which corresponds approximately to two-thirds of the income of workers in Bulgarian state farms.

The factor which most distinctly influences the overall level of income of the co-operative members is, however, the fact that beside a payment in money some of the member's remuneration is received as *payment in kind*. The proportion of this has decreased within the last seven years by about one-half, forming now about 8·5 per cent of the member's total remuneration.[1] Besides payment for work done in the co-operative itself, a large proportion of the farmer's income (about one-third) comes from *personal farming* on a very small scale. We regard both these facts as an anachronism in the present-day highly commercialized Czechoslovak agriculture. Interest in personal farming is motivated today not by the fact that co-operative members would be reluctant to give up their own farms altogether, but mainly by the attempt of the farmer to realize for himself the relatively large difference between the marketing price of agricultural products and the retail prices of foodstuffs.

There is of course a close relation between the payment in kind from the co-operative farm (these consist of plant products, mainly fodder) and the personal farms. The farmers are mainly interested in obtaining fodder,[2] which they use on their personal farms. The products of the personal farm are to a large extent consumed in the farmer's own family, although some animal products (*e.g.* eggs), owing to a traditional intensity of small-scale farming production, still play an important part in the marketing funds.

In the same way as in payment in kind, we again observe a progressive development : the proportion of income from personal small-scale farming in the farmer's personal income has decreased from 42·7 per cent in 1958 to 32·3 per cent in 1962 (decreased by 30 per cent). At the same time, the farmer's income from the co-operative's joint operations increased : the rate of this increase, however, sufficed only to compensate for the decrease, not to surpass it.

After elucidating some structural differences we may now try to compare the income of agricultural co-operative members, described above, with the annual wages of other employees of the socialist sector of national economy. From these comparisons, it would seem that the annual income of a co-operative member, after recalculation from the joint operations of the co-operative, and after inclusion of income from personal farming and calculation of payment in kind

---

[1] Calculated for retail foodstuff prices this difference is of course doubled ; in well-situated co-operatives payment in kind should be about one-half that of economically weak co-operatives.

[2] Whereas due to price relations they prefer to buy bread, flour, butter, and, in some regions, even milk, from state-owned shops.

valued at retail prices, approaches the wage paid to workers on state-owned estates, and equals about 80 per cent of the mean wages paid in enterprises of the socialist sector of the national economy.

These results, however, must be made more precise, or they would not be fully comparable. First of all we must use a comparable length of net working time, *i.e.* 2,298 hours annually. After this modification the co-operative member's income comes to 80 per cent of wages on state estates, and two-thirds of wages in the other branches of the national economy.

VII. SOCIAL SECURITY FOR CO-OPERATIVE FARMERS

A comparison of the level of personal income cannot be carried out in isolation from the level of income stemming from social consumption funds, mainly from social security. Our farmers formerly had no social security. The only form of security in old age were savings, gained by limiting expenditure through life, or an annuity obtained as a condition for the farm being handed over to the younger generation. When joint farming in co-operatives started, this situation altered fundamentally. Today, all co-operative members are insured against sickness, old age, disablement and death ; they receive a contribution in the case of childbirth ; they receive maternity allowances ; many have paid leave ; etc. However, the level of social security still differs between the co-operative and the state sectors. Only some categories of co-operative farmers, especially those who have recently come to work in agriculture from schools or from other sectors of the national economy (agricultural specialists, tractor drivers, young people, etc.) have an assurance of a level of social security absolutely equal to that of industrial workers. This, however, concerns only about 10 per cent of the co-operative farmers. For the rest, a different system obtains, according to which social security payments are made not from state, but from co-operative funds, and mostly according to lower premiums.

VIII. CONCLUSIONS

The differences in the level of income and social security which still exists, only affect, however, the standard of living of the families of co-operative members to a small extent. The income from agriculture is generally only a supplement to the income of an agricultural–industrial family, the main part of whose income is formed by wages from the industry. Nonetheless, it has had a negative influence on the evolution of Czechoslovak agriculture as a whole. It led to the fact that the most efficient and most highly qualified workers left for work in industry, and mostly women and old people remained in agriculture (in the western, industrial region of Czechoslovakia, Bohemia and Moravia, two-thirds of the labour force are women).

Experience of the development of Czechoslovak agriculture in recent years justifies the conclusion that an essential prerequisite of a modern, highly productive agriculture and of its elevation to the level of industry must be the achievement of a high level, we may say a ripening, of the whole national economy. The problem will mainly consist in an intensification of industrial production, in order to avoid a further wide-scale migration of efficient labour forces from agriculture, and on the other hand, to bring about the reverse process — a redistribution of manpower in favour of agriculture. Of course, we do not mean that the aim should be a return of manpower directly into primary agricultural production. It is necessary to strengthen mainly those sectors which are concerned with the production of means of production, and which offer services to agriculture (the so-called 'agribusiness'). An enhancement of the level of the material production base, together with some structural changes, will lead to such an increase in labour productivity, and thus also of the co-operative farmer's income, that it will be possible to solve more easily the present differences between the level of remuneration between the individual sectors of the national economy, and to solve the great territorial discrepancies. Such an outcome — together with a widening of the powers of co-operatives and state estates, and removal of some administrative hindrances which limit the farmer's economic initiative — is the main object and sense of the newly introduced system of management policy of the national economy of Czechoslovakia in the field of agriculture.

DISCUSSION OF DR. FLEK'S PAPER

Professor Badouin said that the picture drawn in Dr. Flek's paper is rather an unusual one. Dr. Flek took his stand on the general statement that Czech agricultural suffers from a lack of labour, and argued that exodus from the land has been the cause of it. Professor Badouin compared this with the more normal situation where the exodus from the land usually lags behind the needs of the economy. It is not clear what effect the lack of population in the agricultural sector has had. Dr. Flek gave a picture of the rural exodus going too fast to safeguard the prerequisites of agricultural growth but he did not discuss the consequences on the rest of the economy. Professor Badouin asked if the shortage of agricultural manpower acts as a break on the economy as a whole or whether it leads to a lack of harmony in the development of other particular sectors.

Professor Badouin said that Dr. Flek drew a picture of an exodus which does not necessarily involve geographical mobility but only professional mobility. This implies commuting of one sort and another. This suggests that in the next few decades we might see greater fusion between urban and rural societies with serious consequences on rural infrastructure. Is this sort of urbanization of the country-side the result of a deliberate government policy or simply a random process ? He contrasted the case in France where the planner seems to be moving towards a dubious solution in making an arbitary distinction between the development of urban and rural space. This involves the creation of regional metropolitan towns — like Parises set in the provinces. The picture of gradual urbanization in Czechoslovakia which Dr. Flek's paper implied, seems to be preferable as long as it receives adequate planning.

Turning to the question of the role of co-operatives in the Czech agricultural structure, Professor Badouin asked if Dr. Flek would try to clarify a little more what the co-operative stands for in Czechoslovakia. The extreme picture is of a purely economic entity with the families which work in the co-operatives dependent on it both as producers and consumers ; but it was clear from Dr. Flek's picture that this is not the case in Czechoslovakia, families are not dependent on the co-operative but merely give it a certain amount of service which appears to be almost residual to their other activities. The individual farm worker has only a subsidiary link with the co-operative and his standard of living does not fully depend on the co-operative. In view of this picture of the co-operative it would be interesting to know, in a little more detail, Dr. Flek's view of its status.

Professor Badouin said that there seemed to be a contradiction in two statements from Dr. Flek. He had stated that the differential trend between the levels of earning in different regions continues to grow but, elsewhere, Dr. Flek said that the exodus from the land is increasing in speed. Did Dr. Flek mean that the flight from the land is basically linked to the rate of growth in other sectors, whatever the situation in agriculture ? In this case, the price policy introduced in 1960 and described by Dr. Flek would seem to be insufficient.

Dr. Flek said that the answer to the question about the consequences of the unbalanced state of manpower distribution in Czechoslovakian agriculture and industry is very complex. This distribution is in large measure the result of the post-war development of Czechoslovakia which has had very special features. Soon after the war there was a heavy transfer of citizens of German nationality into the country and rapid industrialization especially in the years of the cold war. As a result of the cold war there were serious shortages of the basic materials and machines for industrial development. Czechoslovakia was not able to buy on the world market such essential commodities as copper, and was therefore obliged to make heavy investment in high-cost copper mines, etc. This first task of developing industry at great cost made heavy demands on manpower and the government used policy measures to ensure better conditions in industrial life than in agriculture. At that point it was thought that collectivization would make farming possible with far less manpower. But, in the event, mechanization and collectivization proved insufficient, and development was slower than could satisfy the needs for agricultural output. The process of collectivization was also costly in as much as the small-scale capital structure had to be replaced by one of a larger scale. However, in this replacement operation, chances of improvement in technology were largely missed, despite a considerable improvement in the conditions of work in the agricultural sector.

Dr. Flek replied to the question whether urbanization was the result of policy or whether it had been brought about by random process. He said that, again, the historical context is most illuminating. Czechoslovakia, especially Bohemia and Moravia, had been, before 1914, the most rapidly industrializing part of the Austro-Hungarian empire. The main force of this industrialization had been the employment of part-time farmers who then became industrial workers with small pieces of land. Some of these had continued (and even now continue) to be classified as farmers although their land holdings were probably not more than half a hectare, and although a major part of their income came from industry. This general picture could be typified by the family whose husband works in industry while the wife works in agriculture. The kind of urbanization which Czechoslovakia had experienced and is experiencing was a result of this kind of development. Policy had, however, played some part in recent years in changing this line of development. The building of houses in larger villages was encouraged, and building in the smaller villages was discouraged. The results of this policy are not yet clear. There is obviously considerable resistance to moving, even though the standard of housing elsewhere is higher, and villagers continue to improve their houses rather than move to, or build, new houses.

In answer to Professor Badouin's general point about how the income of co-operative farmers can be improved, Dr. Flek said that both improvements in production methods and the use of the price level are necessary, and should be an aim of government policy. Under the present economic situation, which is certainly not ideal, no general improvement in the prices of agricultural produce seems to be possible. An *ad hoc* policy of additional payments to various areas and various lines of production is the more likely development.

The standard of living of farmers has been made worse by the fact that so many co-operative farmers are women and old people. The conditions of work are clearly worse than they are in industry. While young people in agriculture, working with machinery, have incomes comparable to those in industry, the social conditions, hours of work, and the attractions of towns frequently make them decide to leave. The tractor driver will earn as much in agriculture as, say, in public transport, but he is obliged to work a long day, is liable to the jibes of his friends, and perhaps even the derision of the girls.

Professor Obolenski said that he found Dr. Flek's report most interesting, particularly since it was based on practical experiences of agricultural co-operatives. Czechoslovakia is a country with highly developed industry which at the same time has achieved considerable successes in co-operative agriculture. Dr. Flek had been frank in outlining the major problems and had suggested some ways to further the development of co-operative agriculture.

The conclusions which Dr. Flek came to seem to be quite reasonable, especially those which deal with further mechanization of agricultural processes, and those that stress the role of material incentive to co-operative members by the establishing of a guaranteed income minimum and by increasing productive potential of labour. It should be stressed that the problem of raising living standards is closely connected with the productivity of labour. Professor Obolenski asked if Dr. Flek would describe in greater detail what measures are being taken in Czechoslovakia to solve the problem of agricultural income by means of raising the productivity of labour.

Finally, Professor Obolenski said that he would like to hear from Dr. Flek upon what lines the specialization of production in co-operatives is to proceed.

Dr. Jacobi said that he questioned an apparent contradiction in Dr. Flek's paper. On page 313 of the paper Dr. Flek reported that farmers are interested in the local market since in that market they can realize for themselves the relatively large difference between the standard marketing price of agricultural products and the retail price of food. Dr. Jacobi said that these findings are in agreement with his own. However, on the same page, Dr. Flek said that the production from personal plots is consumed by the farmer himself and his family. Does it not seem more likely that the main incentive for intensive work on the personal plots is the possibility of obtaining a high price on the local market ?

On page 306 of his paper Dr. Flek said that in Czechoslovakia there is for the worker no fundamental difference between town and country ; yet on pages 306–7 he reported on statistical research on housing conditions in non-urban areas in which it had been found that the situation in the country-side is less favourable than in the towns. Further on the matter of relative living conditions, Dr. Jacobi said that he was distressed by the implication, drawn by Dr. Flek on page 307, that people living in villages and travelling to towns to work do not mind the generally lower standard they endure. This is an old argument and a dubious one. As anyone who realizes the position of housing shortage in Prague would recognize, there are strong economic reasons which oblige people to endure the lower standards.

Professor Ruttan said that he was interested in the contrasts between the paper by Dr. Flek and that by Professor Bishop. The latter had argued that there is no necessary relationship between the productivity of labour and the productivity of land ; whereas Dr. Flek assumed that the productivity of labour is closely associated with that of land. Such differences suggest different policy implications. While Professor Bishop looked to more efficient functioning of the labour market, Dr. Flek spoke in terms of a better income distribution. It would be interesting to discuss these policy differences.

Professor Bishop said that he was interested to see that Dr. Flek reported that the return to farm labour in Czechoslovakia is approximately two-thirds to four-fifths of that in other parts of the national economy. This is approximately the same as that in the United States. Professor Bishop noted that in Czechoslovakia there had been considerable out-migration from farms, and asked whether this had been brought about by government policy. If so, what rate of out-migration had been aimed at by these policies ? Professor Bishop asked what implications these migrations have had on the returns to farm labour in Czechoslovakia.

Dr. Flek said, in answer to Professor Obolenski, that the potential advantages of specialization are a great untapped reserve in Czechoslovakian agriculture. Agriculture has not developed in specialization up to the present time. The multi-crop multi-stock systems are still basically the same as they were twenty years ago. All the prerequisites for specialization are present, and the new model of management for the economy will be expected to encourage the development of specialization and to increase the initiative of co-operatives in agriculture. In the short run, these changes will probably bring about a rapid increase in the income of farmers. Not only will the incomes of farmers be directly comparable to those of other sectors, but, in so far as those that work on co-operatives also own farms, they will be able to gain great private advantage as a result of the high incomes.

Dr. Flek said, in answer to Dr. Jacobi, that the farmer is faced with two alternatives. Either he can sell extra produce or, more frequently, he can simply avoid being obliged to pay the higher prices for food in the ordinary shops by consuming food which he gets from his co-operative at a lower price, or the food from his own plot. The importance of this activity of selling on the local market is sometimes so great that the income he obtains from his work on the co-operative may be regarded as income from an additional job.

On Dr. Jacobi's second question, Dr. Flek said that it is necessary to distinguish between a *fundamental* difference between town and village, and what is simply a difference! If you met two girls, one from agriculture and one from Prague, you would not be able to tell the difference between them. As a result of the post-war development of villages, approximately two-thirds of the working inhabitants of villages are factory workers in near-by towns travelling by special buses the 5 to 10 kilometres to work. Such workers do not feel the necessity of moving into the town. A survey in social research conducted by a car factory after a vast reconstruction of the town had been completed, came up with the surprising result that workers from villages did not want to move into the new modern houses even though the rents were even cheaper than the cost of normal repairs to a village house. However, it is true that the main reason why so many workers still live in villages, even though working in the towns, is that there was a serious delay in building new houses in the towns after the war, and it was easier simply to improve the housing in the villages. In Slovakia there are even new houses being built for families in villages which are taken up by industrial workers from the towns.

Dr. Flek agreed that differentials exist between different regions as far as farm incomes are concerned. This situation is best dealt with by differential taxes on the regions, and by the provision of credit at low rates of interest, direct income subsidy, or the sale of capital goods at less than the normal price, to those co-operatives which are in bad condition. The state also intervenes, on occasion, by sending agricultural specialists whose salaries are paid by the government. In the case of certain specific items with special distribution among the regions, it is also possible to aid the co-operatives by alterations in the relative prices of agricultural products. It is often the case, however, that the way to avoid the disparities between the different co-operatives is by dealing with underlying political and social difficulties rather than relying on economic methods which are indiscriminate in their effect and costly to execute.

Dr. Flek said that, on the question of state intervention in the migration of rural workers, the original movement of workers out of agriculture had been so fast that it out-stripped the improvement of economic conditions in agriculture. Although mechanization had taken place, so that there is approximately one tractor per 33 hectares, and although fertilizers are used, the modernization has not been completed so that the fertilizer often has to be loaded by hand, and much heavy work remains. With the rapidity of the initial migration and the loss of the younger members of the rural population, this heavy work has to be done by the old. Faced with this situation, the state has had to reverse its policy in favour of agriculture.

Hindrances have been put in the way of people who want to leave agriculture. The would-be migrant has to obtain the agreement of the co-operative he is intending to leave as well as the acceptance of the enterprise he intends to join. However, such administrative barriers are clearly insufficient and the only answer is the improvement of conditions and incomes in agriculture. The young will not return to agriculture until conditions are radically changed.

PART IV

CONTEMPORARY PROBLEMS IN DEVELOPED AGRICULTURE: THE PROBLEMS OF VERTICAL INTEGRATION, INNOVATION AND DOMINATION

M

Chapter 15

VERTICAL INTEGRATION AND DEVELOPMENT OF FARMS: THE PERFECTING AND DIFFUSION OF INNOVATIONS IN INTEGRATED SYSTEMS[1]

BY

JOSEPH LE BIHAN

Institut National de la Recherche Agronomique, France

I. INTRODUCTION

THIS paper proposes to examine certain effects of the vertical integration processes on the pace and methods of development of agricultural plants and, especially, the problem of the perfecting and diffusion of innovations in integrated systems. This is a field of research almost totally unexplored, however strange that may seem, and this paper contains many more working hypotheses or question marks than actual results.

With the gradual industrialization and urbanization of an economy, the growth of farms is more and more conditioned and carried along by the pressures of the ancillary and processing industries. The mechanism of this growth is relatively well known at least from the qualitative point of view and can be broken down into three elements :

(a) *The regular increase of input purchases by farms* and the *technological invasion* of the ancillary industries by a stock of scientific knowledge incorporated in their products (fertilizers, seeds, pesticides, feeding-stuffs).

(b) *The pressure of the final consumer* whose needs are increased and

---

[1] The problems discussed in this paper are studied more systematically in the following three studies : J. Le Bihan *et al.* : 'Vertical integration and development of farms', in the series : *Informations internes sur les structures agricoles*, EEC, Brussels ; J. Le Bihan, 'Research, development and diffusion of innovations in animal feeding stuffs industries, and the processing of animal products', INRA, 1966 ; A. Savag, 'Research on setting up and diffusion of innovations in industrialised farming', thesis at the Faculté de Droit et des Sciences Économiques, Paris.

are transmitted by the processing and conditioning industries and their distribution system.

(c) *Improvement in the co-ordination* between the different processes of the ancillary industries, of farming and of the processing industries and also the setting up of powerful integrated centres of decision possessing the commercial, technical and financial means of orienting and accelerating the development of a vertical unit in the agricultural and food economy. This element simply means the development of a pattern of organization corresponding to the technological possibilities and commercial opportunities arising from the two preceding elements.

The centralization of the power of intervention and, above all, the creation of organic and hierarchical links between the various elements of the production plans (marketing, credit, production methods, inputs), allow the integrating firms to increase the timing and intensity of their pressure on the growth of the farms linked to them by contract or by equivalent ties. In order to understand the importance of this pressure one must take into consideration at the same time :

(a) The influence of the integrating firms which may vary from firm to firm ; and

(b) The resistance of the environment which is likely to be particularly strong in the field of agricultural products because of structural inflexibility, attitudes of non-acceptance or of acceptance, level of education and resources, etc.[1]

In the limited framework of this paper we shall examine first the methods of gaining control over the agricultural production units adopted by the integrating firms, leaving aside any kind of constraint which could diminish their ability or desire for action ; and secondly, the potential for research-development and innovation diffusion existing in the ancillary and processing industries. The first point allows us to give a general and theoretical view of the possibilities of intervention by the integrating firms, whereas the second touches on the study of their *real innovation capacity* (perfecting and propagation).

[1] We should also measure the consequences of these 'pressures' on the internal dynamics of the farm in terms both of direct influence on the integrated sub-group, and of indirect influence on the non-integrated sub-group. However, these aspects of the problem lie beyond the limits of this paper.

II. MEANS OF CONTROL OF PRODUCTION UNITS BY INTEGRATING FIRMS — THEORETICAL POSSIBILITIES

A. Introductory

First we must examine the means of intervention of the integrating firms from a very general point of view, and this without taking into consideration the various internal or external restraints which limit them. In theory, the integrating firms organize the activity of the agricultural production units according to fixed models studied in advance. In order that the running of the integrated unit should conform to the model, the co-ordinating centre must be able to control all the strategic decisions which determine the unit's activity: for example, types and combination of the productive factors used, size of operation, production and delivery planning, etc. The control of these decisions allows the integrating firm to run the integrated unit according to its technical and economic preferences.

Amongst the technical and economic aims of the integrating firm, we can mention first of all the *lowering of the running costs of the whole system* especially by increased productivity in the agricultural production units and a reduced cost of internal communications (transport and collection of the products, transport and distribution of certain production factors, technical advisers, etc.). In this way the integrating firms are able to control an adaptation of the patterns of certain agricultural productions. *The search for a consistent level of quality which corresponds to the needs of the consumers* may be considered as the usual second aim of the integrating firms, especially when they carry on their activity in relatively flourishing circumstances. *The adjustment of the timing of the production programmes* of the integrated units to the required optimum utilization of the expensive fixed equipment controlled by the integrating firms (particularly factories for the processing and treatment of agricultural products) is another aim, the successful achievement of which is a basic condition of the efficiency of large enterprises.

In order to achieve the three aims given in the previous paragraph, the integrating form is induced to control (*asservir*) the agricultural production units by squeezing them more or less tightly in a 'pincer system', within which the production units have a greater degree of self-governing power if they are dealing with a form of production the techniques of which are not precisely laid down, and a smaller degree of autonomy where the techniques are precisely laid down. The achievements of the Findus group in Sweden in the production of vegetables for canning is an illustration of a successful organization scheme of this sort.

In return for the transfer of decision power, the integrating firms grant the integrated units certain securities.

It is therefore according to the two concepts of efficiency and security that we should analyse the means of intervention of the integrating firms at the level of the integrated units. To simplify the analysis we shall consider a product which is part of a vertical chain, and which therefore constitutes *one production unit on a farm*.

B. *Increasing the Efficiency of Integrated Production Units*

In order to improve the productivity of the integrated production units, the integrating firms have efficient means of intervention which are employed to varying degrees, according to the actual circumstances.

(1) *Adjustment, Perfecting and Diffusion of Innovations*. The power to introduce innovations and the control of the diffusion of these innovations, through the pressure they can exert on production units, are powerful instruments of change in the hands of the integrating firms.

The most firmly established concerns have at their command their own research and development centres, able to guarantee them an autonomous production of technological innovations. However, even the most powerful and the most varied concerns appear rather as *importers of crude innovations*, whose role seems to be limited to perfecting and adjustment. This striving after research and development is an essential factor in the growth of the integrating firms.

The farms grouped under the integrating firms are automatically made aware of these innovations, usually by means of technical regulations included in, or added to, the contracts — at least when it is a matter of production techniques in the strict sense of the word. The transmission of some of the other very important innovations is a result, in fact, of the integrated units being obliged to use certain factors of production of high productivity (seeds and young selected animals, balanced feeds, etc.) controlled by the integrating firm.[1]

In theory, the interval between the perfecting of the innovation and its introduction at farm level may always be shortened.[2] It is

[1] Most of the productive resources coming from industry used in farming are really 'carriers of technological innovations'.
[2] Poultry farming is a typical example of the rapid and systematic diffusion of perfected innovations.

therefore probable that the rate at which the innovations are diffused in an integrated system is higher than in an unorganized system, which may help to give a higher profit to the first category of farmers. Moreover, it would seem that the innovations spread in this way are, in general, more perfected and better co-ordinated than those transmitted to the farmers by the classical networks of popularization.

The spreading of innovations in the non-integrated system occurs principally in the form of fragments of information concerning this or that particular point which the farmer must, in theory, combine in a plan of campaign. The fragmentary, incomplete, not to say incoherent, nature of the many pieces of knowledge, transmitted to the farmer through a number of different networks is clearly not favourable to the diffusion of innovations, *especially if the farmers in question have not previously had the advantage of a good technical and economic education.*[1] In any case, the adoption of innovations will, in these circumstances, be left to the opportunities and desires of the individual. Some farmers systematically adopt innovations as soon as they appear ; some — the most common — simply imitate the former ; and some, for various reasons, only adopt the innovations much later. Therefore, the process of adoption of innovations in these conditions cannot but be slow and, above all, conducted in great disorder.

The integrating firms on the other hand will be sources of systematic diffusion of coherent innovations which have been studied and selected and whose adoption will be almost obligatory. The integrating firms also have at their disposal a network of technical advisers who supervise the farmers, thus ensuring the execution of production programmes. In the most advanced concerns, these peripatetic employees have radio transmitters enabling them to convey, in as short a time as possible, any useful information to the central organization. The result of this is almost automatic communication between the centre and the fringe of the system, thus ensuring more rational decisions. This supervision on the part of the agents of the integrating firm makes possible, moreover, the centralization of the results obtained, and greatly facilitates the perfecting of the different innovations.

To sum up, the innovations in the integrated system are transmitted in the form of an annual, or almost perennial production programme

[1] This is the case in most of the countries of the EEC, except for the Netherlands.

adapted to the needs of the integrating firm. This programme will have technical and economic components. The technical component includes not only all the quantitative aspects of the production,[1] but also the qualitative aspects such as the search for the kind of product which corresponds to the requirements of the market. Finally, the production programme must enable the integrating firm to obtain its agricultural raw materials at prices which allow it to establish a good position on the market ; hence the need to conceive the production programme in economic terms.[2]

This propagation of production and organization techniques becomes all the more important as farm management increases in complexity, demanding on the part of the farmers more and more knowledge. Furthermore, in most countries, the traditional networks of popularization are not really adapted to the needs of an agriculture subjected to swift and far-reaching technological changes.[3]

(2) *Allocation of Financial Resources.* The diffusion of innovations in farming more and more comes up against financial restraints. The amount of capital involved in agricultural production is increasing rapidly and regularly and the possibility for internal accumulation of capital is insufficient except for a small number of units enjoying a geographical or technical advantage. The possibility of recourse to an external source of capital, both for everyday transactions and for investments, therefore conditions the application of the majority of the innovations already mentioned.[4]

Thus, *the policy of technological pressure can be supplemented by financial pressure.* The passivity of the financial agents who specialize in the allocation of credit to farmers, and, in some cases, the inadequacy of these means of allocation of credit themselves, have

---

[1] Compare this with the case of popularization by official bodies or by the technical services of the supplier of a particular input. In these cases only one quantitative aspect at a time will be improved.

[2] Some more progressive integrating firms already use electronic computers to plan the economic policy of the farms within their group.

[3] A growing number of investigators are aware of the need to rebuild the networks for the transmission of technical and economic knowledge to the farmers ; for example : 'Cooperatives should consider providing engineering and architectural services plus the sale of buildings and equipment. They can assist farmers in the "systems" approach to agriculture by selling, as an example, *a complete system of efficient beef-production and not merely beef-feed.*' W. G. Leith, 'Changing agriculture demands changing Cooperatives' in *News for Farmer Cooperatives*, April 1965, p. 3.

[4] The amount of capital per agricultural labourer is distinctly greater in an industrial and commercial agriculture than in any other branch of industry.
Cf. G. Muller and Helmut Schmidt, *Invested Capital and Productivity in Agriculture and Industry*, p. 100, I.Fo. Munich, 1959. M. Gonod, 'Input and the capital coefficient in agriculture', lecture given at the École Pratique des Hautes Études (Paris), (cyclostyled), p. 32, Jan. 1964.

caused the active intervention of the integrating firms.[1] These interventions are the basis for the flow of money into agriculture, part of which flow, at least, comes from industry and commerce.

The integrating firms can contribute towards the financing of production units controlled directly or indirectly.

Direct interventions, in the form of credit, are above all meant to facilitate the financing of inputs in integrated units. This form of intervention has developed particularly in the different branches of industrial breeding and especially in the field of poultry-rearing, as, for example, the granting of credit for the purchase of day-old chicks and of poultry feeding-stuffs. The various productions which are less dependent on inputs not of agricultural origin have given rise to less substantial intervention ; for example, in the form of loans for the purchase of seedlings and seeds.

As far as investments are concerned, the intervention of integrating firms is usually more indirect. The existence of a production contract and the various financial securities granted to the integrated unit constitute worth-while guarantees for the financial agents. In some cases the integrating firms can, moreover, expressly guarantee the fulfilment of the loan and assure the fund recovery by means of an automatic levy on the price of the products.

In varying forms there is thus built up a privileged financial network, which has several effects on the future development of the farms which profit from it.

(3) *Reduction of Farm Investment.* The direct provision of productive services (transport, preparation of the land, upkeep of buildings for rearing, harvesting, storage or preserving of produce) saves the farmer a certain number of expensive investment items which do not pay in certain types of farms.[2] In these circumstances,

[1] L. A. Jones and R. L. Mighel, 'Vertical Integration as a Source of Capital in Farming', in *Capital and Credit Needs in a Changing Agriculture*, E. L. Braun, H. G. Diesslin and E. O. Heady, Iowa State University Press, 1961. For a more detailed study of the amount and means of allocation of financial resources to poultry producers in the EEC, see J. Le Bihan, *The Organisation of Chicken Production and Commercialisation in the E.E.C.*, Chapter iii.

[2] Among the operations which may be affected, directly or indirectly, by the integrating firm one can quote as examples :
   (a) *Crop production* : Fertilizer-spreading, crop-spraying, harvesting ; *i.e.* operations which chiefly concern crops, the production of which can be highly mechanized.
   (b) *Animal production* : Decentralized mixing of foodstuffs, preparation of fodder, mechanical disinfecting of animal quarters, hiring of milk refrigerating implements, etc.

See R. L. Mighel and L. A. Jones, *Contract Production of Truck Crops*, U.S.D.A., E.R.S. No. 152, March 1964.

Certain Swedish achievements, notably the organizational policy of Findus, are

one can consider the reduction of investment as an indirect form of allocation of financial aid. The farmer can, therefore, put this capital to another use, or simply keep it to permit greater flexibility in the financial structure of his farm. The advantage to the integrating firm is that it can use this fixed equipment more efficiently because of the volume of production and, because of the greater regularity of its use of such equipment which means an effective increase in the rate of use. Thus certain large integrating firms in the field of agricultural production take over all that can be carried on more profitably on an industrial scale, that is to say all those aspects of production whose cost is decreased if they are carried out according to highly specialized techniques and on a large scale. Carrying out these different operations, either directly or indirectly, by calling upon specialized sub-contractors (which seems to be the most common solution), enables the integrating firm to strengthen its control over the quality of the products. Other advantages may also be obtained, especially as far as the employment of seasonal labour is concerned. The result is to leave under the control of the farmers only those operations most suited to the size of their farms, for example certain cultural operations such as ploughing and sowing.

(4) *Disassociation and Co-ordination of Activity in Contract Production Units : Chains of Satellite Units.* As we have already pointed out, technical progress allows the dislocation of the processes of farm production, *stricto sensu*. In the various branches of crop production, this dislocation has caused the establishment of real industries (selected seeds, etc.).[1] The integrating firm can directly control seed and seedling multiplication, calling upon a certain number of farms in their group. It is, above all, in the various branches of animal production that this striving towards disassociation and re-co-ordination of productive processes has appeared during recent years. The perfecting of special feeding-stuffs and the use of tranquillizers have increased the possibilities of disassociation in the processes of animal production in specialized units of increasing size, permitting a

---

studied in J. Le Bihan, 'Study of Certain Big Industrial and Commercial Concerns in Swedish Food Economy'. Cyclostyled, I.N.R.A., April 1964, p. 75.

[1] In the realm of animal production, selection and breeding have taken on an industrial character only in the poultry branch, by the development of large firms for the selection and multiplication of poultry strains in large-sized industrial plants linked to agents by a system of concessions.

Cf. J. Le Bihan and F. Nicolas, 'Economic Analysis of Operations of Reproduction in Modern Poultry Farming', Fascicule No. 1. *Growth and Coordination of Firms Producing Day-old Chicks,* Travaux de la Station Centrale d'Économie et de Sociologie rurales de l'I.N.R.A.

lowering of costs and an improved standardization of products. In these cases the integrating firms commonly impose a fixed minimum size. Some examples can be given : the creation of production units of started pullets ; integrating concerns in the production of eggs for consumption, allowing a better use of the producer's hen-houses ; the attempts, more or less successful, at least until now, towards the specialization of pig-breeders as breeders proper or as raisers ;[1] the process in specialization in beef-breeding (calf-breeders, producers of calves for slaughter, producers of young rearing cattle). The integrating firms can, at the same time, secure co-ordination between these specialized and complementary links of the chain and create real chains of satellite units. The inadequacy of former structures of co-ordination, giving rise to speculation and the absence of quality standards, and the technical changes creating new and original situations, both favour this sort of behaviour on the part of integrating firms.

C. Various Kinds of Security Granted to the Producers[2]

Thanks to the various securities granted to him, the farmer can accept the policy of growth of the integrated units and the probable reduction in the variety of his activities, without endangering the financial balance of his farm. Vertical integration therefore appears, from this point of view, a substitute for the classical policy of internal diversification and equalization as practised in traditional farming.

(1) *The Reduction of Material Risks in Production.* The material risks in production are, from the start, reduced, thanks to a technical staff of high quality and to almost continuous aid provided by the integrating firm. The control of the quality of certain factors of production and the spreading of adequate technical knowledge constitute, in themselves, an elementary form of risk reduction. Among the interventions which reduce risks can be given the disinfecting of hen-coops and other rearing buildings and the taking of soil samples from fields given over to vegetable cultivation to prevent cryptogamatic attacks.[3]

[1] P. Coulomb and B. Roux, *Industrialization of pork production: attempts at the elaboration of organisation models*, Station Centrale d'Économie et de Sociologie rurales de l'I.N.R.A.
[2] On the analysis of risk and uncertainty in the theory of the firm and integration, cf. Sydney Hoos, *Lectures on Uncertainty and the Firm* (cyclostyled), University of Naples, Centre di Specializzazione e Ricerche Economico Agraria, 1961, p. 220 ; and E. R. Jenssen, E. W. Kehrberg and D. W. Thomas, ' Integration as an adjustment to risk and uncertainty ', *Southern Economic Journal*, April 1962, pp. 378–384.
[3] In the vegetable-producing concern of Findus in Sweden and in France, all the fields intended for the production of peas are subjected to a *chemical analysis*,

These interventions can show initially what is needed and what further interventions will be needed ; the success of the last depending in many cases on the speed with which they are realized. We can give, as an example, soil treatment carried out in time, thanks to the almost daily visit of the integrating firm's advisers and to close links with the organizing centre of production ; the same applies to sanitary treatment in stock farms.

(2) *Security Gained in the Field of Price and Income.* The planned co-ordination of decision in an integrated system,[1] the improvement of the information transmission between the centre of the system and its fringe, reduce, to a large extent, the level of uncertainty and in theory increase the efficiency of the structured economic entity. For example, the integrated system has at its disposal agricultural products of standard quality in quantities large enough to secure commercial bargaining positions and advantageous prices. The running costs of industrial processing equipment are decreased by a correct adjustment of production and delivery planning. Finally, industrial equipment in ancillary industries can also be improved by a more regular output, for example, factories producing feeding-stuffs, incubators and seed-multiplying plants.

This increased productivity and the improved prices are distributed within the system between the centre and the fringe and, in theory, take the form of price and income guarantees granted to the integrated units — although this fixing of the rules for sharing profits may give rise to disputes requiring arbitration. The means of price or income guarantee vary according to the product. Examples of the most common means are : fixed guaranteed price, guaranteed minimum price ; price determined according to the sales results following a mathematical ratio ; price supplement according to a price index observed on a market whose quotations are published ; a guaranteed minimum profit margin considered as normal ; a guaranteed minimum income for each item delivered. Certain means implicitly encourage the improvement in efficiency of the production units. Examples are : differentiated prices according to the quality of the product delivered, or to the size of the production unit ; guaranteed income related to certain technical ratios of production (rate of transformation, for example).

---

making possible a rational fertilizing programme and a bacteriological analysis so as to detect in the soil *Aphanomycea Fungus*, which attacks the root of the pea and which cannot be eliminated economically by treating the soil.

[1] It is, of course, supposed from the beginning that the centre of the system is well enough informed to avoid errors of decision.

III. THE POTENTIAL OF FIRMS IN ANCILLARY INDUS-
TRIES TO CONDUCT RESEARCH-DEVELOPMENT AND
TO DIFFUSE INNOVATIONS

A. Introductory

The innovation capacity of the integrating firms is one of the decisive factors in the dynamism and growth of an integrated system, above all when the branches of agricultural production controlled are capable of fundamental technical changes, and this applies equally at the stage of raw-material production as at that of the industrial processing. Let us concentrate on the problem of innovation in agricultural production itself and on the real possibilities at the disposal of the integrating firms for the introduction and spreading of these innovations. From this angle we can approach the more general problem of research-development in ancillary and processing industries and the diffusion techniques intended to ensure that their results reach the farms. A study such as this raises both conceptual and methodological problems which we must outline first of all. Then we shall turn to the first results of an investigation in progress in France in the feeding-stuffs industry.

B. Conceptual and Methodological Problems

Scientific research into the introduction of innovations and the spreading of new knowledge within the mechanism of an economy is still of recent date, and the concepts used are not always satisfactory, at least for the analysis of the actual situations which arise in the agricultural and food industries.

(1) *The Vertical Transmission of Technical Innovations and the Need for Study of Diffusion Functions.* For several years we have had at our disposal, in a growing number of countries, more or less exhaustive statistical information on the research-development efforts of firms. We need only refer to the annual reports of the National Scientific Foundation published regularly in the U.S.A. since 1951, and, more recently, the standardized nomenclature for the measuring of scientific and technical activities of firms worked out in the course of research at the OECD.[1] The definition of research-

---

[1] Cf. 'The Measuring of Scientific Technical Activity ; Standard Method proposed for Enquiry into Research and Development', Head Office of Scientific Affairs : OECD, Paris, 1962.
For France, the Délégation Générale à la Recherche Scientifique et Technique (D.G.R.S.T.) has published an annual report on research and development in

development used by the writers of these works includes fundamental research, applied research and development, this last meaning, more or less, technical activities whose principal object is the improvement of the product or the process. This definition of research-development does not cover the diffusion of the results of the scientific and technical activity. This last limitation does not allow a full understanding of the innovation processes in those sectors where the autonomous production of technical knowledge is slight and which *depend* on other sectors for their innovations. For these dependent sectors we must try to bring to light the links between the technical-knowledge-propagation centres and the points of implementation. The problems of intersectoral diffusion of new technical knowledge, especially those concerning the dependence relationships which may exist between the innovation-exporting sectors and those which simply import and, in some cases, adapt the innovations, have, to our knowledge, not yet been studied. The communication between two sectors, one providing the driving power, the other being driven or carried along, is achieved by means of a network[1] superimposed on the exchange of materials. We shall notice, in the network linking the sector of agricultural production and the ancillary and processing sectors, two kinds of knowledge diffusion.[2]

(*a*) Explanation accompanying diffusion as information (non-integrated system).[3]

(*b*) Explanation accompanying diffusion but as orders to be obeyed (integrated system).

These two kinds of diffusion can be measured according to the same criteria as those of research-development (number of qualified employees, cost, etc.).

(2) *The Production of Innovations in Ancillary and Processing Sectors.* The statistical results published in the U.S.A. and elsewhere show that, by and large, the amount of research-development

---

industry since 1962 ; cf. the report 'Research and Development in French Industry in 1965', p. 101, *La Documentation Française*, Paris, 1964.

For a fuller analysis see F. Russo and R. Erbes, 'Research Development: Basic Concepts and Problems', Cahier de l'I.S.E.A., Series T, July 1959, p. 77.

[1] F. Russo, 'The Philosophy of Knowledge Diffusion'. Lecture delivered at the European University Centre in Nancy (1964–65). (Not published.)

[2] A policy of up-to-date knowledge-diffusion is one element in the commercial strategy of firms and in market functioning.

Cf. R. L. Cloding and Willard D. F. Mueller, 'Market Structure Analysis as an Orientation for Research in Agricultural Economics', *Journal of Farm Economics*, August 1961. See especially pp. 550–551.

[3] A good deal of the knowledge spread in this way supplements that incorporated in the resources of non-agricultural origin used by the farmers.

expenditure, when compared with the value of sales, is relatively small. A separate evaluation of research-development expenditure for each branch of activity would perhaps enable us to make this judgement a little more detailed and precise. If, in fact, research-development expenditure is small in traditional *technically stationary* branches, it may be considerably greater in new or progressive branches, such as the canning industry (deep freezing), the feeding-stuffs industries, certain animal products processing industries and other industries.

However, it appears that even in those branches where opportunities are greater and where technology is on the move, research-development expenditure is usually limited to *development*. In the agricultural and food industries there has, consequently, been observed a dual system of pressures. The linked ancillary and processing industries promote technical changes in agriculture by their example, mainly owing to efforts towards *development and diffusion*. The ancillary and processing industries are seen, in their turn, to be carried along by other industries having a high research-development level, chiefly by the chemical and applied chemical industries.[1] Thus one finds, within the branches of the agricultural and food industry sector, diffusion alone or development and diffusion, or, more rarely, research-development and diffusion.

When production innovations are introduced in integrated systems (the perfecting of a new process, or especially the improvement of a product or of an existing process) an integrating firm, even though it undertakes little or no research-development in this branch, can be induced to engage specialized sub-contractors for certain operations. It is well known that the big producers of vegetables for canning transmit in this way to firms specialized in seed-selection, improvement programmes for certain varieties of seeds. The most dynamic integrating firms whose main area of activity lies in post-agricultural production industries (processing and sales) can in this way *consciously direct the innovation processes* in preceding branches of activity (*i.e.* ancillary industries) *according to the requirements of the market*, in spite of the fact that the ancillary industries usually have far more advanced facilities for research-development.

[1] We should also note the influence of university research and that of various bodies financed by public funds.

C. Research-Development and Diffusion (R.D.D.) Potential of the French Feeding-stuffs Industry

The feeding-stuffs industry exerts an undeniable influence on the industrialization and organization processes of animal husbandry : poultry farming, pigs, calves for slaughter and to a less extent, young fattening cattle. Its influence varies in degree according to the country in question and according to the stage of technical evolution.[1] The part played in the integration of the production of chickens in countries such as the U.S.A., France and the Benelux countries by firms producing feeding-stuffs is already sufficiently well known.

In France the diffusion network in the feeding-stuffs industry has made up for the quantitative and qualitative inadequacy of official or professional services of agricultural information provision, at least during the years 1950–65.

Let us first of all briefly outline the growth and structure of the industry. The total production of feeding-stuffs has increased from 1,270,000 metric tons in 1955 to more than 4,000,000 tons, 47 per cent of which is poultry feeding-stuffs, in 1964. There has also been spectacular progress in the production of dried powdered milk for calves which increased from 15,000 tons in 1957 to more than 250,000 tons in 1964.[2] In 1963 the feeding-stuffs firms numbered 982,[3] 124 of which were agricultural co-operatives, that is to say 12·5 per cent of the total number representing 16 per cent of the production.

The industry can be sub-divided into four homogeneous sections in accordance with the research-development factor.

(a) The group composed of branches of international firms controlled 15 per cent of production in 1963.

(b) The group of national or regional integrated firms controlled 7 per cent of the production.

(c) The group of satellite or associated enterprises attached to 'service firms' controlled 10 per cent of the production in 1963. The service firms are firms of a special type specializing in R.D.D. and in the production of concentrates incorporated in the feeding-stuffs. They control the producing enterprises by means of a system of special contracts.

---

[1] G. Sévérac and J. Le Bihan, *Study of the Economic Functions and Operations of the Feding Stuffs Industry*, A.F.C.A., Paris, 1965, p. 167.
[2] The technology of powdered milk with added vitamins and adjuvants was first perfected in the Netherlands and in Great Britain, and then in slightly different form in France and in Federal Germany. This is a typically European technological innovation. [3] Excluding those producing molasses.

(*d*) The group of small- or medium-sized independent firms controlled 8 per cent of the total production.[1]

Let us eliminate straight away the last group (*d*), which is supported on the technical side by certain firms in the chemical industry in the production and sale of vitamins and various adjuvants. The diffusion carried out by this category of producer has not yet been studied.

The group of feeding-stuffs producers allied to the service firms, group (*c*), has recently been the subject of a detailed analysis : the nine service firms had, in 1963, a total number of 337 persons engaged in R.D.D., of whom 197 were of graduate level or equivalent, that is, 56 per cent of the total. *Two-thirds of these employees (of graduate level) are engaged in diffusion.* Six per cent of those of graduate level and 10 per cent of the best technicians are employed in applied research. The development activities make use of more than 94 per cent of the total number of scientific workers in this kind of enterprise. To the employees of the service firms themselves we must add the graduates and technicians employed by the basic industries whose entire activity is given over to diffusion and to the supervision of the producers. In 1963 there were 477 graduates or technicians, 12 per cent of whom are of graduate level. Altogether the group formed by the service firms and their galaxy of producers had, in 1963, a total number of 814 scientific and technical workers of whom 251, or 29 per cent, were of graduate level. Diffusion and supervision occupy 187 workers of graduate level, or 74 per cent of the total number of graduates employed by the group.

We have also collected information on the research and experiment facilities of the service firms. The research and experimental stations are, in fact, experimental breeding units where the service firms carry out the testing of their different products. Seven firms have a central research and experimental unit. We have tried to measure the importance of these units in relation to the number of animals permanently kept there. Six firms have given figures which can be analysed in the following manner : [2]

size of chicken units	27,700 birds
size of layer and productive units	10,700 birds

[1] The agricultural co-operatives do not form a homogeneous category as far as the R.D. factor is concerned, at least until now. Some co-operatives can be classed in the second group, others (more numerous) in the third and a few in the fourth.

[2] An enquiry carried out in 1959 by the Nutrition Committee of the A.F.M.A. (American Feed Manufacturers Association) in 74 firms (which is a small sample bearing in mind the size of the American industry) has shown that R.D. had used an annual total of 3,800,000 animals ; the figures are given in O. M. Ray, *On the Farm Milling* (cyclostyled), Feed Production School, Kansas City, Sept. 1960, p. 2.

size of pig units	925 animals
size of sheep units	1,040 animals
size of rabbit units	700 animals

The second group, (*b*) above, includes 25 graduates and 47 technicians of whom 7 graduates and 17 technicians are engaged in diffusion R.D. Finally the first group, (*a*) above, includes 45 graduates and 89 technicians of whom 21 graduates and 22 technicians are engaged in R.D. Altogether, in 1963 the French feeding-stuffs industry had a total number of workers in R.D.D. distributed as follows :

TABLE I

PERSONNEL IN RESEARCH-DEVELOPMENT AND DIFFUSION IN
FRENCH FEEDING-STUFFS INDUSTRY

	Graduates and equivalent	Technicians	Total
R.D.	82	97	179
D.	229	602	831
Total in R.D.D.	311	699	1,010
D. as % of total in R.D.D.	73%	86%	82%

The diffusion and technical supervision network set up during the last few years by the French feeding-stuffs industry therefore seems considerable. Clearly a means will have to be found of analysing the distribution of these employees according to the branch of production and the type of sales system (integrated system or non-integrated system).

IV. CONCLUDING REMARKS

In conclusion, it seems desirable that the problems of innovation in agricultural production should be re-thought and re-studied, bearing in mind the structural changes which have already occurred, which are in progress, or which can be foreseen, in the relationship between the farm and its environment. The traditional micro-economic analyses, based on the farm having almost absolute autonomous powers of decision and receiving a stream of fragmentary knowledge, cannot give a realistic explanation of the breadth and pace of the technical changes which have taken place in the various branches of agricultural production in the course of the last ten to fifteen years.[1] We must stay close to changing reality in the farming and food world and adequately improve our methodology.

Finally, let us suggest two kinds of study which seem particularly useful in the present state of rapid chage in agriculture in industrial countries.

(*a*) *A Stock-taking of the Means of Research-development and Diffusion available in Ancillary and Processing Industries*

A task such as this could be carried out by an international organization such as the OECD. The choice of this kind of geographical unit would have the advantage of bringing to the fore the importance, from the point of view of R.D.D., of large enterprises operating in several countries and the resultant dependence phenomena.[2]

(*b*) *A Critical Study of the Action of Public Authorities to Encourage the Propagation of Technical Progress in Agriculture*

In all industrial countries the public authorities (state or public bodies) finance the diffusion of information, either in the form of the creation of specialized services of technical advice to farmers, or in the form of financial aid allocated to services created and controlled by professional agricultural organizations. In most cases these networks for diffusion work in an official relationship with the enterprises controlling the production and marketing of agricultural products.[3] Such a structure is probably not advantageous for the *rapid* diffusion of the most advanced techniques, at least at the present stage of development.

The action of public authorities might perhaps have more influence if it were concentrated on the improvement of the receptivity of the farmers to the organizational innovations (changes in the inner structure and in the external relations of the farms, etc.). Upon these innovations depends, in most cases, the adoption on an economic scale of the new techniques of production.[4] The improvement of the

---

[1] It is probable, if not certain, that similar phenomena of the systematic transfer of new technical combinations may have occurred historically at the integration of those branches of agricultural production today classed in the technically stationary sub-group ; for example, the sugar industry and the production of sugar beet in France, or the brewing industry and the production of barley. This is a field of research which could be explored by a team of specialists in the historical study of techniques and innovations.

[2] One must not forget that a great many technological innovations are, directly or indirectly, introduced into European countries by American firms.

[3] See Appendix.

[4] For example, the optimum size of agricultural production units is growing as the new technical processes of production are mastered.

general knowledge of farmers and particularly the improvement of their economic backgrounds by means of a constant effort to educate them in economics and to keep them well-informed, can certainly ease the structural changes and, at the same time, modify their social content (increased tendency towards collective action and the rebirth of agricultural co-operation in all its forms).[1]

APPENDIX

INFORMATION ON THE DISSEMINATION OF TECHNICAL KNOWLEDGE
IN FRANCE, ITALY AND U.S.A.

A. Data on the Importance and Structure of Information Dissemination Services in France Financed by Public Funds.[2]

The importance of the information dissemination services receiving financial support from the state has greatly increased since 1958. The cost of this intervention in francs has increased five times since 1958. In 1964 the staff of these state-aided services of dissemination numbered 2,800 ; 810 of these were working in specialized fields ; and 1,890 of them were polyvalent ; a quarter of the total number were graduates in agronomy.

The professional agricultural organizations employ approximately 2,000 persons, or more than 72 per cent of the total number. Within the professional agricultural organizations the economic section (*i.e.* the co-operatives and similar organizations) is of slight importance, being only 36 persons, 8 of whom are of graduate level. Altogether only 22 co-operatives are state-aided in their attempt to spread new techniques amongst their members. This situation, to a large extent, reflects the gradual crumbling of our co-operative system, ill-adapted to the new structures.

B. The Information Dissemination Services of the Federconsorzi and of the Consorzi Agrari (Italy).[3]

With respect to information dissemination the Italian situation is rather different from that in the other countries of the EEC. The Italian agri-

[1] J. Le Bihan and P. Coulomb, 'The Economic Education of Farmers in an Industrial Society : Principles and Methods', *Économie rurale*, 1966.
[2] Source : Ministry of Agriculture. Services financed in part by Public Funds are included.
[3] Information supplied by Dr. F. Catella of the Main Office for Technical Control of the Federconsorzi.

cultural services, the organization of which is centralized, have considerable means for spreading technical information. This dissemination function is part of an economic co-operative organization which is extending its action. In 1964 the organization employed 750 persons, 530 polyvalent and 220 specialized. These are aided by means of information, refresher courses and experiments by the central Technical Bureau of the Feder-consorzi which is composed of 42 technicians, 22 specialized in the improvement of techniques (poultry production, pigs, breeding, production of vegetables for canning, etc.) and 20 concerned with the contact between the technical centre and the fringe.

C. *The Number of Scientific Workers Engaged in the Development of Farming and Food Industries in the U.S.A. in 1962.*

For the U.S.A. we have at our disposal some facts drawn from a report made by Dr. Nathan Koffsky and entitled : 'Agricultural Research : A Technological Resource in the Nation's Economy'.[1] According to Dr. Koffsky, in 1962 there was a total of 26,000 scientific workers in farm and food industries of whom 20 per cent were employed by U.S.D.A., 20 per cent were employed by the individual experimental stations of the states and 60 per cent were employed by the firms.

The 15,500 scientific workers employed by private firms are distributed as follows : 50 per cent in agricultural processing industries, including feeding-stuffs industries ; 25 per cent in the agricultural machinery and equipment industry (production and sales) ; and 25 per cent in that section of the chemical industry directly linked with farming (fertilizers, pesticides, adjuvants contained in the production of feeding-stuffs).[2] All these figures seem reasonable.

It must be borne in mind that the number of workers employed by the processing industries include those engaged in commercial research (consumer research). The nomenclature proposed and accepted by the OECD allows of a less broad definition of research-development that does not include commercial research.[3] The numbers of staff employed by the agricultural processing industries, which in a country like the U.S.A. have commercial services of considerable size, therefore appear larger than they are.

[1] From hearings before a Subcommittee of the Select Committee on Small Business on the Role and Effect of Technology in the Nation's Economy ; in a review of the effect of government research and development on economic growth; United States Senate, 88th Congress, 1st Session, Washington, 1963, Part 2, pp. 139–145.
[2] The author thanks Dr. Koffsky for having been kind enough to give him this supplementary information.
[3] Research-development does not include marketing studies, quality control, expenditure for sales promotion or sociological research.

Chapter 16

INNOVATION IN STOCK FARMING: INFORMATION FLOW FROM THE AGRICULTURAL AND ANIMAL FOOD INDUSTRIES[1]

BY

JEAN VALARCHÉ

Fribourg University, Switzerland

I. INTRODUCTION

INNOVATION is exerting greater and greater pressure on the various sectors of the economy. With the developments of science and the rising standards of living, our societies are becoming both more productive and more exacting. They are constantly calling in question conditions of work and patterns of consumption. This means that our entrepreneurs must conform more and more closely to the Schumpeterian notion of an entrepreneur. The innovations of the present day are both technical and commercial; they concern the distribution of goods as much as their production. Moreover, they affect agriculture as well as industry. But because of the conditions peculiar to the agricultural sector, the new ideas have reached it at a later date, in original forms and with different repercussions on people's lives and work.

The principle is nevertheless the same: the necessity of keeping everyone informed of the up-to-date methods which will enable output to be increased. The procedures are likewise identical: specialization and co-ordination through the instrumentality of the entrepreneur. The broader the division of labour, the greater the need for a single policy-making centre. But in agriculture co-ordination may take innumerable forms. It ranges from mere verbal agreements to the fifteen articles of the Swiss contracts for poultry fattening. The latter merit our attention for a number of reasons. Firstly, they concern stock-raising which had hitherto been the

[1] Translated by Mrs. M. C. Lemierre.

agricultural sector the most averse to innovation. The length of the cycle of reproduction discouraged investors. The differences inherent in natural products obscured the market. The workers were disheartened by the unending attention required and could not understand why so much toil brought in so little money. But a series of innovations has transformed into variables everything that had hitherto been accepted as constants. From its first day to its last, livestock is now reared in a completely different way and consequently yields a much higher profit than ever before.

Of all the branches of animal farming it is the rearing of broiler chickens that has made the most progress in respect of the selection and cross-breeding of stock, the type and manner of feeding, the financing of production and the conditions of sale. It has in fact become an artificial or 'earthless' type of rearing, as independent of nature and as dependent on capital as any industrial undertaking.

The example of Switzerland has been chosen because chicken-farming in that country is governed by a very special type of integration. A financial scheme has been set up whereby the integrating firm is located at both the 'upstream' and 'downstream' (or 'intake' and 'output') ends of the production line. It also organizes the work of the integrated firms and is in addition the main distributor in Switzerland for poultry as for the rest. The reason why the Swiss farmers have been so ready to fall in with this 'Migros' scheme is that information was used as much as coercion for the introduction of the new methods and that the relationship established is sufficiently well balanced to dispel the fear of any repetition of what happened in the United States and in France.

II. INNOVATION BY INFORMATION

A. The Aims of Information

The primary objective of the present-day entrepreneur is to achieve a continuous flow of production and distribution. This aim answers the requirements of an affluent society. The latter demands produce of uniform and standardized quality, fresh all the year round. It is responsive to advertising and disposes of a growing purchasing power. It promotes competition to bring down the price of anything that ranks as a mere subsistence item. It imposes its wishes directly on the salesman, indirectly on the producer. The salesman serves as the connecting link between two multi-unit levels. Whereas the consumers are located in aggregate groups, the producers are widely dispersed. Their production has to contend with physical and biological hazards. They control neither the quality nor the quantity of their produce, nor can they guarantee delivery dates.

There nevertheless exist means of controlling the supply of foodstuffs, including animal products. The chicken of a specific breed, reared on the right food, will at a given age have reached the weight and the maturity which will make it ready for distribution by the salesman and for consumption by the buyer. When poultry is reared in an artificial medium it is possible to time the broods with slaughtering capacity and delivery to the shops. With mass production the poultry farmer is in a position to bear the cost of the new techniques. With mass distribution the dealer can make his supplies available to a far greater number of customers.

B. The Obstacles to Information Flow

Information pure and simple was not enough to change the methods of the chicken farmers. Three reasons may be discerned :

(1) The training of the farmer, usually apprenticed to his father, working in a single undertaking, restricted to the experience of one man and one holding ;

(2) The dispersion of the producers, distant not only from each other but also from the distribution centres, as well as from the agricultural colleges ;

(3) The obscurity of the market. The traditional trade practices do not correspond to the demands of mass distribution. The small dealer cannot afford a market research service to keep him informed of the trend of demand. He possesses neither storage capacity nor preservation facilities. He therefore takes only minimum supplies from the producer and is denied the price guarantees which would encourage him to expand.

C. The Transmission of Information

The suppliers of information distinguish between two types of poultry farmer. One type gets his information from publications and applies the instructions direct. This is the case in Switzerland for a number of suppliers of the S.E.G. (*Schweizerische Eier- und Geflügelgenossenschaften*). Merely by following the advice they obtain in this way, they are able to raise poultry complying with the commercial standards. These, however, are farmers who have had the benefit of scientific training in agricultural colleges and they constitute only a minority.

For most of the farmers the issuing of advice is not enough ; they need the guidance of an 'integrator'. This is the case in Switzerland of the 'Optigal' chicken-farmers. 'Optigal' is a limited company formed by the Federation of the 'Migros' consumer co-operative societies and the 'Provimi' company which manufactures foods for livestock. The initiative thus came from both an 'upstream' and a 'downstream' firm. 'Migros' wished to increase the sales in their shops and 'Provimi' wished to find outlets for their made-up foods. The subsidiary firm they have formed transmits information on the latest poultry-raising techniques to 200 farmers.

The information is supplied in the first place by an 'acquisitor' offering 'fattening contracts' to certain poultry farmers. He describes the treatment to be given to the day-old chicks which they will receive together with the necessary amount of suitable food. He explains the 'work/profits margin' which will remunerate the poultry-rearing (0·50 f. per chicken). He ascertains whether they can obtain the credit needed for the construction of the special 'fattening house' (65,000 Swiss f.). Otherwise the acquisitor will transmit their request for a loan from 'Optigal'. He also acquaints them with the standard cost price which serves as a basis for fixing the purchasing price of the matured fowls. The preliminary information is thus both technical and financial.

Once the fattening cycle has started, a technical expert will generally call once a month on the chicken farmer. This is a matter of routine checking, involving no transmission of information and ceasing after the farmer has produced his first few batches of chickens. In the event of any disease amongst the poultry, one of the company's veterinaries will prescribe a treatment. The farmer will henceforth rely solely on his association for his information. 'Optigal' instigated the creation of a 'Swiss Association of Optigal Chicken Fatteners' which acts as a technical link between the integrating firm and the farmers. It briefs its members on how to prevent the outbreak of disease and avoid waste of chicken food. It also assembles the data which will be used as a basis for the next discussion of the purchasing price of the chickens. The integrator works out the cost price very carefully and adjusts the purchasing price accordingly whenever there is a change in any of the factors of the cost price (chicken food, heating).

Any 'reciprocal' information can come only from the association, for the 'Optigal' technicians take no notice whatsoever of the particulars provided by individual poultry farms. Their attitude is more or less that of Taylor towards his factory operatives.

In addition to information at 'production' level there is information at 'distribution' level, designed more to advertise the products than to provide instruction on technical matters. In Switzerland as elsewhere there is a campaign for the consumption of white meat in preference to red. Doctors assert that chicken can be eaten by everyone and poultry shops are supplied with leaflets developing the same theme. Moreover, red meat is turning pink, if not white, now that cattle are being slaughtered at 18 months to provide 'baby' beef.

One of the most interesting points about the 'Optigal' formula is the light it throws on the link between information and domination. Every operator would rather be given technical information than ordered to perform a task. The chicken fattener is therefore placed in a milieu such as to make it impossible for him not to be rational. Moreover, the availability of information is both the cause and the result of domination. It is the cause in that the farmer is attracted to integration by the services it brings him. It is the result in that the farmer needs guidance for the successful implementation of his programme. It is a financial necessity for him to fatten his chicks up to a kilo and a half in eight weeks : his plant can only be written off with at least five batches of chickens a year.

III. INNOVATION BY DOMINATION

It is obvious that the integrator is the dominant party. It remains, however, to identify the signs of domination, discern its methods and define its scope.

A. *The Criteria of Domination*

Unity of management comes first and foremost. The links between firms located at different stages of the economic process are links not of association but of subordination ; one of the firms occupies the policy-making position at the head of the line. It exercises the various functions enumerated by the theorists: organization of production, assumption of risks, command. All three are essential and they are seen in the 'Optigal' example. A report by the

'Optigal' managing director gives a clear description of the various phases in the organization of production[1] : diagnosis of the situation, plan of action, administrative organization, supervision of implementation of the plan.

It will, however, be necessary to ascertain later in the paper whether all the risks are assumed by the integrator.

Domination is also visible at the origin of co-ordination. In France the initial impetus came from two co-operative societies producing cattle food, but the idea was taken over and developed by a food factory and capitalistic integration greatly outweighs co-operative integration in respect of broiler rearing. In the United States the initiative also came from firms making composite food-stuffs. In Switzerland, in addition to 'Optigal' there is the other co-operative society we have already mentioned, the S.E.G. which makes chicken food and puts out contracts for broilers reared on its product, but its output is only half that of its rival. The fact is that a food factory must above all be assured of a steady rate of distribution. The product does not keep for more than three or four weeks ; the vitamins it contains become decomposed and the fats oxidized. Furthermore, it is a bulky product and the manufacturer could not afford to hold large stocks. The position was neatly described by a French commentator who said, 'The chicken is now merely the residue of Totaliment'.

Domination is visible lastly in the nature of the fattening contract. The parties to it are unequal in size. 'Optigal' is affiliated to two large-scale concerns, one of which dominates Swiss production of cattle food and the other the distribution of food products in Switzerland. It is true that by grouping together, the chicken farmers strengthen their individual contractual position. They are, as it were, the numerous cells of a single production organism which by constituting a group acquires a dimension more comparable to that of its supplier and its consumer. It nevertheless remains a purely technical augmentation. The chicken fatteners' association is not on a footing of partnership with the integrating firm. It has merely a 'gentleman's agreement' with 'Optigal', specifying a standard cost price established by the integrator and approved by the association. As for the purchasing price of the finished products, it is fixed by 'Optigal' after merely seeking the opinion of the association. The

[1] F. Guendet, submission to the 'Group of Experts on the Rationalization of Agricultural Undertakings' of the Agricultural Division of the European Economic Commission, at Geneva on 4 September 1963.

latter was powerless to prevent the fall in the purchasing price in the autumn of 1964. All it could do was to obtain a brief respite : for six months the price would remain at 5 centimes higher than the price imposed by 'Optigal'. The inequality of the contracting parties makes this meagre power of discussion inevitable.[1]

B. Means of Domination

First, financial means of domination which turn on the amount of capital invested. This is shared between the integrator and the farmer. The integrator provides the food and the day-old chick, which represent 85 per cent of the standard cost price. The small contribution made by the farmer is explained, however, by the long period of amortization. The fattening house costs 65,000 f., the major part of which (the actual structure) will be written off over 15 years, another portion (heating plant and ventilation equipment) over 12 years and the remainder (feeding troughs and drinking fountains) over 5 years. Actually, it takes ten years of supplies to 'Optigal' for the farmer to recover his capital outlay. His financial commitments extend even further, moreover. When he receives the day-old chicks and their food from 'Optigal', he is not required to pay for them (their cost will be deducted from the price he gets for the matured fowls) but for his produce he accepts a draft of 10,000 f. which will enable 'Optigal' to obtain cheap credit terms for the payment of the fodder.

Domination is also characterized by a number of signed agreements. 'The fattener shall fatten all the chickens procured for him by 'Optigal' and shall place at the disposal of the latter the totality of the chickens he had fattened [article 2 of the contract]. He shall fatten the chickens exclusively in the standard chicken house designed by 'Optigal' and in compliance with the instructions issued by the latter [art. 3]. He undertakes to house only 'Optigal' chicks in these installations and to refrain from fattening any other chickens on his farm [art. 4]. He shall use only the food supplied by 'Optigal' for the fattening of his chickens [art. 5].' It should be noted, however, that in all countries the poultry-rearing co-operative societies impose more or less the same exclusive conditions on their members. For example, the co-operative societies in the département de l'Ain in France enforce the same technical and commercial restrictions without

[1] See J. Valarché, 'Le Pouvoir de discussion des agriculteurs', *Revue d'économie politique*, 1964, No. 3.

causing their members to complain about integration.

What may be more disquieting are the changes imposed by the integrator. Since 'Optigal' first came into operation, the workload has increased owing to the fact that the fattener now has to rear larger broods ; the 'batch' comprised 3,600 chicks at the beginning of 1963 and 4,200 at the beginning of 1965 (the maximum capacity of the fattening house is 4,400). Also, the series now follow in more rapid succession ; instead of the two weeks at first allowed for cleaning out the fattening house between one series and the next, the interval has now been reduced to one week. It is true that in this way the fattener can raise more chickens and thus earn more money, but the demands of the increased workload will force him to forgo other lines of production and will leave him even more in the power of the integrator. Furthermore, he can neither refuse to take the maximum number of chickens stipulated in the contract (21,000 per annum) nor discuss the date of arrival of the chicks.

C. Scope of the Domination

It is often said that the farmer subjected to these constraints is no longer comparable to the head of an undertaking but is reduced to the role of a man taking piece-work and 'making up' a given raw material for his employer. In both cases the producer is a mere operative performing a specific job. The money he gets for 'making up' the raw materials is not comparable to a regular salary, for the amount depends on the quality of his workmanship. This piece-work does not occupy all of his time. It must also be noted that the utilization of this outside labour reduces the fixed overheads of the undertaking. Nevertheless the position is not altogether the same in the two cases. First, the labour market is much clearer for the chicken farmer than for the piece-worker ; the former knows on what basis his earnings are calculated whereas the latter does not generally know this. Secondly, the chicken farmers are more closely united than the piece-workers, by reason of their adhesion to an association recommended moreover by the integrator. Lastly, there exists a more manifest community of interests between the chicken farmer and the integrator than between the piece-worker and his employer. The 'Migros' slaughter-house and the 'Provimi' company need the 'Optigal' clientele. The prospects of demand are very promising ; the *per capita* consumption of chicken is only 5·3 kilos in Switzerland. It has doubled in the last five years but is still well below the U.S. figure. This explains the signing of a five-year contract with the possibility of renewal for a further five years, whereas no such element of duration is found in the agreements governing piece-work done at home.

What is less reassuring is the refusal of the integrator to commit himself as regards price either for the supplies or for the deliveries. The French law on contractual farming enjoins that the contract shall stipulate the price of the reciprocal supplies. 'Optigal' on the contrary reserves the right to modify the purchasing price of the chickens according to changes in the price of the supplies. Since the scheme was first introduced, the price of the chicken food has fallen from 62 to 60 and then to 59 f. per 100 kilos ; the price of the chicks has fallen from 0·95 to 0·93 f. apiece ; the price of the oil for the central heating has also been reduced and the purchasing price of the chickens has dropped from 2·84 to 2·65 a kilo and within the next few months will have gone down to 2·60 f. This still leaves it slightly higher than the price paid by the poultry-rearing co-operative societies in France in 1964 (2·70 French f.) and very much higher than the price in the Netherlands (2·10 French f.). Will international competition allow it to remain at this level? Nearly three-quarters of the poultry consumed in Switzerland are imported and the 'Migros' company is the biggest importer as well as the biggest 'entrepreneur' of nationally produced poultry. Will the close mesh of the net drawn round 'Migros', 'Optigal' and the chicken fatteners tighten under the lash of the storm like the knots of a wet rope or will it break?

IV. THE CONSEQUENCES OF INNOVATION

A. On the Pattern of Production

(1) *Labour-saving Effect.* Mechanical and chemical innovation has revolutionized poultry production. It is astounding to think that only two and a half hours' work is needed to rear over 4,000 chickens which will be ready for consumption in half or even a third of the time required for their predecessors. The labour-saving methods already familiar in industry are now making themselves felt in agriculture.

(2) *New Division of Labour.* The 'Optigal' fatteners are never engaged exclusively in poultry-rearing. But to devote their two and a half hours to poultry-rearing they have to relinquish another venture. To a certain extent they are deserting traditional cattle-

raising, thus freeing some of their land, and are growing more crops. The difficulty of finding the additional labour prohibits any widespread expansion. The shortage of cow-herds in particular acts as an incentive to crop specialization in addition to poultry-farming. Having less animal manure, the farmer will extend his manure crops and will cut his corn less close so as to leave more to plough back into the soil.

(3) *Isolation of the Work.* The 'Optigal' poultry farmer takes nothing from his natural environment, except water, and gives nothing back apart from the litter of his stock. Whereas the traditional type of stock-rearing was a constituent element of agricultural undertakings, the fattening of broiler chickens emerges as a side-line from the point of view of work although essential as a source of revenue.

B. On the Structure of the Market

(1) *The Progression Towards a Concerted Economy.* In poultry-rearing as in all contractual farming, the mechanisms of the market are being partly replaced by the policy planning of the big firms. On the basis of the incomes policy and the controlled economy now prevailing to a greater or lesser degree in all countries of the West, it is possible for the big firms to foresee the trend of demand and to plan their production accordingly. We are thus reaching the stage of a *concerted* economy.

(2) *The Defectiveness of Competition on the Produce Market.* The price policy followed by 'Optigal' reflects the defectiveness of competition. Three different products are offered on the poultry market : the frozen chicken, usually imported, the fresh 'contractual' chicken mostly an 'Optigal' product or secondarily from the S.E.G., and the farm chicken reared by non-integrated farmers. 'Optigal' therefore banks on the differentiation of the product. There is a material differentiation in that the 'Optigal' chickens are American hybrids for which 'Optigal' holds the monopoly in Switzerland and the methods of rearing them distinguish them from farm chickens, as the consumer very easily discovers. There is also a sales policy designed to convince the customer, through advertisements and the presentation of the produce, that the 'Optigal' chickens taste as good as the farm ones while costing little more than frozen ones. The market is in fact a differentiating 'oligopoly', which explains the price fluctuations of the 'Optigal' chicken. In the course of the year the

price ranges from 4·50 to 5·20 f. a kilo. Neither the volume of production nor the cost price of the other integrator explains the periodical price drop but the price of the fresh chicken cannot deviate to any marked degree from the price of the Danish or American frozen chicken (4·50 f.).

(3) *The Defectiveness of Competition on the Labour Market.* The relations between chicken fattener/worker and integrator/employer are likewise established in conditions of imperfect competition. The market is not directly regulated, it is true, but there is neither freedom of access to it nor rapidity of reaction. The would-be chicken fattener has to find the credit needed for the erection of his broiler-house and access to the market is thereby restricted. He cannot give up his job of chicken fattening until he has been able to write-off a building which would be difficult to use for anything but poultry-rearing. Nor can there be much prospect of working under another integrator ; broiler rearing has to terminate in a carefully timed slaughter and distribution system which is provided only by 'Migros' in the geographical area (French-speaking Switzerland) in which the chicken farmers are located.

C. On the Economy as a Whole

(1) *Reduction in the Costs of Marketing the Product.* With the conventional trade practices, there are very wide margins because of ignorance of the markets, losses during transport and storage, lack of uniformity of the products. Vertical co-ordination has narrowed these margins considerably : the chicken bought at 2·65 f. a kilo from the fatteners was sold at an average price of 5·20 a kilo, ready for consumption, to the consumer in 1964.

(2) *Expansion of Consumption and of National Production.* Per capita consumption was 4·3 kilos when the 'Optigal' company was first launched in 1961. It had risen to 5·3 kilos by 1963, *i.e.* an increase of 23 per cent in two years. The rise in consumption was over 50 per cent due to a rise in imports. But the position was reversed between 1963 and 1964. While the *per capita* consumption rose from 5·3 kilos to 5·9 kilos, there was a slight drop in imports. The increase in consumption was thus due to the expansion of national production, particularly of the contractual type (S.E.G. from 2 to 2·5 million kilos, 'Optigal' from 2·7 to 4·5). This is explained by the reduction of the average price. The price of the integrated chicken remains unchanged but the price of the farm chicken has come down

(from 8 to 6·50 f. a kilo) as a result of the 'Optigal' competition. As there has at the same time been a rise in the price of other meat, poultry is now the cheapest meat. Demand for it is increasing, both as a result of higher incomes (the average Swiss citizen is better off from year to year) and because it is taking the place of other meat.

(3) *Investment in Agriculture of Non-agricultural Capital.* The 'Optigal' chicken-farming system entails new equipment for agriculture : experimental centres, rearing quarters for laying hens, incubators, fattening houses. Equipment on this scale would have been impossible without the assistance of outside capital. The profits earned by the 'Migros' rational distribution system and the chemical innovations developed by 'Provimi', have made it possible to give Swiss agriculture a new look.

(4) *The Productivity of Agriculture.* The 'Optigal' chicken fatteners are well ahead of the average Swiss farmer in respect of their financial and technical capacities. The 'Optigal' statistics show their holdings to cover an average area of 30 acres which is definitely above the average for Switzerland. The acquisitors select only applicants who have had vocational training, not that broiler rearing requires skilled labour — practically all the operations are automatic — but, as with all livestock raising, there is always the biological hazard which only the trained eye can detect quickly enough to avert disaster. On these terms, integrated broiler rearing yields greater returns than any other agricultural production. The 'Optigal' purchasing price ensured a wage of 4·45 Swiss francs per hour in 1962, plus a bonus of 20 centimes per chicken, which raises the remuneration of chicken fattening up to the level of the most highly skilled operative in an expanding sector of the economy.

V. CONCLUSION

In studying the 'Optigal' methods we find industrial principles and insurance procedures being put into practice. Mass production, division of labour, unceasing technical innovation, all these characteristics of modern industry have now reached agriculture. The reduction of hazards, the compulsory alarm system carefully specified in the fattening contract recall the methods of insurance companies. The innovation is total. A great change has thus come over agricultural enterprise ; the bonds of interdependence forged by the traditional farming methods have been severed.

The farmer is confronted with a risk but it is one which may be averted by the constitution of a group and the recognition of complementary interests. Yesterday's farmer used to turn tradesman to sell his poultry at the market ; this is no longer the role of the integrated farmer of today but his group can maintain the contact with the world of commerce. The individual farmer will avoid becoming a mere wage-earner by maintaining activities other than chicken fattening.

From the technical point of view, the farmer is now in a position of greater subordination. But does this necessarily entail the idea of domination? The agricultural sector appears rather to be emerging from its state of dependence. In the old days the farmer would frequently have to ask for an advance from the wholesaler taking his produce and would commit himself to hand over his whole crop, which gave rise to abuses. The 'Optigal' contract does not occasion any such dependence. This is yet another point common to industry and modern farming : the worker is less dependent than he used to be on his employer. The possibility of introducing innovations with a minimum of social upheaval reflects the trend in a world where the sectors of production, like the social classes, are now more evenly balanced.

JOINT DISCUSSION OF PROFESSOR LE BIHAN'S AND PROFESSOR VALARCHÉ'S PAPERS

Mr. Allen said that in discussing these two closely related papers by Professor Valarché and Professor Le Bihan he would treat the term 'integration' in the widest possible sense to refer to situations where a single management unit has complete day-to-day control over consequentive stages in production and marketing of agricultural produce, or, alternatively, where there is a close long-term binding contract between two groups otherwise independent. The topics under discussion are essentially special aspects of the nature and extent of economies of scale in modern agriculture, and of the influence of these on the systems of producing and contract-making to meet the modern needs of the market.

There is a danger that over-generalization might be made from studies of such integration cases as those given in the two papers under discussion, and that integration of this sort might be taken as a typical development to be expected in other sectors. Mr. Allen said that he thought that such generalization would be unsafe. Even within the broiler industry there are considerable disparities : in the United States the early stages of integration in broiler production are undertaken by feed manufacturers, but this has not been the case in the United Kingdom, where the main integrator is a builder, and the second largest integrating group had until recently little connection with feed manufacture. In the U.S.A. the more recent advances in integration have not been associated with feed manufacturers but with such groups as the Cotton Producers Association and Arkansas Valley Industries.

There are clearly many cases in which economies of scale are to be derived, for example, where production has to be brought through geographical bottlenecks or where rigid quality control is necessary. The vegetable-growing industry, the export of fruit from Australia, Argentinian meat production, are all subject to the requirements of canning processes, quality control and special marketing arrangements. However, none of these have shown systems of integration of the kind analysed in the two papers. To take another example, the feedlot system realizes economies of scale no less important than those enjoyed in the broiler-house, yet in this case, as in the cases of fruit and vegetable production, integration of the type under discussion has not taken place. These examples suggested that *static* economies of scale are not the fundamental reason for integration. It might be easy to 'externalize' static economies of scale, along the same lines as those that had been adopted in the past by the British textile industry, as it was to 'internalize' them by integration.

In food production there are many examples of economies which are enjoyed externally by firms within particular industries and which have obviated the need for integration. For example, there has been a development of specialist commercial firms giving technical advice on such matters as the care of animals or the proper way to set up a feedlot, and there has been the growth of specialized futures markets in such commodities as pork bellies providing security to the unintegrated operator through hedging ; it is possible that the recently introduced futures market in beef cattle will be developed to play the same role.

One of the reasons commonly advanced to account for integration is that the integrator, for example the feed firm, wishes to maintain his share of the market. This argument might explain the original integration but it does not explain its continuance. In fact, in some cases feed firms which had been the original integrators have moved out of broiler production. The reason for these withdrawals is that particular economies of scale often turn out to be smaller than expected. Also, it might be found that there are cheaper ways of replacing the economies of scale externally rather than internally by integration, even after the integrated organization has been set up.

Mr. Allen said that he strongly agreed with the emphasis that Professor Le Bihan had placed on the importance of innovation and rapid technical change as an integrating agent. Integration is a search for a more effective method of consuming the result of technical change than is provided by the normal market mechanism. The market mechanism is often unable to turn rapidly into external economies those advances which the integrating firms are able to secure as internal economies. However, the incapacity of the normal market structure to reap these benefits is not necessarily permanent. Nor did it necessarily follow that this reason for vertical integration will apply in other lines of business, such as pork and beef production. In the United States pork production is being conducted in larger-scale units, since more and more the health risks are being overcome. But pork production on this scale requires a high degree of management ability, and the rate of technological advance is not expected to be as dramatically rapid as it has been in the case of broilers. These factors meant that the same sort of integration is not likely to appear. Furthermore, the credit services offered by county banks, which are now often backed by large city banks, remove the threat to producer of not being able to raise the funds required to keep his operation up to date.

The case of cattle production in the United States is similar to that of pig production with the added complication of special grading requirements which place a yet higher premium on good management. In these industries, economies are enjoyed externally and the firms which already operate the modern beef and pork production units are capable of adapting to rapid technological change.

Mr. Allen said that he had four specific questions to ask Professor Le Bihan. Would he say something about the structure of the feed industry in France ? There seemed to be an unusual breakdown in the statistics on page 338 of his paper referring to the branches of international firms as opposed to national and regional firms ; it was stated that the branches of international firms controlled 15 per cent of production in 1963, while the national or regional integrated firms had 7 per cent of production. Secondly, what significance does Professor Le Bihan attach to the fact that, as he points out, United States firms spend little money on research and development in relation to total sales revenue ? Surely these firms are able to draw on extensive university fundamental research ? Moreover, is it not misleading to measure this expenditure in relation to sales, since these are large-volume industries with quite small profit margins? Third, Mr. Allen noted that Professor Le Bihan argued for improvement in theory and methods of analysis to deal with the problems of integration, and he asked what Professor Le Bihan thought is required. Is not the traditional theory of the firm adequate ? It is simply necessary to recognize that the cost curve could shift because of technological advance, and to forget the frequently held view in agricultural economics that small-scale enterprise is more efficient than large. Finally, Mr. Allen asked if Professor Le Bihan would elaborate on the question of the role of extension work in the framework of increasingly integrated modern farming.

Discussion of Integration and Innovation

Mr. Allen referred to Professor Valarché's point that the integrator refuses to commit himself to price guarantees to the operator. Is this situation for the farmer worse than his condition before the integration had taken place ? Surely the integrator could not be expected to divorce himself from the forces of market competition, and guarantee prices when the price for the final output might not be stable ? Mr. Allen also asked for clarification from Professor Valarché on his views about the defectiveness of competition of the product market. Certainly the market does not conform to the perfectly competitive model, but when considered by less abstract and less rigorous standards it does not really seem that competition is seriously defective or unworkable. Finally he asked Professor Valarché whether the reader would be right in thinking that the paper is largely concerned with emphasizing the social as opposed to economic considerations in the process of integration. There did seem to be a latent dialogue between Professor Le Bihan and Professor Valarché on this point, with Professor Valarché influenced by the 'way of life' arguments, while Professor Le Bihan presented an apology for integrators.

Professor Valarché said that he had emphasized in his paper that the case he had studied is indeed a special one. However, it has many important aspects. It represents one solution to the problem of unstable incomes in farming which attend the unstable prices for agricultural products. It also shows how wealthy societies, which are prone to serious difficulties in the co-ordination of meat supply, can deal with these difficulties by means of invention and innovation ; that is, in ways which traditional husbandry methods, limited by land and manpower, cannot manage.

Innovation in stock-farming is indissolubly connected with information and domination. The social problems that arise from these issues are as important as the purely economic issues. Under the capitalist system, information is provided for the farmer at the same time that constraints are set upon him. Constraint is an essential part of the system, just as it is in co-operative forms of farming. The French farmer co-operative employs similar constraints on production, supply and delivery as are used by integrators, and in both cases success depends largely on the elimination of personal considerations. The main difference in the introduction of innovation by co-operatives rather than by means of the integrated capitalist structure, is that the co-operative gives more attention to the dissemination of information. In the co-operative framework it is more difficult to eliminate inefficient producers and far more information has to be provided so that producers work properly. Moreover, the flow of information in a co-operative system is reciprocal. Under the integrated capitalist system the flow of information from the bottom to the top is officially excluded. A constraint has to be viewed both from the economic and the sociological standpoint. Confronted with a major economic problem of modernization, social constraints have to be accepted.

Professor Valarché cited the case of farmers who made contracts with firms to supply produce in return for technical help, and who after one year offered their improved production to another firm. Social constraint is necessary if disruptive action of this type was to be avoided. In any case, there are limits to the constraints that could be placed on a farmer. In Switzerland, for example, the range of alternative lines of production and the state support of prices mean that a farmer can change to a different range of products without heavy loss, and might even move from one sector to another without difficulty. Finally, many integration schemes can be conducted both by downstream and by upstream firms ; this implies that the farmer has the opportunity of playing one end of the industry off against the other, and so of reducing the threat of domination.

Professor Ruttan said that people seem to regard integration as though it was a new concept in economic organization. In fact the subsistence farm, (for example, the medieval manor) had shown a high degree of integration. Among the reasons for disintegration were the emergence of wider geographical trade and the development of specialized commodity markets. Under these circumstances information was spread by the single dimension of prices, whereas under integration, the communication system became multi-dimensional and personal. This permits greater precision in the transmission of information and ideas.

Professor Gale Johnson said that if the Swiss, French and West Germans are concerned with the social problems of integration in the broiler industry, there is a simple solution : import more broilers from Denmark and the United States !

In Professor Valarché's paper there seems to be an implied criticism of the integrating group for the prices they charge for inputs. Might these high prices be due to the policy of the Swiss government on import duties ? Certainly the price of feed that goes into the chicken in Switzerland is higher than it is in the Netherlands and in France. Also the importation of chicks is probably hampered by import duties or health requirements.

Professor Bishop said that he was glad that Professor Le Bihan had pointed out the effects of changes in industrial development on research and educational systems. However, he thought that the changes in educational services and in the nature of contact with researchers are not unique to integrated firms, but reflect the growing size of firms in general and the growing cost of making wrong decisions.

Professor Bićanić asked if Professor Le Bihan could explain how the formation of vertically integrated contract systems is co-ordinated with French agricultural planning.

Professor Gulbrandsen said that an important question was why the integrator does not buy the necessary factors of production and run the enlarged production units himself, instead of incurring the costs of teaching the new technology to the farmers. The main answer is that there might be external effects such as the avoidance of tying up more land than is necessary and breaking into profitable rotation patterns which it would not be worth the integrator to undertake for his own purposes. It might also be very difficult to get enough land in one place in order to reap full economies of scale.

Professor Robinson said that it might be valuable to widen the discussion to ask the general question : By what means does innovation enter the agriculture of advanced countries ? In the case of the United Kingdom, advance in agriculture had come from rich farmers such as Coke and Arthur Young. This advance spread from these farmers, who were rich enough to withstand the costs of experiment, to the general run of farmers. Professor Robinson asked whether he was right in thinking that this process is not dead. There seemed to be important cases in the United Kingdom of men who make money in other activities and who then try out advanced ideas in farming, with the capital they had amassed.

The role played by university farms and farms of the Ministry of Agriculture is analogous to that played by wealthy amateurs. The ideas developed on these farms are absorbed by the braver farmers and so spread to the farming community generally. In some parts of England an important part of the dissemination of technical information is achieved by conducted farm walks in which farmers examine each other's methods critically. Professor Robinson asked whether the methods of dissemination of information differ widely from country to country. Is the impression correct that in the United States the bulk of new knowledge is obtained from rich university departments ? In general, from where does knowledge disseminate and to whom ?

Dr. Jacobi said that he thought that Professor Robinson was right that a great deal of technical education goes on by farmers looking over the fences of other farmers. But the technical education which Professor Valarché was dealing with in his broiler industry example is not of the same sort. Technical information is disseminated, but it is scarcely technical education since there is no need for the farmer to understand what he is doing and there is little chance for him to experiment or do other than follow precise instructions. Thus the farmer applies progress but is hardly technically educated in Professor Robinson's sense.

Mr. Allen said that he would like to amplify Professor Robinson's point and relate it to what has happened in the case of vertically integrated concerns. There are a large number of ways that innovations can be disseminated — by visits to extension farms, by the use of trade journals which are widely read in the United States, and by any other methods. The problem is to choose that method of dissemination which suits the industry and innovation in question. Innovations today typically involve very large economies of scale and heavy levels of risk taking, which was not the case previously. The broiler industry development itself has been a colossal undertaking, and, although agricultural colleges and universities have played an important part, very big risks still had to be borne by individual firms. A major factor in determining which form the dissemination of innovation takes is the very rapid pace at which technology advances. The ordinary market arrangements are unable to create the special organizations necessary to perform the dissemination. In the broiler industry know-how has advanced so swiftly that specialist advisory firms cannot be set up fast enough, and organizations which can secure these advances as internal economies have flourished.

In the United States there is extensive fundamental research carried out by agricultural experimental stations, and commercial firms are concerned to adopt the latest ideas from these stations and from the university. In many lines of agricultural production — the hog and beef industries being the outstanding examples — there are men of very great ability and educational calibre, who already run large operations, and who often know far more than the people actually engaged in research. In the two lines of business mentioned, technology is not changing so fast that people cannot adapt or create a system of services through normal market development. Vertical integration in these cases might well prove unnecessary and even relatively unprofitable.

Professor Le Bihan said that from the outset he had imposed two restrictions on his analysis : only technologically progressive production processes were considered ; and the model he used left on one side the distinctions between internal and external constraints. Professor Le Bihan said that the impact of integrating firms on a number of agricultural units is very complex in its effects, and there is no literature that has yet been published on the question. The interactions that occur upon integration are not susceptible to classical analysis and no satisfactory analytical study has been made. The difficulty is that each individual case has to be taken separately since the size of the reactions and the techniques adopted follow no one pattern. Also the impact of the intervention of an integrating firm on the development of an agricultural undertaking is not permanent so that there is never any one definitive and unchanging situation that can be taken as the norm.

Professor Le Bihan said that he agreed with Professor Ruttan that integration is by no means a new process. Apart from the subsistence cases, there are a number of agro-industries that have been integrated for a long time, such as the sugar-beet industry and much of the brewing industry.

In answer to Professor Allen, Professor Le Bihan said that the structure of the feed-stuffs industry in the Common Market countries is very different from that in North America and Britain. The major concerns in France, Italy and Belgium have been launched by Dutch firms ; this gives the industry a very specific international structure. The main feed-stuff firms which produce concentrates which supplement the diet of the animals, supply a constellation of small manufacturers who are tied to these main firms by five-year contracts so that their production policies are controlled, and so that information is provided for them.

Discussion of Integration and Innovation

Professor Le Bihan said that it had been argued that an important part is played by universities and other research institutions. He said that he had been dealing mainly with the European case and, in Europe, neither the universities nor the public research bodies have shown the dynamism that is typical in the United States. In fact Europe is an importer of the discoveries made in the United States. The major innovating firms are disseminators of inventions, but were seldom initiators in invention. These firms are to be found both upstream and downstream from the farmer, although it seems that the fertility of firms in making technical innovations declines as one left the upstream firms and examined the downstream final distributing enterprises. Thus the great distributing organizations such as 'Migros' run on research and development services and play only a minor role in the diffusion of real technological change. In contrast, the integrated complexes which are linked to upstream factor producing industries show, at the farm level, the consequences of greater technological stimulation. The industry that has been integrated by a downstream organization does provide a stimulating effect on the farmer, but this is as a result of the price stability that it can offer rather than as a result of technical stimulation.

In reply to Professor Bićanić, Professor Le Bihan said that in his view French indicative planning is largely a matter of words rather than of actions. There are provisions for planning information to be relayed to major contractual systems, but these are scarcely used by the planners. If anything, it would seem that France was in stage of 'deplanning'.

Professor Valarché said there are government regulations concerning Swiss agricultural production as Professor Gale Johnson had said. The government insists that importers take over a part of the indigenous production where that is thought necessary for the health of Swiss agriculture. Thus 'Migros' sells more imported broilers than Swiss broilers, but it cannot import more than it does without being liable to the law which requires it to absorb internal production. On the matter of feed importation, Professor Valarché said that the duty paid varies according to particular commercial agreements with the exporter. The issue is complicated by the fact that poultry is in competition with veal, and if the duty on skimmed powdered milk were reduced veal might become more profitable than poultry.

Professor Allen had expressed surprise that the paper showed concern about the integrator's refusal to commit himself on prices. Professor Valarché said that he was not arguing that the integrator should resign his function as the integrating agent, but that there are indications of a movement away from healthy profit-sharing. He agreed with Professor Allen that there is some price competition and therefore fluctuation in prices under the present integrated system, but that this is appreciably less than in the past.

It is often argued that integration brings the peasant into the wage economy and that it reduces him to the status of a wage-earner in his own home. Certainly it is true that the producer of chickens under the present scheme is concerned only with his physical productivity and that the value productivity is entirely in the hands of the integrator. However, it is important to remember that the chicken-raising activity is usually only one part of his range of farm activities. The financial commitment of a peasant in entering chicken production was 65,000 Swiss francs for a 16-acre farm and this is only a small part of his whole investment in the farm — usually not more than 25 per cent. Thus chicken-raising is typical of 'workshop' cultivation while the rest of the farm activities continue as before. It is perhaps in the context of his work that the largest change is to be seen ; the farmer no longer uses local veterinaries and technicians for many of his requirements, but works rather on a regional or national basis.

Chapter 17

THE PROBLEMS OF VERTICAL
INTEGRATION IN AGRICULTURE:
THE HUNGARIAN CASE[1]

BY

LÁSZLÓ KOMLÓ

Budapest, Hungary

I. INTRODUCTION: THE NEED FOR THE
INDUSTRIALIZATION OF AGRICULTURE

VERTICAL integration in agriculture involves the organization and co-ordination of the successive stages of agricultural and foodstuffs production. As it is a stage in the level of industrial development of productive forces, its principles are valid for agricultural sectors which have reached an industrial stage of production in countries where this evolution has taken place. Its effect on production relationships is to provide and develop those of a capitalist type in a capitalist economy and those of a socialist type in a socialist economy.

In theory, therefore, the principles of vertical integration will be adaptable to Hungary as a socialist country. But its concrete forms and the phasing of its development will be determined by the economic and social conditions of our agriculture and our country.

The first question that arises is, therefore, to ascertain whether the factors already exist that call for the industrialization of the country's agriculture.

The export of foodstuffs and of agricultural produce was already playing an important part in Hungary's economy before the war. It has since shown a steady upward trend and is now four times the pre-war figure. The products are exported to countries of the West and of the East. The Western markets of today constitute a demand which can only be met by an industrialized agriculture. Before long the same will be true of the socialist markets. If, therefore, we wish to maintain and even increase these exports, our agriculture will have to be industrialized as soon as possible.

[1] Translated by Mrs. M. C. Lemierre.

The Problems of Vertical Integration and Innovation

It is known that *the concentration of the food trade,* linked with urban and industrial concentration, also plays an important role in the industrialization of agriculture. It so happens that there has been a change in our previous policy of specialization and for the last few years we have been developing a network of supermarkets and multiple stores. Unfortunately self-service is not yet fully established and this detracts somewhat from the advantages of this concentration. The more widespread this system of commercialization becomes, the more keenly we shall feel the inconveniences resulting from the incomplete realization of its potentialities. This too will contribute to make the industrialization of our agriculture inevitable.

Our *food industry* was even in pre-war days the most highly developed in all Central Europe and, since then, particularly in the last ten years, it has been modernized and concentrated and has been accompanied by the construction of new high-capacity factories. Its development is continuing on the same lines and at an increasingly rapid rate. The trend is already significant and will become even more marked in the coming years, developing it into an industry in which the share of permanent capital will be a high one, thus further stressing the need for continuity of operation. Needless to say this goal can only be reached if we achieve a corresponding industrialization in agriculture. Can it be said that our agriculture is ready to meet these new requirements?

II. THE STATE OF AGRICULTURE AND OF GOVERNMENT CONTROL

(1) *The Structural Situation*

Before the war agriculture was in many places in a condition that can only be described as semi-feudal. Following radical measures of agrarian reform, we first made good the very considerable damage caused by the war and then set about remodelling its structure on socialist lines. On account of the mistakes that were made, collectivization was only completed in 1961. We have now reached the point at which about 14 per cent of the total area of cultivated farm land is occupied by the state farms and 80 per cent by the co-operative production associations. The average size of the state farms is 4,800 hectares (nearly 12,000 acres) and that of the production co-operatives 1,500 hectares (approximately 3,700 acres). This is a very favourable factor for our agriculture and will facilitate its industrialization.

366

The position is less encouraging in respect of the mechanization of agriculture. Despite large-scale investment, the desired degree of mechanization has not yet been attained everywhere. We reckon 66 hectares (163 acres) of arable land per tractor in the state farms and 93 hectares (230 acres) in the production co-operatives. There is almost total mechanization in the production of grain crops whereas for other vegetable crops and above all for the production of livestock, mechanization is still only in its early stages. The industrialization of our agriculture thus calls for a much faster rate of mechanization than in the past. Furthermore, this need is made still more urgent by the structure of our agricultural population. Although its active elements still represent 30 per cent of the country's total active population, some 60 per cent of that population is over 50 years of age whereas only 6 per cent come from the under-27 age groups.

A last but most important point to be noted is that by the collectivisation of our agriculture we have actually achieved only the concentration of landed property, without as yet progressing to the concentration of agricultural production itself. Our state farms and, above all, our production co-operatives still follow the pre-industrial pattern of multi-crop and multi-stock farming.

(2) *The Price System and Government Regulation*

This survival of the multiple farming system is partly attributable to the current agricultural price system. It is a problem which affects the production co-operatives more particularly, for prices are of much greater importance to them than to the state farms. Our price policy is still, in fact, based on the principles of traditional farming, designed to ensure a specific income to a type of agriculture comprising multi-crop and multi-stock farms, without paying very much attention to the costs of production per sector and taking the view that it is the gross income of the undertaking that matters. As I see it, an agricultural price system of this kind constitutes an impediment to the specialization and concentration of production, *i.e.* to the industrialization of agriculture.

The other peculiarity of our present agricultural price system is that, in addition to its basic function in the exchange of goods, it also serves to transfer a portion of the agricultural income to the centralized state funds, from which the necessary share is then redistributed through a system of loans and aid to the farms and in particular to the production co-operatives. It is by this system of redistribution that the state attempts to remove the differential rent and at the same time ensure the balanced development of farms which may not all enjoy the same economic or natural benefits. Although this aim is fully justified, it is important not to lose sight of the fact that this role deprives prices of their original function ; it disrupts and impedes the normal evolution of the production of goods and to remedy this state of affairs we have to resort to a multiplicity of financial or even administrative measures. I think it would be preferable to restore their original function to prices and, in order to obtain the desired revenue from agriculture, to apply an improved system of taxation, especially as the farms have to keep their accounts.

The requirement that the production co-operatives pay their land taxes in corn (a survival from the age of compulsory deliveries), and the requirement that the farms tender raw materials (cereals) in payment for the compound foodstuffs obtained from the factories, both, in my view, retard the disappearance of the multi-crop, multi-stock system. There are other factors that have very much the same effect, such as the fact that many producers live largely on their own produce because of the inadequate foodstuff distribution networks in rural areas, or the regulation forbidding exchanges between agricultural undertakings, with the result that each has to produce certain staple elements needed for its own production (corn, grain crops for animal fodder, etc.).

Yet another factor contributing to the continuance of the multi-crop, multi-stock methods is our agricultural planning system. It was designed within the framework of the old type of agriculture and based therefore on the horizontal interdependence of the various branches, made inevitable by the inadequate development of the forces of production. On the basis of the national requirements, the agricultural production plan is drawn up by the Ministry of Agriculture which sets the goals with respect to areas to be sown and species of livestock to be produced. These objectives are then divided in an empirical manner between the regions and districts.

Furthermore, the allocation of the objectives between the various administrative areas is effected by somewhat oversimplified methods. The planning agencies are content with obtaining a more or less uniform distribution of the various lines of production and go to no trouble to encourage any regional specialization.

(3) *The Lack of Vertical Co-ordination*

As for the production plan for the food industry, it is drawn up and allocated separately by the Ministry of Food Industries. Vertical co-ordination between these two branches of the national economy takes place only at national level and is the responsibility of the Planning Office. If it occurs at lower levels this is merely by chance.

Vertical co-ordination between agricultural production and the food industries is no better with respect to investment. This is at present in the hands of two authorities, the Ministry of Food Industries and the Ministry of Agriculture. Each is responsible for the investments in its own sector and the task of the Ministry of Agriculture is further complicated by the fact that it must take into account the production co-operatives' desires to invest. These two authorities in fact pursue an investment policy corresponding to their own horizontal requirements, as a consequence of which the location of the processing factories rarely coincides with the needs of the farmers, while agricultural investment on the other hand often takes little or no account of the needs of the processing factories. It is obvious that the continuation of this system of investment might have regrettable results. Collectivization has just been completed ; the major investments are still to be made, and, if in so doing we cling to our present methods we may jeopardize, perhaps for many years to come, the possibility of achieving the effective industrialization of our agriculture.

It is the survival of the multi-crop, multi-stock system and the accompanying lack of vertical co-ordination that are responsible for the inconsistencies between our food industry and our agriculture. The food industry has to collect its raw materials in small batches scattered all over the country which involves high transport costs. Furthermore, it cannot count on the uniformity of these raw materials, coming as they do from so many different farms dispersed over the whole country and largely escaping adequate inspection by its technicians. When we also remember that by far the greater part of our agricultural production is still strictly seasonal, it is not difficult to understand why our food industry is in so unfavourable a position.

Although there is a contract system for agricultural production, it is a very loose arrangement and the food industry has, as yet, no decisive say in the production pattern of the farms. On the one hand, the production co-operative associations, representing co-operative ownership, retain considerable independence with regard to the pattern of their production and, on the other hand, the planning of the food industry's production is not sufficiently co-ordinated with that of the Ministry of Agriculture for agricultural production. This explains why, in spite of the system of production under contract, there is not sufficient vertical co-ordination between the factories of the food industry and the farms which, moreover, are not compelled to produce under contract since there is not, as a rule, any relative surplus production from the farms. It is obviously the concentrated food industry which feels the lack of vertical co-ordination most strongly and it was not without purpose that it made itself heard so forcibly over the question of the vertical integration of our agriculture.

III. STEPS TOWARDS CO-ORDINATION AND INTEGRATION

It has just been shown that the essential anomaly between agriculture and its economic environment stems from the fact that agricultural production is at a lower level of industrialization than that which has been reached both by the food and agricultural industries and by the food trade.

The problem therefore is to replace the old system of agricultural production by a new one, bringing it closer to the level of industrialization of its economic environment. The new system will have to attach greater importance to specialization in farms. There will have to be greater flexibility, for instance, in the link established under the old system between livestock production and the animal food crops grown on the farm. One of the most valuable results of the new system would be to lead to specialization by regions, in the light both of their natural suitability and of their relative position in the national economy. These two types of specialization are essential for the successful introduction of production techniques of an industrial character. The vertical co-ordination of production and investment planning must therefore be achieved by methods which will encourage such specialization. These methods are well-known ones which have long been in use elsewhere. They are adopted wherever production arises, whether in capitalist or socialist countries. They are the methods of which we have spoken under the heading of vertical integration. It has, however, been seen that in our system of agricultural policy planning there are a number of obstructions to the change-over from the old methods. The necessary transformation can only be achieved if we first overcome these 'superstructural' obstacles.

One of the prerequisites is, of course, to change the agricultural price system in order to do away with the inconveniences we mentioned earlier ; in particular, all payments in kind would have to be abolished. But this is a problem of political economy which lies outside the scope of the present paper. There would then have to be far-reaching changes in the planning methods followed for agricultural and food production. Vertical planning should come first, with second place given to horizontal planning according to the vertical sectors of the food economy. The adoption of a planning system of this kind and also the solution of various other delicate problems of the food and agricultural economy would be facilitated by the merging of two ministries, namely the Ministry of Food Industries and the Ministry of Agriculture. If the new joint ministry was organized with departments for each sector of the food economy, these would be able to draw up their respective plans on the principle of vertical interdependence, leaving the task of horizontal co-ordination to the general planning department. I also think that if all the agricultural produce made available for the market passes through the food factories or through the purchasing centres, which will have previously placed production contracts with the farms, there will be no need for the ministry to draw up a detailed plan of agricultural production. It would possibly suffice merely to set the production goals for each processing factory and for the amount of produce to be handled by each purchasing centre. It would be for the factories and purchasing centres themselves to determine their requirements in agricultural produce and to place the necessary production contracts with the farms.

I am convinced that with these changes, the food and agriculture planning system would become much simpler and more effective than it is at present ; the economic responsibilities and activities of the vertical units (processing factories and purchasing centres) would increase considerably and they would be able to concentrate agricultural production more efficiently and more expeditiously ; the farms, having likewise acquired greater independence as economic units, would be able to negotiate with the pole of integration and determine their patterns of production in the light of their economic positions, on which the pole of integration would exert a favourable influence.

The necessity of achieving vertical and horizontal co-ordination of agricultural investment would likewise call for the merging of the two Ministries and their mechanisms. It would be the only way of ensuring full co-ordination not only at a national level but also at regional and even district levels. At these lower levels the processing factories and the agricultural produce purchasing centres could assume the task of co-ordinating agricultural investment, taking into account vertical and horizontal interdependence on the one hand and the requirements of the necessary concentration of agricultural production on the other hand. A system of this kind would, I think, make co-ordination and implementation of the investment plans both simpler and much more effective than under the present system.

After the removal of the 'superstructural' obstacles, it would remain for us to establish the integrated vertical units.

IV. METHODS OF INTEGRATION

Despite almost general agreement on the need for vertical integration, there is much difference of opinion as to the methods to be adopted for its achievement and in particular as to the allocation of the role of *the pole of integration*.

To understand the origin of this disagreement, it must be remembered that before the war our agriculture was of small farm type and had not attained the level of development at which it could have formed part of a co-operative food industry. The processing factories of that period were in the hands of the capitalist firms or of the large landowners and as such were subsequently nationalized by the socialist government. This industry is now being directed and developed by the Ministry of Food Industries, and agriculture itself is not allowed to build its own processing factories. This explains why our food industry is quite separate from agriculture; it is entirely state-owned, whereas in agriculture it is co-operative ownership which predominates. It is from this situation that the discussions about the pole of integration have arisen. Three main standpoints may be discerned.

According to the first, the pole of integration should be in the hands of the food industry, which would otherwise be unable to organize and concentrate the production of its necessary raw materials according to its requirements. The opponents of this school of thought argue that agriculture would thus be completely subordinated to the food industry and would lose all independence.

According to the second standpoint, which is that of the opponents of the first, *the factories of the food industry should be handed over to the agricultural undertakings* which would organize the agro-industrial combines. The advantages of this solution, according to its advocates, would be as follows : the production of the raw materials and their industrial processing would be part of the same undertaking and under the same management ; the earnings of agriculture and of the agricultural population would be increased by obtaining the added value accruing from the processing of the produce ; this method would also make it possible to eliminate seasonal unemployment in the agricultural labour force.

From the third standpoint the solution would be to organize *agro-industrial associations, comprising on the one hand the processing factory and on the other hand the state farms and the production co-operatives supplying its raw materials.* This solution would not interfere with the present forms of ownership and the associations would operate as 'trusts'.

The choice of the pole of integration is thus seen to raise very complicated problems and for all three solutions there are arguments which cannot be lightly set aside. Without wishing to appear biased, I should like, however, to make one or two comments. First, I do not consider it possible to find uniform solutions for all the sectors of the food and agricultural economy. They are at the present time at varying stages of development for which the appropriate solutions would have to be found. It would be impossible, for instance, to install a modern processing factory within one of our agricultural undertakings, for in most cases the latter would be incapable of supplying it with sufficient raw materials. I do not therefore think the agro-industrial combines can be set up until agricultural production has been concentrated and specialized, *i.e.* until agriculture has reached the final stage of industrialization. Even then, it would perhaps be preferable to have agro-industrial associations, for it would scarcely be possible to concentrate agricultural production to the point at which a single undertaking could keep a large processing factory supplied. Consequently, the combined solution would appear better suited to processing factories of smaller capacity and on these lines a start could be made even now in certain areas entirely devoted to vineyards or orchards.

The creation of agro-industrial associations would likewise require a fairly intensive concentration of agricultural production, otherwise we should find ourselves in a situation where practically every farm would be a member of each agro-industrial association. For the moment, our agricultural production is still too widely dispersed and it is therefore only in the few areas devoted exclusively to vines or market gardening that associations of this kind could be set up immediately.

It would, at the present time, appear to be ascending vertical integration and the traditional production contracts which might be adapted for the majority of our agricultural sectors. The former could be applied to the production of chickens, eggs, bacon, veal, dairy produce, green peas, asparagus, tomatoes, pimentoes and spices, and the latter for the production of grain crops, dried beans and peas and all the usual plants grown for industry.

The traditional type of production under contract is the general practice in practically all branches of agriculture in Hungary and it is the food industry (under the authority of the Ministry of Food Industries) which places the contracts with the farms. I see no valid reason for upsetting this system, which has given good results ; it should, on the contrary, be improved and orientated towards vertical integration by entrusting the role of integrating pole to the food industry. In my view, it is only along these lines that we shall be able to achieve sufficient concentration of agricultural production and of the food industry to pave the way for the subsequent introduction of agro-industrial associations and agro-combines. At all events, I do not think we can accept the arguments of those in favour of placing agricultural industry in the hands of the producers in order to raise the level of their earnings and provide a remedy for seasonal unemployment.

As regards the question of agricultural earnings, it must be remembered that they are not determined by the food industry. Most of our agricultural prices are fixed by the government which in so doing takes into account the proportion of a particular price which goes as earnings to the co-operative farming sector. We could thus have economic prices for agriculture even without the intervention of the processing factor and conversely they might be non-economic even with the assistance of the latter. It would always depend on the incomes policy followed by the government.

I also fail to see how the installation of processing factories within the farms could resolve the problem of the full employment of the agricultural labour force. The food factories of today likewise require permanent manpower. It was only when the processing of agricultural produce was a small-scale cottage industry that it could be carried on with seasonal labour borrowed periodically from the farms. I therefore consider that the problem of employment will have to be solved within the framework of agriculture itself and the more industrialized it becomes the more capable it will be of ensuring full employment for its manpower.

In these conditions, urgent attention will have to be given to the reorientation of training for management posts and of scientific research.

V. TRAINING AND RESEARCH

In the old days, farming could be left to the type of man who had been trained exclusively in rural economy on the horizontal plane. He was familiar with the methods of growing all kinds of crops and rearing all species of livestock, he had studied the organization of production and of labour but he had little knowledge of the 'vertical plane'. Industrialized agriculture, on the contrary, requires specialists who have received vertical training in one of the industrialized branches of agriculture, *i.e.* who are familiar with all phases of the vertical chain of production for a specific product.

If we examine the make-up of the staff of a vertical unit we see that it includes the economic planner, the agronomist, the processing factory engineer, the commercial specialist, the planning specialist and the qualified accountant. This shows that when agriculture is industrialized it can no longer be run solely by men who have specialized in the various branches of rural economy but will need the services of those who, in addition to their basic training, have studied the particular sector of food economy with which they will be concerned.

In Hungary the training of farm managers is still proceeding on traditional lines. However, the first steps have been taken to reorganize it by arranging for finishing courses of a specialized nature. It will be readily understood that for the time being we cannot do more, as our agriculture is only just beginning to move towards industrialization. The complete modification of the training system will obviously not be an easy task, nor can it be achieved very rapidly for, as in other countries, it will first be necessary to provide adequately qualified teaching staffs.

In view of the fact that scientific research and higher education are kept fairly separate in our country we shall probably have to rely on the findings of research to bring about a change in the attitude of the teaching staffs. But in fact research itself is still tied to the traditional type of agriculture : the scientific research institutes are organized and work on a 'horizontal' plane, as do the food industries and agriculture itself. To change the content or the aims of higher education, we should perhaps begin by reorganizing the framework of research, that is to say, by setting up research institutes corresponding to the vertical sectors of the food economy or else by providing for a central institute with its departments organized by sectors.

Furthermore, any such reorganization of scientific research will be required primarily by the need for the industrialization of agriculture itself. There are indeed no examples for us to follow with regard to the methods of organizing and developing an industrialized agriculture in a socialist economy, and it is for this reason that I think it should be for our own research to determine the phasing, the methods and the procedures which would lead us along this path that has hitherto remained unexplored.

VI. CONCLUDING REMARKS

It has not been possible in this short paper to make more than a passing reference to the basic problems now confronting our food industry.

I should like to add, however, that it is not by chance that our agriculture and our food industry have reached the point at which there are such glaring inconsistencies between the superstructure and the level of development of the productive forces. They are merely undergoing the same difficulties as our economy as a whole, which is at the present time engaged in moving from a phase of extensive industrialization to a phase of intensive industrialization to which the old superstructures are no longer adapted.

Everyone is fully aware of this state of affairs and this year a Committee for the improvement of the economic machinery has been set up, to enquire into ways and means of renovating the superstructures. This Committee includes *inter alia* an Agricultural Commission which intends to speed up the industrialization of agriculture, in the knowledge that only in this way can the productivity of labour be increased in this sector and a faster rate of development achieved for the rest of the national economy.

DISCUSSION OF PROFESSOR KOMLÓ'S PAPER

Professor Fauvel said that the burden of Professor Komló's paper was the presentation of an acute problem which is simple to grasp. Agricultural production has to be integrated to satisfy the demands of modern industry. In Hungary the structure inherited from the past, and the measures of centralization that had been a feature of collectivization had worked against rational integration. The situation is one of confusion and wastefulness, with products that are not standardized, being gathered in small quantities, at great cost, within a system of regions in which specialization is not encouraged by present policies. Ministerial control is divided between two ministries so that administrative initiative towards integration is hampered.

There is little doubt that vertical integration is as necessary in socialist countries as in other countries. However, the planning structure means that different methods would be used in the socialist countries. Where so large a degree of control is exercised by the government, the question of ministerial control and co-ordination is most important. On this matter, Professor Komló restricted his recommendations to the two ministries at present directly involved in agricultural policy. But the position of Hungary as an exporter of agricultural produce means that the Ministries of External and Internal Trade will also have to co-ordinate their policies with those of the ministries (or — after Professor Komló's recommendations had been realized — ministry) directly concerned. Professor Fauvel said that he thought Professor Komló overlooked the whole question of external trade policy which impinges heavily on the formulation of agricultural planning.

Although the solutions which will be, and are being, realized in socialist countries are different from those in the non-socialist world, the underlying problems are the same. The whole process of food production has become a kind of public service the maintenance of which is assured by a system of contracts, arranged either by commercial bodies or by centralized agencies. The concern with vertical integration and joint control is partly a result of problems encountered in realizing this public service, such as the need to stabilize production at a desirable level, the need to give an equitable level of income to farmers and the need to ensure the efficient fulfilment of other objectives which were thought to promote public welfare. The other main reason for concern with vertical integration is that it means the realization of large economies of scale which have important repercussions on the status of farmers and the growth of monopolistic power. From neither point of view are the problems new. At many points in the Conference it has been accepted that the forces of integration have to be employed, and it is clear that if farmers are not able to make their peace with the new arrangements they will in the long run have to withdraw from the market.

The Problems of Vertical Integration and Innovation

Mr. Allen said that he would like to take up the general question of vertical integration as it had been posed by Professor Fauvel. This integration is simply a question of economies of scale and has been covered by economic analysis since the time of Marshall. However, the form in which these economies are realized is not always the same. They are realized either as economies which are internal to a large integrated organization, or as economies which are largely external to the separate links in the marketing chain. Thus in Yugoslavia it seems that the agricultural marketing system has been adapted to securing these economies externally. This could be contrasted with the internal economy methods envisaged by Professor Komló. It would seem that the Yugoslav method implies a greater degree of competition in the marketing system.

Mr. Allen said that on the question of the role of innovations, there often seems to be the implicit assumption that agricultural marketing remains a relatively unsophisticated industry, in which innovations do not involve imagination and the readiness to bear risks. Mr. Allen said that he questioned this view. It is clear from United Kingdom and United States experience that vertical integration and improved marketing methods result from the high profits that are attainable by innovation. The readiness to bear heavy risks is the crucial factor. Similarly, if it proves possible in the next few years to adopt broiler methods in hog production, it would be the result of a very few American farmers being prepared at the present time to bear the very heavy risks in hog management.

Thus, while at an unsophisticated stage of basic food requirements a bureaucratic and purely administrative marketing system may be able to perform the job effectively, it is likely to prove impossible to develop a successful system of vertical integration in socialist economies, unless, in some way, reward for innovation can be introduced. A marketing system of some flexibility is essential so that it becomes worth while for the enterprisers to innovate.

Dr. Jacobi said that the parallel drawn by Professor Fauvel between the system of vertical integration in non-socialist countries and the system of government contracts in socialist countries raises one problem that should be examined with attention, namely the question of bargaining power. Whether it is large industrial enterprises and retail chains which negotiate the contracts with the farmers, or whether it is the government which establishes these contracts, the system might be a useful tool for the farmers in that it ensures a certain price. However, its usefulness to the farmers may well change. As long as demand outstrips production, so that bargaining power is largely in the hands of farmers, the usefulness is assured ; but when over-production — as in Switzerland and Italy — becomes a typical condition of agriculture, the contract system may become a serious threat to the bargaining power of the farming sector.

Chapter 18

JOINT DECISION-MAKING PROCESSES IN PRESENT-DAY AGRICULTURE

BY

SIDNEY HOOS

University of California, U.S.A.

I. INTRODUCTION

THE architects of the Conference have assigned to me a broad subject, but with constraints of time and space. Within such limitations, I sketch a limited number of market-structure situations in which joint decision-making processes prevail in present-day agriculture. They apply to the United States, although they are not limited to it ; broadly similar joint decision-making processes operate in other countries. A *caveat* is that my experience and views are flavoured, likely biased, by my observation and study of developments in my own country.

I first set forth a taxonomic outline into which one may cast the variables and the types of decision-makers participating, then are noted some market structures in which joint decision-making processes prevail. With such background, I shall consider some implications and inferences of joint decision-making processes in present-day agriculture.

II. DECISION-MAKING: OF WHAT AND BY WHOM

Economic and institutional variables are subject to joint decision-making processes. The major economic variables include production functions, cost rates of inputs, prices of output, annual volume and seasonal flow of output, and product characteristics. Not all of these are necessarily included in all of the joint decision-making processes to be considered ; some are oriented to only one economic variable, while others encompass several.

The institutional variables include legal, legislative and political

379

ones. They reflect the extent to which government and agriculture, or politics and farming, are intertwined.[1] But on this phase of the subject I shall touch tangentially ; others here will deal with it in depth.

The participants in joint decision-making processes in present-day agriculture may be structured, in a simplified sense, into three broad categories : (1) farmers, (2) those with whom farmers do business and (3) government. When either farmers *or* those with whom they do business participate jointly in the decision-making process, I refer to that as joint decision-making of the unilateral type. When farmers *and* those with whom they do business participate jointly in decision-making, that is referred to as bilateral decision-making. When government participates jointly with *only one of either side* (farmers or their opposites), we again have bilateral decision-making. And finally, when government participates with *both* farmers and those with whom they exchange, we have trilateral joint decision-making. All four categories prevail in the market structures with joint decision-making in agriculture.

Among the several examples involving joint decision-making, I comment on co-operative marketing of outputs and of inputs, co-operative processing, co-operative bargaining, contract farming, marketing agreements and order programmes, supply control involving output and price programmes, and international commodity agreements.

III. MARKET STRUCTURES WITH JOINT DECISION-MAKING

A long-established and widely used form of unilateral joint decision-making in agriculture is the conventional co-operative marketing association. Here marketing includes either or both the sale of farm output and the purchase of farm inputs. Farmers band together in terms of horizontal integration, generally provided for under national legislation which permits a special status for co-operative organization by stipulated exemptions from cartel or anti-trust laws. Multiple ownership and control of the co-operative may prevail along the lines of the Rochdale principles or a variation of them.

[1] Dale E. Hathaway, *Government and Agriculture*, New York, Macmillan Co., 1963, p. 412.

In such a conventional marketing co-operative, the day-to-day or operating decisions are usually under the jurisdiction of employed management. But the broader policy decisions are made by the board of directors elected by the farmer membership. Such joint decision-making is unilateral in the sense that only farmers participate in the formulation of the policy decisions.

With *marketing* as its prime focus, the co-operative marketing association is in the main concerned with the temporal flow of output and/or inputs, their pricing, and product characteristics. In a pure sense, the marketing co-operative conceptually is concerned with demand functions for the product output and for the factor inputs. Farm production functions and control of farm output do not fall under the joint decision-making area of the pure marketing co-operative; each farmer member makes his own decision as to his seasonal output volume and his method of production. It is in the sale of the product output or the purchase of the factor inputs that this type of co-operative marketing association is relevant for joint decision-making.

The restriction of joint decision-making to sale of output and purchase of inputs, however, need not prevail; other areas of decision-making may enter. In the latter cases, the enlarged dimensions of decision-making include aspects which so closely impinge on marketing that the co-operative finds it necessary or advisable to include them in its joint decision-making area. Yet, unilateral joint decision-making continues with only farmers participating in the function.

Another form of unilateral joint decision-making in agriculture is reflected in co-operative processing associations.[1] This is a type of co-operation, greatly expanded in recent years, in which individual farmers join together in the establishment of facilities for processing their crops. This is a combination of horizontal and vertical integration. Farmers integrate horizontally to establish a co-operative for the marketing of their farm product in conjunction with their own processing facility, which involves vertical integration. The decision-making area, in comparison with conventional co-operative marketing, is now more encompassing and includes more variables. But the decision-making remains unilateral since only farmers, or management on their behalf, make the decision and do so jointly through the farmer-elected board of directors.

[1] Sidney Hoos, 'Cooperative Canneries', *California Agriculture*, Vol. 14, No. 1, Jan. 1960, pp. 4 and 6.

The types of decisions involving joint decision-making under co-operative processing associations include among others the range of raw products to be received from farmer members, the quality of the raw product received, the production functions or processing methods and layout in the processing facility, the nature of the end product manufactured, the pricing of the end product and the imputation of returns earned by the processing facility to the farmer-owner members of the co-operative.

Generally, although there may be exceptions, the co-operative processing association does not limit the farm output to be received from members and processed by the facility. In other terms, the processing co-operative does not control the volume of output produced and delivered by farmer-owners of the facility. A major purpose of the co-operative facility is to stand ready to receive the raw product delivered by the co-operative members ; through their joint decision-making they establish the policy of the processing facility to be one of providing a home for the farm output of the members. Such unilateral joint decision-making by the farmers forces upon the processing facility management the responsibility of selling the end product manufactured from the farmer raw-product deliveries. The unilateral decision-making of the raw-product producers — the farmers who control the processing facility — may engender strained relations with its management.

A type of co-operative marketing, which is long established but in recent years has received increased attention, is referred to as co-operative bargaining.[1] It reflects horizontal integration and may be either of the unilateral or bilateral joint decision-making type.

A group of farmers — for example, canning fruit growers — join together horizontally in a marketing co-operative. Each farmer member turns over his processing fruit to the co-operative for its sale to canneries and other processors. The individual farmer member of the bargaining co-operative relies wholly on its board of directors (or a select committee) for the bargaining on price with the customers of the co-operative. The purpose of the bargaining co-operative is to participate in the determination of price and other terms of trade on behalf of the farmer co-operative members.

If the bargaining occurs under a term contract, the specifications of which (other than price) were jointly developed and accepted by

---

[1] Peter G. Helmberger and Sidney Hoos, *Cooperative Bargaining in Agriculture: Grower-Processor Markets for Fruits and Vegetables*, Berkeley, Division of Agricultural Sciences, University of California, 1965.

the bargaining co-operative and customer, bilateral joint decision-making occurs. The decision-making exists in an institutional and legal environment established bilaterally ; the seller and buyer participants play a role in the development of the contract and its consummation. But it may occur that the bargaining co-operative sells to buyers with which it has no term contract. In such cases, the joint decision-making is on the part of the co-operative's directors and is of the unilateral type.

Another form of market structure where joint decision-making plays an essential role is that which operates under the name of contract farming.[1] That form of business organization is widespread in industry but has made substantial inroads into agriculture only in recent years. The participants in contract farming usually include two parties : the farmer, as the producer and seller of the farm product, and his customer as the purchaser of the farm product. A special type of vertical integration is introduced through the term contract binding the farmer and the purchaser of the farm product.

The comprehensiveness of the contract binding the two parties is not unique but may vary in detail. In poultry-raising, an area where contract farming has made its greatest impact in United States agriculture, the terms of the contract may specify nearly all details of the farm operation including the supply of chicks and their quality specifications; the layout and operation of the hen-houses or the environmental conditions under which the chicks are raised; the source, volume and type of feed and its rate of consumption; the time-schedule of the marketing of the broilers to the purchaser; and the price terms for the farm inputs and output. Thus, contract specification of production functions, supply and demand schedules of the inputs and product, and the temporal dimensions of the farming operation are all involved. Some observers of contract farming have evaluated its state of development as one leading to the farmer's degeneration from an independent entrepreneur to a 'hired man' for the feed company or the broiler contractor. Yet, in terms of joint decision-making, the farmer operating under that type of contract farming has participated jointly with the broiler taker in the specification of the terms incorporated in the contract. The extent to which the farmer has a voice in establishing the specifications in the contract

[1] Ewell P. Roy, *Contract Farming, USA* ; Danville, Illinois : Interstate Printers & Publishers, 1963, p. 572 ; Michel Labonne, 'Prix et contrats de production en agriculture', *Économie rurale*, Vol. 61, July–Sept., 1964, pp. 81–92.

depends on his bargaining power. But in terms of a market structure model, contract farming reflects joint decision-making of the bilateral type.

Still another type of market organization with joint decision-making is defined as marketing agreements and orders.[1] In some countries, broadly similar institutions are referred to as marketing boards or schemes. In our classification they may involve unilateral, bilateral or trilateral joint decision-making.

In the United States, a marketing agreement is an agreement between an agency of the government and handlers of a farm product. The agreement is binding only on those handlers who voluntarily agree to abide by it ; other handlers are not subject to the marketing agreement and its terms. In contrast, a marketing order is binding on all handlers who market the product or all farmers who produce it. Once in effect, a marketing order and its provisions apply to those who voted against it as well as those who favoured it. Marketing orders may apply only to handlers, only to farmers, or to both jointly. Since marketing orders are more prevalent than marketing agreements and their participant coverage more complete, our comments are limited to such orders.

Marketing orders are established within enabling legislation and are operated under the authority of a government agency. But before a marketing order can be made effective, it must receive affirmative majority vote on the part of farmers if it is a producer order, an affirmative majority vote by handlers if it is a handler order, and an affirmative majority vote by both farmer producers and handlers if it is a joint producer-handler marketing order.

The marketing order includes one or more of the following provisions : control of total product volume to be marketed during the marketing season ; regulation of intertemporal distribution of the marketing volume during the season ; the financing and operation of advertising and market promotion ; the financing and sponsoring of research and investigation ; the establishment of quality, packaging and grade standards for the product marketed ; and the specification of unfair trade practices to be prohibited. Control of farm production is not provided for by marketing orders ; they regulate only the marketing of the product as specified in the order's provisions.

The joint decision-making in marketing orders occurs at several

[1] Sidney Hoos, 'The Contribution of Marketing Agreements and Orders to the Stability and Level of Farm Income', *Policy for Commercial Agriculture*, report of the Joint Economic Committee, 85th Cong., 1st Sess., Washington : Government Printing Office, 1957 ; pp. 317–327 and 799–825.

points in their promulgation and operation. First, farmers or handlers, or both in the case of a joint order, make a joint decision through the procedure of voting. They accept or reject a marketing order and the provisions to be included in it. Secondly, the government agency, under whose authority the order operates, participates in the decision as to the establishment or continuation of the marketing order. The government agency may veto or discontinue the order on the grounds that its operation is not consistent with the enabling legislation. Thirdly, the order is operated by farmer and/or handler participants who compose an advisory board or administrative committee for the government agency. The decision-making is joint in that the board or committee proposes to the agency, which approves or rejects ; neither party can unilaterally make effective one or more of the provisions. The marketing of the product is subject to the bilateral or trilateral joint decision-making.

Joint decision-making in present-day agriculture is not limited to forms of co-operative marketing or special situations as in contract farming and marketing orders. Joint decision-making occurs in a broader context also, in government production or acreage controls and in price support operations.

Wherever under general enabling legislation, the introduction and use of such controls and programmes are subject to majority affirmative voting of the farmers, they jointly participate with government in affirmation or rejection of the regulations.[1] In that sense, the joint decision-making is bilateral with both government and farmers participating in the process. In some countries the joint decision-making may be unilateral, determined solely by the government or by the farmers simply acting jointly.

Finally, reference may be made to international commodity agreements for agricultural products.[2] Here the joint participants are governments which are signatories to the commodity agreement. The joint decision-making applies to the source and destination of the product and its rate of flow from producing-selling to purchasing-receiving countries, and to the price and other terms of trade. In the formulation of the agreement, the selling and purchasing countries are arrayed in a form of oligopoly-oligopsony model of market

---

[1] Hathaway, *op. cit.* ; M. R. Benedict and O. C. Stine, *The Agricultural Commodity Programs: Two Decades of Experience*, New York : The Twentieth Century Fund, 1956, p. 510.
[2] Food and Agriculture Organization of the United Nations, *International Commodity Arrangements and Policies*, by Gerda Blau, Commodity Policy Studies, Special Studies Program No. 1, Rome, 1964.

structure. The terms of the agreement emerge from the negotiations between the seller and purchaser categories. But the management and operation of the agreement reflects joint decision-making of the bilateral type. Here, as in some other market structures, unilateral joint decision-making underlies the bilateral or trilateral form of joint decision-making.

IV. SOME IMPLICATIONS AND INFERENCES

The preceding brief and limited survey of market structures reflecting joint decision-making suggests some features common among them. Production, supply and demand functions — the economist's basic conceptual figments — are not disposed of ; they continue to play their traditional roles. But organizations and market structures of various types are introduced, with legal and institutional forms superimposed. The individual cell of decision-making gives up some of its autonomy and merges its decision-making with that of other decision-making cells. The individual farmer, for example, joins a co-operative or votes in favour of a marketing order because he believes, rightly or wrongly, that the result of joint decision-making emerging from such an institution is more favourable for him than if he were not to have the framework. Implied is a notion of increased returns from greater numbers of decision-makers jointly participating in the decision-making process. Yet, where is the logic or empirical evidence supporting such a notion? It seems we must seek elsewhere to account for the increasing amount of joint decision-making occurring in present-day agriculture.

It is in the search for improved returns that more farmers are favouring organizations and institutions in which joint decision-making plays a role. Yet, it is not because the decision-making is joint that the farmers join ; rather, it is in the belief that higher incomes will be the result.

It is not the jointness itself which is of essential relevance in joint decision-making in agricultural market-structure. Rather, it is the aggregation of economic resources which together with institutional forms lead to market structures susceptible to monopoly power and bargaining power for the farmers as a group. The individual decision-making farmer, as a price taker, strives to become a price-maker. Recognizing his inability to achieve that goal by himself, he joins with others of like mind. The decision-making of the resulting organization may or may not be the joint outcome of the individual farmer members, other than their decision to join such an organization. There can well be a large gap between the farmer members of a marketing co-operative and its management which makes decisions, often blurred between operational and policy. Such a situation is not unique to private corporations.[1] The concept of monopoly power as a vehicle for increased returns to farmers need not be examined here ; it is well established in the literature, with attention given to its analytical aspects and social implications.[2] The phrase 'monopoly power' is seldom if ever explicitly referred to in the rationalization of farm programmes and institutions with or without joint decision-making aspects. Yet, the concept and its implementation is embodied in many farm programmes and their complementary institutions.

In recent years, in the United States at least, relatively increased attention is being given to 'bargaining power' for farmers.[3] Such bargaining power is presumed to emerge from the unilateral joint decision-making of commodity groups of farmers. In the speeches, reports and documents from the highest echelons in agriculture to the grass-roots leaders, the attainment of increased bargaining power is stressed as necessary for augmented farm incomes. Yet the concept of bargaining power is not explained or examined. Its meaning is presumed to be self-evident, and its measurement is not articulated ; as used, it remains largely void of analytical content.

Without fully developing the argument here, it is proposed that a negotiator's bargaining power be viewed as his ability in a negotiation process to change a term of trade — let us say, price — in his favour. The ability reflects a composite of cardinally measurable and non-measurable influences such as economic resources at each negotiator's command, the will to bargain on the part of the participants, their relative degree of willingness to reach a bargain or forgo a deadlock, and the relative bargaining skills of the participants. The degree to which the price, or other terms of trade, changes in the bargainer's favour reflects his bargaining power. It is measured in relative rather than absolute units.

[1] R. A. Gordon, *Business Leadership in the Large Corporation*, Washington D.C. : Brookings Institution, 1945, p. 369.

[2] A. P. Lerner, 'The Concept of Monopoly and the Measurement of Monopoly Power', *Review of Economic Studies*, Vol. i, No. 3, June 1934, pp. 157–175 ; Joe S. Bain, 'The Profit Rate as a Measure of Monopoly Power', *Quarterly Journal of Economics*, Vol. 55, No. 2, Feb. 1941, pp. 271–293.

[3] Helmberger and Hoos, *op. cit.* ; George W. Ladd, *Agricultural Bargaining Power*, Ames, Iowa State University Press, 1964, p. 163 ; Center for Agricultural and Economic Adjustment, *Bargaining Power in Agriculture*, Report No. 9, Ames, Iowa State University Press, 1961, p. 107.

The measurement of bargaining power as such has not been attempted in the accessible literature, to my knowledge. An analytical formula has been proposed,[1] but after experimentation I view it as not applicable in empirical work. For that reason, I have developed for my own use, and offer here, another measurement of bargaining power that emanates from price theory and is empirically operational. Here it is indicated only.[2] With this formula, measurement of bargaining power in various situations in agriculture is being investigated.

Bargaining power is relevant for joint decision-making because even within the unilateral type it is a force in the decision-making process, as it is in bilateral and trilateral joint decision-making. Yet, the concept and measurement of bargaining power require further theoretical and empirical investigation. This also applies to the economic theory of joint decision-making. In the vanguard of development in decision theory, relatively little attention is given to joint decision-making. This is reflected in the empirical research on joint decision-making processes in present-day agriculture.

Where joint decision-making occurs in agriculture, *ad hoc* procedures largely prevail. The recently developed methods envisaged in decision theory have yet to be adopted to any significant extent by joint decision-makers.[3] Although such approaches as linear and curvilinear programming, game theoretic structures and simulation and heuristic programming have gone forward in research at the firm level, they remain largely outside the province of joint decision-making in agriculture. With their adaptation to handle jointness in decision-making, along with the progress in the economic theory of joint decision-making, participants as well as researchers may gain a

---

[1] John T. Dunlop and Benjamin Higgins, 'Bargaining Power and Market Structures', *Journal of Political Economy*, Vol. 50, No. 1, Feb., 1942, pp. 1–26.

[2] Let $p_s{}^*$ =seller's desired or target price — one which is either profit-maximizing, satisficing, or established by some other procedure as full cost mark-up ;

$p_b{}^*$ =buyer's desired or target-price — one which is either profit-maximizing, satisficing or established by some other procedure ;

p' =realized price at completion of bargaining ;

BP_s =seller's bargaining power ; and

BP_b =buyer's bargaining power.

$$BP_s = \frac{p' - p_b{}^*}{p_s{}^* - p_b{}^*} \; ; \; BP_b = \frac{p_s{}^* - p'}{p_s{}^* - p_b{}^*} \; ; \; BP_s + BP_b = 1.$$

[3] R. D. Luce and H. Raiffa, *Games and Decisions*, New York, John Wiley & Sons, Inc., 1957, p. 509; H. A. Simon, 'Theories of Decision-Making in Economic and Behavioral Science', *American Economic Review*, Vol. xlix, No. 3, June 1959, pp. 253–283; Robert L. Bishop, 'Game-Theoretic Analyses of Bargaining', *Quarterly Journal of Economics*, Vol. 77, No. 4, Nov., 1963, pp. 559–602.

more adequate understanding of the joint decision-making processes in agriculture.

————

DISCUSSION OF PROFESSOR HOOS'S PAPER

Professor Mossé said that Professor Hoos had put forward a formal analytical framework which took into account the number of groups that took part in a bargaining discussion, and the aim of the decisions to be taken. But, there is little to be found on the question of the content of the decisions and on the actual process of the making of joint decisions. Professor Mossé said that he would therefore like clarification on two groups of points.

First, what are the true objectives, motivations and behaviour of the people, both as individuals and in groups, who take part in the making of a decision ? For example, what are the respective positions which they attribute to price, to income, to quantity, etc. ? Are they rational and are they well informed ?

Secondly, what are the tensions which emerge either within a group or between groups ? How are such tensions resolved ? In particular who tends to dominate a co-operative group ? Are farmers likely to be deprived of their power of individual decision by salaried management, or are they likely to be deprived of it by certain kinds of leaders ? To what extent is discipline to be expected within the group ?

Professor Mossé said that Professor Hoos's discussion of the effects of the joint process seemed to him to be too short and somewhat misleading. The question of monopoly power was treated tangentially, almost as though he were trying to dodge it. It was subsumed under the all-embracing discussion of bargaining power.

Professor Mossé said that he would like to ask a number of questions on the probable effects of special agreements, such as cartels, in the taking of decisions. First, will such special understandings enable the income of farmers to be increased ? Second, will such agreements make possible the reduction of the margins taken by intermediaries or will it result in these becoming more rigid ? Might they not help to establish a vertical coalition between producers and intermediaries especially in bilateral types of decisions ? Thirdly, will these agreements make it possible to stop the destruction of produce as a way of affecting the market, or will they rather favour such practices with the common piling up of stocks being followed by a co-operative destruction of surpluses ? Fourthly, what are the results of these agreements on consumers ? Are consumers (or can they be) sufficiently protected by the intervention of government ?

Professor Mossé said that to translate the measure of bargaining power given by Professor Hoos into literary language, he would say that it took as its basis the spread between the initial aspirations of the participants and the final price arrived at. However, this spread is not given any absolute value but has a relative value with respect to the initial spread between the different aspirations of the parties which confront each other ; or, what comes to the same thing, bargaining power is represented by the rebate (or the premium) obtained, as a percentage of the initial spread of prices.

Professor Mossé said that he had two reservations about this formula. Is it really possible to take as a measurable reality the initial aspirations of participants ? To what extent do such desired prices really exist ? Secondly, is the loss (or gain) that is said to be obtained over and above the desired prices really based on a satisfactory notion ? If the seller starts with an initial high price and revises downward this high 'desired' price, the buyer is likely to appear to receive a high gain by Professor Hoos's measure but this did not prove that the *buyer* has great bargaining power or even has made a good deal. When, in Africa, a seller demanded 50,000 f. for a statuette of ebony but in fact sold it for 5,000 f. the buyer might none the less have been taken for a ride. It is true that in value relative to the initial spread, the gain of the buyer seems high, but — and this was the point he wished to emphasize — this is really a false appearance of advantage.

Professor Mossé said that to return finally to the question of bargaining power, it seemed to him to be discussed in too narrow a manner by Professor Hoos, and he hoped that the discussion would explore the matter more widely in attempting to provide answers to the questions he had posed.

Professor Mundlak noted that Professor Hoos had said that his paper went into an area where economics stopped. That made things difficult. He thought that Professor Hoos's measure of bargaining power should take account of the shapes of the curves involved. Professor Mundlak argued that the cost and demand curves delimit the area in which bargaining power might be exercised, and so, to some extent, limit the possible outcome that will be measured by Professor Hoos's formula.

In the California peach industry, with which Professor Hoos had had long experience, when bargaining takes place before the harvest there is very little freedom for the grower. Will this be reflected in the measure and how does such a situation compare with a situation where growers have alternative markets ?

Professor Le Bihan asked whether Professor Hoos or others are developing measures of bargaining power in the selling of animal products in the United States. Professor Hoos had developed the measure for fruit and vegetables and it would be interesting to know whether it had also been developed for animal products.

Professor Le Bihan asked whether there are available statistics to allow a comparison over ten or fifteen years between the growth of co-operative food industries on the one hand, and non-cooperative food industries on the other. As far as he knew, such data had never been published. The time has surely come to try to find out the truth about the dynamism of co-operative processing of agricultural products.

Mr. Allen said that he was not happy about the concept of bargaining power put forward by Professor Hoos, particularly as it was suggested that it should be generalized outside the area in which Professor Hoos had used it.

Mr. Allen said that he did not like the concept because it does not get at the roots of the problem. It does not ultimately tell us whether the group has, or has not, got strong power. Surely the proof of bargaining power lies in the ability to change the market structure within the framework of normal competition. This can only emerge over the long run. For example of such a real proof of bargaining power, we need to look at such successes as the reduction of imports obtained by bargaining, or the inception of special legislation on behalf of the bargainer. Such actions will raise the overall price level over a long period, but will not show in the formula. Mr. Allen said that it also concerned him, that if both sides come forward with the same desired price, the measure would indicate that neither side has any bargaining power, whereas it ought only to state that there has been no bargaining. The context within which the agreement takes place might reduce the possibility of bargaining to a minimum. This is particularly the case where the seller has been protected.

Professor Robinson said that he supported Professor Allen. Professor Robinson asked whether Professor Hoos thought that his paper was talking about the important problem. The big issue, in the United Kingdom context at least, is at what level support prices are to be fixed. It is on this issue that the strength or absence of strength of farmers is shown. In the negotiations on this question three parties are involved : the National Farmers' Union ; the government, who are slightly inclined to favour the monopoly ; and the consumer, who is not directly represented and whose only power lies in refusing to consume. Professor Robinson asked whether Professor Hoos would regard these negotiations as matters of bargaining in his sense. If so, are the interests of consumers to be regarded as having no relevance in what happens between the negotiating parties ?

Mr. Allen said that he was concerned by the abuses that arose when someone invented a concept and others applied it uncritically. He said that the Lerner article (R.E.S. 1934) on the measurement of monopoly power had produced a rash of studies using historical data, attempting to measure the 'degree of monopoly'. These studies usually give no consideration to the complex of factors involved. He feared that with what Professor Hoos had put on paper, people will think only in terms of what Professor Hoos is talking about, and forget the other things that Professor Robinson is talking about.

The Problems of Vertical Integration and Innovation

Professor Ruttan asked if it would help if Professor Hoos stated how he had observed the variables in his function.

Professor Hoos said that Professor Mossé asked for additional information and ideas. He was certainly sympathetic with this request but said that when he had been invited to prepare the paper and was given a limitation of space he had been sufficiently naïve to take that limit seriously.

Professor Hoos said that he had given much thought to the problem of decision-making with the farm groups that he worked with. He had come to the conclusion that to say much on certain aspects of the process we need to call in the help of the psychiatrist. In economic theory we can think about and discuss the basis and the results of decision-making, but some dimensions of the process itself seemed to him to lie outside economics. As far as the true motives of joint decision-makers are concerned, he had assumed that the motive is increased income, though not necessarily maximum income. Professor Hoos said that his views were based on the actual decisions of various farm groups. For twenty years he had been working with joint decision-making processes by agricultural and processor organizations, and his ideas had come from direct observations in this work.

Professor Hoos said that the question about who will dominate a group he took to be an organizational problem. Clearly, in a joint decision-making process it is not usually true that the effect is such that one man has one independent vote. Leadership may take one of many forms and is modified by coalition or the creation of cells within the whole group, each of which are dominated by one or more persons.

Professor Mossé had felt that he had slighted monopoly problems. Professor Hoos agreed that monopoly power plays an important role, but he had turned attention to the problem of bargaining power because it has unique aspects and it is, he thought, a concept which needs greater clarification. In the United States, bargaining power is a popular and fashionable phrase, but very few have asked what we mean by it. He had argued that bargaining power consisted in the ability to change the terms of trade, and he had tried to propose a specific measure. It certainly can be used in practical terms with farm groups since he himself had so used it for several years. It can be used within a model of profit maximization or one dependent on other target prices as satisfying prices. Professor Mossé had given a somewhat esoteric example where a falsely high price had been quoted, and a high measure of bargaining power resulted. Professor Hoos said that he had assumed that fictitious prices are not employed; he thought that such stratagems are easily detected, and such prices are not the same as the target prices he was considering.

Bargaining power in United States agriculture has important consequences for agricultural policy. People believe that one of the problems of the depressed situation of certain crops is that farmers are in an unfavourable bargaining position *vis-à-vis* large purchasing groups. Thus,

in fruits and vegetables for canning relatively few firms buy a large proportion of output in a single year. This raises the important question of
whether depressed farm prices are, in fact, the result of unequal bargaining
power between producers and those to whom they sell. That is one of the
major problems to which the recently appointed National Commission on
Food Marketing is addressing itself.

Answering Professor Mundlak, Professor Hoos said that he had been
working on how he could take the shapes of the relevant economic curves
into account. He believed that the formula could be refined and functions
used rather than particular points on them as he was doing at present.

Answering Professor Le Bihan, Professor Hoos said that co-operative
bargaining organizations are being developed for various products. The
National Farmers Organization in the United States is organizing farmer
co-operative bargaining associations in animal products and grains.

On the question of the comparative growth and dynamism of co-
operatives and non-cooperative industries, there are at present, as Professor
Le Bihan had said, no official comprehensive comparative statistics. Professor Hoos said that he had hoped that the National Commission on Food
Marketing would yield something. His impression, in the absence of
adequate statistics, was that the co-operative movement is not growing in
terms of the share of the market to the same extent as private business.
The co-operative groups obviously suffer from organizational problems.
They are plagued by problems of internal politics and often dominated by
particular groups or individuals. The co-operative movement is not in
fact stagnant, but its growth is not a relatively strong one.

Professor Hoos thought that Mr. Allen was falling into the confusion of
monopoly power with bargaining power ; they are different concepts even
though they share in common some elements. Monopoly power involves
a situation on only one side of the market ; whereas, bargaining power
involves a relationship between the two sides. In particular negotiations,
monopoly and bargaining power may both be involved. It is for many
purposes important to distinguish between the two. In further reply to
Mr. Allen, Professor Hoos remarked that he did not fear that a rash of
measurement studies on bargaining power might emerge. In fact, there
is need for much additional empirical work on the existence and use of
bargaining power in agriculture. For such work to be sound, there is
required a conceptual framework to support the empirical work, and it
was to that end that he proposed his measure of bargaining power which is
based on price theory.

In response to Professor Robinson who asked what the issues involved
are, Professor Hoos commented on the rapid growth in the co-operative
bargaining movement in the United States during the past decade, the
reasons for the establishment of farmer co-operative bargaining groups,
and the problems they face in negotiating with processor-buyers.[1] In-

[1] Peter G. Helmberger and Sidney Hoos, *op. cit.*

The Problems of Vertical Integration and Innovation

volved are the establishment of terms of trade through negotiations and the resulting price. The horizontal and vertical integration which occurred in food processing industries resulted in fewer buyers from farmers, and the fewer buyers had larger resources than formerly. One issue is whether such a change in the structure of the processing industry and the changing grower-processor markets results in terms of trade less favourable to the farmers because the processors have acquired augmented bargaining power. In the belief that such is the situation, more and more farmers in the United States are turning to co-operative bargaining associations as a countervailing device to balance the bargaining power of processors. With the issue being the existence and use of bargaining power in the negotiation of the terms of trade, economists are faced with clarification of the bargaining power concept and its measurement in order to analyse the process and outcome of joint decision-making in specialized farmer co-operatives established for bargaining with processor-buyers. The effectiveness of farmer co-operative bargaining associations has implications for national farm price policy as evidenced by the investigation of the National Commission on Food Marketing in the United States.

PART V

CONTEMPORARY PROBLEMS IN DEVELOPED AGRICULTURE: CASE STUDIES IN AGRICULTURAL POLICY

Chapter 19

THE INCOME OBJECTIVE IN
AGRICULTURAL POLICIES

BY

H. ASTRAND

Swedish Agricultural Credit Association, Stockholm, Sweden

All animals are equal, but some animals are more equal than others.

George Orwell, *Animal Farm*

I. INTRODUCTION

OUR starting-point for this study is a country whose national policy aims at the well-being of all its citizens, and simultaneously at raising their standard of living as rapidly as possible. We shall also assume that the more important decisions made in this country are arrived at by majority vote on democratic lines, but that at the same time great attention is paid to the opinions of minorities : to this end, there is readiness to make compromises. We assume moreover that a policy of great restraint is observed in the exercise of economic pressure, not to mention compulsion, on the common citizens when their interests as individuals conflict with the interests of the general public. This ideal democratic state must have an agricultural policy. Just what its agricultural policy shall be, is a question that is perhaps not easy to decide.

What is the actual situation at the point where our discussion begins? We might say that it is approximately as follows : Agriculture is no longer the main industry of the country, for it now employs less than a fifth of the population, maybe no more than a tenth. Most of the enterprises engaged in it are family ones, with a manpower contribution of 3,000 to 7,000 hours, but there is also a growing number of commercial agricultural companies and a considerable number of undertakings in which the farmer works only part-time. Most agricultural properties are owned by the farmers themselves. Farms are often inherited. Production, at least for some commodities, exceeds the capacity of the domestic market, so that normally a certain amount of goods are exported. Application of modern techniques is bringing about a rapid increase in production

397

per hour of work ; this increase is in fact faster than the depletion of labour, so that the total value of production shows a clear tendency to increase. The share of agriculture in the national product is considerably less than the proportion of the population engaged in it, partly because age-distribution is strongly affected by the falling away of manpower in the younger age-groups, but primarily because the yield per working hour is appreciably lower in agriculture than in the rest of trade and industry. Farmers have relatively strong professional organizations, and their political influence also, as a result of time-lag in parliamentary representation, is greater than their proportion of the population entitles them to.

Our political system aims at the welfare of *all* citizens. Expressed in these terms, this means that there is heavy pressure in favour of immediate action. The worst paid, the handicapped, those who are queueing for housing accommodation or medical treatment—all these demand that something must be done quickly. They have no desire to bide their time indefinitely or to submit to a process of gradual adaptation, to wait for an allocation of the production factors that will make for a more effectively functioning social order. Their demands are mainly concerned with incomes. Higher incomes mean more freedom of action for the individual, better social standing, greater security. And as far as work is concerned, is not the labourer worthy of his hire ? Is not the production of food at least as important as the manufacture of motor-cars or television sets ? Trade organizations and political parties constantly seek solidarity with *all* their members, not merely with some of them. If they pursued the latter policy, they would risk defections. Public consciousness of belonging to an important social group — to an indispensable group, in fact — is therefore kept very much alive and maintained in its full vigour by the work of these organizations and parties. What organization conscious of the urge to self-preservation would say to any substantial group of its members : 'You have nothing to do with us' ? None of course would do so, except nation-wide organizations which have become so large and have their managing staff of permanently employed officials so firmly rooted that they can bend before a local storm without being shaken to their foundations.

II. VARIOUS KINDS OF INCOME OBJECTIVES

So we ought not to be astonished to find that income objectives play a central role in agricultural policy, and they will certainly always do so. If we try to formulate such objectives, however, we find a copious collection of samples to choose from in various countries. Without making any attempt to compile a complete list, we mention some of the different kinds below.

The simplest type does not primarily involve a comparison with groups outside agriculture, but with agriculture itself at an earlier period. The United States 'parity prices' thus aim at giving farmers the same purchasing power as they had obtained at an earlier base-period. The system works, however, only for the prices of particular commodities and necessities, and has quite varying effects on farm incomes, depending on the way their production is concentrated and the capital behind them. The Agricultural Adjustment Act of 1938 admittedly lays down a principle of equality of status between farmers' incomes and incomes in non-agricultural industries as the target of agricultural policy. However, it has not been possible to realize this ideal. In Sweden, the aim throughout most of the 1930's was to restore prices to the same level as before the agricultural crisis, envisaging that this would entail a general restoration of the income level. The Swedish system of reviewing farm prices, developed in detail later (1941 to 1956), which applies to a great many items of income and expenditure including compensation for the work of the farmer's family, is also based on a fixed period used as a reference point. The relative incomes of the farmers were to be maintained by means of an equally large percentile increase in incomes and expenditure. Except for a few years, the system did not give satisfactory results.

The agricultural policy programmes of most countries nowadays contain income objectives framed in general terms. The British Agriculture Act of 1947 mentions 'prices which give a reasonable return for work, a fair standard of living for farmers and a satisfactory rate of interest on invested capital'. Article 39 (para. 1) of the Rome Treaty speaks of 'securing a reasonable standard of living for the agricultural population'. The members of EEC, however, have in most cases made rather more far-reaching statements in their national agricultural programmes. The French law of 1960 proclaims equality — in the economic and social sense — between agriculture and other sectors of industry. The Italian Five Year Plan promises to 'work for improved living conditions and higher incomes for the rural population'. The aim of Netherlands agricultural policy as it was formulated in 1962 was 'to bring about an acceptable standard of living for the agricultural population'. The West German 'Green

Plan' spoke in 1963 of 'allowing the farming population to share in the general rise in the standard of living'. The German agricultural programme also mentions comparisons of income with the earnings of certain groups of industrial workers. This is also true of the Swiss programme.

These pronouncements on income objectives do not provide many criteria by which the level of income aspired at might be defined and measured. But several political programmes for agriculture do contain demands for rationalization as a condition upon which better incomes for farmers must depend. It may perhaps be of interest to see what a very ambitious income objective would lead to, and what problems the pursuit of such an objective would bring with it. The following section is therefore devoted to Swedish income policy, which has been carried very far in this respect — preposterously far, many people will think.

III. SWEDISH-TYPE INCOME OBJECTIVES

The agricultural policy programme of 1947, which is still in force, says among other things that the aim of the governmental steps taken shall be to ensure that the population working in agriculture enjoys the same possibilities as those engaged in other industries to earn a reasonable income and to share in the general increase in prosperity which may occur in the future. An agricultural committee had investigated the difference in nominal wages between farm workers on the one hand and certain groups of workers in rural areas on the other, such as forestry workers, road-building labourers and general workers in industry. It was found in 1946 that the difference was between 15 per cent and 20 per cent of the farm worker's wage. It was considered that owners and operators of small family farms had about the same level of earnings as that of the farm workers. The 1952 enquiry into agricultural prices resulted in a far more precise definition of comparisons between incomes. As a group for purposes of comparison on the agricultural side, those farms of 10 to 20 hectares in the lowlands were chosen which are included in the enquiries into the agricultural economy carried out every year with government assistance. These investigations include highly detailed balance sheets. The farms were admittedly not representative of all farms of the same size in Sweden, for they were better managed and more intensively conducted, but this fact was regarded almost as an advantage, as it was desired to make the groups used for comparison submit to more rigorous standards of efficiency than those applicable to ordinary farms. Both economic accounting and labour book-keeping figures were available for the farms selected. Interest on the farms' own capital was deducted from the incomes, but the capital which formed the basis for calculating interest charges was first reduced by 'inflationary profit'. This profit was defined as the difference between increases which had occurred in real estate values and investments made. It was intensively discussed whether comparisons should be on the basis of hourly wages or wages per year. The working hours of farmers were then about 10 per cent longer per year than those of industrial workers with their eight-hour day and three weeks' holiday. Parliament resolved, however, that the comparison should be made on an annual-income basis, the farmer's personal income per year being worked out after employed labour and work done by the other members of the family had been paid for in accordance with the collective agreement in force for farm workers. As a non-agricultural comparison-group, industrial workers over 18 years of age in the two lowest of five then existing cost-of-living index localities were chosen. These two groups mainly comprised industrial undertakings in rural areas and in less densely populated districts in southern and central Sweden. The result of the comparison was that for the six years 1946/47 to 1952/53 the farmer's average income amounted to 4,850 kr. if the capital expenditure was estimated from the market value of the property, and 5,300 kr. if it was calculated on the considerably lower taxation-value. The pay of the industrial worker was 5,850 kr. A correction for the price of agricultural payments in kind reduced the difference by about 300 kr. and the correction for inflationary profit reduced it by a further 300 to 400 kr.

However, the enquiry was not restricted to these comparisons : two other methods were also tried. One was to compare the average taxed income in *all* farms of 10 to 20 hectares with the average wage for all industrial workers in rural areas. For both parties, these incomes include a certain amount of income from capital, and the declared earned income of the wife and of children living at home. In view of other circumstances also, for example the valuation of agricultural payments in kind, a comparison made by this means may be misleading. The other method tried was to make direct comparison between the standards of consumption in the two groups. Expenditure per consumption-unit turned out to be quite similar for farmers and non-farmers in rural areas. An enquiry of this kind, however, makes a supplementary investigation into savings essential. During certain periods in the lives of the farmer's family, especially at the start, there is probably very often a certain amount of consumption of capital, while on the other hand saving towards the close of the active period seems to be at a very high level.

On the basis of all these investigations, the enquiry arrived at the conclusion that in and around the year 1952 there was not very much difference in incomes. However, it was evident that the incomes of farmers in some years were higher than usual because of good extra earnings from forestry work and felling. It soon became obvious that the income gap was quickly widening, and by the end of the 1950's and the first part of the 1960's it was certainly of considerable size.

New calculations made in the great agricultural enquiry which has been going on since 1960 have been mainly carried out with the aid of 'standard farm price calculations', as they are called, for farms of 15 and 25 hectares respectively. They are based on standardized accounting results. Information from tax-declarations has also been used. In Sweden, declarations to the income tax authorities are obligatory and extremely detailed. Great attention is devoted to the family income, in other words to the aggregate income of the farmer, his wife and their children under 16 years of age. Information on such family incomes has also been worked out for industrial workers. Inflationary profit in agriculture has been calculated afresh, and the interest claim of capital has been corrected for that purpose. The results for 1960 are stated in Table I. It should be observed :

(a) that the 'standard calculation' material, unlike the information from the tax declarations, is not representative ;

(b) that the basis for the standard calculations has been simplified, so that it only includes farmers fully capable of work and does not comprise farms of an extremely specialized character ;

(c) that housing expenditure and housing capital are not included in the standard calculations ;

(d) that the standard calculations are based on normalized quantities of capital goods and harvests, while the declaration material refers to actual quantities for the year 1960 ;

(e) that the wages of industrial workers have been supplemented by the expenses incurred by employers for certain types of social insurance which must be paid for the employees by the farmers themselves.

Table I

INCOME COMPARISONS IN SWEDEN FOR THE YEAR 1960

(in Swedish kronor)

	15 hectares arable land	25 hectares arable land
A. *Farmers*		
1. *According to standard calculations*		
Total income of family including adults, and work done by members of the family on the farm	12,907	15,663
Farmer, wife and children under 16	11,201	13,125
Wife's earned income	2,086	1,974
Earned income of children under 16	299	474
Family's capital income	2,139	2,682
Farmer's earned income (balance)	6,677	7,995
2. *According to tax-declaration enquiry*		
Farmer, wife and children under 16	10,742	13,599
Family's capital income	1,839	2,392
Farmer's earned income	6,637	8,882
B. *Industrial workers*		
Total income of husband and wife according to tax declaration	15,060	
Industrial worker's wages according to official statistics	11,767	
Wife's earned income and family's capital income (balance)	3,293	

It is evident from the table that the difference between the comparable figures for farmers and industrial workers is very large. In spite of supplementary increases in prices it does not seem to have diminished much during recent years. The most important reasons why it has not done so are the following : Industrial prosperity has made it possible for workers' wages to be very much increased. The wages have reacted on the wages of farm workers and on agricultural expenditure. The rate of rationalization in farming has not been as high as it has been in industry. Exports of food surpluses at low prices have helped to keep down commodity-prices.

IV. PROBLEMS OF MEASUREMENT

Let us revert to the question of income objectives. If agriculture lags far behind, we can of course content ourselves with a formula which accepts that the incomes of farmers ought to increase more rapidly than those of other groups. But we cannot avoid measurement problems by this method either. What principles should be used as a basis for the selection of comparison groups? It can hardly be reasonable, for instance, to treat all those working in the farming industry as a single collective entity and to calculate an average income or mean income for them. Among them there are employers and employees, whole-time and part-time workers. Moreover, it is very difficult to make a sharp distinction between agricultural work and forestry work, etc. We must have a sharply defined group of employers or employees for whom reliable statistical data are available. This requirement alone means that the possibility of making comparisons is very restricted.

What selection should be made for the non-agricultural sector? The same objections can be made against taking them as a whole, and using the national accounting figures, for example, to estimate income conditions. The distribution of age-groups alone, in and outside agriculture, may strongly influence such a comparison. It would seem advisable to take for comparison with the farmers some group of employers in another sector of trade and industry who are subject to similar conditions as regards competence and capital needs. Despite a keen search, however, it has not been possible — in Sweden at any rate — to discover such a group. When it was decided to make a comparison with industrial workers, the choice was made, among other reasons, because they constitute a very large group with great political influence, and one which has demonstrated that it can obtain its share of the results of production. The requirements for competence demanded from a small farmer have not been thought to be such as to render a comparison of this kind invalid. Moreover, in all probability the income statistics for industrial workers are comparatively reliable.

V. LEVELLING OF INCOMES AND THE RISE IN THE STANDARD OF LIVING

When choosing groups to be compared, we soon find ourselves confronted with the question of what standards of efficiency the selected undertaking must conform to. Consider what was said in our introduction : one of the objectives aimed at in our ideal democracy is the most rapid increase possible in the standard of living. The economists have told us that we will get it in the long run by allocating resources in such a way that their marginal productivity is everywhere the same. Can we bring about such a proper distribution of resources if there is a rigorous levelling of incomes? Is there not a considerable risk that far-reaching levelling will have a preservative effect on the structure of the undertakings and reduce the mobility of labour? In all probability the answer largely depends on the way in which the additional income is provided. It is well known that higher incomes in democratic states are usually lopped by progressive income tax. We shall not at this point go into the much-discussed question of how this affects the rate of development in the national economy; we merely draw attention to it.

VI. VARIOUS FORMS OF ADDITIONAL INCOME

Additional income in agriculture can be acquired in principle either by (a) increasing the prices of agricultural products, (b) lowering the cost of agricultural capital goods or means of production and (c) adding to income without regard to production. The first of these methods has been applied for a long time and to an excessive degree, and it must now presumably be evident to most people how dangerous it can be. Supported prices mean over-payment of the means of production in agriculture and lead to the retention of resources in the industry and thus to over-production. The cost of marketing the surplus must to a considerable extent be paid for by the domestic market's customers, as the producers would otherwise find their incomes reduced and would have to be compensated by new increases in prices. There are, however, many motives behind the demands for higher incomes by means of higher prices for commodities: Production costs are so high — can't we even cover *them*? Vague notions of a 'fair price' play a great part in the debate. Let the consumer pay for what he consumes — that is much simpler and less controversial than giving subsidies via the national budget. In wealthy countries the proportion of the cost of food in the total household expenditure is falling, and the farmer's share of the price the consumer pays is probably often less than 50 per cent. In such countries, however, it is possible to carry budgetary price subsidies quite a long way without arousing any very strong reactions from the consumers. Differentiated prices for the benefit of farmers in areas with inferior natural conditions, for example mountainous regions and those situated far north, are appreciated and accepted there, relatively speaking without difficulty, as socially justifiable measures. In poorer countries, which usually have a proportionally large agricultural population, it is too expensive to give any major part of the support via budget. A policy of high import duties is generally the adopted instead.

Efforts have been made to reduce the disadvantages of increasing income through raising the prices of commodities, by fixing the price level so that it gives a reasonable profit only to very efficiently run farms. If the demand for efficiency is set so high that the undertakings in question must satisfy the demands stated above for marginal productivity there will be very few farms, however, which will be profitable to run, and total production will fall steeply. The majority of farmers will be exposed to heavy economic pressure, and that is something which we in our ideal democracy do not find socially acceptable — so we raise prices.

The same objections on grounds of principle could probably be raised against a cheapening of agricultural capital goods, for example by means of subsidized commercial fertilizers or interest subsidies, as the objections which apply to an income support given by means of increased price for products. There remain additions to income which are neutral as far as production is concerned. We have examples of these in Sweden in the 'acreage allowance', which is payable in varying amounts to a large number of farms of 2 to 10 hectares, and which have to pass a means test taking account of both income and property. They are not paid if the taxed income exceeds a certain minimum sum. As long as such allowances are tied to the ownership of agricultural property, they must in principle have the same drawback as the methods previously mentioned, namely that they have a restrictive effect on the mobility of the means of production. But they have the advantage that they do not offer a direct stimulus to increased production. Income support which even helps to reduce production is conceivable. An example is the United States agricultural programme. There is no guarantee, however, that the income allowances will not be used for new investments in the enterprises.

VII. LABOUR MUST BE MORE MOBILE

Our discussion seems to lead to the conclusion that income objectives in agricultural policy are absolutely inevitable but very difficult to define more precisely. Realizing them by means of income assistance in various forms leads to conflicts with the demand for optimum allocation of the resources in the community.

Is there, then, no way out of this dilemma, which as a matter of fact is not confined to agriculture but applies to any weak or slow-paced industry? To find a solution, we must return to the demand that everybody in a democratic society must be able to earn a reasonable standard of living. This is not to say that he must be able to find it in the particular place where he happens to have been born or where he lives. Nor does it mean that his income must be derived from just the trade or profession that he may have chosen for various reasons. The community fulfils its obligations towards him if it can offer him an alternative occupation on reasonable terms. In the country that was quite briefly described in the introduction, there is room for free enterprise, but there is not unlimited room. If we believe that free enterprise leads to an optimum allocation of resources, the task of the community is to take away as far as possible the obstacles to the mobility of productive resources — above all of the most important of all the productive resources the country has, its people. The income objective in our agricultural policy is merely part of a general income objective for all sectors of trade and industry. The programmes of agricultural policy should include a statement to this effect: 'Our aim is that the population now working in agriculture shall achieve a reasonable standard of living — either in or *outside* agriculture'. The programmes must, in addition to proposals for making the remaining agricultural enterprises more efficient, also include proposals for

(*a*) a very comprehensive general information service, specially concentrated on explaining what alternative possibilities of employment exist for the agricultural population;

(*b*) vocational advice at an early stage for all young people;

(*c*) an economic advisory service which makes severely realistic calculations for investments in agriculture;

(*d*) re-training of farmers for other trades and professions;

(*e*) release of farmers' tied capital assets. In an inflationary economy — and what economy is not inflationary nowadays? — it is reasonable to suppose that this could be done by offering some kind of facilities for guaranteed-value investment, for example in guaranteed-value bonds;

(*f*) early-retirement on pension for farmers who cannot be retrained.

Only by means of a long-term agricultural policy of this kind, wholly integrated in a general economic policy, will it conceivably be possible, sometime in the future, for the population now working in agriculture to attain satisfactory income-levels.

Chapter 20

SWEDISH EXPERIENCE IN AGRICULTURAL POLICY[1]

BY

ODD GULBRANDSEN
Agricultural College, Uppsala, Sweden

I. INTRODUCTION

WHEN I was invited to write this paper I was told that the case of
Swedish agricultural policy seemed to be one of the few interesting
ones. By the expression 'interesting' may, as I see it, be understood
a radical and/or successful policy, which can serve as a model for
other countries. If we look merely on the strictly agricultural part
of Swedish economic and social policy, I am afraid I have to be
disappointing.

The development of Swedish agriculture with regard to production
volume, efficiency and income has been highly influenced by general
economic development and policy. The conditions for realizing
certain goals for agriculture are chiefly to be found outside agriculture. For example, the outflow of labour from agriculture, which
plays a strategic role in the realization of agricultural goals, has been
determined mainly by outside forces and policy. It is probable that
the most interesting part of policy for agriculture lies in the radicalism
of Swedish labour market policy, education policy and pension policy.

Consequently, I start my review with an analysis of the general
forces and policies and their effects on agriculture. After this follows
a short presentation of the special agricultural policies and their
recent development. The paper ends with some aspects on the future
policy problems.

---

[1] The judgements made in the paper are entirely my own and are not necessarily
officially agreed with or agreed with by other Swedish experts.

Case Studies in Agricultural Policy

II. THE ROLE OF GENERAL FORCES AND POLICIES

(1) The Labour Force in Agriculture

The Swedish economy has been characterized during the last two decades by a not very fast, but continuous growth. Development has been concentrated on the engineering and service industries, which have a relatively high labour consumption. The annual number of working hours has decreased markedly due to shorter daily working time, longer vacations and more free Saturdays. These developments have created a large demand for labour. The sources of labour have primarily been agriculture, a number of young people entering active age, the reserve of housewives and some immigration. The high demand has been promoted by a full-employment policy and the movement of labour stimulated by an active labour market policy. This policy has been intensified in many respects in recent years and adult education has been introduced. The possibilities for the young to choose the best job have been broadened by a prolonged general education which includes more vocational training.

The high demand for labour and the freedom of choice have attracted labour out of agriculture and reduced recruitment, and account for the fact that about 10 per cent of the workers left agriculture annually during the fifties. During recent years also young farmers have moved out at an increasing rate. To a very large extent the shift in job has been carried out progressively with part-time farming as an intermediate stage.

Increases in old age benefits and the introduction of a general pension scheme have increased the rate of retirement for the older farmers. Reduced recruitment, the growing outflow of young farmers and earlier retirement have accelerated the reduction of the number of farms from 1 per cent per year in the beginning of the fifties to about 5 per cent per year now. Over all, the volume of the labour force has decreased by 3–5 per cent per year in the last decade. On the push side may be noted low wages, seasonal unemployment and profitable substitution for labour by other inputs, such as machines and fertilizers. As the difference in income is great among farmers, there have been large groups pressed by low income, despite the fact that as a mean the income level and development have not been bad.

(2) *Agricultural Policies and Their Effects*

The fast outflow of labour could have permitted rather radical changes in production volume, agricultural structure and efficiency if agricultural policy had made a national use of the factors involved. It is true that the volume of production has stagnated and even decreased somewhat during recent years, but this is no proof that agricultural policy has been radical and successful.[1] On the contrary, with slower increase in the use of industrial production means, *e.g.* less over-mechanization, a marked decrease in production could have appeared.

If the constant production is divided by the declining labour input the mathematical result is a rapid increase in labour productivity, but this need not mean an optimal path in the rationalization process. For example, the structural changes have been slow: the average size of a farm increased from only 12 to 14 hectares between 1944 and 1961. It is true that the amalgamation and withdrawal of the smallest farms have accelerated, but as the number of the largest farms has also decreased and the total arable area has contracted, the net effect turns out to be limited. As Sweden has a certain reputation for its efforts in administrative rationalization work, it may look surprising that the effect of it has been so limited. To understand the reasons for this we have first to take a short look at the general features of agricultural policy in Sweden.

III. *MAJOR FEATURES OF AGRICULTURAL POLICY*

(1) *Restricted Buying of Land*

The three most important fields of policy in Swedish agriculture are rules for establishment, for price regulations and for rationalization activities. The rules for establishment are the oldest and date

---

[1] The changes in production and self-sufficiency in Sweden are illustrated by the following figures for 1954 and 1962, years which are comparable with respect to weather conditions.

	1954	1962
Total crop production, billion crop units	9·71	9·56
Imported feeding stuffs, billion crop units	0·17	0·37
Total crop supply (incl. fodder)	9·88	9·83
Agricultural production, 1,000 billion calories	7·89	7·05
Consumption of agricultural products, 1,000 billion calories	7·62	7·65

[2] The means used by the agricultural boards in the work of rationalization are : expropriation ; right to buy before anyone else (excluding near relatives) at the seller's offer on the market ; active buying ; intermediation ; loans and subventions ; technical service ; assistance by economic advice. The farmers' own offers and demands at the boards for additional land have played the most important role in the work of amalgamation.

from the time of the settlement of wood industries and speculation in land during the wars. The rules impede companies from buying land. Individuals also may not be allowed to buy if speculation can be assumed or if the purchaser is judged to have enough land and if the offered land can be used to complete another smaller farm. These restrictions on the rights of buying land, somewhat relaxed for the first time in 1965, have had a negative influence on the rate of the amalgamation of land, and are the reason why the efforts of administrative rationalization may have been counteracted. As it is the same administrative boards which have handled both the control of buying land and the rationalization work, this counteraction may seem curious. The explanation is that the directions for the rationalization work were given at a time, 1947, when it was not obvious that structural transformation induced by general economic forces could take place. In reality a transformation, hidden in the ordinary interpretation of statistics, was already under way. When the transformation accelerated in rate during the fifties, the directions for rationalization became obsolete and served to obstruct instead of promote progress. Gradually critics of these effects of policy have forced more liberal directions to be taken. For example, it is now permissible to make up farms to a two-family farm size.

(2) Limits on Administrative Rationalization

Instead of leading the intended drive towards structural rationalization the efforts of the state have the character of an activity adjusting itself to development.[2] The compulsory powers, with which the boards were provided at the beginning, have been used to a very limited extent. It is estimated that only one-sixth of all amalgamations are handled by the boards. The net effect, measured by the number of farms amalgamated with the boards in existence is markedly less and, as said before, probably negative as compared with the number there would be without the existence of the boards.

As a consequence of the lagging structural rationalization the other parts of the rationalization work have also somewhat lost their intended effects. This refers especially to state support of long-term investments such as buildings and land improvements. A rather large percentage of the support to building may be regarded as obsolete in the light of continued structural transformation, as the support was exclusively given to farms with less than 20 hectares. To avoid such errors it has been insisted since 1959 that investments shall prove to be profitable to get state support.

(3) *Price Policy : Income and Production Goals in Conflict*

The primary goal for price policy has been income parity between certain groups of farmers (so-called basic farmers with 10–20 hectares, and from 1965 so-called standard farmers with 20–30 hectares), and rural industrial workers. In the negotiations this goal has been central both for the determination of the price level and for the relations between prices. The attainment of the main aims has, however, been limited by the rule that export surpluses in principle have to be paid for by the farmers themselves. Although this rule has not been effective for small surpluses (due to consumer-paid regulation funds), it has probably functioned as a restriction against large surpluses. Since home market demand has stagnated, this restriction may be at least one important reason for the stagnation of production volume. The restriction has, however, also caused an autarky for every product. Since the most profitable branches naturally first attain a surplus situation, they have to pay the penalties of being in surplus. On the other hand a deficit in a product leaves room for increases in prices to achieve the income goal. This means that support is given to branches with low comparative advantage. Consequently, the support stimulates a misallocation of resources within agriculture and gives a lower income effect than would more equally distributed support.

From these viewpoints the conclusion can be drawn that the exclusive concentration on income goal has distracted attention from price as the only effective means (besides production quotas) of regulating the volume of production. The result has been an over-allocation of industrial resources to agriculture and a misallocation of resources within agriculture. The misallocation within agriculture is partly dependent on the over-allocation, since at a lower total volume the risk of surpluses of individual products diminishes ; and both effects together must have restrained the possible income development. As the price guarantees have been extensive — such as guarantees against world price falls, cost increases and compensation for income increases in industry — the income development ought to be very favourable. In fact, the income gap for manual work has remained and the farmers' income disposable for living expenses has stagnated, since a large part of the income increase is ploughed back in property values realizable only at the disposal of the farm.

The technical price system in Sweden is said to have served as a model for the regulation of the common market of the EEC. I am not sure that all Swedes are grateful for this plagiarism, especially if Sweden remains a commercial outsider to the EEC. At the start of a new price system in 1956 the change-over from quantititative restrictions to import duties was regarded as a step forward. The intended prices were called middle prices and the actual prices allowed to vary between certain boundaries (about $\pm$ 15 per cent) without change in import duties. This implied an adoption of short-term variations of the world market prices, as long as they were less than the interval between the boundaries. On the other hand the boundaries could cut the effect of long-term changes in prices, *i.e.* those which should govern the adjustments in composition of output. A positive quality of the system is, however, that short-term large variations in prices are truncated.

An important intended property of the duties was that they should be given sizes, which would lead to the same relations between the profits of different production branches as at world market prices. This was interpreted so that the price relations should be unchanged by duties which were uniform percentages of world market prices. This, however, can only give a correct answer when then are no intermediate products and the percentage of costs for industrial inputs is the same for different productions. It can be shown that the required duty has to be calculated by the following formula to keep profit relations unchanged :

$$d = p\left(1 - \frac{c_i}{1 - c_a}\right)$$

(d = duty ; p = profit increase ; c_i = percentage of revenue referring to industrial inputs ; c_a = percentage of revenue referring to agricultural inputs from other sectors of agriculture.)

In the practical setting of duties the idea of uniformity has, however, been relinquished on several important points and replaced by autarchian or emergency price setting with the consequences already mentioned.

In the current operation of price regulations obtaining since the 1930's, strongly organized farmers' co-operatives and unions play an important role. The large percentage of the food supply processed and distributed by farmers' co-operatives (*e.g.* 99 per cent of milk deliveries, 90 per cent of slaughtered animals, 60 per cent of cereal sales) give these organizations powerful bargaining power. This power provides at least one important reason why it has been possible, despite the fall in the international food prices, to keep the national farm prices in step with the enhancement of the general price level in Sweden. Effective organization and fast structural rationalization of the co-operatively owned food processing industries here certainly influenced price margins favourably. On the other hand, interest in development of new processed foods has been low and has even been impeded by strict price regulation techniques. Furthermore the so-called equal price principle, applied on the deliveries from members, has limited the benefits of large-scale farming and consequently impeded the structural transformation of agriculture. During the last few years, the equal price principle has been relaxed in some co-operatives in order not to lose members with large deliveries.

(4) *Conclusion*

As a general verdict on Swedish agricultural policy it may be said that in the fifties the aims and means for the different activities were well co-ordinated. The leading goal was to equalize the income of 'basic' farmers to income in other comparable industries and to transform smaller farms to the size of basic farms. But technical and economic development had in the meantime made this size unreasonably small, and therefore the goal lost its reality.

Changes gradually had to be accepted by the policy, first in rationalization work (1959), then in the control of buying land (1965) and finally in price policy (1965). The new policies imply liberalization of the use of support to different sizes and types of farming, and of the buying of land by different categories of purchasers. Strict rules in price policy to cover income equalization have been submitted and the general aim for income level has been shifted to the 'standard' farmers. The policy for agriculture can be said to have disintegrated in the last few years. However, a committee, set up in 1960, exists which has the duty of proposing new lines of agricultural policy.

IV. FUTURE POLICY PROBLEMS

I am of the opinion that we in Sweden have come to a point where it is necessary, or at least desirable, to find a new comprehensive policy for agriculture. The problems have changed rapidly in many respects and they need more effective solutions, to avoid a lagging agriculture, expensive for the general economy and yet unsatisfactory for the farmers. Though Sweden, as a small, neutral nation with a highly developed economy, is a special case, with the solution there having limited validity for other countries, some principal viewpoints may be of interest.

In the discussion of agricultural policy reference is often made to the slow adjustment path of resources, *e.g.* of labour input, as an argument against the price weapon to solve a surplus situation. This is naturally correct in the short run, but in the long run, at such high yearly losses of labour as in the case in Sweden, this standpoint is untenable. For example, with a reduction of the labour input of 5 per cent per year and an increase in labour productivity of 3 per cent per year, production decreases by 2 per cent per year. This means that the production level in Sweden may reach a volume that will be insufficient to meet requirements in cases of emergency, such as a blockade, within a decade. The point is that in a country with a prosperous economy and with full employment and an active labour policy, the power of action in the long run for agricultural policy has become strong. The necessary adjustments in production volume may be governed by moderate shifts in the price level.

(1) *Arguments for a Low Price Level*

The price system in Sweden is based on rather complex market regulation, in which import duties are only one feature. Not only the sale of agricultural products but also the activities within the principal sectors of food industries are regulated. This has forced these industries to work exclusively as home-market industries, restricted in competition and hampered in the development of quality products. As food industries and distribution take an increasing part of the consumer's payment for food, now about 70 per cent (50 per cent if only expenses for home-produced products are considered), while the agricultural sector recedes both absolutely and relatively, a system with market regulation becomes unsuitable. At the high protection level, about 50 per cent for recent years, the risks of there also being misallocations in the food sector outside agriculture are obvious. Even within the agricultural sector singular effects arise, since two sectors exist, one of which uses the products of the other ; for example pig production could be profitable without being subsidized if coarse grain could be bought at world market prices, but with the high prices for feeding stuffs a surplus of bacon has to be exported with heavy losses. To avoid such peculiarities subsidies ought to be given as early as possible in the production process, for example only to crops on delivery or in the case of integrated production (of both feeding-stuffs and animal products) as a lump sum, proportional to the calculated fodder production. Such a low price level reduces the demand for the introduction of compensatory duties on competing products in different stages of the food production process (*e.g.* corn, margarine).

It would seem that I am pleading for an English subsidy system for Sweden. This is quite consistent with the general change in the problem of Swedish agriculture. Since the world market prices seem to remain constantly low, there is no comparative advantage to be found in a large agricultural domestic output. The only reason in the long run for keeping a level of production which cannot compete with the prices at which Sweden can buy food from abroad, is our safety in time of emergency. Consequently, the necessary support to keep sufficient resources in agriculture for this task is to be regarded as a part of the military expenses. There is naturally a social problem during the adjustment time, but as already pointed out this is transient at the existing high demand for labour.

One argument for maintaining a large agricultural output is world hunger. Sweden ought to use its surplus resources in agriculture to reduce hunger in under-developed countries. Such help would, however, be very expensive for the Swedish economy and relatively ineffective compared with technical assistance. The comparative advantages of Sweden giving help with technical education and capital are surely great and the multiplicative effect for the economy of the receiver ought to be greater than an increase of food consumption based on Swedish food. If food assistance is to be given, it would be cheaper to buy the food on the world market than keep the present factors of production in Swedish agriculture.

In the present discussions in Sweden voices are also found advocating the connection of Swedish agricultural prices to EEC prices. This is meaningful if a close association were to be realized at some future date. That implies, however, that Sweden relinquishes the right to govern the volume and composition of agricultural production with unknown consequences for its economy and safety. Realistically speaking, this means that Sweden changes its line of strict neutrality, a change which obviously depends on factors outside agricultural policy.

(2) *Political Implications of Vast Structural Changes*

The central question in the work of rationalization is what size of farms ought to be promoted. The mechanical revolution in agriculture has changed the requirements for farm organization in a drastic way. In land-free sectors, as hog and broiler production, great structural changes have already taken place, but in the land-bound sectors of production they proceed slowly. This does not mean that the advantages of large size are less in the last-named sector. Careful analyses of the profitability of different sizes of farms show that even within the frame of the family farm the optimal size in the long run exceeds 200 hectares. Probably further profits are to be made if the restriction of labour input per farm imposed by the family principle is relieved. This is indicated by the enormous capacity of the largest existing tractors, combines and other field machines, which make it possible for one machine to manage some hundreds of hectares during a harvest period. The idea of specialization is difficult to realize further within the frame of one family, as biological and economic factors call for combination. The real advantages of this idea are carried into effect only at a scale of production so large that both combination and specialization are possible within the same enterprise. Another problem for the optimal family farm is the need for a large amount of capital, which implies considerable fortunes to keep the farmer independent and safe against risks. Such fortunes are not consistent with the aims of a socialistic country as is Sweden.

It seems to be apparent that the family farm is no long-term solution, and therefore some method to bridge the transition of slow amalgamation into large farms would be preferable. Naturally politicians recoil at the severe encroachments upon farm structure and social life that such a step would mean. In the present discussions they are considering some reasonable size of family farm as an intermediate step, and it is very intensely argued that no returns of scale are found beyond the size of the optimal family farm.

Whether the structural transformation continues at the same rate as at present or whether it is accelerated, it implies a fast depopulation of agriculture and deep changes in rural life and outlook. The population concentrates in towns and the earlier growth of small villages has now turned into a decline. Investments in roads and communications are concentrated to serve inter-town traffic, and therefore the standard of rural communications lags behind. On the other hand there is an increasing flow of population from towns to the countryside as leisure time is extended. It is estimated that the number of leisure houses, which was already larger than the number of farm houses, has in a decade grown to several times the number of farm houses. Leisure houses are naturally localized in other places than the farm houses, and therefore we have to meet an enormous rearrangement of the landscape. Consequently a planning policy for the country-side is needed to avoid the destruction of nature. It is possible that this planning may be co-ordinated with the localization efforts, which were started in 1965. The aims of the localization policy are, however, as they are now formulated, to try to brake the accelerating movement of people from the northern part of the country to the southern urban districts and to keep the outflow of rural population, which now only passes through the towns in the northern districts, in these northern towns.

To sum up, the development and welfare of agriculture in Sweden have to a large extent depended upon the effects of general economic and social policies and this relationship seems likely to be strengthened in the future. This creates greater possibilities for the formation of agricultural policies with means which are capable of realizing the goals. Interest may be focused on the time horizon to reach the goals and on the dynamics of goals and means.

———

JOINT DISCUSSION OF MR. ASTRAND'S AND PROFESSOR GULBRANDSEN'S PAPERS

Professor Knoellinger said that Mr. Astrand's anonymous society might be recognized as Sweden or as any of those countries in the Western World — those welfare states — that present the highest degree of national economic and social integration. All these integrated countries have, to cite Myrdal, 'gradually found it appropriate to take vigorous measures to even out differences in income and wealth'. Agricultural policy in most of these countries has brought about an income redistribution in the interest of the farming population, usually by means of some price parity arrangement.

Professor Knoellinger said that in the historical perspective, the Great Depression of the 1930's is something of a watershed. It spelt loss of income and of security to a great many people, and compelled governments to fight the destructive forces of the market. In Sweden, for example, representatives of farmers and workers rallied in support of legislative measures which lifted farmers' incomes enough to provide their labourers with a fair share. Even in America the government had to intervene, despite the nation's individualistic past. He said that it was true that before the Great Depression, incomes had not been formed in a vacuum. The whole framework of society influenced the process of pricing from which incomes originated, and sometimes there occurred direct governmental protection of sectional interests — often the wealthy interests. But this had not much in common with the modern idea that, since a farmer's efforts and sacrifices are not less than, say, those of a non-agricultural worker, he should be entitled to a remuneration that equalled the worker's wage, and that, if this is not ensured by the market, it should be brought about by law. In the wake of this egalitarianism we have had, as Myrdal put it, 'to accept the fact that to an increasingly large extent prices are no longer . . . a simple function of free market competition but are manipulated and directed and are, indeed, results of "political" group actions'.

The whole fabric of price supports, subsidies, etc., obviously does not work as it is supposed to do. Agricultural incomes — on average, at least — persistently deceive policy-makers. This Mr. Astrand showed with figures that were, of necessity, rather fragile in spite of all refinements. Further, as Mr. Astrand also stressed, there is a grave risk that those kinds of income assistance to farmers which have now been practised for about a generation are likely to delay a rational allocation of resources in the economy at large. However, in view of the arguments of such men as Professor Galbraith it should be conceded, firstly, that part of the remarkable advance in agricultural productivity may have been the result of price support legislation which reduce price and income uncertainty and induce farmers to invest in new technology, and, second, that farm support prices are a useful part of the so-called built-in stabilizers of the economy.

In spite of the thesis of Professor Galbraith that the problem of income inequality has ceased to preoccupy men's minds, there are still worried policy-makers. There is the example of the problem of regional income disparities — quite apart from the problem of income disparities in the world at large. 'Incomes Policy' has become a modern catch-phrase ; and while the primary aim of such policy might be monetary stability, it also implies difficult problems of equity and equality. Agricultural incomes are an important issue in this debate.

The hard-boiled 'liberal' (of the European type) would tell marginal farmers to go out of business if the demand for their products is flagging. In Sweden, Professor Eli F. Heckscher in the early thirties had suggested something like this and had been promptly posted as hostile to farmers. He had conceded that some alleviating measures might be a necessary concomitant to the exodus. However, he had been quite confident that the migration process would be a smooth one, whereas Professor Myrdal had pointed out a number of obstacles that had to be taken into consideration. Although Myrdal also advocated fundamental structural changes in the economy, he pleaded, at the same time, for far-reaching governmental interference. In Heckscher's opinion any subsidizing of agriculture should at least be subjected to a predetermined time limit. This idea was not accepted. After all, farmers are not watch-makers, hatters or hair-dressers. They are, as a rule, much more numerous, and much more influential in society. Thus they, and many whom they have influenced, reject the ultra-liberal solution of their income problem, because they deem it too risky and too full of hardships. This includes the hardship of giving up the particular 'way of life' to which a good many of them are attached. On the other hand, prompted by the agricultural income gap, which no parity price plans seems to be able to eliminate, the flight from the land continues. In practice, therefore, we have a mixture of the liberal solution in the American sense, and the liberal solution in the European sense : *i.e.* government income support schemes, as well as voluntary migration to other pursuits.

Professor Knoellinger said that he did not find the solution a bad one in every respect. However, it cost a great deal in terms of resource mis-allocation ; and Mr. Astrand's arguments for acceleration of the migration process should be taken in earnest. The co-operation of farmers them-selves is an important condition of success.

Professor Knoellinger used a comparison between the experience of Finland in agricultural policy and the experience of Sweden, to show how difficult it is to disentangle the problem of agricultural income objectives. He said that in Finland the political spokesmen of farmers are still some-what ambiguous in these matters. Although Finland is a smaller country with a climate less suitable for agriculture, the economically active agri-cultural population is as large as, or perhaps even somewhat larger than, that in Sweden (in absolute numbers). Finnish farmers, therefore, play a much more important part in their country's politics than Swedish farmers do. The subsidizing of Finnish agriculture is an extremely heavy burden on other members of the community. The income parity policy applied in Finland is modelled upon the Swedish pattern. There is a continuous outflow of people from Finnish agriculture, although the annual rate of this depopulation is lower than in Sweden. Lack of skill, information and housing facilities are major obstacles to farm people who want to move to other occupations. Moreover, non-agricultural enterprises complain that they are being deprived of sorely needed investment capital because of the heavy taxes imposed on them to support, among other things, low-productivity farming, and that consequently they are unable to offer new employment rapidly enough to help accelerate the stream of people leaving the land. This being the case, farmers have to stay on their small plots and, in order to make both ends meet, have partly to rely on subsidies ; this completes the vicious circle. Furthermore, there is the other serious, vicious circle of inflation. Industrial expansion in Finland is hampered by high interest rates and by a wage level which, although lower than the level in Sweden, has been rising under the influence of the recently created common labour market in Northern Europe. Thus when industrial wages rise, subsidized agricultural incomes have to follow suit, and if this rise affects food prices, wages have to be adjusted again.

In his résumé of the paper, *Professor Nussbaumer* commented on Professor Gulbrandsen's point that price support policies have increased the capital invested in farms rather than increased the disposable income of farmers, so that the differences between the urban and rural standards of living continue. This clearly shows the unfavourable distortions brought about by a system of production and price stabilization reinforced by compulsory contributions designed to compensate for losses of some producers. The example stands as a lesson to other countries.

The price-fixing arrangement with its provision for 15 per cent variations and sliding import duties is akin to the system applied by countries in Europe which, between the wars, used a sliding tariff. To what extent has unfavourable speculation, accompanied by artificial influence on internal price levels, appeared in Sweden as it did in these countries between the wars ? What is the justification for the choice of 15 per cent variation limits ?

Turning to the policy recommendations in the paper, Professor Nussbaumer asked what Professor Gulbrandsen thought of the policy of giving direct aid to farmers as a means of regional development and income support, to be used along with other policy recommendations that he had made.

Professor Nussbaumer asked whether Professor Gulbrandsen would comment further on his point that the size of farms would have to increase to 200 hectares in the foreseeable future. Should we not be a little more careful with such estimates, and differentiate between farms which produce different kinds of products ? Is it because of specific Swedish conditions that the figure of 200 was chosen ? On the matter of hired versus family labour to be used on these larger rationalized farms, Professor Nussbaumer said that the problem is not simply a straight economic choice, but rather a question of the price that society is prepared to pay if, for social reasons, it wishes to preserve the family farm in its traditional form.

Professor Nussbaumer asked what alternative means for financing the larger farms could be employed in Sweden. He asked what part is played, or can be played, by rural credit co-operatives ; how much financing is done directly by the government ; and whether there is available any special access to capital such as that enjoyed by farmers who can use woodlands to finance occasional large investments.

Professor Nussbaumer asked for clarification of the definition of town and village used in the paper. In a densely populated country (such as Austria) towns of 15,000–20,000 inhabitants are easily achieved.

Professor Nussbaumer said that he had three general comments on the problems confronted by Professor Gulbrandsen. First, there is the difficulty with price support policies that they make it hard to see what in fact should be done. There are other methods of achieving many of the same ends ; for example, direct income subsidies can be used. The purposes of agricultural policies need greater specification for any policy to be fruitfully applied. For example, the probable increases in productivity per man-hour and per acre, and the number of people to be retained in farming and in rural employment have to be established. Simply to continue aid to farmers at the same time as subsidizing consumer prices, involves the government of many European countries in paying both aid and subsidies. This seems to be the worst possible choice since it means that the markets are almost entirely politically determined, and since it means that all economic restraint of the costs of further production and selling of agricultural goods is removed from the immediate environment of the farmer.

Second, even if the political problems are successfully solved, the use of price subsidization may have very damaging effects. As long as farmers work primarily for their own consumption needs, and try to maintain, as their main motive, a traditional standard of living instead of working for the profit motive, it is possible that there will be an inverse relation between prices and the quantity supplied. In any case there is likely to be only a slight stimulation to production from an increase of prices. In such cases a positive price policy will not lead to over-production, but will involve an increasing amount of money in subsidization. On the other hand, in developed countries, where the profit motive is effective, it would seem better not to interfere with the market mechanism, both because of the threat of over-production in agriculture, and because of the possible misallocation of resources in other sectors. Direct support of agricultural incomes from government funds, would be a simpler method, both more easily administered and less liable to produce distorting effects.

Third, Professor Gulbrandsen had wondered whether the agricultural over-production of developed countries could not be used to alleviate food shortages in developing countries. Professor Nussbaumer asked whether the agricultural imports of some developed countries could not be used to relieve the balance of payment difficulties of some of the developing countries.

Professor Astrand said that Professor Knoellinger was correct in saying that farmers have a larger influence on agricultural policy in Finland than they do in Sweden. The farming population in Sweden is only 10 per cent of the total ; the farmers' votes no longer count in the same way ; and the Labour Party has provided a climate for agriculture which has been rather cold in the last few years.

Professor Astrand described some of the methods used by the Swedish government to encourage rationalization. In order to obtain greater mobility of labour, the labour exchange service is considered essential in disseminating information to everyone who is interested in moving. There is a need for wider economic advisory services to farmers in order to help them make realistic calculations for investments. Farmers are obliged to show economic calculations before government investment aid is available. Government boards are available to help the farmer and to provide state guarantee through the rural banks. At these banks a lower interest rate would be charged for that type of credit. The system of credit guarantee replaces all kinds of direct loans from the government. A government training scheme is used to teach small farmers to run larger farms, and this makes them capable of changing farms. A government pension scheme for early retirement, which rests heavily upon the Dutch model, is at present being formulated.

Professor Gulbrandsen said that it had been the policy of the Swedish government up to 1959 that farmers with sufficient farm size could not buy additional land. After 1959 this restriction had been relaxed in cases where the additional land adjoined the farm, and where the land, by being incorporated in the farm, raised the farm income to a proper level.

The questions on subsidization are not really relevant to Sweden, since subsidies to consumers have not been used since the 1940's.

On the question of the 15 per cent variance allowed for prices — this figure was given simply as a representative one, since the limits in fact vary between different products. The use of such variances to avoid rigidity in price setting has developed because of the experiences of the 1930's to which Professor Nussbaumer had referred.

Professor Gulbrandsen said that the estimate of 200 hectares as the size of family farms in the future was based on programming calculations of optimum size for a typical farm. Clearly if regions are highly specialized in the production of certain products, or if they combine forested areas with other products, the optimum size will be different. Moreover, if the restriction that they should be family farms is lifted in the calculations, the optimum size may well be far larger.

Professor Gulbrandsen said that the size of town which he had in mind in his paper was to be defined with respect to the different services that it could offer, rather than with respect to the number of people alone. The required size would vary from region to region, but populations above 10,000 would usually prove acceptable.

Professor Bishop said that he had been concerned throughout the Conference at the tendency to treat agriculture as a separate part of the economy, and so to think of policies used within the agricultural sector as 'self-contained'. Even in Sweden the problem seems to be looked at too narrowly. In the cases where specific incentives are provided to encourage migration from farms, is sufficient consideration given to the effects this migration will have on those who remain in the rural areas, in particular on the small businessmen in rural towns and villages ?

Agricultural Policy in Sweden

Mr. Tracy said that the subjects discussed in these papers were of great interest to him personally and to his organization, the OECD, which has done a great deal of work in this field. It is now fairly clear what a rational agricultural policy should look like. For example, it should not include too rigid a guarantee for agricultural incomes, otherwise it ran into the difficulties indicated by Professor Astrand, and by Professor Knoellinger in his introduction to the discussion. It should not seek to implement an income guarantee solely through prices, but should leave a certain degree of flexibility to prices, and support farm incomes, if necessary, through more direct means. On the other hand, it should seek to create farms large enough to provide an adequate income for a farm family, and it could do this by helping some people to move out of agriculture, by helping elderly farmers to retire, by putting together different pieces of land into units of sufficient size, and so on.

Various countries had adapted their policies in these ways. Besides Sweden, there had been important developments in the Netherlands, France and, very recently, in the United Kingdom. The trouble is that the appropriate policies are not adopted in all countries, or if they are adopted, they are applied only half-heartedly. Economists can probably agree on what ought to be done, but the resistance lay elsewhere, particularly among farmers' organizations which need to be convinced of where their real long-term interests lay. Economists have an important task in helping progress to be made in this direction.

Professor Robinson said that discussion of the withdrawal of people from agriculture and their transfer to other sectors is of major concern to the general economist and, in particular, to those involved in policy discussions in the United Kingdom. It is frequently asked why the United Kingdom appears so low in the league table of economic progress, with a growth rate of about $3\frac{1}{2}$ per cent. One of the main reasons is that, for a long time, the agricultural population has been as low as 4 per cent of the total working population so that by now there is very little opportunity to reduce this population further. This denies the country the annual increase of national income — enjoyed by other European countries with their large agricultural labour resources — to be obtained by moving people into industries with higher productivity. The United Kingdom has arrived at the point where the net value added in agriculture is 90–95 per cent of that in other parts of the economy. This situation is not to be found in most of the countries of Europe. This process of shifting population to sectors of higher productivity is not one that should be retarded, and agricultural policy should encourage the release of rural populations in order to increase the output of other sectors.

Professor Robinson said that he was concerned by the circular arguments which so easily arise in discussion of returns to capital. The selling price of a farm is the capitalized value of what someone can get out of farming it. With any system of support policies aimed at making farming profitable, higher valuations would be placed on farms than if these policies were not in force. However, what can be gained from farming the land is also dictated by the gains of scale. The farmer whose farm is too small will pay a high price to increase the size of his unit in order to reap these gains. This is as it should be if the aim is to increase the efficiency in agriculture. In the United Kingdom the consolidation of farms had been encouraged and land values partly reflected these gains in scale.

Professor Glenn Johnson said that evidence from the United States is pertinent to the points made by Professor Robinson. The bulk of the empirical evidence indicates that in the United States the marginal products of factors are not far from in line with the capitalized values of land prices. The buying of small farms so that they can be added to other units does not seem so much to be an attempt to obtain returns to scale, but rather an attempt to bring a farm which is out of balance with respect to other factors, back into a state of balance. In fact, calculations indicate that more or less constant returns to scale exist after a certain size has been achieved, and that land purchases are made to achieve balance rather than to achieve size.

Professor Zemborain said that the farmer has to be seen as a manager of an enterprise, and not as a simple worker. Even to be in business he has to have managerial capabilities. He did not think it appropriate to compare the farmer with the industrial worker. These managerial capabilities are the main determinant of how large a farm can be. A farm does not usually need very heavy investment. If well managed, a farm with one tractor in the Argentine can be as efficient as a farm using three or four.

Dr. Odhner said that since he was involved in reshaping Swedish agricultural policy, he had found the papers and discussion most interesting.

He said that the problem brought up by Professor Bishop, of the effects on other parts of the rural population of the exodus of agricultural manpower, is not special to agriculture. The same policies of mobility have been applied to all other sectors of industry. It is only for political reasons that they have not been applied to agriculture until very recently. The threat of loss of votes has retarded the application of the policy. It is now to be hoped that a policy of stimulating mobility can be carried out.

Swedish agriculture suffers from many of the problems raised in Professor Glenn Johnson's paper. The structure largely depends on family farms which show the same general lack of efficiency from an overall point of view. The figures, which show a greater increase in productivity in agriculture as a whole in comparison to the rest of industry, are mainly due to the fact that small farmers who produce almost nothing for the market have moved to other occupations in increasing numbers. The productivity in family farms is in fact lower than industry in general. This inefficiency is due to lack of education, paucity of technical know-how and the small size of many units. Increases in size are held back both by the threat of surplus production, which makes intensification hard to achieve, and by the rigidity of land ownership. Dr. Odhner said that the structural policy pursued in Sweden since the Second world war has been, in large measure, a deception. The mobility of land, which it was thought the policy would promote, has not been realized. Farms have stayed much the same size and it is now to be doubted if administrative methods can achieve structural change of the kind required. The great increase in land prices has also retarded land mobility. Productivity increases are not the only reason for these high prices. Speculative pressures and the demand for land for roads, housing, airfields and for the leisure pursuits of town dwellers, have affected prices and required the farmer to compete with other means of deriving profit from land.

Dr. Odhner said that he would like to know what can be done about this problem of land prices. Mr. Tracy had said that we know the general lines of a rational agricultural policy, yet, in this matter of land prices there is the dilemma that, if they are controlled, rigidity is induced into the land market, but, if they are left entirely uncontrolled, the prices soar. What should be done ?

Professor Bishop said that although he agreed with Professor Odhner that the labour market in agriculture, if left free to function, works as well as that in many other sectors, the question he had originally asked concerned the external effects produced by specific incentives to increase mobility. The movement of farmers affects many others ; it changes their economic environment, and in some cases obliges them to move as well. Whether we are analysing a capitalist, socialist or mixed economy the nature of these changes is similar and the direction of these changes the same. The need is to increase the level of economic literacy, so that people will move, with full information, in the direction in which economic forces direct them, and so that they will make full use of the incentives provided to encourage them to move.

Professor Mundlak referred to the question of scale and adjustment of resources brought up by Professor Glenn Johnson. There are two ways in which a farmer, who thinks that his use of factors is out of balance, can get them back into balance : either he can increase the size of his farm, or he can reduce the use of other factors. If he chooses the former, as appears to be the case, he is effectively realizing economies of scale.

The situation is not perhaps as bad as some participants seem to argue. Adjustments to economic forces takes place very rapidly in many economies. Farm income support policies have never achieved the objective of income parity. If this objective were achieved, there would be little if any movement out of agriculture.

Professor Glenn Johnson said that if we take the case of the small farmer contemplating the purchase of more land, it is likely that with his present organization some assets earn more in their present use than he could get for them if he sold them, yet less than it would cost to acquire more of them. Consequently such assets are fixed. If, moreover, the factors were out of balance, the loss-minimizing adjustment can often be the purchase of extra land, quite independently of any increasing or constant returns to scale that might result.

Professor Hathaway said that it appeared that Sweden had achieved some degree of structural reform without any increase in total agricultural production. This is a unique situation for an industrial country. Has this been achieved with a stable level of absolute and relative prices or with a rising level ? It is hard to believe that the device of obliging farmers who produce an exportable surplus to pay for those surpluses, can alone be responsible. This policy has not been effective in the United States as a restraint upon over-production.

Professor Robinson took up Professor Bishop's point about the effects on other rural occupations of the contraction of the size of the agricultural population. A recent meeting of the Royal Economic Society, addressed by Douglas Jay, then President of the Board of Trade (which is responsible for this), had discussed the problem of location policy and raised the same point. The general effects of contracting industries on the whole infrastructure of a region had been discussed. In the United Kingdom the authorities take account of this contracting multiplier effect, in deciding whether to try to take work to the workers or to move workers away from the declining regions. There are strong arguments favouring taking work to the workers ; for instance, the Pigovian arguments about the difference between public and private social product are applicable.

Professor Zemborain said that studies of land prices in the Argentine revealed that land prices, in relation to price of agricultural output, had remained almost stable for a long period of time. However, the composition and the direction of investment in the rest of the economy had changed. From 1960 to 1965 land prices relative to agricultural output prices had not changed ; however, in the stock market the price had reduced from 100 to 30. It was clear that land prices should be studied in relation not only to agricultural prices but in relation to prices in the whole economy.

Professor Astrand said that there had been no direct regulation of land prices in Sweden. However, if a new law, allowing companies to buy land, comes into operation it will certainly mean heavy increases in the price of land. This presents a serious problem. Inflationary pressures are already at work and these would be aggravated. One of the difficulties is the traditional reluctance on the part of farmers to part with their land in view of the rising land prices. The advice of the old farmer is always to hold onto land. 'Sell as much as possible, but never sell land.' The state does buy land, but not with the aim of holding onto it, or conducting any rationalization itself ; it is sold as fast as possible to private enterprises or companies. Forest land comes under a different category ; the government has funds to purchase forest land, and does so on an increasing scale.

Professor Astrand, in reply to Professor Zemborain, said that he thought that in Sweden it is quite valid to compare the small farmer and the industrial worker. The industrial worker is very well paid and highly skilled.

Professor Astrand said that he appreciated the difficulties involved in calculating capital values of land, and the income of farmers. His practice had been to take a plausible fixed price for land, to calculate the interest that could be earned for that sum and to regard that as part of the farmer's income.

Professor Gulbrandsen said that the difference in land values in Sweden and the Argentine mean that whereas the effective constraint to size in the Argentine may be management ability, in Sweden it is rather the problem of acquiring high-priced land.

The high price of land hampers profitable adjustments for many farmers. However, the state is aware of this, and it is from state encouragement to enlargement that rationalization would come. There are also many ways in which farmers could themselves overcome the difficulties of financing their operations. Moreover, the farmer would contribute to state schemes to aid him in this respect, in the same way that at present the farmer pays one-third of the costs of financing crop insurance schemes.

In answer to Professor Nussbaumer's question about the possible value of woodlands as alternative sources of finance, Professor Gulbrandsen said that in the major forested regions of the north, agriculture cannot compete with the agriculture in other regions. It is not intended to retard the rapid disappearance of these farms by restructuring them. In fact, farmers at present do obtain much of their incomes, as well as some capital requirements, out of forestry.

Professor Gulbrandsen said that he agreed with Professor Robinson about the importance of the movement of population out of agriculture in the growth of European countries. Between 1950–1960 the contribution to the increase of gross domestic product from this source in Sweden, was some 15–20 per cent of the total increase. Despite the fact that only about 9 per cent of the total population remains in agriculture, it provides a reservoir which could be drawn upon at an increasing rate. For example, in 1964 the decrease in total agricultural labour force was about 8 per cent.

Professor Gulbrandsen said that Professor Hathaway was correct in thinking that it is not only the fact that farmers have to pay for exportable surpluses themselves, which accounts for the static production. Relative prices have remained pretty stable. The income parity objective has not been achieved largely because incomes in other parts of the economy have increased so fast that they outstripped the rise in productivity in agriculture. Finally, the high demand for labour in other sectors has meant a very high rate of out-migration.

Chapter 21

PURCHASING CONTRACTS AND PRICE POLICY AS MEANS OF PLANNING AGRICULTURAL PRODUCTION [1]

BY

M. POHORILLE

Central School of Planning and Statistics, Warsaw

I. INTRODUCTORY

SINCE the end of the Second world war, the contract system has been gaining ground steadily in the agriculture of most of the economically developed countries. There are several reasons for this.

(a) The development of food processing and transformation industries leads to concentration both in the demand for agricultural products and in their supply, for these industries are dependent upon regular supplies of farm produce of a given quality.

(b) Production of agricultural inputs is an expanding industry. Anxious to widen their markets, firms in the industry often take a hand themselves in organizing both agricultural production and marketing and so initiate a process of vertical integration in agriculture. The contract system is one of the basic forms of such integration.

(c) A revolution has taken place in food distribution. Chains of supermarkets and the multiple outlets of wholesale dealers are handling more and more of the food supplies to the public, and these chains need regular supplies of standardized products just as much as the food industries do.

(d) In agriculture, as in industry, people are becoming more and more aware of the need to plan production.

[1] Translated from the French by E. Henderson.

430

II. THE DEVELOPMENT OF THE CONTRACTUAL SYSTEM IN THE PLANNED ECONOMY

(1)   In a socialist system, all these factors act with greater force than they do in the capitalist economy.   In socialist countries, where industry and wholesale trade are nationalized and the economy as a whole is planned, the contract system is a sort of link between the collective (or, in the transition period preceding socialism, individual) agricultural units and nationalized industry.   At the same time, the contract system is a form and an instrument of planned agricultural production.   As such it assumes all the more importance since it has been learned, by trial and error, that any attempt at central indicative planning with respect to the areas to grow different crops is doomed to failure.   In these circumstances, the best way of fitting agriculture into the central plan is to use voluntary contracts with farm enterprises as a means of arriving at the desired composition and volume of agricultural output.   These contracts stipulate the sale of pre-determined quantities of produce of defined standard to the state trading organizations ; they are concluded *ex ante*, that is, before production decisions are taken, and for this reason exercise a highly effective influence on the planning of agricultural production itself.

The essential point is that the contract system can reconcile general social interests with the individual financial interest of agricultural producers.   Thus production plans are transmitted to agriculture not in the form of administrative orders, but in the form of economic incentives, and farmers are completely free to choose whatever pattern of production seems most advantageous to them.

(2)   The contract system certainly is no panacea which will automatically resolve the problem of balancing demand and supply with respect to farm products.   This depends upon the general conditions of economic growth, and requires the most diverse means of influencing agriculture.   The desired composition of output demands the prior establishment of an appropriate system of relative prices.   Should rapid expansion of demand for agricultural products necessitate an acceleration of the rate of increase in supply, other measures have to be applied, such as stimulation of investment in agriculture and its ancillary industries, intensive promotion of technical progress, higher expenditure for infrastructures, etc.   But none of this diminishes the part which contracts can play as a means of mobilizing the supply of agricultural products and adapting it to the country's needs of consumption goods.   On the contrary,

whenever the government considers it indispensable to apply certain other, non-market, methods of influencing developments in agriculture, it appears reasonable to link these other measures with the contract system, for this creates between the state and the farmers reciprocal relations resting on the very sound principle of *do ut des*. Farms are furnished with part of their resources from outside (credits, inputs, technical assistance, etc.), and in exchange they assume an obligation to produce and sell a given quantity of produce, in accordance with social demand. Price policy, too, takes on new features when associated with the contract system. It is often argued that planned influence on production is more effective when it does not run counter to market forces and is based on incentives which work through the intermediary of the price mechanism. This is no doubt true enough, but the reverse is equally true. Prices can become an effective instrument of control over agricultural production when price incentives rest on broader state intervention involving direct influence on the conditions of production and sale.

(3) We are to conclude that in socialist countries the contract system tends toward general application in agriculture, that is, toward encompassing all agricultural production and all farm enterprises. It is indeed only as a general system that it can gain its full usefulness as an instrument of planning agricultural production.

(4) However, the achievement of this desired model of a contract system comes up against a number of obstacles, as follows:

    (*a*) The methodology of planning in the field of agricultural production is still inadequate;

    (*b*) the further extension of the contract system creates serious organizational difficulties;

    (*c*) overhead costs are often higher in contract purchasing than on the free market.

It will therefore need long preparation before the contract system can be quite generally applied. To direct this process consciously, it will be necessary to form very clear ideas on two points:

    (*a*) the order in which the contract system is to be applied to different groups of agricultural products, and

    (*b*) the negative and positive aspects of the transition period.

(5) There are certain methods of stimulating the supply of agricultural products which are very close to the contract system, even though they do not involve formal agreements of the purchase contract type between farmers and purchasing agencies. I have in

mind, for instance, the purchase of milk and eggs by co-operatives : a co-operative dairy establishes continuous relations with the farmers of its districts, who supply the dairy regularly with a fixed quantity of milk of a specified fat content ; the dairy, in return, guarantees the farmers a fixed price in advance, as well as certain other benefits (*e.g.* skim milk for animal feed). This form of purchase might be called *paracontractual* and it might be argued that, with its much lower incidental expenses, it meets the case very well indeed and limits the scope of the contract system properly speaking.

One case where no question arises at all is that of industrial crops. These always have been subject to the contract system, given that industry needs a regular supply of raw materials and that farmers would not grow these without being sure of being able to sell them. This leaves such groups of agricultural products as livestock products, cereals, vegetables and fruit ; these are all within the contract system, though the contracts themselves vary.

(6) The general logic governing the development of the contract system is as follows. Apart from industrial crops, the agricultural products to be drawn into the system first are :

(a) those to be exported, which must meet certain quality standards and delivery dates ;

(b) those the production of which particularly needs to be stimulated ;

(c) those which help the contract system itself to develop its functions in agricultural production. The organizations which award the contracts perform certain activities which influence the very process by which the goods subject to these contracts are produced. In effect, wherever the contract system is most developed, the bulk of the organizational functions in the farms concerned falls to the organizations which award the contracts. It is they which supply the farms with seed and pesticides, insist that the specifications and instructions for production are met, etc.

(7) The development of the contract system is not free of contradictions. The first of these is of a general nature. In countries where it is difficult to find outlets for agricultural produce (that is, in the conditions characteristic of certain European countries and of the United States), the contract system has its attractions for the farmers but is of little use in counteracting surplus production. Fundamentally, the value of the contract system certainly is that it stimulates, and not that it restrains, an increase in agricultural output.

Where, on the other hand, the market is open in the sense that there is no difficulty in selling what is produced, the contract system is much less attractive for the farmers, who need to be afraid neither of a fall in prices nor of a lack of markets. In such cases the farmers' interest in the contract system needs to be stimulated by including in the contracts, apart from the price guarantee, additional clauses conferring certain other benefits on the farmers, such as credit facilities, priority in the allocation of inputs in short supply, etc.

It might of course be argued that the best way to resolve these contradictions would be to eliminate any disturbances of market equilibrium and to forestall any situation of either surplus or shortage of agricultural products. But this is quite another problem. The question is not what conditions might be most propitious for the development of the contract system, but how the contract system can best be used as an instrument of planning and developing agricultural production. The answer is simple enough. The contract system has no independent function ; it is merely one of the elements in planning the economy — an element, it is true, which is very important, but which needs to be associated with many other means of influencing agricultural production.

(8) Another major problem which we encounter in our system is how to co-ordinate the activities of the numerous institutions concerned with contracts for farm produce. In 1961 we had fifteen organizations in Poland. While they all work to one single national plan, they have their own network of agents and their own raw material departments.

It is no doubt all to the good to have a certain amount of competition between the contract agencies.[1] On the other hand, dispersion of effort in the organizational field along often quite divergent lines diminishes the efficacy of the contracts as instruments to raise agricultural production and pushes up trade expenses (contract costs, inspection, on-the-spot checks of cultivated area, etc.). Recently we have tried, in Poland, to assign the duties deriving from the contract system to the agricultural development plans at municipal level, and to charge the People's Councils with the function of co-ordination.

Insistence that farmers strictly meet the pre-established delivery dates, grow more industrial potatoes, as well as vegetables and fruit ripening at different seasons, and differentiation of prices according

---

[1] Generally, there is no price competition, but competition only in the service sector, that is, the organization of a speedy delivery system for the goods subject to the contract, their transport, etc.

to the time of year — these and many other similar measures can help to ensure a more regular flow of contract produce to the purchasing centres. This would greatly reduce expenses.

Improvements in planning methods with respect to agricultural production are, to a large extent, the key to the effective use of the contract system as an instrument of planning.

III. CONTRACTS AND THE ECONOMIC PLAN

(1) Agriculture obviously does not lend itself as easily to planning as industry. Contrary to what happens in industry, forecasts of the course of agricultural production are more reliable for several years ahead than for short periods. The same is true of forecasts concerning agricultural incomes and prices. In the short period, fluctuations in output are bound to cause income and price fluctuations.

The key problem is how to determine plans and purchasing methods as well as prices for agricultural products such as to create the necessary conditions for a steady increase in agricultural production and for stable food prices at retail level, and to do all this without detriment to the indispensable flexibility of the planning system as a whole.

(2) The purchasing plans for agricultural products are worked out on the basis of the overall agricultural balance and of commodity balances.

The overall agricultural balance shows the aggregate volume of agricultural production, of which part goes to cover the requirements of farms (seed, animal feed, household consumption and losses) and part is marketed.

The commodity balances include both those agricultural products which are bought domestically by the state trading agencies (and the co-operatives) and those which are imported. For each product, the pattern of its use is laid down in terms of the proportions to be allocated to transformation, consumption, export, stockbuilding, etc.

(3) State purchasing of agricultural products takes three main forms : (*a*) contracts ; (*b*) purchases on the free market ; and (*c*) obligatory deliveries.

In contract purchasing, prices are fixed every year in close connection with the general purchasing plan. If, for example, the government wants a larger acreage to be given over to sugar beet, it increases the contract-purchasing plan and offers better prices to sugar beet growers (or makes other sales conditions more attractive for them). Nevertheless, prices are normally fairly stable. Contract terms are changed only at intervals of several years.

Agricultural products not purchased under the contract system are bought on the free market, where price fluctuations are much sharper. Price policy with respect to free-market purchases is designed to keep these fluctuations within pre-established limits (the purchasing agencies buy within a given price range).

The third form of state purchasing, obligatory deliveries, applies to part of the output of cereals, to meat and to potatoes. At present, it encompasses about 17 per cent of marketed production and is of minor importance. Delivery is taken at fixed prices, which are lower than those on the free market. The difference between the two prices is paid into the Agricultural Development Fund, which provides farmers with money to buy tractors and other farm machinery and also finances collective investment (for example, storage facilities for fruit and vegetables, grain cleaning units, drinking water supply, etc.).

(4) In the current system of agricultural purchasing the contract system occupies a front rank position. This is evidenced not only by the growing share of contract buying in total state purchasing of agricultural products, but also in the steadily increasing part the system plays in creating the conditions of balanced growth in agricultural production.

In this last connection, four important points need to be made.

(*a*) The contract agencies are increasingly active in the planned supply of basic inputs (*e.g.* fodder, fertilizers, etc.) to farmers — a matter of decisive importance to an expanding agriculture.

(*b*) The contract system relieves farmers of all the risks of price fluctuations on the market not only in the short, but also in the medium run. In a free market economy, the contract system in the medium run intensifies the process of the elimination of marginal farms. The contract buyers introduce technical progress into farming methods and generally tend to deal with the more efficient farmers. The process of selection becomes more brutal as outlets become scarcer, and eventually the least efficient farmers not only fail to get their contracts renewed, but lose their hold altogether and have the greatest difficulty in getting back to where they started. In the planned economy, the contract agencies in any given branch sometimes arbitrarily limit the application of the contract system in certain regions or with respect to certain groups of farms, but this is done deliberately for economic reasons. Any change in the contract system has to be discussed with the local People's Council responsible for the direction of the whole contract system in the municipality's territory, and the Council can put forward another solution.

(*c*) The contract system is used as a means of adjusting the structure of output to the social and economic structure of agriculture. The contract plans are designed to develop labour intensive crops in farms having unutilized manpower reserves, to develop the spirit of economic co-operation among farmers, and to adapt the geographical pattern of production to the quality of the soil in different parts of the country. Thus the contract system plays an important part both in resource utilization and in putting a given social policy into effect.

(*d*) More and more frequently, the contract agencies undertake projects designed to found relations with farmers on contracts stretching over several years. This means that particular areas can be given over permanently to certain crops, though in order to encourage farmers to conclude such contracts, the government needs to provide investment credits.

(5) Thus there is a diversity of purchasing methods coupled with a diversity of prices, of which some are fixed, some subject to very small and others to larger fluctuations. In practice, we thus strike a sort of compromise between two contradictory claims which farmers habitually advance, namely, stabilization of prices and stabilization of income. At the same time, we also manage to reconcile the interests of agricultural producers with those of consumers. Experience shows that in spite of price fluctuations on the free market food prices can be kept fairly stable and uncontrolled changes in the distribution of national income can be prevented.

IV. THE GENERAL PRINCIPLES OF PRICE-FIXING FOR AGRICULTURAL PRODUCTS

(1) The contract agencies are not free to decide themselves at what price to buy agricultural products. The prices of basic products are fixed by government. Price policy takes account not only of the

current market situation, but also of the immediate and later consequences of price changes and their influence on production and the market as well as on the real income of different social groups. It must be stressed than when prices are fixed by government, the problem of appropriate price relations assumes key importance. When prices form freely on the market, the price system is always more or less coherent — which does not mean that it always meets the requirements of economic growth. But price intervention by government always entails a risk of infringing the logic of price equilibrium. In the absence of close co-ordination of decisions concerning price changes for different agricultural products, there is a danger of internal inconsistencies and disproportions, which would weaken the effectiveness of the whole set of measures to influence agricultural production and the market. Hence equilibrium of prices must be seen as a whole, and its internal logic respected. This can be done by what I would call the production/distribution formula in price fixing for agricultural products.[1]

(2) The point of departure is the distinction between the problem of the general level of agricultural prices and the problem of price relations as between separate agricultural products. Farmers have to make a choice among various possible ways of using the factors of production at their disposal, and they base their decision not on absolute profitability, but on the relative profitability of the farming branch under consideration. At given production cost, the relations between the prices for different agricultural products are of decisive importance in arriving at a suitable structure of agricultural production.

The general level of agricultural prices (and its relation to the prices of services and of industrial goods bought by farmers) determines the level of real income in agriculture and the volume of farm household consumption and accumulation. In its turn, the rate of increase in agricultural production depends upon the level of accumulation. The general level of agricultural prices must therefore be determined in close connection with the planned increase in national income, its distribution and the actual methods of its distribution.

(3) When I speak of the distribution of national income, I have in mind its allocation to the accumulation fund and the consumption fund just as much as distribution among social classes (workers and peasants). In mentioning the actual methods by which national income is redistributed, I want to stress the close link which exists

---

[1] This formula constitutes an attempt at theoretical generalization of the practice of price-fixing for agricultural products in socialist countries.

between the level of agricultural prices and the various forms of income transfer as applied in any given period. Suppose, for example, that, in accordance with the general principles of economic policy, the greater part of investment in agriculture is made by the state (*e.g.* in the form of tractor stations) ; in that case, the prices paid to farmers should be lower than they would have to be if farms took care of their own investment. It follows that any changes in the system and scope of agricultural finance must be accompanied by well-defined changes in the general level of agricultural prices.

(4) The reasoning behind the production/distribution formula is as follows. The plan provides for a specified increase in agricultural production ; the achievement of this increase needs a given amount of investment in agriculture and ancillary sectors. Part of this investment is financed by the state, part by the farmers. But this decomposition of investment expenditure according to sources depends not alone upon the planned increase in production. It depends upon a whole series of economic, political and institutional factors, which also govern the rate of increase in farm household consumption, such as it figures in the general economic plan.[1]

(5) Increase in agricultural production, prices, incomes, consumption and accumulation are interconnected magnitudes. It is obvious that the level of agricultural incomes depends upon the volume of output and upon prices, and that, on the other hand, the achievement of a given volume of output requires a corresponding level of farm incomes (during the preceding period), of which a well-defined portion must be devoted to productive investment.

The government influences the distribution of farm income (as between consumption and investment) by means of a system of tax reliefs on invested funds, low-interest credits and assistance from the Agricultural Development Fund.

(6) Once a feasible rate of increase in farm consumption and accumulation has been determined, and once practical arrangements have been put in hand to make sure that the accumulated funds will be used for investment such as to raise agricultural production by a given percentage, then the basic data for defining the general level of agricultural prices are at hand.

---

[1] In Poland we usually adopt rather similar rates of increase for the household consumption of farming and for the working population. But this rule is not obligatory in all circumstances. In the presence of major divergencies between the standard of living in the towns and in the country, it may be necessary to raise rural consumption more quickly than working-class consumption. Throughout the twenty years since the war incomes have been growing much faster among the agricultural population than among the workers.

Given this general price level, the relations between the prices of different agricultural products must be so shaped that they discriminate in favour of certain crops and induce changes in the structure of production to match as far as possible the changes in the structure of society's consumption needs.

(7) In this connection the calculation of production costs plays an important part. Attempts at fixing prices directly on the basis of unit cost have been disappointing ; what is needed is to follow the dynamics of price and cost changes, so as to gain a picture of variations in the profitability of the different lines of production and of the farmers' likely reactions thereto. Cost calculations are equally indispensable for fixing agricultural incomes. Thus the production/distribution formula does not exclude the calculation of unit costs in agriculture, but merely divests it of its absolute significance and avoids oversimplification of the cost-price relationship.

(8) In the view of some economists, the complex study of correct price relations is completely redundant, since the market resolves this problem perfectly. It is true that information obtained from market analysis is very helpful in the consideration of the problem under discussion, but the question is not as simple as all that. First of all, it is not merely a matter of identifying the equilibrium conditions of the market at any given moment, but of forecasting future changes with accuracy. Secondly, prices which are subject to volatile and transitory factors can give no proper guidance to farmers in their choice of the right line of production.

Unexpected price changes generate uncertainty. By contrast, expected changes play a constructive part, in that they are an incentive to farmers to alter production in the desired sense. Price policy acts like a sieve : it lets through certain price changes and holds back others.

(9) This brings me to the fundamental question as to how we can link up the theoretical system of farm prices (as defined on the basis of the formula outlined above) with the actual price system for end products of agricultural origin, which latter system comes about under the influence of the factors which determine market equilibrium (supply and demand). The two price systems can be combined by manipulating five factors :

(a) the margin between the retail price of the end product and the producer price ;
(b) taxes levied on farmers ;

(*c*) subsidies and credits to agriculture ;
(*d*) the prices of agricultural inputs ;
(*e*) the prices of industrial consumer goods which farmers want.

By manipulating these factors the plan can be made internally consistent, that is, its targets for an increase in agricultural production and incomes can be harmonized with an economically justified system of retail food prices.

(10) Any effort at ensuring general market equilibrium in the course of a process of economic development depends decisively also on wages policy, in so far as effective demand for consumer goods should increase in proportion with the rate of increase in their supply.

V. CONCLUSION

The principles underlying the determination of agricultural prices, as discussed above, can aptly be named production/distribution formula. These principles indeed rest on the premise that prices should be fixed in close relation with the planned proportions of income distribution and with the long-term development targets for agriculture.

This formula constitutes sufficient foundation for a rational price policy in the purchase of agricultural products. In this field, much scope is left for *ad hoc* decisions. Such decisions indeed intervene in every price formula, as we know from experience. Contrary to appearances, no formula provides a hard-and-fast answer to the question of the desirable level of prices for different agricultural products. It is undoubtedly one of the virtues of the system described above that it establishes a more flexible link between prices and production than other systems, such as, for instance, the system of parity prices or those based on the so-called cost of production formula.

DISCUSSION OF PROFESSOR POHORILLE'S PAPER

Professor Papi, in his comments on Professor Pohorille's paper, said that his first criticism was that it only took into account a few of the problems of the market and of price stabilization. Professor Pohorille had underlined the fact that purchasing contracts relieve farmers of the risks of fluctuations in price that are prevalent in all market economies. It is true that Professor Pohorille pointed out that any modification in the system of government contracts has to be discussed in advance with the local 'People's Council'. None the less, the paper completely ignored the structural problems of agriculture in socialist countries. It ignored the problem of the most efficient combination of productive factors, upon which the successful development of all internal and external economies of agricultural income turned. On page 435 of the paper Professor Pohorille had given what he took to be the principal problem : 'The key problem is how to determine plans and purchasing methods as well as prices for agricultural products, such as to create the necessary conditions for a steady increase in agricultural production and for stable food prices. . . .' Professor Papi said that it is difficult to see how this problem can be attacked or resolved without the least mention of fundamental problems of agricultural productive processes. Thus, the paper was reduced to a simple description of one of the many methods of static market stabilization. Given this approach, one could not say, as Professor Pohorille did, that the method of contracts 'creates the necessary conditions for the steady increase in agricultural production'. This method might be used in an attempt to create a certain stability, but Professor Pohorille had given a description of a static approach to the problem which did not take into account any conception of the dynamics of equilibrium growth. In such equilibrium growth the size of undertakings, the size of the agricultural sector in relation to other sectors and the size of the population on the land, would have to undergo profound and planned changes. These were not discussed in the paper.

Professor Papi said that the second point that he wished to make about Professor Pohorille's paper was yet more fundamental. It concerns the sale and purchase of agricultural products on the market which Professor Pohorille called a free market. Professor Papi asked whether there really are markets — not to say free markets — in socialist economies. It is a crucial point which cannot be passed over in silence if full understanding is to be achieved. The point at issue is the process of the formation of prices in a socialist system with decentralized production.

In order to explore this point more fully Professor Papi cited *The Economic Theory of Socialism* written by F. O. Taylor and Oscar Lange in 1939. Professor Papi enumerated the major tenets of socialist economic theory given in that book, emphasizing, in particular, that the productive activity centres on two pivots ; first, the determination by the central authorities of the prices of factors of production, and, secondly, the special nature of the competitive behaviour of all those who take part in the productive activity of a decentralized enterprise in a socialist economy. The results are that the price of product or service can never depart from the cost of production. Under the conditions set up by these tenets, neither in the case of factors of production, including labour services, nor in the case of final products, can a true market be established. Once the central authority becomes a monopolist, in the supply of all factors of production both from inside and from abroad, and so arbitrarily decides the price of the factors of production, it may not be said that a market can be established. As Taylor and Lange made clear, the accounting price of the factors of production has to be established by trial and error. A true market cannot be established for labour services, the price of which — even before any exchange activity has taken place — is fixed from the moment that the central authority chooses one or another criterion of income distribution and fixes the prices of factors. A true market cannot be established for final products and services, since their sale price cannot diverge by much from the cost of production and these costs of production flow straight from those prices which the central authority has fixed for the factors of production. Under these conditions, any market that is said to be established is purely figurative.

Professor Papi concluded that to speak of markets and even of free markets seemed to him to be very difficult. The economy of any socialist country is governed by the fundamental power of the central authority.

Professor Georgescu-Roegen said that he wanted to take up Professor Papi's point about the problems of the kinds of prices that existed on a market. Professor Georgescu-Roegen agreed that prices are the regulators which governed the quantities of real distribution in the current economic process. But at the present juncture we face some different types of problems, such as what criteria should be used in deciding on a policy of vertical integration and how far should vertical integration go in a given situation. Professor Georgescu-Roegen said that he suspects that a particular case of vertical integration which has been quite profitable in one country might be unprofitable in another. Upon what criteria should we decide issues such as these ? Current prices are only superficial, external aspects. The problems such as that of vertical integration are radical problems : they pertain to how much we should work and how much we should get out of work. Professor Georgescu-Roegen illustrated his point with a problem which, he thought, represents an actual situation. For the sake of vertical integration, a country having a sufficient output of beef, may decide to construct a single central packing station. This decision implies that all beef be distributed deeply frozen. But would not the local consumer be then in a worse situation than he was before, if he happens to prefer the fresh to the frozen meat ? Might not the fever of vertical integration cause us to move against our own preferences ? In the opinion of Professor Georgescu-Roegen price-cost analysis is incapable of dealing with such fundamental problems.

Dr. Jacobi said that he did not agree with Professor Papi on the matter of the market situation in Eastern Europe. The local market in socialist countries is one of the only free markets left in Europe. Consumers look for additional food from this market, and the demand is considerable ; the farmer is interested in supplying as much as he can to that market, since the prices are relatively high.

Professor Stipetić said that before any pricing policy can be effectively used for planning production patterns, it is essential to know what supply responses to these price changes would be. It may well be that price increases will simply mean higher returns to farmers without an increase in total production; or it may be that more land, labour and capital will be applied to particular crops at the expense of other crops. Much work needs to be done on this. At present there is little knowledge on supply response to changes in agricultural prices. In backward economies there is even the possibility that increases in the relative prices of agricultural products may lead to decreases in market supply if peasants eat more of the products themselves. Even in the advanced countries — for example, the United States — it seems that supply response is very low for agriculture as a whole. Professor Stipetić asked what the total supply response is for agricultural products in Poland. Are there differences in the supply reactions of peasant holdings as opposed to socialist enterprises? If so, what reasons can be put forward to explain this?

Professor Tepicht said that Professor Papi's criticisms of Professor Pohorille's paper raise the question as to how far the system of contracts is like a watch movement repeating the same pattern again and again or how far it follows, or prepares the way for, developments in the national economy as a whole. In many if not all socialist countries, the contract system had been started at times when there was insufficient supply of agricultural products of all kinds. Production contracts are then a necessary device of governments in the race to obtain the essential production levels and in the efficient allocation of the means of production among farmers. The situation has radically changed in many of these countries. Certain products have already reached their ceiling; thus there is little chance of increasing the production of sugar beet, tobacco and some other products, although the ceiling of production in meat and cereals is still distant. If certain sectors of the market are thus becoming saturated, the relationship between the interested parties is changed. The central authorities who used to run after the peasant to get his agreement are now pursued by the peasant in search of contracts. The contract system, under these circumstances, can be used more explicitly as a tool for structural reorganization and regional changes.

In Poland, at present, purchasing conditions for a particular commodity are uniform throughout the year and, in most cases, throughout the country, to farmers of different types and in different regions. In these conditions, prices are a rather blunt instrument of policy. In any case they form only one of the instruments at the disposal of the government. Production contracts can become one of the main tools for the modernization of agriculture — and are particularly important as long as the agricultural sector remains predominantly one of peasant agriculture. Thus

444

the introduction of compound feeds and the dissemination of all kinds of advice and improvements are connected with the production contract system. Such changes are not peculiar to socialist countries ; as in the case of vertically integrated organizations in the non-socialist countries, the system of contracts goes hand in hand with improvement and specialization. Contracts are linked with complementary undertakings, so that state industry and the farmers are brought into close and flexible *rapport*.

Professor Plotnikov said that he had been greatly interested by Professor Pohorille's investigation of the contract system as applied to planned agriculture. The system facilitates the stimulation of agricultural production, the encouragement of farmers to produce specific kinds of agricultural products and the creation of organized and closely integrated relations between agriculture and industry.

A system of contracts in agricultural production had been in use in the Soviet Union for more than forty years. It was applied on a large scale particularly in industrial crops, milk and some other products.

Professor Pohorille referred to expenditures and purchases of certain products through the contract system that are greater than purchases through the free market. Would Professor Pohorille provide figures showing these differences ?

Professor Pohorille had argued that between the many contracting organizations there exists a form of undesirable competition. Had the attempts to co-ordinate the activities of these organizations with the help of local councils achieved any worth-while results ?

Professor Plotnikov said that he would like to hear in more detail how the formula of 'production-distribution' worked in establishing prices for agricultural products ; and, finally, a more detailed explanation of the role of production costs in setting prices.

Professor Robinson said that he wished to return to the problem of prices for agricultural products which had been brought up by Professor Papi. Professor Pohorille's paper — along with the other papers from socialist countries — raised questions that were familiar to many non-socialist countries in the administration of their price support policies. Professor Robinson said that he agreed with Professor Pohorille that the point of departure is the distinction between the problem of the general level of agricultural prices and the problem of price relations as between separate agricultural products. Farmers have to make a choice among various possible ways of using the factors of production at their disposal, and they base their decision not on absolute profitability, but on the relative profitability of the farming branch under consideration. A number of European countries had run into the problem of increasing regional specialization in milk, meat and other products. Around such specializations had developed specialized marketing facilities which generate external economies for the major regional agricultural activities. If a surplus in the product of such a specialized region emerges it creates an acute policy problem. Use of changes in relative prices to eliminate the surplus is bound, in these circumstances, to have a profound effect on all the farmers of particular regions who are ill-adapted to move into alternative lines of production. Certainly the desired result cannot be achieved overnight. Only by large price differentials and powerful incentives would such farmers be persuaded to reallocate resources away from the regionally specialized productive activities. The use of price support policy to achieve such changes would have a strong impact on the incomes and general welfare of a region. Since price support policy is in part aimed at realizing income and welfare objectives, there would be a continuous and serious conflict between the objectives of policy. Furthermore, if prices are used primarily as incentives to get a desirable balance of different products, serious political pressures are likely to emerge ; and to these politicians are highly sensitive.

Professor Robinson asked whether these policy conflicts arise in socialist countries. Are the present farms so 'multi-product' that they can easily switch from surplus to deficit products ? Have the planners in these countries faced these problems ?

Professor Pohorille said that Professor Papi had asked whether the paper and the system it described are to be taken as only dealing with price stabilization problems or whether they are also to be regarded as including attempts to stimulate agricultural production. He did not overestimate the role of the contract system. The contract system is not a panacea which could solve all the problems of imbalance between supply and demand. These major problems of balance are a function of the general conditions of economic growth and change and would require a variety of policy measures to deal with them. If an increase in demand called for an acceleration in the rate of growth of supply, measures would be needed to stimulate investment in agriculture and to promote intensive technical improvement. However, none of this diminishes the importance of the part played by the contract system in mobilizing supply and adapting production to the needs of consumers. There are three major instruments of planning : investments in the agricultural sector ; the system of contracts ; and the system of prices. The first is as important as the last two are. Professor Pohorille said that he thought that his paper had made clear some of the important ways in which the three instruments are linked.

Referring to Professor Papi's question whether a market for agricultural prices really exists in socialist economies, Professor Pohorille said that the arguments of Hayek and von Mises concerning the impossibility of setting a rational price system under socialism were proved wrong a long time ago.

It was theoretically established that it is possible — by successive approximations — to set up a price system in accordance with the plan and serving its optimization. He did not want to examine these problems in a more detailed way because it would take him too far away from the topic of his paper. He accepted that the market he referred to is not 'free' in the classical sense. He said that he wondered whether Professor Papi could, in fact, point out a country where there is such a free market for agricultural produce.

The free market in Poland has been mentioned in one sense only — the prices are not fixed by the state. This market is a marginal one and of very little significance. It consists of direct sales to consumers made up of the non-agricultural rural population. Sales on this market consist of less than 20 per cent of total transactions in agricultural products. This local free market has peculiar features : there are no middle men ; it is entirely local ; and all the transactions are retail.

A more important question is the possibility of domination of the market through which most of the produce moved — that is the market of government purchases. It had been asked whether the farmers are dependent on the government and its arbitrary decisions. A number of facts rebut the view implied in this question. Despite the decisive part played by the state, it does not hold an absolute monopoly, as the existence of the local free market testifies. If the prices that are offered by the government do not enable the farmers to make a reasonable return there is no guarantee that planned production will be forthcoming. In any case producers are carefully consulted by the government and persuasive pressure is used by both sides. Under perfect competition the farmer would as a matter of fact have to sell at prices dictated by the market. It is, after all, true of all Western countries that it had been the farmers dislike of the free play of market forces which had first led to the imposition of guaranteed prices. The farmer in Poland had an alternative choice open to him as to what he should produce. In fact farmers in Poland are in a strong position because of rapid increase in demand for farm produce.

Professor Pohorille returned to the question of the precise role of agricultural prices in the socialist system, and of the effect of the fluctuation of prices on the distribution of national income. In general the price of consumer goods is aimed to correspond to market equilibrium, and price is regarded as the balancing factor between supply and demand. However, planning in his opinion is not to be based on the arbitrary decisions of market forces under a particular condition of income distribution. The planning commission has to take account of individual preferences in a more complete sense, and prices and wages are instruments of the consumption plan. If the state modifies the prices of particular industrial goods, this has an insignificant effect on the real income of workers, and does not prevent the planners from establishing a position of global balance between supply and demand as indicated by the price mechanism. The case of changes in agricultural prices is significantly different. Changes in prices might mean not only a change in the flow of essential commodities which are universally bought, but it would mean a significant change in the distribution of income. Agricultural prices, therefore, hold a special place in the planning decisions. They are the main economic instrument to inform farmers about the planned objectives and the conditions of plan fulfilment.

Professor Pohorille said that Professor Robinson's question raised problems which are fortunately not of practical importance in Poland, since the country has little regional specialization. However, the problem of co-ordination of the different roles played by price is fundamentally significant. Professor Pohorille said that in such cases the only solution seems to be the direct influence on income either by differential taxes or by direct income subsidies.

Professor Pohorille said that he was unable to give Professor Plotnikov the precise figures of distribution costs on the free market and by contracts. The percentage varies from commodity to commodity.

Professor Pohorille said that there is evidence of competition between contracting activities as they affect the farmers. However, researches conducted through People's Councils suggested that the contract system can be co-ordinated to prevent this having undesirable consequences.

Linking the central plan with regional and district plans of agricultural development as well as with the investment programmes prepared on the basis of these plans becomes at present more and more important. The central plan sets up definite frames for the detailed plans covering a region and 'gromade' (the smallest administrative unit in rural areas). Development programmes of 'gromada', besides productive investments and social and cultural facilities, also include long-term contract plans.

With regard to the question of Professor Plotnikov on the role of production costs in setting the prices Professor Pohorille explained why, in his opinion, the Cost of Production formula is of no use in socialist economies. With a system of prices which is mechanically tied to costs there is no possibility of using prices as an operational instrument in influencing agricultural production. Besides, since the differences in costs of production in various groups of farms are very important, the Cost of Production formula does not provide a clear answer how to set the prices. The prices are largely determined by considerations of income distribution and income security. In capitalist societies the guaranteed prices based on the Cost of Production formula have one advantage — they create some feeling of security among farmers; in Poland the farmers do not fear a gradual fall in agricultural prices. The calculation of production costs plays an important but *indirect* part in setting up agricultural prices. Following the dynamics of price and cost changes the authorities may get a picture of changes in the profitability of the different lines of production and of the farmer's likely reactions thereto.

Professor Pohorille said that he agreed with Professor Stipetić that the question of the elasticity of supply is crucial. He said that in Poland research is being undertaken to estimate supply responses. At present it is thought that supply is price elastic, and that some use of price policy would bring forth the expected results. There is little more that can be said at this stage of research.

Chapter 22

AGRICULTURAL PLANNING IN U.S.S.R.

BY

K. P. OBOLENSKI

All Union Research Institute in Agricultural Economics, Moscow, U.S.S.R.

I. INTRODUCTION

THE possibility of planning agricultural production by the state is a considerable advantage of a socialist economy. As is well known, there is practically no country that today pursues a *laissez-faire* policy in the area of agriculture. In almost all countries agricultural production is controlled by the state to some extent, as examples, we can cite the system of the Commodity Credit Corporation in the U.S.A., the policy of a *dirigement* of farming in France, the extensive system of subsidizing agricultural work in England, etc.

As the level of agricultural production is liable to frequent fluctuations while the demand for its produce is not elastic, even the most active advocates of free enterprise are bound to recognize the necessity of state interference in agriculture. Perhaps the most striking example of such a transformation of ideas is given by the current programme of state regulations for agriculture developed by the U.S. Department of Agriculture. Under the conditions of private ownership of land and other means of production the state regulations on agriculture bear, however, a limited nature, and cannot exert a decisive influence upon the trends and the rate of growth of agriculture. Only under socialist conditions can the state take a decisive part in determining the direction and rate of growth of agricultural production.

II. THE BASIS OF AGRICULTURAL PLANNING IN THE U.S.S.R.

In planning just as in other areas of economic activity the principle of democratic centralism is realised, *i.e.* the coupling of state guidance with the broad activities of the working masses. The main direction

449

in the development of agriculture is determined by the state through the establishment of plans regulating the sales of farm produce necessary for the state, plans for capital investments, plans for production and sales of equipment (*i.e.* tractors and implements) necessary for agriculture. The state fixes the prices for farm produce, which is sold by collective and state farms to the state, and prices for manufactured goods acquired by the collective and state farms.

An indispensable condition for efficient organization of planning in socialist agricultural economy consists of planning from below, *i.e.*, directly within the collective and state farms, and in enrolling the broad masses of collective farms and sovkhoz workers for production planning.

It must be admitted, however, that the exceptionally favourable opportunities inherent in socialist agricultural planning have not always been taken full advantage of in recent years. Sometimes purely administrative, economically unfounded measures were taken instead of economic stimuli of planning; principles of socialist planning were violated; and economically erroneous advice and recommendations were given. This manifested itself in a gap between the production plans and the material and technical basis, in the discrepancy between the plans for stock-raising and the forage reserves available, in the incompatibility of prices for some farm products with their production costs, and in the imperfection of remuneration methods (remuneration insufficiently connected with productive results), etc. All this has caused a slowing down of the development of the agricultural economy.

The Central Committee of the Communist Party of the U.S.S.R. brought in, at its March 1965 plenum, fundamental changes in agricultural planning, following from the requirements of the economic laws of socialist economy. They are as follows :

(1) A firm plan was established regulating procurements of agricultural produce over a period of several years ; the plan provides for optimal conformity between national interests and the internal interests of a given enterprise. Thus, the plan for grain purchases has to be firm and stable throughout the planned five-year period, including 1970. A firm and invariable plan for grain purchases was established for each year of the next five-year period and for all republics, regions, districts, as well as collective and state farms. Firm plans were set up for purchasing livestock products and other products in each year of the coming five-year period.

Establishment of firm plans for the purchase of agricultural products in the next five years enables the collective and state farms to determine more precisely the prospects for the development of different branches, the structure of areas to be sown and the composition of the herds, to introduce proper crop rotations, systems of agriculture and livestock farming, systems for management of the whole enterprise, and to provide stability in the development of agricultural production.

(2) With the purpose of further agricultural development and with a view to securing the fulfilment of the established plans for procurement of agricultural products and thus to raise the living standard of the people, it was decided to increase capital investment considerably and raise the output of tractors, trucks, combines and other farming machinery. The established capital investments to be made by the state and by collective farms in the next five-year period (1966–70) totalled 71 billion rubles; state investment in construction of projects for production purposes and purchase of technical means amounted to 41 billion rubles. Approximately this amount was invested in agriculture in the course of the past nineteen postwar years. In the next five years agricultural supplies will total : tractors — 1,790,000 units ; farm machinery — 10·7 billion rubles ; trucks — 1,100,000 (as against 394,000 supplied in the last five years). To fulfil this extensive programme, about eighty new plants and work departments must be constructed and over 4 billion rubles invested in this work over the five-year period.

Thus, a stable material and technical basis is to be created for agriculture, providing ample possibilities for a systematic and balanced development of the agricultural economy.

(3) Large alterations were introduced in purchase prices for agricultural products, which provide for reimbursement of production costs and profitability of agricultural production. These measures of economic stimulation will help to speed up the growth of labour productivity in the agricultural sector, and enable the collective and state farms to put themselves broadly on a self-supporting basis.

But in what manner are the above-mentioned planning principles implemented in the U.S.S.R.? First of all, it must be pointed out that planning from the centre (centralized planning) is coupled with an active participation by the enterprises themselves and by their workers in drawing up a plan. The general instructions and recommendations concerning means and directions for the development of economic life on the basis of the state plan of national economy, and prices for farm products, are worked out by central organizations with due regard to proposals received from the periphery. Formulation of plans for production development takes place immediately at enterprise level with the wide participation of the workers. This manifests itself specifically in the fact that only plans for purchase of farm produce are fixed by the state ; the sown areas, crop yields, livestock and its productivity and gross output of the farm are planned by the enterprises themselves.

Such a division of responsibility in planning between enterprise and state allows a minimum of interference by the state into the activities of the enterprises, helps to develop in all ways the initiative of the collective and state farms, giving simultaneously to the state a powerful means of influencing the economic structure of agriculture and a means of securing a certain fixed amount of foodstuffs for supply to the population.

While formulating plan targets for selling agricultural produce, the following factors are taken into account : economic and natural conditions of farm production ; requirements of its proper specialization and rational organization ; availability of labour, material and technical resources ; and the necessity of avoiding long transport distances for perishable goods. The marketable production of those crops which are used as raw materials in the processing industry, is concentrated, if possible, at a short distance from processing enterprises. The extent of such concentration must correspond, on each collective farm, with the requirements of a rational agricultural management, timely fulfilment of agricultural work, state of forage reserves and the possibility of their further extension. Thus the planning of state purchases appears as a means of introducing the most rational allocation of agricultural production.

In contrast to the countries where the specialization of agricultural production is implemented under the influence of market relations, and where the specialization of areas with commercial agriculture and cattle-breeding proceeds slowly and is complicated, for instance, by the elimination of a great number of small enterprises, in the U.S.S.R. this problem is being solved much more easily through the system of plant purchases, which takes into account production costs, transport possibilities, availability of labour, etc.

Development of scientific bases for such specialization plans constitutes one of the most important tasks of the scientific research institutes of farm economics, in particular of our All-Union Research Institutes of Agricultural Economics. In spite of the fact that the role of the state is limited to a narrow range of the most important indices, the principle of planning runs through the whole agricultural system of production from top to bottom. The achievement of this is, to a considerable extent, due to the fact that enterprises themselves participate broadly in internal planning. In the enterprises there are worked out long-term plans (for five-year periods), annual production and financial plans, annual production targets for sub-units of the state and co-operative enterprises, for brigades and livestock farms. So-called operative plans are drawn up for separate periods of farm operations and work orders are drawn up for separate jobs.

While developing long-term and annual production and financial plans the following types of balancing calculations are applied : balancing of demand for agricultural products ; balancing of labour ; balancing of forage ; balancing of power resources and other technical means; balancing of demands for fuel, lubricants, fertilizers and other materials and of resources to meet them ; balancing of money incomes and expenditures.

Plans within the enterprises are drawn up through broad discussion by workers and active participation by key economists, agronomists, animal scientists, engineers, etc. The principle of democracy in planning is reflected in the fact that agricultural production plans are drawn up for each district, region or republic on the basis of plans of separate enterprises. The planning on a country-wide scale is accomplished by the Ministry of Agriculture of the U.S.S.R. and its local bodies. Another function of these bodies is the settling of differences with local planning organizations, if discrepancies are revealed between local and state planning. As can be seen, the general system of planning is a flexible tool securing observance of state demands on the whole and the wide variety of interests of enterprises in question. It would be erroneous, however, to believe that a system of planning, however perfect it may be, can work well without the use of powerful economic stimuli, which ensure material incentives for each enterprise and its workers, in fulfilment of plans and in obtaining required agricultural produce ; these stimuli include economically sound prices and a rational form of remuneration for agricultural labour.

III. PLANNING OF AGRICULTURE AND PRICES FOR AGRICULTURAL PRODUCTS

In agricultural planning great weight is given to steady improvement of prices for agricultural products, as well as for industrial products which are destined for agriculture. We are far from considering prices as being a universal economic factor of state influence upon agriculture, but their significance is undoubtedly increased through development of market relations in our economy. An essential advantage of a planned economy — in contrast to economic systems with spontaneous price formation — is the possibility of building a price system rationally, differentiated for seasons and zones of the country. All economists realize quite clearly the negative consequences that result from sharp fluctuations of prices for agricultural products that are generally characterized by an insufficiently elastic demand, and from worsening of the terms of trade, *i.e.* from a decrease in the price ratio between articles sold and purchased by agricultural producers — a situation which obtains in almost every capitalist country today.

In the last few years in the U.S.S.R., almost all grain needed by the state, as well as sugar beet, sunflower seeds, raw cotton, fibre-flax, and the most part of vegetables, milk, meat, potatoes and other products were purchased from collective and state farms.

Collective and state farms sell their agricultural products to official bodies in charge of purchases at purchase prices (collective farms) or at fixed prices (state farms). When delivering the products direct to the trade network the farms obtain retail prices minus a trade rebate. These prices are centrally determined and adjusted. For a number of products (vegetables, fruits, berries and several others), if they are used to meet local needs, the delivery prices are fixed by local bodies on the basis of demand and supply analysis.

Of decisive importance for the economy of collective and state farms are the planned delivery prices ('purchase' and fixed prices) for which they sell more than 90 per cent of their marketable produce. The purchase prices are determined in the U.S.S.R. in accordance with objective requirements of price building and state policy in the price area. This price building is done with due regard for the interests of production in agriculture and industry as a whole. The main factors taken into consideration in building up the system of delivery prices are as follows :

(a) Costs of production born by collective and state farms in various branches and zones ;

(b) budgetary allotments by collective farms (income tax) ;

(c) level of net income required to provide for planned rate of growth of production funds and to cover expenditures for social and cultural needs of collective farms.

454

The prime costs of production give an initial basis for determining delivery prices. As used for setting up prices, the production costs for collective and state farms are calculated by means of average indices for a period of several years (3–5 years) since the prime costs are liable to significant deviations in some years.

In short, prices must always cover prime costs and provide for some accumulation. In contrast, under conditions of the capitalist market, with its spontaneous price building, prices often do not cover production costs. For instance, in the U.S.A., if the wages of the farmer himself are imputed at the rate of hired workers, production costs are in most farms higher than sale prices, and the established guaranteed prices are in most cases also below prime costs.

For several products the system of price planning has particular peculiarities. They manifest themselves through different levels of net income that should flow to agricultural producers. The figures given in Table I illustrate the production profitability for some products in collective farms. It can be seen that the prices which were valid in the period 1961–63 provided the greatest profitability for grain production :

TABLE I

PROFITABILITY OF SELECTED CROPS

	Average Costs for 1 Metric Ton 1960–63	Sale Price of 1 Metric Ton (rubles)	Profitability (per cent)
Grain	42	86	205
Sugar beets	16	29	181
Potatoes	38	72	189
Vegetables	65	82	126
Raw cotton	226	389	176

The highest profitability is that for grain. This is because of the importance of this product for the national economy and by the necessity to stimulate its production in ever greater quantities. But for several products the most important consideration is not in the need to differentiate between the levels of stimulation to be given to particular lines of production but the need for different degrees of territorial and seasonal differentiation.

Because of the fact that agricultural production is carried out over a large territory with a great variety of soils and climatic conditions, the prices for such products as grain, sugar beet, sunflower, milk, meat and several others are differentiated by zones. The zonal differences in prices between the highest and lowest levels are : for sunflower seed, 36 per cent ; for sugar beet, 48 per cent ; for beef, 37 per cent ; for milk, 44 per cent ; for wheat, 94 per cent, and for rye, 117 per cent. In view of the fact that with such differentiation it is impossible to take into account fully the natural and economic differences in the production conditions of some areas, where it is economically profitable to cultivate some particular crops, the planning bodies of the union republics and large regions are granted a right to differentiate the purchase prices (within the average prices established for them) by groups of districts and, with respect to cotton, by groups of enterprises as well.

To achieve redistribution of rents generated within the agricultural sector the zonal prices system is supplemented with a taxation system. The two systems serve as a means to create more equal economic opportunities for all collective and state farms ; first and foremost to step up the profitability of those in less favourable production conditions.

For flax, potatoes, pulse and several other crops, common prices are applied for all areas. It is considered that the marketable production of these crops must be carried out preferably by enterprises which dispose of favourable conditions for their cultivation. *i.e.* where a relatively high profitability is secured with common prices.

Delivery prices not only stimulate the raising of the volume of production and the cutting of the production costs but due to their structure they also stimulate the expansion of the area growing the crop and improvement in the quality of agricultural produce. The quality requirements for products are the reflection of standards and various basic norms, which are established with due regard to the use value of products, world standards and production levels achieved.

The main criterion for differentiating the delivery prices, which depend ultimately on social usefulness of agricultural and animal products, is the effect on the national economy of production and consumption of different quality products. Because of the difficulty of measuring this precisely, the degree to which this index is taken into account is different for various products. The delivery prices for wheat are, for example, differentiated according to two basic groups : soft wheats, and durum and hard wheats. To the latter belong wheats containing in their grains more than 16 per cent of gluten (albumen). In view of the bread-making qualities of durum and hard wheats the delivery prices are fixed 40 per cent higher than those for soft wheats. In calculating the payments for grain such factors are taken into account as how well sorted it is, moisture content, quality for storage and some other quality indices. The grain of valuable top-quality types of cereal crops is paid 10 per cent above the prices fixed for corresponding cereal crops. The price for brewing barley is fixed 20 per cent higher than that for soft wheat. The purchase prices for raw cotton are differentiated by two quality groups which are determined as upland (American long-staple cotton) and fine-fibred brands. In its turn, each group is sub-divided in accordance with quality and technological properties of the fibre into types and production brands. Similarly, there is a differentiation of prices for a number of other agricultural products.

In our country there are direct links in development between agricultural producers and trade organizations. The specialized state and collective farms adjacent to large cities and industrial centres supply their produce directly to the retail trade network. Such a method of delivery for farming products is profitable both for the enterprises and the trade network.

In the U.S.S.R. the agricultural producers (collective and state farms) are large enterprises and they enjoy equal rights when entering in contractual relations with trade organizations. Their share in retail prices is very large, up to 80 per cent. This contrasts with the case in capitalist countries where 'vertical integration' is in process and agricultural producers fall into a heavy dependence on trade and industrial companies. Thus, they become the weakest link in the chain of contractors and their share in retail prices decreases.

For the material and technical factors that are supplied to agricultural enterprises the selling prices are established in similar ways to those used to determine prices for agricultural products, *i.e.* with a view to compensating the production expenditures and to providing average profitability enjoyed by industrial enterprises. In the course of establishing prices for industrial products such factors as the quality of products, the technical and economic data of machines, content of active substance in fertilizer, etc., are considered as well. In addition, preferential prices are to be expected for the most economically effective machines, fertilizers and other materials.

The prices for industrial products used in agriculture (machines, fertilizers, fuel) are relatively low. This embodies the state policy directed at raising the profitability of agricultural economy. The achievement of this is partly due to the fact that the turnover tax for this sort of production is low, and for some articles is not imposed at all. Owing to high prices for agricultural products and low prices for factors of production the ratio between them is far better for agricultural producers in the U.S.S.R. as compared with many other countries.

TABLE II

RELATIONSHIP BETWEEN PRICES FOR AGRICULTURAL INPUTS AND OUTPUTS IN U.S.A. AND U.S.S.R.

(volume of agricultural products in metric centners which must be sold to pay off one unit of input)

	Wheat		Milk		Beef	
	U.S.A.	U.S.S.R.	U.S.A.	U.S.S.R.	U.S.A.	U.S.S.R
Wheeled tractor up to						
20 h.p.	227	160	178	77	35	15·3
50–59 h.p.	738	280	577	135	115	26·8
Caterpillar tractor						
35–50 h.p.	1,044	300	816	145	164	28·8
Self-propelled combine	949	512	741	248	147	49·2
Petrol	9·9	9·6	7·7	4·6	1·5	0·9
Diesel oil	5·9	4·2	4·6	2·0	0·9	0·4
Ammonium sulphate	39	23	31	11	6	2
Simple superphosphate	30	11	23	5	5	1
Potassium chloride	13	3	10	·1	2	0·2

Note : For the U.S.A., prices for fertilizers are taken free at the station of destinati (fertilizer mixing station), and for the U.S.S.R. the prices are f.o.c. (free on car) at the stati of departure.

In 1965, as a result of the need to speed up the development of agriculture and livestock farming, the purchase prices for wheat, rye, buckwheat, rice, millet, milk and meat were raised considerably in the U.S.S.R. This took place simultaneously with a substantial expansion of production and an increase in the supply of machinery and fertilizers to agriculture, and realization of other measures aimed at strengthening the productive basis in collective and state farms. Moreover, the most important rise in prices for the crop products named above was made in areas with relatively poor soil-climatic conditions of production.

With a view to increasing, on collective and state farms, the material incentives to increase production and supply of wheat and rye to the state, a 50 per cent price supplement will be paid for each ton of wheat or rye sold over a purchase plan beginning with this year (1965). The economic significance of a rise in purchase prices for farm products consists not only in the considerable increase in the output of these products and the elimination of the previous failure to make profits in livestock farming, but also in the fact that favourable opportunities are created for planning agricultural production on the basis of greater specialization, both within the main natural-economic zones and over the country as a whole.

It must be noted that this rise in purchase prices for agricultural products was carried out without a rise in retail prices, hence without decreasing the real income of the population.

IV. AGRICULTURAL PLANNING AND MATERIAL INCENTIVES

In our country the systems of remuneration of labour, however various they are, proceed in general from the following principles :

(*a*) an objective appreciation of work quality with due regard to the professional skill of the worker, his experience, the intensity of work and concrete conditions of work in each particular case ;

(*b*) scientifically founded rate setting for labour, *i.e.* determination of the amount of work required to carry out one or another operation, or to produce a certain amount of a product of given kind and quality ;

(*c*) guaranteed wages for workers, which meet his needs and the needs of his family, provided that his output-standard is carried out ;

(*d*) equal pay for equal work, *i.e.* for a unit of the same work, or for producing the same product, other things being equal, without regard to the nationality, sex or age of the worker ;

(*e*) incentive remuneration for more productive and efficient labour ;

(*f*) side by side with funds for private consumption, funds for social consumption are also specified ; these being used to improve the living conditions of the population.

In this way, labour wages in our country are related to the amount and quality of work, and to the amount and quality of produce yielded. At present, particular attention is being concentrated on how to improve the ways of relating workers' wages to the fulfilment or over-fulfilment of state plans, the growth of output, decreases in production costs, while ensuring that the interests of workers and those of the state do not remain in conflict with each other but coincide ; the common purpose is to increase aggregate production and cut its prime costs.

The principles listed above underlie the remuneration of labour in collective and state farms. At the same time, both the common features inherent in the social nature of these enterprises, and the essential peculiarities in their productive organization, are taken into account. The difference consists generally in the fact that the workers on state farms receive fixed wages, similar to those in industrial enterprises, but the labour of the collective farmers is paid for by means of an income distribution, based on the quantity and quality of the work of each.

In state farms a single scheme of labour remuneration is applied — a common tariff scale with wages ranging in proportion 1 : 2 for the highest and lowest categories of machine-operators, and 1 : 18 for such categories as manual labourers and labourers who work with horses. Output standards (and standards for work with animals), and estimation of pay for yields and for operations carried out are developed on the state farms and approved in co-ordination with the local trade union committee.

Since 1962 a piece-work and bonus wage system has been generally applied in state farms, the essence of which consists in the pay being determined for each centner of produce or for each 100 rubles of product cost. Pay for each centner of produce or for each 100 rubles of value is determined through division of the planned wage fund (which consists of the tariff fund plus 25 per cent supplement for agricultural and 15 per cent supplement for animal products) related to the planned output for each brigade, detachment and team. Therefore the payments themselves proceed from a plan and are closely connected with it.

When the production plan is over-fulfilled workers receive increased wages. Thus they are materially interested in better production results, in fulfilment and over-fulfilment of the plan. Side by side with basic payment for fulfilment and over-fulfilment of production plans, a bonus scheme exists to ensure that workers are rewarded if agricultural operations are carried out in proper time and are of high-level quality. This bonus can amount to up to 1·5 months' wages of a worker. In addition, there is a bonus scheme to remunerate workers for products yielded over the established plan and for saving of direct costs as compared with the plan. For an over-fulfilment of the plan for gross produce, brigade, detachment, work team or worker receive up to 20 per cent of the cost of the extra produce ; for saving direct costs as compared with planned costs per unit of produce they receive 25 per cent of cost saving in crop production and 40 per cent in livestock production. The bonuses for workers are made proportionally to their wages in brigade, detachment and team. A worker can receive bonuses amounting to the sum of his wages for a six-month period.

In the state farms managing personnel and specialists receive monthly pay, the rate being determined in accordance with the volume of production. They receive bonuses for over-fulfilment of planned output and for yielding an over-plant profit. Managing executives and specialists of farm departments and production sections receive bonuses for over-fulfilment of plans and cutting of cost prices for produce. The bonuses are distributed among executives and specialists of state farms and of their production units in proportion to their salaries.

Such a system of work remuneration stimulates growth yields, raising of crop yields, productivity of animal husbandry, cutting of production costs and hence raising of the profitability of enterprises. For instance, in 1963 when the plan for gross output was under-fulfilled, all extra pay and bonuses for plan over-fulfilments made up only 7 per cent of base earnings in the state farms, but in 1964, as the plans were over-fulfilled by most enterprises these extra payments amounted to up to 26 per cent of the base earnings.

In collective farms a scheme of guaranteed monthly pay is being widely introduced which is based on tariff plans and tariff rates similar in their sizes to those of the state farms ; a piece-rate and bonus system of labour remuneration is used. The earnings of collective farms become more and more closely related to the results of production — to fulfilment and over-fulfilment of plans.

By 1964 about one-third of all collective farms had adopted the scheme of guaranteed monthly pay to remunerate labour of collective farmers. A substantial part of these farms set up the payments close to the level of wages received by state farm workers. There is in progress a continuous levelling in forms and standards of labour remuneration in collective and state farms in the U.S.S.R.

Quite a number of artels remunerate the labour of collective farmers using wage rates based on a unit of production (on a centner of yielded produce), especially in livestock production, and thereby connect the wage rates with final results of production, with fulfilment and over-fulfilment of planned targets. Many collective farms still remunerate the labour of farmers on the basis of work-days with monetary payments and payments in kind on account, in accordance with procedures established in these farms. When remuneration of collective farmers is carried out on the basis of work-days they are still interested in realizing better results in their work and in carrying out obligations to the state, as their earnings grow with the growth of production, with cutting of production costs and raising of profitability of the farm.

For instance, in 1963 the collective farmer received as a supplement for fulfilment and over-fulfilment of plans, an average 8 kopecks on each ruble of base payment ; in 1964 since production results were better, they received 15 kopecks on each ruble of base payment.

The rise in the earnings of state farm workers and collective farmers will be continued as before on the basis of a rise in labour productivity. Simultaneously the tariff rates of the higher categories of state farm workers will be increased, and in all collective farms the guaranteed monthly payment will be raised up to the level of the corresponding wage rates of the state farms. The remuneration of agricultural workers on the basis of final results of production, fulfilment and over-fulfilment of plans, will become relatively more important.

V. CONCLUDING REMARKS

Thus the system of labour remuneration in agriculture will be developed in the direction of higher material incentives for production results, with the help of closer dependence between work results and remuneration of immediate producers. The characteristic feature of labour remuneration in our country consists in the fact that it creates for the worker a direct interest both in efficient operation of his enterprise, and in fulfilment of the national-economic plan as a whole.

To fulfil successfully the plans of production and state purchases of agricultural products, great importance is attached in the U.S.S.R. to cost accounting in collective and state farms. Its essence consists of a commensurability between production costs and production results, aimed at increasing production. To provide for sound cost accounting, planning must include all units of an enterprise (brigades, detachments, cattle-breeding farms, etc.).

Thus the following requirements are realized in the system of agricultural planning in the U.S.S.R. :

(*a*) combination of state leadership with a broad activity of the workers ;

(*b*) combination of state interests with the interests of each agricultural enterprise ;

(*c*) continuous technological progress, and consolidation of material and technical bases of the agricultural sector ;

(*d*) extensive use of economic stimuli (such as prices, systems of labour remuneration and cost planning) with the aim of fulfilling agricultural production plans, raising the output of high-quality agricultural produce and bringing down prime costs.

DISCUSSION OF PROFESSOR OBOLENSKI'S PAPER

Professor Gale Johnson began by pointing out some obscurities in the presentation of Professor Obolenski's paper. He noted that the coverage for the tables in the paper is not complete ; some data were missing, there was no documentation and some of the variables or concepts, such as average cost, were not defined.

Professor Johnson then referred to the statement on page 451 of Professor Obolenski's paper : 'in the next five years agricultural supplies will total 1,790,000 tractors etc. . . .' Professor Johnson said what he assumed is meant by this is that the present plan for that period has set the indicated quantities as targets or goals for 1966 to 1970. Professor Johnson thought that it would be accurate and more believable if stated that way. Anyone who has read in the field of Soviet agriculture over the past decade can remember how many times some of these statements have been made and how frequently they have been wrong. Professor Johnson said that he was referring to some of the statements about agricultural production as well as to the statements about farm inputs.

This example from Professor Obolenski's paper led to Professor Johnson's first main question to Professor Obolenski. He asked what reasons there are for believing that the gaps between the actual planning process and the ideal will be any less than in the past. It is clear from Professor Obolenski's paper that we are not to be left in doubt that socialist planning is superior to the type of interventions engaged in by capitalist countries in the field of agriculture. Professor Obolenski described what might be called an ideal type of planning. The question is whether this ideal can ever be achieved. Professor Johnson noted that Professor Obolenski referred to the possibility of the agricultural plan falling short of this ideal when he said 'it must be admitted, however, that the exceptionally favourable opportunities inherent in socialist agricultural planning have not always been taken advantage of in the last few years'. Professor Johnson cited the constant criticism of Soviet political leaders of the agricultural planning policies of their predecessors. Their repeated critical and damaging evaluations cover all but one — the last — of the thirty-six years of centralized agricultural planning in the U.S.S.R. Professor Johnson said that he did not want his remarks to be interpreted as saying that nothing has been accomplished by the Soviet agricultural system. Substantial output gains had been achieved though a large fraction of these gains were concentrated into the five years from 1953 to 1958. However, little progress had been made in reducing the real cost of Soviet agricultural output. It is in view of these considerations that he asked his question.

On page 455 of Professor Obolenski's paper there was a table to illustrate one aspect of the principles for setting purchasing prices. The relative priorities for commodities were reflected in differential rates of profitability. Professor Johnson said that although there was a date given for average costs, there was no information about the period used to provide the purchase price figures. Professor Obolenski had stated privately to Professor Johnson that the purchase prices were for 1965, however, the introduction to the table refers to 'prices which were realised in recent years . . .' and this surely implies more than the current year. Professor Johnson said that he would be glad to hear more about this.

Professor Johnson said that this table on page 455 was used to illustrate the following important principle : '. . . prices must always cover prime costs and give a certain accumulation' and it was added 'but under conditions of the capitalist market with its spontaneous price building, prices often do not cover production costs'. Professor Johnson said that he did not know if we were meant to assume that socialist planning always leads to prices covering prime costs and more, or that this should be the case, or that it will be the case. Since Professor Obolenski spoke of conditions that have prevailed in the United States, it seems reasonable to assume that it was a factual as well as a normative statement for the U.S.S.R. Professor Johnson said that he wished that the author had included data for beef, pork and milk. As late as 1962, Mr. Krushchev presented information that indicated that prior to the price reforms of that year, the sales prices did not cover the costs of producing most animal products. During the winter of 1964–65 a high Soviet official again noted that livestock prices generally did not cover costs and that it should be no surprise that many collective farms were reluctant to increase meat production since when they did so, net incomes were reduced. As noted by the author, milk and meat prices were again increased this year, but with the increase in grain prices, Professor Johnson said that he wished to know whether, even now, the quite high prices for livestock products would cover average national costs.

Professor Johnson said that his final points and questions related to the table on page 458 of Professor Obolenski's paper, which, once again, lacked data and provided no documentation and which omitted what would have been interesting information on the periods before and during the important price reforms of the 1950's. However, from this table it is clear that the $\frac{\text{output}}{\text{purchased input}}$ price ratios in the U.S.S.R. are high in relation to those in the United States. This implies quite substantial differences in the efficiency with which these inputs are used in the agriculture of these two countries. But from studies made by his colleague, Professor Arcadius Kahan, Professor Johnson was able to state that the relative real income of the farm and non-farm population in the U.S.S.R. and U.S. are approximately the same (*i.e.* the ratio $\frac{\text{average farm income}}{\text{average non-farm income}}$ is approximately equal in the two countries). Professor Johnson asked whether Professor Obolenski anticipated that it would be necessary to have such high $\frac{\text{output}}{\text{purchased input}}$ price ratios indefinitely in order to achieve moderately satisfactory income levels for collective farm members. Professor Johnson said that if Professor Obolenski's answer is in the negative, he would like to know when purchase prices of final output are expected to decline relative to purchased inputs.

Referring again to the table on page 458, Professor Johnson noted the substantial differences in the relative prices of wheat and other grains and of fertilizer in the U.S.S.R. and the U.S.A. Professor Johnson asked how Professor Obolenski explained the differences in the amount of fertilizer applied per hectare in the two countries. He asked whether this has been due to limited supplies of fertilizer, differences in climatic conditions, a lack of complementary factors or for other reasons.

In answer to Professor Johnson *Professor Obolenski* said that at the present time the prices of all farm produce without exception cover production costs (*i.e.* the cost price) both in collective and state farms. Moreover, they are sufficient to yield a clear profit which provides for planned increases in production funds and meets the cost of social and cultural requirements. Up to 1965 animal husbandry was either unprofitable or slightly profitable on most farms ; but, at the time of writing, animal husbandry has become universally profitable because of higher prices paid by the government.

Professor Obolenski said that the discrepancies between the U.S.S.R. and the U.S.A. will not disappear as rapidly as one would wish because of the greater productivity of agricultural labour in the U.S.A. However, this process is progressing as can be seen from the higher rate of growth of the national revenue in the U.S.S.R. From 1950 to 1964 the national revenue of the U.S.S.R. has increased 3·35 times as compared with 1·56 times for the U.S.A.

Professor Obolenski said that Professor Johnson expressed doubts as to the feasibility of fulfilling the agricultural development plan for 1966–70. Professor Obolenski said that it seemed to him that references to the pronouncements of politicians or criticism of the activities of governments have no place in the functioning of this Conference. As for the essence of this question, as an economist, he was personally convinced that the plan can be realized. Never in the history of the agriculture of the U.S.S.R. has a sum of 71 milliard rubles been allotted to agriculture for a five-year period (this sum equals the total amount spent on agriculture over the past nineteen years). Furthermore, measures have been taken to raise material incentives (farm prices have been raised, taxes lowered, pensions granted, etc.) and to strengthen democracy in the administration and planning of agriculture. In addition, it should be pointed out that the U.S.S.R. has had a very rapid rate of development in the past. For example, in 1954–58 the rate of growth of agricultural production was 7·6 per cent per annum. All this gives reason to differ from Professor Johnson's pessimistic views.

Professor Obolenski said that with respect to the planning and administration of agriculture in the U.S.S.R., he had spoken of these sufficiently frankly at the beginning of his paper. But these shortcomings can all be completely remedied, especially once they are clearly understood. They must not be allowed to overshadow the great virtues of a planned economy such as exists in the U.S.S.R.

Professor Dandekar referred to the paragraph on page 450 in Professor Obolenski's paper where it was stated that from March 1965 a 'firm plan was established regulating procurements of agricultural produce over a period of several years'. Professor Dandekar asked what exactly is involved in this purchase plan, and what commodities are specified in it. He also asked whether the purchase plan determines which collectives are to deliver the specified commodities. If not, how are the decisions on the contributions of quantities of commodities from each collective made ?

Dr. Hsai said that Professor Obolenski on page 450 of his paper implied that, up to March 1965 (before which purchase plans had been confined to one year), farmers had found it difficult to make long-term plans and the Soviet agricultural sector had been in a state of uncertainty. However, in answer to Professor Johnson, Professor Obolenski had referred to a growth rate of 6–7 per cent per year between 1955 and 1958. Professor Hsai said he wished to know how this growth rate had been achieved in view of the uncertain conditions endured by farmers.

Dr. Hsai said that the zonal price system and the taxation system amount to a subsidizing of inefficient farm enterprises. Would Professor Obolenski explain how these systems operated ?

Professor Bishop said that he had been intrigued by the figures given on page 458. They suggested the establishment of very favourable terms of trade for agriculture. Are these related to a scheme for the transfer of labour from farms ? Are the high ratios between the price of farm products and the prices of farm machinery expected to increase the rate of substitution of machinery for labour ? What rate of movement out of agriculture does Professor Obolenski expect in the future ?

Professor Georgescu-Roegen said that the table on page 455 of Professor Obolenski's paper showed very high rates of profitability for the agricultural products. If such high rates existed in a free enterprise economy for any commodity, it would be the unmistakable sign of the fact the corresponding sector was either under a monopoly or that there was an acute temporary shortage of output from that sector, which is virtually the same thing. Professor Georgescu-Roegen said that he would like to learn from Professor Obolenski what these high rates of profit represent in the case of a socialist economy such as U.S.S.R.

Professor Campbell asked whether the bonuses of which Professor Obolenski had written on page 460 of his paper are paid in a good season ?

Professor Obolenski in answer to Professor Dandekar said that in the matter of planned purchasing each farm has to make its own balance between different payments. These include payment to members of the collective itself, payment for the quantity of animal feeds it requires and payments for seed grain and other input needs. On the basis of these calculations each farm informs the regional authorities how much of the different outputs it might be able to sell. Higher level planning organs (working on the basis of regional and district estimates) balance the estimates derived from the figures from collectives and the requirements of national economy. From this emerges the precise calculations for each farm and these amounts are embodied in the final plan.

Professor Obolenski referred to Dr. Hsai's questions. The lowering of the growth rate of between 6–7 per cent after 1958 is explained by other reasons than the fact of single-year purchasing plans. However, the five-year planned purchasing period gives farmers a longer perspective for his own planning. On the matter of the tax system, Professor Obolenski said that net income (profit) is the taxed quantity and the tax varies with the level of net income. The poorest farms are exempted from paying taxes. In general, the demands of taxation on agriculture in U.S.S.R. are not great and do not play any decisive role.

Referring to Professor Bishop's questions Professor Obolenski said that he did not think it would be desirable to increase the prices for agricultural machinery. It would worsen the conditions of agricultural production. If prices of machinery were raised, incentives to increase the productivity of labour would be reduced. On the question of rural population migration into the towns Professor Obolenski said that this, in general, developed on a planned basis and the few occasions where it occurs spontaneously are exceptional and not very important.

In answer to Professor Georgescu-Roegen, Professor Obolenski said that in the U.S.S.R. agriculture taken as a whole, incomes are about 50 per cent of those outside the sector ; this provides the opportunity for developing agriculture on a still larger scale. Meanwhile the lowering of prices for food, clothes, etc., are all a consequence of lowering production costs in agriculture. Lowering of such production costs in agriculture is one of the major aims of present policy.

Speaking of the bonus payments Professor Obolenski said that the system was only established at the beginning of this year. They are expected to have favourable effects on agriculture as a whole, and in particular would play a part in raising production and standards in those commodities which are most needed.

Chapter 23

THE U.S. SOUTH AS AN
UNDER-DEVELOPED REGION[1]

BY

WILLIAM H. NICHOLLS
Vanderbilt University, U.S.A.

I. INTRODUCTION

ECONOMIC development (whether national or regional) is a long-run historical process which cannot appropriately be examined without reference to the past. For this reason, the present paper takes 1930 as the historical turning-point between the U.S. South's long period of economic lag and the more recent period in which it has closed much of the gap in material well-being between itself and the rest of the United States. We shall therefore begin by examining the relative status of the Southern economy in 1930, followed by a discussion of the historical, social and political factors which help to account for the South's lag behind the national economy. We shall then turn to a statistical documentation of the substantial spurt in the South's rate of economic development during 1930–60, concluding with a discussion of the job that remains if the South is finally to achieve full economic parity with the other regions of the United States.

II. THE SOUTH IN 1930

By 1930, the United States had already clearly established itself as the world's richest nation. None the less, the Southern states — stretching from Texas and Oklahoma to Virginia and Florida — remained a relatively under-developed region.[2] The extent to which

[1] This paper is a revised version of an article published as part of a recent book, Avery Leiserson (ed.), *The American South in the 1960's*, Praeger, New York, 1964, pp. 22–40.
[2] Where possible, I have used statistics which define 'the South' as the thirteen-state region embracing Virginia, North Carolina, South Carolina, Georgia, Florida, Kentucky, Tennessee, Alabama, Mississippi, Arkansas, Louisiana, Texas and Oklahoma. From an economic point of view, West Virginia might also

the South was disadvantaged relative to the rest of the nation (or non-South) may be illustrated by the following statistics. In 1930, nearly half (43 per cent) of the South's gainfully employed population still remained in agriculture, almost three times the non-South's 15 per cent. On the other hand, only 14 per cent of the South's work force was employed in manufacturing, little more than half of the non-South's 26 per cent. The South's relative urban population (34 per cent) was also only about half of that (66 per cent) of the rest of the nation. With crude birth rates 39 per cent above the non-South and nearly identical crude death rates, the South's crude natural increase (15·1) was twice that of the non-South's (7·7).

Thus, the South in 1930 was characterized by a much heavier dependence on agriculture, a relatively rural population, and a relatively high natural increase of population. All this would not have been so bad if the South's agriculture had been efficient and prosperous. Unfortunately, 46 per cent of its gross cash farm income was derived from cotton, which was produced by techniques which (like most other Southern farm products) made very ineffective use of human resources — a reflection of the relative plenty and cheapness of labour in an over-populated agrarian economy. As a consequence, Southern agriculture contained most of the nation's low-income farm families. Typically, its farm families were poor because each had so little land and capital to work with. This perennial problem was intensified by high birth rates which exerted a constant downward pressure on farm-labour productivity and incomes.

To some extent, population pressure in the rural South was relieved by large-scale migration of its farm people to other regions where non-farm job opportunities were more plentiful. Thus, during 1910–30, the thirteen-state South had lost 2·0 million people by migration to the rest of the nation. Omitting Florida and Texas, which enjoyed considerable in-migration during that period, the remaining eleven Southern states had actually lost 2·7 million by net out-migration during the two decades prior to 1930. In so far as the South had been industrializing, some further relief had been provided for its rural over-population. But the South's own rate of industrial-urban development had been far from sufficient — even in combina-

---

appropriately be included in the South while, for some purposes, Oklahoma — and more recently Florida and even Texas — might appropriately be excluded. Where Delaware, Maryland and the District of Columbia are included here, it is only because of the vagaries of Census classification.

tion with a high rate of out-migration — to solve the region's low-income problem. Furthermore, such manufacturing as the South attracted was largely in such low-wage, low-productivity industries as textiles, furniture and garments which required less capital and lower labour skills than did the more desirable, higher-wage industries.

As a consequence of all these factors, the average Southerner was in 1930 *less than half as well off* as his counterpart in the rest of the nation. In that year, *per capita* income in the South was $322, only 47 per cent of the $624 enjoyed in the non-South.

III. WHY THE SOUTH LAGGED BEHIND

As I have argued at length elsewhere,[1] the South's great lag behind the rest of the United States in rate and level of economic development was largely attributable to historical and cultural factors. Prior to the Civil War, the South prospered on the basis of a dominant cotton economy, associated with a plantation system and slave labour. The consequences were several :

(1) The South embraced an agrarian philosophy which positively opposed industrial-urban development as an inferior way of life.

(2) The South took over from England the 'aristocratic ideal' of a society dominated by large land-holders, with a carefully stratified and rigid social structure in which first the Negro, and later the typical white, had his place.

(3) The South's political structure was based on a narrow electorate which gave disproportionate weight to the economic interests of large planters.

(4) The South's dominant socio-political leadership gradually lost its sense of social responsibility, reflected particularly in its opposition to the advancement of public-school education and industrial-urban development.

With military defeat and reconstruction, the South reacted by re-examining its distinctive value sustem. It made frenzied efforts to industrialize and for the first time established a basis for the public financing of common school systems. However, these efforts were aborted by the modest results of its belated industrialization campaigns and by the use of the battle-cry of 'white supremacy' to defeat the threat of agrarian radicalism, under which the interests of the

[1] William H. Nicholls, *Southern Tradition and Regional Progress*, University of North Carolina Press, Chapel Hill, 1960.

rural masses of poor whites and freed Negroes were beginning to coalesce. Thus, by 1900, the old aristocratic agrarian values were reinforced and restored to their original dominance.

The results were highly unfavourable to the South's general economic development. The large planters were insulated from competing economic forces. The *rural* middle class, which contributed so much to the democratization and economic development of the North and West, was abnormally subordinated in the South's rigid social structure and undemocratic political structure. The South's *urban* middle class was handicapped by the slow growth of Southern cities and by a political system which discriminated strongly against industrial-urban interests. The South's rural leadership accepted a disproportionate number of low-income people as normal and inevitable. It promoted its self-interest in a cheap labour supply by diverting the attention of low-income whites through numerous devices which at least clearly supported their claims to superiority over the Negro race.

This rural leadership also held strongly to the belief that the South's low-income people, whether white or Negro, were poor because they were innately inferior. Thus, it rationalized a policy of inaction towards the improvement of schools and other public services, towards social and political reform, and towards the promotion of industrial-urban development. In such a static and stagnant environment, only by migration to other regions could most low-income people better their lot.

IV. SOUTHERN ECONOMIC PROGRESS, 1930–60

Despite these severe handicaps, the effects of which were still strong in 1930, the South enjoyed a substantial acceleration in its rate of economic development thereafter. The two decades following 1940 were a particularly dynamic period. Millions of surplus Southern farm people were attracted to other regions in which they could substantially improve their economic status. Still other millions were finding it possible to obtain better non-farm job opportunities within the South itself. This new Southern industrial development received its first strong impetus during World War II when, labour becoming almost impossible to obtain at any price in other regions, the South's relatively plentiful labour supply emerged as a major asset.

Since industries now sought out Southern location, not because Southern workers were 'cheap' but because they were of higher quality than those then available elsewhere, the quality of Southern industry was substantially upgraded. In the process, old prejudices (which held that Southern workers were unsuitable for the more skilled industrial jobs) were dissipated, encouraging further movement of industry to the South even after labour supplies elsewhere became less tight. The South's relatively unexploited water, forests and other natural resources also helped greatly to attract chemical, paper and other higher-quality industries to the region. Finally, Southern industrial-urban development was, to a rapidly growing extent, reinforced as a prospering South became increasingly attractive as a market for industrial goods. Let us look at some of the evidence on the impressive extent of this recent regional progress.

(a) Industrial-urban Development

First, the growing relative importance of manufacturing employment in the South may be shown by the following figures :

TABLE I

RELATIVE MANUFACTURING EMPLOYMENT, TOTAL

Year	South %	Non-South %	South as % of Non-South
1930	14·5	25·7	56
1940	14·8	27·2	54
1950	18·4	29·0	63
1960	21·3	29·4	72

Source : Based on U.S. Census of Population data, which includes Delaware, Maryland, District of Columbia and West Virginia in 'the South'.

During 1930–60, as a percentage of the total gainfully employed, manufacturing employment grew from 14·5 to 21·3 per cent in the South but only from 25·7 to 29·4 per cent in the rest of the country. In the process, the South closed a major part of the gap between itself and the non-South in relative manufacturing employment — standing at 72 per cent of the non-South in 1960 as compared with only 56 per cent in 1930.

In this connection, it is also interesting to break down manufacturing employment into non-durable goods (which in general tend to be produced by lower-productivity, lower-wage industries) and durable goods (which on the whole are higher-productivity, higher-wage industries). For 1950 and 1960, relative manufacturing employment was as follows :

TABLE II

RELATIVE MANUFACTURING EMPLOYMENT

Year	Non-Durable Goods *			Durable Goods		
	South %	Non-South %	South as % of Non-South	South %	Non-South %	South as % of Non-South
1950	11·1	12·5	89	7·3	16·5	44
1960	12·5	11·6	108	8·8	17·8	49

* Includes industries not otherwise specified.
Source : As for Table I above.

Here, we see that during 1950–60 the South actually surpassed the non-South in the relative importance of employment in the manufacturing of non-durable goods, the relative of South to non-South increasing from 89 to 108 per cent. Furthermore, while the South still lagged far behind in durable-goods manufacturing, even here it showed a gain from 44 to 49 per cent relative to the non-South, indicating a continuing process of upgrading in the composition of Southern manufacturing.

Second, the South passed during 1930–60 from a predominantly rural to a predominantly urban region, as shown by the following data on relative urban population :

TABLE III

URBAN POPULATION AS PERCENTAGE OF TOTAL

Year	South %	Non-South %	South as % of Non-South
1900	18·0	50·0	36
1930	34·1	66·0	52
1940	36·7	65·7	56
1950	44·0	65·8	67
1960	57·7	74·4	78

Note : Excludes Delaware, Maryland, District of Columbia and West Virginia.

Source : Based on Calvin Hoover and B. U. Ratchford, *Economic Resources and Policies of the South*, Macmillan, New York, 1951, pp. 23–24 ; and U.S. Census of Population, 1960.

Whereas the rest of the nation's population became more urban than rural shortly after 1900, it was only half a century later (during the 1950's) that the South's urban population exceeded its rural population. The rapid acceleration of Southern urbanization may be indicated in the fact that its relative urban population increased during the single decade 1950–60 nearly as much as during the preceding thirty years 1920–50. At the same time, the South's relative urban population moved up sharply from 52 to 78 per cent of the non-South's during 1930–60.

(b) Decline of the Rural-agricultural Sector

As the South was becoming more industrial and urban, it was of course becoming less agricultural and rural. Thus, the percentage of the Southern labour force employed in agriculture — which had slightly exceeded 50 per cent in 1920 — declined as follows during 1930–60 :

TABLE IV

RELATIVE AGRICULTURAL EMPLOYMENT

Year	South %	Non-South %	South as % of Non-South
1930	42·8	14·6	293
1940	34·9	12·8	269
1950	22·9	9·0	254
1960	10·4	5·5	189

Note : Same exclusions as for Table III.

Source : As for Table III above.

It will be noted that the relative importance of agriculture in the South's total employment dropped from 43 per cent in 1930 to only 10 per cent in 1960. While relative agricultural employment was also dropping sharply in the rest of the nation, the decline in the South was sufficiently greater so that, relative to the non-South, the South's agricultural sector employed less than twice as many in 1960, as compared with about three times as many thirty years earlier.

Meanwhile, Southern agriculture had experienced a remarkable change towards greater diversification into farm products of higher productivity. Thus in terms of the composition of the South's total gross cash farm income, the following shifts took place (see Table V).

Case Studies in Agricultural Policy

TABLE V

RELATIVE SOUTHERN AGRICULTURAL INCOME FROM :

Year	Cotton %	Other Crops %	Livestock and Products %	Total %
1929	46·0	26·4	27·6	100·0
1939	26·7	36·6	36·7	100·0
1949	27·2	34·6	38·2	100·0
1960	18·0	39·3	42·7	100·0

Note : Same exclusions as for Table III.

Source : Hoover and Ratchford, *op. cit.*, p. 103 ; and U.S. Dept. of Agriculture, *Farm Income Situation.*

Here, the most striking change was the precipitous drop in the relative importance of cotton in Southern agriculture during 1930–60. In 1930, after reigning for well over a century in the South, 'King Cotton' was still a tyrant, accounting for nearly half (46 per cent) of the region's agricultural income. By 1960, when he accounted for only 18 per cent, 'King Cotton' had clearly been reduced to the status of a constitutional monarch. In substantial part, the production of cotton had become much more efficient, thanks to mechanization and other improved techniques associated with a more productive use of farm labour. Furthermore, by diversifying into other crops, Southern agriculture had become less dependent upon the vagaries of a single crop. Finally, by its notable expansion of livestock and livestock products (particularly cattle and poultry), from 28 to 43 per cent, the South had also found ways to utilize its plentiful labour supply more fully throughout the calendar year, to follow better soil-conservation practices and to raise annual farm incomes significantly.

With its rapidly increasing urban population, the South's relative

TABLE VI

RURAL POPULATION AS PERCENTAGE OF TOTAL

Year	South %	Non-South %	South as % of Non-South
1900	82·0	50·0	164
1930	65·9	34·0	194
1940	63·3	34·3	185
1950	56·0	34·2	164
1960	42·3	25·6	158

Note : Same exclusions as for Table III.

Source : As for Table III above.

rural population also declined sharply, as shown by the figures in Table VI.

These data are simply the converse of those on relative urban population, but are of interest for purposes of comparison with relative agricultural employment, which fell much more rapidly, particularly in the South.

To a large extent, the fact that rural population dropped substantially less than did agricultural employment reflects the general trend towards suburban and rural living by those employed in urban centres. Especially in the South, it also reflects the large numbers of persons who have changed from farm to non-farm occupations without giving up their original residences. While many of these latter persons have considerably increased their annual incomes, substantial numbers (again especially in the South) have probably simply shifted from employment at low wages in agriculture to employment at little higher wages in some marginal non-agricultural activity still in the rural sector. Thus, in recent years, part of the problem of human poverty in the South may well have shifted from its rapidly declining rural-farm sector to its rural-non-farm sector as well as (via migration and unemployment) to the urban sector. Even so, for the most part, declining agricultural employment in the South — whether due to out-migration to other regions or to its own industrial-urban development — has undoubtedly been favourable to raising the incomes of both those who left agriculture and to those who stayed behind.

That a declining agricultural labour force is absolutely necessary for the South's continuing economic development can be illustrated by turning our attention to the available data on the productivity of Southern agriculture, on rates of crude natural increase, and on out-migration both actual and potential.

(c) Productivity in Southern Agriculture

The South's relatively low agricultural productivity in a year as recent as 1949 can be illustrated by the comparisons in Table VII, between one of the best agricultural regions of the South, the Mississippi Delta, with one of the best agricultural regions of the non-South, Central Iowa.

According to these data, the average Mississippi Delta farm had slightly more labour but only one-sixth as much cropland and one-fourth as much land and capital as did the average Central Iowa farm. Accordingly, the average Mississippi farm had a net value of product

TABLE VII

RELATIVE AGRICULTURAL PRODUCTIVITY

Index 1949	Agriculture of		Mississippi as % of Iowa
	Mississippi	Central Iowa	
Average per farm:			
Gross value of product *	$3,029	$9,603	32
Net value of product †	$2,421	$6,052	41
Man-years of labour ‡	1·62	1·42	114
Value of land and capital §	$7,321	$46,197	16
Acres of cropland	36·7	136·7	27
Average per farm worker:			
Gross value of product	$1,870	$6,763	28
Net value of product	$1,494	$4,262	35
Value of land and capital	$4,519	$32,533	14
Inputs per $1,000 of net product:			
Man-years of labour	0·67	0·23	291
Value of land and capital	$3,024	$7,633	40

* Includes all farm products produced, including an estimate of the value of product consumed on the farm.
† Gross value less selected cash production expenses.
‡ Available labour was converted to 'man-years of labour' actually used by adjusting for farm operator's work away from farm, age of operator, and seasonality, quantity and quality of family-labour inputs, defining a man-year as 2,500 hours of actual labour input.
§ Includes value of land, buildings, power and machinery, and productive livestock.

Source : Data derived from Jackson V. McElveen and Kenneth L. Bachman, 'Low Production Farms', *Agri. Info. Bul. 108*, U.S. Dept of Agriculture, Washington, 1953.

only 41 per cent of that of the average Iowa farm. Perhaps in more meaningful terms, the average farm worker in Mississippi, having only one-seventh as much land and capital to work with, produced only about one-third (35 per cent) as much net product as the average farm worker in Iowa. Finally, in order to produce a given amount ($1,000 worth) of net product, Mississippi agriculture used almost three times as much labour, but less than half (40 per cent) as much land and capital, as did Iowa agriculture. Thus, in 1949 — despite decades of large-scale shifts of Southern farm workers into non-farm employment, either in other regions or in the South — Southern agriculture still suffered from an excess labour supply and a deficit of capital, with the consequence of relatively low labour productivity and income.

(d) Major Demographic Trends

In large part, the reason was that the rate of natural increase of the Southern population remained sufficiently high to keep its human reservoir rising in spite of a very large outflow of migrants to other regions. The following vital statistics will amply illustrate the problem :

TABLE VIII

POPULATION FIGURES SOUTH AND NON-SOUTH

(Per Thousand of Total Population)

Year	South	Non-South	South as % of Non-South
Crude Birth Rates			
1930	26·6	19·2	139
1940	25·1	18·4	136
1960	24·6	23·4	105
Crude Death Rates			
1930	11·5	11·5	100
1940	10·4	11·1	95
1960	8·6	9·5	91
Crude Natural Increase			
1930	15·1	7·7	204
1940	14·7	7·3	200
1960	16·0	13·9	115

Source : Hoover and Ratchford, *op. cit.*, pp. 23–24, and U.S. Bureau of the Census.
Excluding Delaware, Maryland, District of Columbia and West Virginia.

These figures indicate that, in both 1930 and 1940, crude birth rates in the South were more than one-third higher than in the non-South, largely the effect of the South's more rural population. At the same time, crude death rates in the South were as low or slightly lower than in the rest of the country, the effects of the South's poorer medical and public health facilities being more than offset by the lower average age of its population. With substantially higher birth rates and approximately the same death rates, however, the South as late as 1940 had an excess of births over deaths per 1,000 population (the crude rate of natural increase) fully *twice* that of the rest of the nation. By 1960, birth rates in the South and non-South had moved close together but, having gained a slightly more favourable relative death rate, the South still had a crude natural increase 15 per cent higher than that of the non-South.

479

If, then, one looks only at the relatives of South to non-South, the results look very favourable to the South's economic development since most of the inter-regional difference in natural increase had disappeared during 1940–60. Such a result would not be surprising since, having become much less rural, the South might have been expected to reduce its birth rate towards the lower level of the more urban non-South, thereby relieving the historic pressures to accommodate the South's excess rural-farm population by out-migration and by internal occupational shifts. Unfortunately, in this instance, the gap in the rate of natural increase was actually closed by an opposite tendency — *i.e.* by birth rates in the non-South rising to the levels which had characterized the South in the pre-war period. This latter tendency may perhaps be attributed primarily to the post-war 'baby boom' in the non-South and to a normal time-lag between urbanization and reduced birth rates in the South. With some recent indications that birth rates in the non-South are beginning to move back towards their lower pre-war levels, plus the probability that natality patterns in the urbanizing South will increasingly resemble national norms, large inter-regional differences in natural increase may well have disappeared for good. Meanwhile, however, the effects of the 'baby boom' in the non-South have been unfavourable to Southern economic development in the sense that the non-South, now providing a larger part of its own labour force, offers fewer job opportunities for the less skilled migrants from the South than the latter have historically enjoyed. As a consequence, the South's further economic progress must depend even more on its own industrial-urban development than in the past.

The extent of the surplus farm labour force in the South as late as the 1950's may be illustrated by the data in Table IX, on replacement ratios of rural-farm males in the working age group 20–64 for low- and high-income farming areas of the United States. The 'replacement ratio' indicates the number of rural-farm males expected to enter the age group 20–64 during 1950–60 — *assuming no net migration* into or out of the given area and no shifts to non-farm employment within the area — per 100 rural-farm males expected to leave the same age group through death or retirement during the same decade. Thus, the figure 168 for the United States means that, for every 100 rural-farm males of working age dying or retiring during 1950–60, 168 were expected (in the absence of migration or local occupational shifts) to become farmers of working age. It further means that, for the farm labour force merely to remain

TABLE IX

REPLACEMENT RATIOS OF RURAL-FARM MALES

Farming Area by Level of Farm-Family Income	Replacement Ratios, Rural-Farm Males of Working Age, 1950–60	Rate of Net Migration, 1940–50 %
United States	168	– 30·9
Medium- and high-income areas (largely non-South)	143	– 28·1
Low-income areas (largely South) :	(200)	(– 33·7)
Moderate	169	– 27·2
Substantial	206	– 34·9
Serious	221	– 37·1

Source : These figures are taken from a Special Report of the Secretary of Agriculture, *Development of Agriculture's Human Resources — a Report on Problems of Low-Income Farmers*, U.S. Dept. of Agriculture, Washington, 1954.

constant rather than increase, about 40 per cent (68 out of 168) would have to find non-farm jobs either locally or (via migration) at a distance.

The above figures are very important, making it clear that, the lower the average family income of a farming area, the higher its natural increase and the greater the excess of its farm labour force, in the absence of out-migration or increased non-farm employment, over previous high levels. While they also indicate that, during 1940–50, net out-migration rates were higher, the greater the extent of rural poverty, they also emphasize the very large magnitude of the downward adjustments still needed in the farm labour force. Thus, even in the medium- and high-income farming areas of the non-South, about 30 per cent of the rural-farm males had to find non-farm jobs for the farm labour force just to hold its own. In low-income Southern agriculture, the corresponding figure was around 50 per cent (100 out of 200). Since Southern agriculture could improve its low productivity substantially only by an even much larger reduction in its labour force — thereby permitting the consolidation of small farms into larger, more mechanized and more efficient units — the need for both increased out-migration and increased local non-farm jobs was obviously enormous. While most Southern states actually succeeded in reducing their agricultural employment by 35–60 per cent during 1950–60, further substantial reductions will continue to be necessary if Southern agriculture is finally to achieve levels of income comparable with its counterparts in the rest of the nation.

Case Studies in Agricultural Policy

(e) High Out-migration Rates

Under these circumstances, while it would have been preferable for the South to have enjoyed a sufficient expansion of non-agricultural employment to keep all of its farm youth within the region, continued large-scale out-migration throughout 1930–60 did serve as a 'safety valve' which made for more rapid regional development than would have been possible in its absence. The extent of this net migration may be indicated by the following figures :

TABLE X

NET MIGRATION PER THOUSAND OF POPULATION

Decade	Whites	Negroes	Total
Florida and Texas			
1930–40	+ 153	+ 55	+ 208
1940–50	+ 704	– 60	+ 644
1950–60	+ 1,657	+ 74	+ 1,731
Total, 1930–60	+ 2,514	+ 69	+ 2,583
11 Other Southern States			
1930–40	– 620	– 459	– 1,079
1940–50	– 1,018	– 1,261	– 2,279
1950–60	– 1,327	– 1,586	– 2,913
Total, 1930–60	– 2,965	– 3,306	– 6,271
All 13 Southern States			
1930–40	– 467	– 404	– 871
1940–50	– 314	– 1,321	– 1,635
1950–60	+ 330	– 1,512	– 1,182
Total, 1930–60	– 451	– 3,237	– 3,688

Note : In this Table, 'the South' includes, in addition to Florida and Texas, the states of Virginia, North Carolina, South Carolina, Georgia, Kentucky, Tennessee, Alabama, Mississippi, Arkansas, Louisiana and Oklahoma.

Source : Data compiled from Everett S. Lee *et al.*, *Population Redistribution and Economic Growth*, Vol. i, Philadelphia, 1957 : and U.S. Bureau of the Census, *Current Population Reports*, No. 247, April 1963.

According to these data, the thirteen-state South lost by net out-migration 3,688,000 people (89 per cent Negro) during 1930–60. However, Florida and Texas differed distinctly from the other Southern states in that they enjoyed persistent net in-migration (gaining a total of 2,583,000 people, almost entirely white) during the same thirty years. The eleven remaining Southern states suffered persistent out-migration, losing 6,271,000 people (53 per cent Negro) during 1930–60 and 2,913,000 during the 1950's alone.

Since less than 25 per cent of the South's population was Negro during 1930–60, the *rate* of out-migration of Negroes was obviously much higher than that of whites. This continuing phenomenon accounts for the gradual reduction in the South's relative Negro population during the last half-century or more :

TABLE XI

NEGRO POPULATION AS PERCENTAGE OF TOTAL

Year	South %	Non-South %	South as % of Non-South
1900	34·3	2·4	1,429
1910	31·6	2·3	1,374
1920	28·6	2·7	1,060
1930	26·1	3·5	746
1940	25·0	3·8	657
1950	22·5	5·2	433
1960	21·0	6·6	318

Note : Same exceptions as for Table III.

Source : As for Table III above.

While the South in 1960 still had three times as large a relative Negro population as did the non-South, this index had been cut by more than half since 1940 and by more than four-fifths since 1900. With this radical redistribution of the Negro population, many Northern and Western cities had by 1960 relative Negro populations greater than that (21 per cent) for the South as a whole and even some of its own major cities. Given its racial attitudes, the South's large loss of Negro population had probably served to ease somewhat the many race-related problems which have so long plagued its general economic development. In the process of even welcoming Negro out-migration, however, many Southerners have too easily overlooked the concomitant heavy loss of white population and have too cavalierly ignored the high cost to the region of losing people of either race when — after having received substantial private and public investments in their rearing, education, etc. — these people have chosen to move elsewhere as they entered their most productive years.

R

(f) Per Capita *Incomes*

In reviewing the changes during 1930–60 in some of the major indexes of Southern economic development, we have found some highly favourable and others much less so. None the less, if we look at the best overall measure of economic development, *per capita* incomes, the very favourable balance becomes clear. In this connection, let us look at the data on real *per capita* income payments :

TABLE XII

REAL *PER CAPITA* INCOME PAYMENTS

(in 1957–59 Dollars)

Year	South $	Non-South $	South as % of Non-South
1929	623	1,335	46·7
1939	665	1,290	51·6
1948	1,200	1,860	64·5
1959	1,652	2,391	69·1

Source : Hoover and Ratchford, *op. cit.*, p. 48 ; and *Survey of Current Business*, August 1962. Data in current dollars deflated by Consumer Price Index.

According to the above table, *per capita* real income increased during 1929–59 by 165 per cent in the South, as compared with only 79 per cent in the non-South. In the process, the South closed a remarkable part of the income gap between itself and the rest of the country. Thus, while the South's *per capita* income was less than half (47 per cent) of the non-South's in 1929, it had risen to 69 per cent by 1959. However, despite these great strides, there has been some tendency in recent years for the South's relative progress to slacken. Upon what, then, does the South's further economic development depend?

V. THE JOB THAT REMAINS

(a) The Income Gap by Residence and Race

In order to understand more fully the problems which still remain, if the South is at least to achieve economic parity with the rest of the nation, it is desirable to break down our inter-regional income comparisons to allow for differences in race and place of residence.

Although somewhat out of date, the following 1949 family-income data, based on the 1950 census, are very instructive :

TABLE XIII

FAMILY INCOME DATA

	% of All Families in Region with Incomes Under $1,000 in 1949			Median Family Income, 1949		
Type of Family	South	Non-South	South as % of Non-South	South $	Non-South $	South as % of Non-South
Rural and urban families :						
All	24·3	11·5	211	2,248	3,330	68
Non-White	43·8	n.a.	n.a.	1,168	n.a.	n.a.
Urban and rural-non-farm families :						
All	17·9	10·5	170	2,622	3,419	77
White	n.a.	n.a.	n.a.	n.a.	n.a.	88 *
Non-White	34·9	n.a.	n.a.	1,389	n.a.	60 †
Rural-farm families :						
All	45·0	19·4	232	1,284 ‡	2,480 ‡	52 ‡
Non-White	69·2	—	n.a.	n.a.	n.a.	n.a.

* Mean rather than median family income, based on an estimate by D. Gale Johnson.
† The author's estimate, based on a U.S. median income of $1,658 for non-white families.
‡ Families of farm operators only. If the families of farm labourers are also included, the median farm-family income for the United States as a whole was lowered from $1,867 to $1,729.
Note : In this Table, 'the South' includes Delaware, Maryland, District of Columbia and West Virginia.
Source : Data compiled or computed from the 1950 Census of Population and from U.S. Bureau of the Census, *Farms and Farm People*, Washington, 1952.

The foregoing family-income data indicate that in 1949, for all families rural and urban, the proportions of families having net cash incomes of under $1,000 were 24 per cent in the South and 12 per cent in the non-South.[1] For the Southern families, however, 45 per cent of those in rural-farm areas, but only 18 per cent of those residing in the non-farm (urban and rural-non-farm) sector, fell into this low-income category. For Southern non-white families, the correspond-

[1] According to preliminary data from the 1960 Census of Population these percentages had fallen to 10·1 and 4·0 per cent, respectively, during 1949–59. This undoubtedly reflects substantial improvement, even though the common $1,000 limit represented (in 1949 dollars) only $825 in 1959.

ing figures were much higher — 69 and 35 per cent, respectively. While the South had about twice as many low-income families (in relative terms) as the non-South, the actual gap was much wider. Typically, the South's low-income families had small children and able-bodied male heads in their most productive years, while their counterparts in the non-South consisted more largely of the widowed, the disabled and the aged, with fewer dependants and lighter financial responsibilities, frequently having savings or other capital resources from which to supplement their low current incomes. Thus, the South's poverty stemmed largely from poorly utilized but potentially more productive human resources, the non-South's poverty representing primarily a welfare rather than a production problem.

Let us now turn to the data on median family incomes in the preceding table. We see first that, for all urban and rural families, the South's median family income was 68 per cent of the non-South's in 1949.[1]   However, Southern non-farm white families (with incomes at 88 per cent of their counterparts in the non-South) had nearly closed the inter-regional gap, particularly if we take into account the higher cost of living associated with the non-South's typically larger urban centres.   On the other hand, Southern non-farm Negro families were only 60 per cent, and Southern farm-operator families were only 50 per cent, as well off as their non-Southern equivalents.   Thus, the above data make clear that, as late as 1949, the South's low-income problem was largely centred in its non-farm Negro population and in its farm population, both white and Negro.   (As to the latter group, it should also be emphasized that — while a much larger percentage of Negroes than whites was in the low-income group — 68 per cent of all low-income farm families in the South were white.)   While changes during 1950–60 may well have ameliorated somewhat the relatively low incomes of Southern farm families relative to those of the non-South, it probably remains true today that further Southern economic progress will largely depend upon further improvement in the South's rural sector.

(b) Towards a Solution

If we are to find a solution for the low-income problems of the South's urban Negroes and its rural people of both races, more

---

[1] According to the author's estimates from preliminary 1960 Census data, this percentage had increased to 74 per cent by 1959.

vigorous public policies will be required. It is largely migrants from these groups who, because of low levels of education and skills, constitute our nation's present alarmingly 'hard core' of unemployment. Among these Negroes and rural whites are additional millions who, although officially recorded as employed in agriculture or other low-productivity occupations, are seriously *under*-employed. Hence, continued emphasis on monetary-fiscal policies to assure expanding non-farm job opportunities and greater public outlays for general and vocational education are essential if the South is to complete its economic renaissance.

Continued efforts to industrialize low-income rural areas will also be necessary. Improvements in public job-information and job-placement services are very much needed to facilitate migration from the rural South. Finally, the resources devoted to supervised public farm credit must be greatly increased to help those who remain in Southern agriculture to reorganize their farms into larger-scale, more mechanized, more productive enterprises. Only through a three-pronged attack — directed at further industrial-urban development, even greater labour mobility and increased farm-capital resources — can the Southern economy at last become a full partner in the nation's further economic progress.

Such a programme will be costly. But, if a modest fraction of the billions of dollars being wasted by our extravagant and ill-conceived agricultural price-support programme were diverted to such uses, the problems of the South's low-income rural people would be far more quickly solved. If the South's Congressional delegations were (like their constituencies) less reluctant to recognize the problems and needs of the new industrial-urban South, more satisfactory federal policies would undoubtedly be forthcoming. But few Southern Congressmen have yet abandoned the old rural traditions which have so seriously hampered the region's economic progress. None the less, under the leadership of President Lyndon Johnson (himself a Southerner), it is to be hoped that they will increasingly support his large-scale programmes to alleviate poverty and to promote education-programmes under which the South will be a major beneficiary.

Clearly, sharp shifts in both racial customs and the rural–urban balance of political power have become essential to the South's further economic development, and are actually under way to a much greater extent than Southerners themselves yet realize. Recent Supreme Court decisions on school integration and legislative reapportionment — while roundly damned by many Southerners —

will probably appear a generation hence to have been landmarks in Southern progress, as will current civil-rights and voting-rights legislation. The principal support for segregating the races has come from the South's tradition-bound rural minorities. The significance of the reapportionment decision is profound, since it will at last give the South's increasingly urban-industrial majorities and its Negroes a voice more nearly proportional to their numbers. As a result, the forward-looking, progressive and dynamic forces which Southern industrial-urban development has already created now promise to erode away the blind sectionalism, the negative and defensive states'-right doctrines, the disinterest in general social and economic betterment, and the race extremism which have so long diverted Southern energies from constructive channels.

The non-economic factors which historically have shackled the South's economic progress are at last in full retreat. The recent sound and fury emerging from the South can easily be misunderstood. It clearly represents the death throes, not the renaissance, of those Southern traditions which are inconsistent with the region's industrial-urban development. In the process, the South is finally creating the environment needed for it to achieve full economic parity with the rest of the nation. I am now confident that, if appropriate public policies are forthcoming, the South can and will achieve this goal.

DISCUSSION OF PROFESSOR NICHOLLS'S PAPER

Professor Jöhr began by asking whether the South of the United States is, in fact, an under-developed region. It is clearly an under-developed part of the United States but we have to bear in mind that the rural income per head in the South of the United States lies above the corresponding values in highly industrialized regions such as Germany and France. Noting that Professor Nicholls took the year 1930 as the point when the South began to close the gap which existed in relation to the economy of the North, Professor Jöhr pointed out that in 1930 the United States was on the threshold of the great depression from which the South also suffered. Other dates could be taken, such as 1922 at the beginning of the so-called golden prosperity period of the 1920's, or 1936 after the great depression, or even 1940 at the beginning of the war boom.

Professor Jöhr summarized the factors which Professor Nicholls had taken to account for the improvements, but asked why the author had not emphasized the attraction for Northern industry of the cheap labour in the South.

In the body of his paper, Professor Nicholls did not make use of the instruments of growth theory, such as the capital-output relation, to explain why the South lagged behind, but rather made use of terms which one might qualify as historical or sociological ; for example, the use of such terms as 'agrarian philosophy', 'aristocratic ideals of society', etc. These arguments seem convincing but they raise a fundamental question for economists : these sociological and historical factors are here given a causal significance, and presumably similar factors may also influence the future development of the South as well as of the North and the future of other countries. Clearly, these factors should be incorporated in a theory of growth, and if economists insist upon using only mathematical methods in any progress in theory, it suggests that economic theory will be refusing to face vital questions for economic development.

It would have been interesting if Professor Nicholls had dealt with the question of causation at greater length. Professor Jöhr suggested, the following lines of discussion. First, is there a mechanism which tends to equalize, in the course of the growth process, the average economic welfare of the inhabitants of two regions with different starting-points ? Second, if there is such a mechanism, of what factors is it composed, under what conditions will it work and under what conditions will there be a process of widening of the gap in relative terms ? Thirdly, can such a theory also be applied to other countries such as Italy in which a similar problem had existed for many decades ? May it also be applied to the world as a whole ? Professor Jöhr said that we had already learned from the paper of Dr. Ojala that we cannot expect that the developing countries will be able to increase their real income per head as fast as the industrialized countries. Can we devise a theory on the basis of Professor Nicholl's paper to help us with this problem ? Finally, it would be interesting to work out the differences in the conditions of economic growth which account for the fact that the developing countries will not, or, at least, are not expected to be able to, follow the example of the American South.

Professor Nicholls seemed to place optimistic reliance on the regional policy he advocated for the South, when he argued that it might achieve 'full economic parity with the rest of the Nation'. Professor Jöhr said that he was not so confident that this result would be achieved if the author really meant by economic parity the equality of the averages of the real income per head of the two regions, and if he excluded the possibility of a radical policy of income redistribution in favour of the South. Professor Jöhr said that he thought there would always remain considerable differences in the average *per capita* income between different regions because of the following factors : the differences in natural resources, the differences of location with regard to transportation systems and the limited opportunities of transforming them, the uneven distribution of entre-preneural energy, the uneven distribution of the quality of labour, the obstacles to migration and finally the differences in climate, which not only affect agriculture and business directly, but also affect the psycho-logical drive of the inhabitants and the specific preferences of rich people as to the region where they would like to live (an important source of demand and taxation). Professor Jöhr said that for these reasons parity would seem to be an unrealistic aim. An approximation to parity might be aimed at, but it might be that all the South could expect at any given moment would be to develop at a pace not slower than the North.

Professor Georgescu-Roegen said that a paper of this type about historical facts is an excellent substitute for a laboratory experiment. Consequently, he would like Professor Nicholls to elaborate a little on the object lessons that we can derive from the rapid development of the South in the United States. A fundamental question is how many other places in the world would fit into the same special category as the South of the United States, given that, at the time when its development began, that region was part and parcel of the most developed country in the world. This spectacular development may be a unique historical event because there already existed a supply of capital internal to the same economy so large that it had been looking even for placement abroad.

Professor Dandekar said that he had found the paper most interesting and had not appreciated before how much experience the United States had had of development within her own boundaries. From this he would expect the United States agricultural economists to have more sympathy with the problems of Asia. Professor Dandekar said that Professor Nicholls had stressed the role of out-migration in spite of the advanced technology available within a short distance. The Southern states had lagged behind in technology and only mass migration had allowed them to pick up. Asian countries with population excesses have no opportunity for out-migration anywhere in the world. Professor Dandekar asked for a more sympathetic understanding of these problems.

Professor Nicholls, in reply to Professor Jöhr, agreed that the concept of under-development was a relative one, but within any large country one had to be concerned with different rates of economic development. The experience of the South of the United States should be of value to any other relatively large country. Professor Nicholls drew the parallel with French Canada where great cultural differences exist between that area and the rest of British Canada. He listed southern and north-eastern Brazil, southern and northern Italy and southern and northern India as other areas where the experience of the South of the United States might be of value. It is true that in absolute terms the development in Arkansas and Mississippi looked good in comparison with India or Brazil, but it is the relative comparisons that are important.

In answering Professor Jöhr's question as to whether cultural differences would continue to effect the development in the South and elsewhere in the future, Professor Nicholls said that cultural differences seem to be fast disappearing. For example, as his statistics show, different parts are becoming more alike in race mixture.

Professor Nicholls agreed that cheap labour had been effective in attracting industry to the South. Historically, this had been a factor since the 1870's, while during the 1920's the cotton industry very largely transferred from New England to the South where wage rates were low. Furniture manufacturing had also moved into the South. So before 1930 cheap labour had been a factor, but thereafter, other aspects of Southern conditions, including the shortage of labour in the non-South, the improvement of marketing possibilities in the South and the higher level of *per capita* income in the South which attracted market-oriented industries, had been more important in development.

Professor Jöhr had been sceptical about an optimistic view of full economic parity between South and North. Professor Nicholls said that he believed that parity is possible because the South itself is so large a region. Even now, urban white population in the South has achieved parity with the urban white population in the North as far as *per capita* income is concerned. The problems which remained are the lower *per capita* income of urban Negroes and of workers employed in agriculture.

Replying to Professor Dandekar, Professor Nicholls said that, although it is clear that the factor of out-migration acts as a safety valve, it is, in itself, not sufficient to make the situation better. He said that in the case of the South, out-migration began about 1810. The South had been one of the earliest settled areas and out-migration from it had been started very early. Thus it is clear that out-migration alone will not close the gap between the under-developed and the developed regions. Professor Nicholls said that the question could be broadened and related to Professor Bishop's paper. The important policy consideration was to choose between the alternatives of moving men to jobs as against moving jobs to men. His own reading of Southern history was that one could not rely on out-migration but had to bring industry to the region and so raise *per capita* incomes.

Professor Dandekar referred to Professor Nicholl's calculations in the Table VII on page 478 on average productivity figures of agriculture in Mississippi. He said that the figures gave comparisons between Mississippi and Iowa, but did not include a comparison between the two regions with respect to output per acre of crop-land. Professor Dandekar said that by his calculations based on the land per working man given below the table he referred to, output per acre of crop-land is higher in Mississippi than in Iowa. By this measure, Mississippi agriculture is more productive than that in Iowa. Professor Dandekar said that he would prefer to see Professor Nicholls call agriculture in Mississippi more productive and interpret the labour productivity figures as resulting from an excess supply of labour.

Professor Nicholls said that while he accepted Professor Dandekar's calculation on the higher Mississippi value yield per acre, he still thought that high yield per man is the ultimate goal of development and that this is where India, for example, failed. It might be that, for countries such as India, the only route to achieving an increased yield per man would involve raising yield per acre in the process but the two did not necessarily go together, particularly if there were enough non-farm job opportunities.

Professor Delivanis said that one of the great merits of the paper was that it drew attention to development within a free trade area. We might, therefore, get indications about the eventual development in Europe if it ever became a completely free trade area. Professor Delivanis thought that Professor Nicholls was right in stressing the importance of emigration from the South as a safety valve. It seems quite clear that any local shortages of manpower can easily be made up when increased capital investment arrives. Meanwhile, the easing of the over-population burden by emigration can make room for effective planning of a labour policy and render labour legislation more efficient.

Professor Delivanis said that he wanted to know more about the effects on rural population and agricultural employment of the changes that occur in the economic structure of regions when, for example, people in Iowa live permanently outside cities. In the U.S.A. the tendency to live out of town has clearly developed greatly. What were the implications of this for development ?

Mr. Tracy said that Professor Nicholls had spoken about the contribution made by agricultural price support programmes to the development of the South. This seems to be in contradiction with his statement near the end of his paper about the 'billions of dollars wasted by our extravagant and ill-conceived agricultural price-support programme', and his statements about the desirability of diverting this money into other uses, for example, for promoting industrialization in low-income areas. Mr. Tracy said that he would be grateful if Professor Nicholls would elaborate on this apparent contradiction.

Professor Bishop asked if it was not correct that total government outlays for agricultural support had been relatively greater and greater per family farm in other regions than in the South. He also asked whether Professor Nicholls really felt that income parity in the South could be achieved within agriculture *per se*.

Professor Nicholls said that he did not really feel ready at this point to speak on the question of rural residences raised by Professor Delivanis. The development of these is a result of automobile use and of well paved roads. It is, in part, connected with the development of part-time farming — the combination of factory jobs with small holdings — which may lead eventually to the abandonment of work on the land so that then the farm will become simply a rural residence. An interesting consequence of rural residence is that it may involve a lower total demand for many services, because people spend their leisure under rural conditions. He said that this applies east of the Mississippi.

Professor Nicholls said that Mr. Tracy's question and Professor Bishop's question were related and he would answer them together. There was no inconsistency in what he had said on price support policy since any transfers are bound to help the region. However, he was arguing that there are far better ways for the Federal government to help, and more important ends to serve. The fact that price supports *per capita* and per family are lower in the South than in other parts of the United States reflects the size of operating unit in the South, which is smaller than elsewhere ; this is one of the objections to price support policies ; the benefits are uneven, and the top ten per cent of farmers in terms of farm size receive over 50 per cent of the benefits from price support.

Perhaps his paper was a bit misleading in suggesting that the Federal government had little to do with closing the gap in *per capita* income between North and South. However, he thought that the role of government could be over-emphasized ; for example, the Tennessee Valley Authority contributed considerably to the Southern economy but less than foreigners were prone to think. However, it is true that, of the six crops given special preference by the Federal government in price-support policies, four of them (rice, cotton, tobacco and peanuts) are Southern crops. Also large income transfers in education and public health have certainly been of assistance to the development of the South.

On the question of the applicability of the study to other parts of the world, Professor Nicholls thought that there was much that could be generalized, as long as one looked at such inter-regional problems within countries of large geographical area. As Professor Georgescu-Roegen had pointed out, the United States South has substantial advantages over under-developed countries ; it is part of a rich and dynamic nation, part of a developed Federal system, and can benefit from other richer regions which can help to supply non-farm jobs to its surplus labour. Other advantages could be listed ; it enjoys a common language and laws with the rest of the nation, there are no barriers to income transfers, no tariffs, no immigration barriers and so forth. Certainly the United States South is in many ways, unique, but the same principles could be applicable to similar differences *within* under-developed countries.

PART VI

CONCEPTUAL PROBLEMS IN THE ANALYSIS OF AGRICULTURAL DEVELOPMENT

Chapter 24

PROCESS IN FARMING VERSUS PROCESS IN MANUFACTURING: A PROBLEM OF BALANCED DEVELOPMENT

BY

NICHOLAS GEORGESCU-ROEGEN
Vanderbilt University, U.S.A.

I. INTRODUCTION: THE ASYMMETRY OF A SYMMETRY

1. It is Wicksteed who first pointed out the 'fascinating' analogy between the laws of satisfaction and those of production.[1] A trivial idea by now, the formal identity between consumption and production theory comes from the fact that in both cases the main problem is one of maximizing an ordinary function of several independent variables subject to a budget constraint. In consumption, it is the utility,

$$u = \varphi(\xi, \eta, \zeta, \ldots),\qquad(1)$$

that must be maximized within a given income ; in production, it is the maximization of the output,

$$q = f(x, y, z, \ldots),\qquad(2)$$

for any given outlay. And there seems to be no greater delight for the author of a text-book than to comb every formal symmetry between indifference curves and isoquants on the one hand, and between the marginal utility and the marginal productivity theorems on the other. Yet, should one compare the march of ideas in the micro-theory of consumption and that of production, one would not fail to be surprised by the marked contrast between the two.

An unending series of contributions have taken us from the Gossen–Jevons conception of utility as a cardinal magnitude depending separately on the amount of each commodity, over an ever-changing panorama of consumer's theory. There came in succession

[1] Philip H. Wicksteed, *The Co-ordination of the Laws of Distribution* (Reprinted as No. 12 of the Scarce Tracts in Economic and Political Science, London School of Economics and Political Science, London, 1894), pp. 8 f., 48.

Edgeworth's complementarity, Pareto's ordinal index of ophelimity, Samuelson's revealed preference, then a new cardinal utility, and in more recent time the view that the preference field cannot even be completely represented by a Dirichlet function such as (1). Certainly, that is not the end of this most elusive problem.

On the other hand, ever since the essential idea behind Walras's coefficients of production found a more adequate expression in (2), this form of representing a productive process has constituted the most solid corner-stone of production theory. To be sure, our present notion of a production function differs from that of Wicksteed when he first introduced the concept in the work cited above. But the difference is rather superficial. As a result of the contributions of Pareto and others after him, we have become aware of the fact that (2) does not cover all situations : joint products and limitational factors require amended models. But these models too involve only Dirichlet functions. And though careful writers discriminate now between the production function of a unit of production and that of an aggregate of such units, the distinction actually tends to strengthen the position that (2) is the basic analytical representation of a productive process.

2. But no student of production theory, it seems, felt the need for raising the same kind of *epistemological* questions about the production function as those that have continuously tormented the students of consumer's behaviour. This is the contrast to which I have referred above. It has, I submit, a very natural explanation. The servants of quantitative economic analysis, especially the mathematical economists, have bent their efforts to defending that sector of their theoretical citadel where the opposite camp has always concentrated its attack ; for the suggestion that human actions can be analysed into laws of mathematical rigidity had been decried long before economics came out with a mathematical theory of utility. On the other hand, even the fiercest enemies of mathematical economics have never challenged the use of the mathematical tool for studying production : after all, production being a physico-chemical process must run according to the rigid laws of nature. And from what one can infer, economists too have concurred in that production theory, for one, cannot possibly upset the apple cart of mathematical economics. There is no need to search further for an explanation of their lack of interest in a critique of formula (2) as an analytical representation of a productive process.

It is precisely such a critique that led the author to the results presented in this paper. These results include, first, the conclusion that there are two analytically distinct types of productive processes, neither of which is described completely by the traditional production function. Second, they will expose the fallacies of the rather widespread practice of conceiving the economic process as a circular flow. As an immediate and perhaps a most eloquent application of these results, they will be used in analysing, from a yet untried viewpoint, the balance between agriculture and industry in economic development.

II. THE PRODUCTION FUNCTION: A CRITIQUE

1. One difficulty of a student approaching for the first time the subject of consumer's behaviour is the vast literature covering the various meanings of the function (1) as well as the experiments by which one could conceivably construct the particular 'utility' function of a given person. On the other hand, the discussion of the production function often amounts to no more than a repetition of Wicksteed's initial theme: 'The Product being a function of the factors of production we have $P = f(a, b, c, ...)$.'[1] In some of the not too recent works one finds a more or less comprehensive discussion of the nature of the production factors, that is, of the elements supposed to be represented by the symbols $x, y, z, ...$ But ever since 'output' and 'input' displaced the classical terminology, most economists, it seems, have felt that the immediately obvious etymology of the new terms renders superfluous any further explanation of what corresponds in actuality to these symbols. Indeed, search as one may, even in the works of the most respected modern authorities one finds hardly any discussion of the empirical scaffold of the production function. All seem satisfied with some purely formal definition, some variation on Wicksteed's theme.

For a representative sample, we may cite, first, Boulding's definition:

the basic transformation function of an enterprise is its *production function*, which shows what quantities of inputs (factors) can be transformed into what quantities of output (product);[2]

[1] Wicksteed, *op. cit.* p. 4. This cavalier treatment is characteristic of most textbooks nowadays. Cf. A. W. Stonier and D. C. Hague, *A Textbook of Economics* (2nd edn., London, 1958), p. 119, or Richard H. Leftwich, *The Price System and Resource Allocation* (rev. edn., New York, 1960), p. 108.
[2] Kenneth E. Boulding, *Economic Analysis* (3rd edn., New York, 1955), p. 585. This definition represents the most popular variant. Cf. A. L. Bowley, *The*

second, Stigler's :

> A production function may be defined as the relationship between inputs of productive services *per unit of time* and outputs of products *per unit of time.*[1]

and, finally, Carlson's, according to which the production function is

> the relationship between the quantity of input bought at the beginning of the production period and the quantity of output turned out by the process at the period's end.[2]

We find no attempt to supplement such formal definitions with some operational instructions about how to determine a production function at least under hypothetically ideal conditions. Boulding likens a production process to a recipe from a cook-book. Just as one such recipe says that 'if we put 4 eggs, 2 cups of milk . . . in a waffle mixture and obey the instructions, we should expect to get ten waffles out of it, no more and no less', so the cook-book of the iron manufacturer tells him that if he 'mixes so much ore, so much lime, so much coke, and so much heat for so many hours', he will get 'so much iron'.[3] But the widely shared position that it is not the business of the economist to know more about the relationship between output and inputs is more directly expressed by Stigler, who says that production functions are 'taken from disciplines such as engineering and industrial chemistry : to the economic theorist they are data of analysis'.[4]

2. But is it not strange to think that the economist can analyse these data without knowing more about their nature than that q is 'output' and $x, y, z, \ldots$ are 'inputs'? In fact no sooner have we proclaimed that all we need to know is that a relation such as (2) exists, that we

---

Mathematical Groundwork of Economics (Oxford, 1924), pp. 28 f. ; J. R. Hicks, *The Theory of Wages* (London, 1932), p. 237 ; Erich Schneider, *Theorie de Produktion* (Vienna, 1934), p. 1 ; A. C. Pigou, *The Economics of Stationary States* (London, 1935), p. 142 ; Paul A. Samuelson, *Foundations of Economic Analysis* (Cambridge, Mass., 1948), p. 57.

In view of the argument to be developed in the following section, we should note that some of these authors — Boulding (p. 201), most explicitly — identify the quantities of inputs with the *services* of the production factors.

[1] George J. Stigler, *The Theory of Price* (New York, 1949), p. 109 (italics added). The treatment of the production function in the 1952 revised edition is lowered to the level of the ordinary text-book and thus is quite uninteresting for the present discussion.

[2] Sune Carlson, *A Study on the Pure Theory of Production* (New York, 1956), p. 12. Carlson uses 'input' as a plural noun.

[3] Boulding, *op. cit.*, pp. 585 f.

[4] Stigler, *op. cit.*, pp. 109 f. For the general agreement on this point see Hicks, *op. cit.*, p. 237 ; Pigou, *op. cit.*, p. 142 ; Samuelson, *op. cit.*, p. 57 ; Carlson, *op. cit.*, p. 10.

turn to discuss such specific structural properties of the production function as the convexity of isoquants or the laws of returns to factors and scale. The space occupied in economic literature by the problem of whether or not there is an optimum size of the unit of production plainly proves that economists have never been consistent as regards what they need to know about the production function. All the harder is it then to understand their manifest indifference toward the foundations of this concept.

A rather piquant consequence of this indifference is the fact that no one seems to have noticed that the Boulding and the Stigler definitions cannot be equivalent unless all production functions are homogeneous of the first degree. Yet, the point is most elementary : Let q, x, y, ... denote the flow rates of output and inputs so that (2) should correspond to Stigler's definition. Clearly, then, during any time interval t the quantity of product is tq and the quantities of inputs are tx, ty, ... According to Boulding's definition, we must also have $tq = F (tx, ty, ...)$. Hence, $f \equiv F$ is a homogeneous function of the first degree, as said.[1]

3. A series of issues — some immediately obvious and familiar to all of us, others brought to light as a surprise by a highly interesting study of Chenery [2] — show how foolhardy is the belief that the production function which the economist needs for his special job comes ready-made from the technical 'cook-books'.

3. 1. The first issue, a very familiar one, hinges upon the qualitative variations of some inputs, especially of the human inputs. Technical blue-prints hardly ever include the administrative, often not even the technical personnel. No chemical formula includes the chemist himself, not to mention the bursar from whom he gets his salary cheque. Engineering blue-prints, therefore, can throw no light upon the problem of the influence that the quality of the entrepreneur, of plant managers, or even of foremen, has upon the amount of output obtainable from the same set of material inputs. The engineer's concern is the efficiency of machines, not the efficiency of 'human inputs'. But in the actual process it is man who operates the machines, who responds actively to their signals and, above all, who keeps the chain of the process going. And at least the economic theorist should know better than to be satisfied with the answer that technical blue-prints pertain to the results obtainable with the

[1] We shall see later on (*IV*. 4) the reason why a great confusion, closely related to the point just made, reigns over mathematical economics.
[2] Hollis B. Chenery, 'Engineering Production Function', *Quarterly Journal o, Economics*, lxiii (1949), 507–531.

'average' quality of human inputs.[1]   What this average is in each activity sector is one important outcome of the general economic process ; hence, for the economic theorist it is an item of explanation, not a datum.

But one may interject that non-quantifiable qualities — such as those, especially, of the persons who do not directly operate machines— cannot in any case be included in the function (2) because all variables must first of all be measurable.[2]   This is indeed correct, but if and only if we insist that (2) should be a Dirichlet function.   But why should the economist bide by this restriction — legitimate though it may be in engineering — if the economic problem involves also qualitative variables?

For a suggestion of how the restriction may be avoided, let $q = F (x, y, z, ...)$ represent the engineering production function of a plant and let us assume that the relation pertains to an *ideal* process. (An illustration of such an engineering production function is the theoretical formula $2nH + nO = n$ molecules of water.)   Let now $q = E (m_i ; x, y, z, ...)$ be the corresponding *economic* production function of the same plant.   Here, m_i, $i \epsilon H$, denotes a particular qualitative unit of 'plant manager'.   There is no need for H, the set of all such qualities, to be a linear continuum and, hence, for i to be a number.   But since there is no *ideal* manager, the difference $F - E$ is necessarily positive.   The smaller this difference, the better is the managerial quality of m_i.

True, one could no longer speak of the partial derivative of E with respect to m_i, but the differences between the values of E for m_i and m_j, $i \neq j$, would still have a meaning useful in analysing the pricing process.   Equally true is the fact that from the operational viewpoint E has an important shortcoming : there is no way to determine exactly and *ex ante* what is the result of substituting m_j for m_i.   But the construct can serve the purposes of economic analysis in the same way in which the system of arbitrary (unspecified) functions do, or the circular concept of 'the fittest' benefits biology.[3]   We should not

[1] More often than not, however, the technical blue-prints pertain to what happens in the 'test tube' or, at most, under laboratory-like conditions.   That is why, if statistical data are sufficiently reliable, the input coefficients computed from them describe more adequately the actual situation than those supplied by the blue-prints.

[2] This opinion is emphatically expressed by Samuelson, *op. cit.*, p. 84, and R. G. D. Allen, 'The Mathematical Foundations of Economic Theory', *Quarterly Journal of Economics*, lxiii (1949), 116, n. 6.   Pigou (*op. cit.*, p. 137), however, argues that leaving out of account the qualitative elements of the problem does not affect the relevance of economic analysis.

[3] One example of the analytical service of the economic production function

overlook the fact that the theorem that maximization of output with constant outlay requires that the marginal product of the 'dollar' be equal for all inputs has been proved without implying that inputs are cardinally measurable. And if the issue of the indivisibility of m_i is critically raised, we should observe that the same issue calls for amending the above theorem even in case all factors were measurable.

3. 2. The other issues, though less familiar, carry more weight because they do not depend upon how one views the role of economics. To begin with, an engineering production formula — as Chenery reminds us [1] — describes an isolated process (say, a refrigeration system or an electrical transformer), not the full process of a plant. Secondly, in such a formula most outputs and inputs are not commodities, that is, entities with which economics is concerned. Instead, they represent physico-chemical properties : pressure, temperature, stress, specific gravity, pH, viscosity, etc. etc.[2] More-over, an engineering formula may not be available for every individual process within a plant. But even if there is such a formula for every process in a given plant, to obtain the engineering production function for the whole plant may be well-nigh impossible. To transform an engineering formula into a functional relation between commodities is not an easy task. Nor are we certain that it is always feasible. There is also the staggering problem of covering all the innumerable combinations of engineering processes by which a given product may be produced.

More important still is the point that the choice of the processes studied by engineering — and for which alone there exist engineering formulae — is narrowly determined by the price constellation of the production factors. As a result, engineering formulae are available mostly for processes that can be used economically only by economies rather advanced in development. Consequently, even if one could derive an engineering production function for every product, the information would not cover all possible factor combinations.

4. The preceding critique is not meant at all to deny that the physico-chemical laws constitute the backbone of any material production. Nor should one interpret it as a belittling of the scientific value of engineering economics — a field which is now open to econometricians thanks to the pioneering work of Tjalling C.

---

suggested above is offered in the author's paper, 'Measure, Quality, and Optimum Scale', *Essays on Econometrics and Planning Presented to Professor P. C. Mahalanobis* (Oxford, 1964), pp. 231–256.

[1] Chenery, *op. cit.*, pp. 529 f.   [2] *Ibid.*, pp. 510 f.

Koopmans — or of all special works concerned with determining an economic production function of the type (2) for some particular industry with the aid of statistical data. The purpose of the critique is to point out that in actuality there is also a great deal of purely economic flesh around the engineering backbone. The economic theorist cannot ignore this economic flesh. There are also good reasons why the offerings of engineering are not of much help to him. Algebraically simple though most of engineering formulae are, because of the difficulties mentioned in the preceding section they can provide hardly any basis for those principles without which the edifice of economic analysis would completely collapse : the principle of decreasing marginal rate of substitution and the laws of returns to factor and scale. As a matter of fact, from the gas pipeline formula [1] we should conclude that output increases more than proportionally with the size, and hence with the quantity, of pipe. And though no one, it is true, can tell what output corresponds to every factor combination yet untried by engineering, economic analysis must arrive at some idea, at least, about how output varies in general. The economic theorist himself, therefore, has to construct the proper tool for the analysis of economic production, and he must tend this delicate job with all the sophistication required by it.

III. THE PARTIAL PROCESS AND ITS ANALYTICAL CO-ORDINATES

1. If the few remarks scattered in literature are pieced together, they reveal that what we want in the very first place to represent by the concept of production function is a catalogue listing every optimal process by which a certain product can be obtained from each possible factor combination.[2] But then it follows with impeccable logic that before we can say anything about the structure of such a catalogue we must have a clear picture of the individual category to which the catalogue pertains. That is, we must first arrive at an analytical representation of a productive process.

No other word is used in economics with such careless ease as 'process'. Yet if one pauses to think about it, no other concept is as full of epistemological thorns as that of process. The reason is that we cannot talk about process without getting entangled in the

---

[1] Chenery, *op. cit.*, p. 515.
[2] 'This catalog of possibilities is the production function', Samuelson, *op. cit.*, p. 57.

difficult problem of Change. Ever since Herakleitos intrigued his contemporaries by teaching that 'you cannot step twice into the same rivers' — because the second time neither you nor the rivers are the *same* — the problem of the opposition between Being and Becoming has tormented the mind of every great philosopher. Science, however, has long since decided to embrace the viewpoint of 'vulgar philosophy', which viewpoint is that there is both Being and Becoming. From all we know, to abandon this dualism is to renounce *analysis* ; and to renounce analysis is to do away with *theoretical sciences*. However, we must not expect that analysis can remain entirely immune to the epistemological ills inherent in any dualism. It does not take much to see this truth.

2. Formidable though the notion of the whole universe in eternity certainly is, at least it does not raise one particularly difficult issue specific to the concept of *partial* process. For the universal process we need not specify what is included in it : the Whole is self-sufficient. That is not true for a partial process. To speak of such a process we must first of all determine its boundary with respect to both time and substance : *no boundary, no partial process.*

But as the entity from which we wish to carve a slice is a *seamless* whole, it contains no lines already traced to guide the carver. For the economist nothing could illustrate this point more convincingly than the still unsettled controversy concerning where the *natural* boundary of the economic process lies. And this situation is not the consequence of the complexity — as one may say — of the economic world itself. There is no natural boundary either between physics and chemistry, or between chemistry and biology. The issue is present everywhere we turn.

For a very elementary example from economics, let us take the case of a truck carrying over a highway a load of lumber from a mill to a furniture factory. Is there any objective criterion for deciding whether the travelling truck is part of the mill process or of the factory process? Perhaps one could answer that it depends upon which enterprise operates the truck. This, indeed, is the practice of book-keepers or cost analysts. But general economic analysis can hardly accept a criterion which becomes idle in case the mill and the factory belong to the same enterprise. And only harm can result from restricting that analysis to processes delimited by the money boundaries of enterprises.

3. Precisely because reality is a seamless whole one can draw a process boundary anywhere one pleases. However, an arbitrary slice

of reality does not correspond to our conception of process. On the other hand, it is hard to say exactly what sort of boundaries determine processes. To wit, every special science draws process boundaries where it suits its own *purpose* best. Without an intimate knowledge of this particular purpose, therefore, one could not say what a process means from the viewpoint of thermo-nuclear physics, for instance. In other words, any analytical process presupposes a purpose, and consequently is essentially a *primary* notion, that is, it may be clarified by discussion, but never reduced to other notions. Almost every important problem in social sciences supplies a striking illustration of these points. So, let us turn to economics.

The purpose of economics, to recall Say's definition, is to study the production, the distribution and the consumption of riches. Nothing is more natural, therefore, than that all partial processes in economics should be delimited in strict reference to riches, viz. commodities. The analytical definition of industry is referred back to the notion of commodity; a market process too is defined in a similar manner. Still more telling is the fact that the economist is not interested, for instance, in separating the process of plate-glass manufacturing at the point where the melted glass pours into the rolling machines: melted glass, under present technology, is not a commodity. From this comes the principle, heard now and then, that what happens inside a productive process is the engineer's, not the economist's, concern. But if strictly followed, the principle becomes a fetter. Without a broad understanding of the internal articulations of a productive process, it seems difficult to arrive at an analytical picture of such a process adequate for the economist's general task.

4. In economic literature we find two entirely distinct modes of representing a productive process; each one embodies one of the two traditional conceptions of such a process.

In one model we easily recognize the idea behind Boulding's and, more directly so, Stigler's definition (*II.* 1): process is represented by a set containing only *flow co-ordinates*. This mode of representing the process of an industry made its conspicuous entrance into the literature through the Leontief 'static system'.[1] But we owe to Koopmans its general formalization: a process P is represented by a vector

$$P (a_1, a_2, ..., a_n),\tag{3}$$

where each a_i indicates 'the rate of flow per unit of time of each of the

[1] Wassily W. Leontief, *The Structure of American Economy, 1919–1929* (2nd edn., New York, 1951), pp. 12 f., 37.

n commodities involved' in the process, negative values corresponding to inputs, and positive to outputs.[1]

The second model, which corresponds to Carlson's definition — and to a certain extent to Boulding's as well — was explicitly formulated for the first time by John von Neumann.[2] In this model, a process *P* is represented by a two-row matrix

$$P\begin{bmatrix} B_1, B_2, ..., B_n \\ A_1, A_2, ..., A_n \end{bmatrix}, \tag{4}$$

where the vectors (*A*) and (*B*) represent the quantities of commodities existing at the beginning and at the end of the period during which the process is completed.

It is puzzling though that the same concept, that of a process, should be represented by two models which are not wholly equivalent. Indeed, (3) contains only import-export type of data, *i.e.* data reported by an observer concerned with *what crosses the boundary of the process within an infinitesimal period of time*. On the other hand, the data in (4) are census type data : both (*A*) and (*B*) represent *what exists within the boundary of the process at an instant of time*. Or, to put it in a still more instructive way, (3) ignores what crosses the *time* sector of the process boundary and takes into account only what crosses the *physical* sector of the same boundary ; (4) proceeds the other way round. Clearly, each model — as I observed in an earlier paper [3] — tells only one-half of the whole story. For one thing, if (*A*) = (*B*) we have no means of knowing whether (4) represents a uniformly going process or a totally frozen state.

5. In order to avoid irrelevant technical aspects and long notations, let us consider the *scientific report* of a very simple process, that of growing maize by a method requiring only a spade for preparing the ground, all other operations being done by hand. A complete list of the elements involved in the process is easily drawn :

1. Land space.
2. Solar radiant energy.
3. Chemicals from the air.
4. Soil with chemical nutrients.
5. Labourers.
6. Spades.
7. Maize grain.
8. Stubble, cobs, etc.

[1] Tjalling C. Koopmans, 'Analysis of Production as an Efficient Combination of Activities', *Activity Analysis of Production and Allocation*, ed. Tjalling C. Koopmans (New York, 1951), p. 36.
[2] John von Newmann, 'A Model of General Economic Equilibrium', *Review of Economic Studies*, xiii (1945), 2. (The original article, in German, appeared in 1935).
[3] 'Aggregate Linear Production Function and Its Applications to von Neumann's Economic Model', *Activity Analysis of Production and Allocation* (*op. cit.*), pp. 100 f.

Let us choose the time boundary so that the process *begins* with the first act of spading and *ends* with the shelling of the last kernel. Let this interval — or the time of production, as Marx called it [1] — be denoted by $(0, T)$, where T is measured in full days. Let us also use the terms 'input' and 'output' with their basic meaning, that is, to denote elements crossing the boundary *into* or *out of* the process, respectively. This is a most natural thing to do. It also has the merit of bringing two important issues into plain view.

First, most of the categories listed above appear both as inputs and as outputs. Second, the spade that goes into the process is *qualitatively different* from that coming out of it. And, as alluded to above, qualitative change does not fit into an analytical representation. To circumvent this incompatibility at least in part, we may follow the common practice of identifying each spade by the total time it has been in actual use since it was 'new'. The report then may look as follows :

Category	Inputs	Time	Outputs	Time
Land-space	One acre	0	One acre	T
Solar energy		$E(t)$		
Air chemicals		$C(t)$		$C_1(t)$
Soil chemicals		$S(t)$		$S_1(t)$
Labourers	One rested man	$0, 1, T - \tau$	One tired man	$\tau, 1 + \tau, T$
Spades	One new	0	One used	τ
Maize grain	One bag	1	Eleven bags	T
Stubble, etc.			X	T

The time co-ordinates represent either the instant a finite entity crosses the boundary or, in the case of a 'continuous' exchange, the total amount of an input (output) up to the instant t. For the sole purpose of avoiding notational complications the length of the working day is assumed constant and is represented by τ.

6. Complete though it is, such a report can hardly serve the purpose of the economic theorist : it includes many items that are not commodities in the usual sense of the term.

The case of solar energy and air chemicals is simple. Land-space not only provides a spatial base for the process at hand, but it also constitutes a *catching net*, as it were, for solar energy and chemicals from the atmosphere. Within a homogeneous climatic region, the average amount of solar energy and exchange of oxygen or carbon dioxide between the atmosphere and some given crop is proportional with the area of the land-space. And since even over a very long

[1] Karl Marx, *Capital* (Chicago, 1933), ii, 272 ff.

time period the two elements just mentioned vary little in intensity, they can be regarded as the 'original and indestructible' attributes of land-space. Hence, solar energy and air chemicals can be lumped together with land-space as 'Ricardian land'.

The problem of the tired labourer or of the used spade, however, is far more complex. By definition a used tool must be an output of some productive process. Yet in no sense can we say that it is the aim of economic production to produce used tools. Consequently, they have no cost of production. Moreover, with the exception of used automobiles and used dwellings, no used capital equipment has a regular market and, hence, a price in the same sense in which new equipment has. Used equipment, therefore, is not a commodity proper, and yet no report of a productive process can be complete without reference to it.

To be sure, in practice one may adopt one of the numerous conventions used in computing depreciation. But such a solution, besides involving some arbitrariness, is logically circuitous : it presupposes prices and interest rate to be already given. Economic theory has endeavoured to avoid the Gordian knot altogether by building its foundation only upon a process in which *all capital equipment is continuously maintained in its original efficiency.* The idea underlies Marx's diagram of simple reproduction as well as the neo-classical concept of static process. But not all its analytical snags have been completely elucidated.

7. To transform our illustrative process into a static one, we should include in it the activities by which the spade is kept sharp and its rivets, handle or blade are replaced when necessary. These activities imply additional labour power, additional inputs and, above all, additional tools. And since the efficiency of these tools must be, in turn, kept constant, we are drawn into a regress which may perhaps stop only after the whole production sector has been included in the process at hand. Moreover, if strictly interpreted, a static process must also maintain its labourers — *i.e.* the variable capital of Marx [1] or the personal (human) capital of Walras.[2] Thus, in the end, we have to include the consumption sector as well — a glaring illustration of the observations made earlier concerning the analytical difficulties of the concept of partial process. To avoid the regress to the whole economic process, we may assume without fear of being unrealistic

---

[1] Karl Marx, *Capital* (Chicago, 1932), i, 232 f.
[2] Léon Walras, *Elements of Pure Economics*, tr. W. Jaffé (London, 1954), pp. 214 ff.

that in every partial process part of the capital equipment and all human capital are maintained by outside processes, each one in turn to be analysed separately. After all, analysis cannot proceed without some heroic abstractions at one stage or another.

8. Once we have removed the qualitative difference from the picture, the analytical representation of a partial process no longer presents any difficulty. The elements participating in a process now fall into two simple and distinct categories. The first includes the elements that figure in the scientific report only as inputs or as outputs ; the second comprises the elements that appear both as inputs and as outputs, more exactly, the elements that enter and come out of the process in an economically, if not also physically, identical form and in the same amount. It is appropriate to refer to the elements of the first category as *flow* elements and to the others as *fund* elements.[1]

The participation of any factor in the process is easily described by a function of t. In case of a flow item, the function, say, $A(t)$, shows *the total amount of the corresponding flow during the time interval* $(0, t)$, this interval being open or closed to the right according to whether the item is an input or an output. By convention, $A(t) \leqslant 0$ for an input flow and $A(t) \geq 0$ for an output flow.[2] For a fund element, the function, say, $B(t)$ shows the *amount of the corresponding fund effectively participating in the process at the instant t.*[3]

To determine the analytical co-ordinates of a static partial process according to the method just outlined is a very simple matter, except for one particular category. In our example this category is represented by maize. Maize figures both as input and as output, but the output is greater than the input. The case therefore presents a difficulty which does not exist for Ricardian land, labour power or equipment. The solution, however, is straightforward and, moreover, inescapable : *one bag of maize is a fund element, whereas ten bags of maize constitute a flow element.*

One point may now be stressed : the analytical co-ordinates of any

---

[1] The reason why here 'fund' is to be preferred to 'stock' should be obvious ; but see p. 524, n. 1 below.

[2] Apart from being non-decreasing in absolute value, the function $A(t)$ may have any form whatever. However, for 'continuous' items, we may safely assume that $A(t)$ has a derivative function $a(t)$, which then represents the instantaneous flow rate at the instant t. In the case of a discontinuous input, for example — $A(t)$ is ordinarily such that $A(t) = 0$ for $t < t_0$ and $A(t) = A_0$ for $t \geqslant t_0$, A_0 being the global input at t_0.

[3] As a mere matter of consistent formulation, the function $B(t)$ for a fund B_0 going in at t_0 and coming out at t_1, should be defined so that $B(t) = 0$ for $t < t_0$ and $t \geqslant t_1$, and $B(t) = B_0$ for $t_0 < t \leqslant t_1$.

process are strictly determined by its boundary. Therefore, it would be a mistake to believe that commodities can be divided into two distinct classes, funds and flows : the same commodity may be a flow in one process and a fund in another.[1] Thus, the clover seed in a process the *purpose* of which is to produce clover seed is a fund, but in a process aimed at producing clover fodder, it is a flow.[2]

9. There are certain implications of the preceding argument on which one would perhaps welcome additional discussion.

9. (*a*) The first such implication is that gross output is never an analytical co-ordinate of a partial process. This does not imply, however, that gross output data are meaningless. After all, the farmer of our illustration brings home eleven bags of maize, an amount which corresponds to gross output. As a purely descriptive co-ordinate of non-analysed facts gross output raises no problem. But facts must ultimately be analysed and analytical categories, once established, must always be kept separate. And the point is that two analytically distinct categories, a fund and a flow, are added arithmetically to arrive at gross output.

One is likely though to ask why in a farming process the fertilizer input, for instance, should be treated differently from the seed input : both undergo a qualitative change inside the process. The issue is easily clarified if we refer to a more stringent illustration, namely, a process in which one hammer is used to hammer (shape) an additional hammer. According to all classifications ever proposed in economics, the first hammer is classified *as capital* (fixed or constant), *not as output*, whereas the reverse is true for the second hammer. The physical identity of the two hammers is no reason against recognizing their distinct roles in the productive process.[3] Nor is the lack of

[1] One may note in passing that this relativity bespeaks the impossibility of analysis to fare well and indefinitely with the dualist separation between Being and Becoming.

[2] The point that the analytical co-ordinates of a process depend strictly upon its boundary has some bearing upon such discussions of the consolidation problem in input-output *analysis* as offered by Leontief, *op. cit.*, pp. 14 ff., and R. Dorfman, P. A. Samuelson and R. M. Solow, *Linear Programming and Economic Analysis* (New York, 1958), Chapter 9 and 10.

[3] Nothing illustrates better the imbroglio produced by the failure to keep fund and flow co-ordinates separate than the celebrated problem of the transformation of values into prices in Marx's economic theory. It all comes from this : in his diagram of simple reproduction Marx fails to distinguish analytically between the hammering hammer, c_1, and the hammered hammer, $v_1 + s_1$. The error is made elementarily obvious by the cacophony used by Paul M. Sweezy, *The Theory of Capitalist Development* (New York, 1942), pp. 76 f., to explain that diagram : gross output (total value) is obtained by adding 'the constant *capital engaged* [in production with] the income of the capitalist [and] the *income* of the worker'. (My emphasis.)

physical continuous identity of the maize seed a reason against treating it as a fund in our analysis.

9. (*b*) The flow-complex that seems to dominate the current economic thought is the only reason why we should also discuss the point that the analytical representation of a partial process necessarily includes both flow and fund co-ordinates. The necessity, however, should be immediately obvious : the *fund* co-ordinates represent the material base of the process, the *flow* co-ordinates describe the change (transformation) achieved with the aid of this base. A framework based upon both Being and Becoming must necessarily include one analytical category for each, fund elements to represent *the unchangeable agents* and flow elements to represent *the object changed by the agents*.

There is no need nowadays to insist upon the analytical difference between *stock* and *flow*.[1] But a great deal of uncertainty still exists concerning the difference between *stock* and *fund* and, especially, between *flow* and *service*. To recall, it was Walras who incorporated in production analysis J. B. Say's old idea that services constitute the fundamental element in a productive process. But while he carefully distinguished between capital funds and their services, he failed to note the difference between these concepts on the one hand and those of stock and flow on the other. Somehow the lacuna has perpetuated itself through all subsequent writings. For a most convincing illustration in point we may choose such an authority as Pigou :

> In a stationary state factors of production are stocks, unchanging in amount, out of which emerges a continuing flow, also unchanging in amount, of real income.[2]

The physical picture becomes even more perplexing as we are further instructed, first, that the services of factors 'become embodied in other goods', and, second, that 'what is directly demanded is the flow of [such] services'.[3] Clearly, one root of the confusion is the improper use of the term 'flow'. Once this term is assigned to convey the idea of *some material substance crossing the process boundary*, it is utterly misleading to apply it also to services. The inevitable trap is that, because a flow can be stored up, we come to speak of services being 'embodied in the product'. The fact that, under certain

---

[1] On which we have been instructed first by S. Newcomb, *Principles of Political Economy* (New York, 1886), and then by Irving Fisher in a series of contributions from 'What Is Capital ?' *Economic Journal*, vi (1896), 509–534, to *The Nature of Capital and Income* (New York, 1919), Chapter iv.

[2] Pigou, *op. cit.*, p. 19. Cf. Walras, *op. cit.*, Lesson 17, especially para. 169.

[3] Pigou, *op. cit.*, pp. 20, 117.

circumstances, the value of services is 'embodied' in the value of the product cannot alter the physical side of a process : only the elements that flow in the process can be *physically* embodied in the outflowing product.

But perhaps we fear that by overtly recognizing that an inflow can be transformed into an outflow only with the aid of services we would implicitly commit ourselves to the view that Ricardian land and capital equipment *must* earn an income. Be this as it may, the fact is that these fund co-ordinates are ordinarily deleted from the analytical picture of a static process.[1] The procedure, in addition to prejudicing the analysis of income distribution, constitutes an analytical distortion full of pitfalls — as we shall see later on.

The familiar leitmotiv that there would be double counting if both the maintenance flow of capital and its service are included in the analytical picture of a partial process is grounded on presumptions as regards value distribution. Marx alone endeavoured to justify the principle that even in the case of labour maintenance flows take care of services. And although Marx painstakingly avoided mentioning service by using instead such expressions as the work performed by a machine or the life-activity of the labourer,[2] there can be no doubt that he treated the labourer as a fund in our own sense.[3] Now Marx does start out by admitting that labour power is

> the aggregate of those mental and physical capabilities existing in a human being, which he exercises whenever he produces a use-value of any description.[4]

But in the end he reduces the participation of the labourer in a production process to 'a definite quantity of muscle, nerve, brain, etc. [which] is wasted' during work.[5] By this volte-face Marx set the pattern for ignoring the brute fact that any productive process requires the participation of the worker's entire fund of muscle, nerve, brain, etc. Why, nature is so constructed that no professor can discharge his duties by sending to class only that part of his nervous or muscular energy which ordinarily he spends during a lecture.

---

[1] The most illustrious example is the Leontief static system. Koopmans, *op. cit.*, pp. 39 f., does include land in his model ; but he does not distinguish between flow and service co-ordinates.

[2] Cf. *Capital*, i, 589, or his 'Wage Labour and Capital' reprinted in K. Marx and F. Engels, *Selected Works* (Moscow, 1958), i, 82.

[3] *Capital*, i, 189 f., especially p. 622. But the most incontrovertible statement is that of F. Engels, 'Marx's *Capital*' in *Selected Works, op. cit.*, i, 464 : 'Labour power exists in the form of the living worker who requires a definite amount of means of subsistence for his existence as well as for the maintenance of his family.'

[4] *Capital*, i, 186.   [5] *Ibid.*, p. 190.

And it is absurd to think that one can cross a river on the maintenance flow of a bridge : a non-existent bridge cannot possibly render a service ; nor can it be maintained.

10. To sum up, the analytical picture of a process by which some product — whether a bushel of maize, a ton of coal or a piece of furniture — is obtained, consists of the following broad categories of co-ordinates :

FLOW CO-ORDINATES

Inputs :	From nature	$R(t)$
	From other processes :	
	(a) current inputs	$I(t)$
	(b) maintenance	$M(t)$
Outputs :	Product	$O(t)$
	Waste	$W(t)$

FUND CO-ORDINATES

Ricardian land	$L(t)$
Capital	$K(t)$
Labour	$H(t)$

Over the production time interval $(0, T)$, each of these co-ordinates is defined as explained in section *III*. 8 above. Their most likely forms are shown in Fig. 1. To retain Marx's convenient terminology, the total time $T_H = \overline{Ot_1} + \overline{t_2 t_3} + \overline{t_4 t_5} + \overline{t_6 T}$ will be referred to as the working period. Needless to add, in most cases $T_H < T$, either because some phases of the process require no labour services whatsoever, or because the process is intentionally interrupted for some reason or other.

A catalogue of all possible partial processes pertaining to the same product is thus contained in the following general formula

$$O(t) = \mathscr{G}\ [R(t),\ I(t),\ W(t),\ M(t),\ L(t),\ K(t),\ H(t)], \qquad (5)$$

which being a functional mapping is a far cry from the Dirichlet function (2). To be sure, among the arguments of $\mathscr{G}$ there normally exist additional relations corresponding to some technical restrictions. They permit the elimination of the flow co-ordinates from (5), in which case we have

$$O(t) = \mathscr{F}\ [L(t),\ K(t),\ H(t)]. \qquad (5a)$$

This formula, we should note, shows how misleading is the familiar proposition that output is a function of the *services* of the

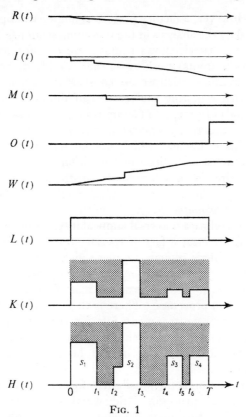

FIG. 1

production factors. Indeed, according to the consecrated meaning of the term, the services of labour, for example, are represented in Fig. 1 by the sum of the areas s_i. Clearly, this sum, though determined by $L(t)$, does not determine this co-ordinate.

IV. THE FACTORY PROCESS

1. A factory is such a familiar object nowadays — especially, to those who are the progeny of an industrialized society — that we are apt to lose sight of two essential facts : first, that the factory system of production represents one of the greatest *economic* innovations in history, and second, that the system is not (and, most likely, will never be) applicable to all production sectors.

Now, any factory represents a partial process in the broad sense of

s

this term. However, a factory process is not a partial process in the sense adopted by the analysis of the preceding section. True, in both processes some material flows (inputs) are transformed into other material flows (outputs) ; otherwise they would not be processes at all. But whereas a partial process consists of a *temporal sequence* of operations, each requiring the services of different factors and for different periods of time, in a factory process all these operations are performed *simultaneously and in a special arrangement*. To the analysis of this difference, which has far-reaching economic consequences, we must now direct our attention.

One point, already hinted, needs to be stressed at the outset : *the partial process constitutes the basis of all production, whether in agriculture, mining or manufacturing*. As an analytical device, the diagram of Fig. 1 has therefore a universal applicability. Among other things, it makes perfectly clear a fact of special importance for the present argument : a farmer's plough, a miner's pick, or a carpenter's plane, do not *continuously* participate in the partial process in which each happens to be used. This is true for most fund factors, including labour, and for all processes. There is then an inherent reason why fund factors may have to remain idle for varying periods of time. Whether this idleness, which definitely represents the most relevant form of economic waste, can in fact be avoided depends upon several conditions.

One such condition is the possibility of using the same fund factor in another partial process. And whether this is economically feasible depends in turn upon demand, joint or simple. For a clear illustration of the role of simple demand — which alone is relevant for the topic of this section — let us refer to the production of tables. If only one table is all that is demanded during a time interval greater or equal to the corresponding period of production T, then obviously production must be carried out by partial processes *in series*, *i.e.* such that none overlaps in time. In this case, there is no way of avoiding a rather long idleness of the plane, of the carpenter himself, etc., unless these same factor funds can be employed in the production of other goods for which there is sufficient demand. But if during an interval equal to the same production period more than one table is demanded, then there are two alternatives : (1) a sufficient number of partial processes are started *in parallel*, *i.e.* at the same time, and (2) a sufficient number of processes are started *in line*, *i.e.* a different process is started at each time instant of a chosen sequence.

Clearly, the analytical picture of the processes run in parallel leads to a diagram identical to Fig. 1, except for the fact that all ordinates are amplified by the number of processes. Consequently, idleness (represented by the dotted areas) also is amplified. The diagrammatical representation of processes run in line, however, is entirely different and depends upon the way their sequence is arranged in time. We need not bother about the general case. Nor is it necessary here to go into the mathematical niceties by which the following two theorems are proved :

A. *Given the number of necessary partial processes, they can be arranged in line so that the idleness of some arbitrarily chosen fund factor is minimum.*

B. *If the number of necessarily partial processes is sufficiently large and end points of all periods during which each fund factor renders a service are commensurable with T, then there is a minimum number of processes that can be arranged in line so that no fund factor used in this arrangement is idle.*[1]

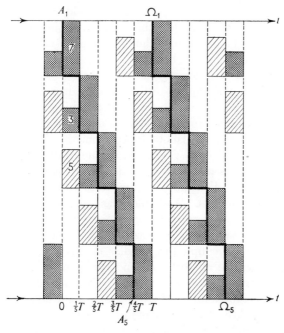

Fɪɢ. 2

[1] If m is the smallest integer such that the ratio between the service period of each fund factor and $t' = T/m$, is an integer, then the arrangement consists of m processes spaced in line at an interval equal to t'.

The diagram of Fig. 2 illustrates the last theorem for a partial process simple enough to avoid unnecessary complications. This process involves only two funds, say, two types of labour. The services of one of these factors are marked by shaded areas. Within the interval $(0, T)$, there are five processes starting in line at 0, $T/5$, $2T/5$, $3T/5$, $4T/5$, respectively. They are delimited by the step lines A_1A_5 and $\Omega_1\Omega_5$. Ten labourers of one category and five of the second category are periodically shifted from one process to the next. Needless to add, it is this sort of arrangement that characterizes the factory system.[1] The view expressed at the beginning of this section concerning the emergence of the factory system is thus vindicated.

2. There is some advantage in considering first a factory process that goes on uninterruptedly, as is actually the case in steel or glass production. For the analytical description of such a process, it is wholly irrelevant that the *same service* is rendered during one shift by Mr. A and during another by Mr. B. The corresponding labour fund, like any other physical fund, never leaves the process. But this is no justification for the flow-complex which would have us ignore the existence of factor funds in describing a steady-going process. After all, a factory is not a phantom without material basis any more than a bridge or a working human being is.

No analytical description of a factory process, therefore, could be complete without the fund co-ordinates. But in the case of an uninterrupted process, we can no longer determine these co-ordinates by following the same procedure as for a partial process. One could determine, it is true, the labour funds from the line of employees passing through the factory gate at the beginning of a shift. But this simple solution does not work for other funds, for they never pass through the gate. The only solution is to take a *census* of what exists inside the process. And since the process is steady-going, it does not matter at what moment this census is taken.

Such a census will reveal the existence of several broad categories of funds. There are first the familiar ones : Ricardian land, L, capital equipment of all sorts, K, and labour of various kinds, H. But two new categories now show up. The first covers the stores (inventories) of commodities which appear in some flow co-ordinates. Let it be denoted by S. The second new category consists of 'goods in process', though 'goods' is here a misnomer : melted glass or half-tanned hides, for example, can hardly fit the term. It is none the

[1] Incidentally, the aggregate of the minimum number of processes arranged in line so as to occupy continuously all factor funds may provide a basis for defining a unit-plant.

less true that *at any time* there exists inside the boundary of a factory a *process fund* consisting of the successive phases through which the input flow passes to become ultimately the product flow. From the analytical viewpoint it is highly instructive to note that this fund is a picture of the qualitative change achieved by the factory process, just as a movie film contains a whole drama at once. The co-ordinate corresponding to this Becoming frozen into a Being, is therefore a complex entity. Let us denote it by $\mathscr{C}$.[1]

The flow co-ordinates are the same as in the case of a partial process. However, a few additional observations seem in order. First, the factory process offers further support to the view that gross output is not a proper analytical co-ordinate of a process : the number of hammers used in producing hammers will never appear as a flow co-ordinate of the hammer factory. *Instead, there will be a fund co-ordinate of hammers.*[2] Second, all input flows can be lumped together as a general maintenance flow : the role of the current input and of part of the direct input from nature is to maintain $\mathscr{C}$. There is nevertheless some advantage in abiding by the old classification.

To simplify the analytical picture, let us make the customary assumption of absolute continuity. In this case, every flow factor can be represented by an instantaneous flow rate, and the analytical representation of a factory process assumes the following general pattern :

FLOW CO-ORDINATES

Inputs :	From nature	r
	From other processes :	
	(a) current input	i
	(b) maintenance	m
Outputs :	Product	q
	Waste	w

FUND (SERVICE) CO-ORDINATES

Ricardian land	L
Capital proper	K
Labour	H
Process fund	$\mathscr{C}$
Stores	S

[1] For practical purposes, one can divide $\mathscr{C}$ into a finite number of sections such that within each the qualitative variations could be ignored in the first approximation. Then $\mathscr{C}$ can be approximately represented by a vector (C). However, one must not overlook the fact that C_1 is not necessarily a commodity.

[2] Surprising though it may seem, the only way of finding out how many hammers are used during a given interval is from the waste flows. However, 'worn out hammers' may not necessarily appear as an item of waste : the heads might be remelted.

3. A few points about this table need special emphasis. First, *every co-ordinate is a point-coordinate measuring some intensity*. The flow co-ordinates measure the intensity of inputs and outputs ; their dimension is (substance)/(time). The fund co-ordinates too measure intensities, of services ; their dimension is (service)/(time). The fact that this last measure coincides with the measure of the corresponding fund — which has the dimension (substance) — should not obscure the dimensional difference between the two concepts. The upshot of these remarks is that there is absolutely no sense in which one can speak of price in connection with any co-ordinate in the table. A water bill is for the water consumed, not for the gauge of the connecting pipe.

Second, let us observe that we would arrive at exactly the same co-ordinates even if the factory would not work around the clock.[1] The table thus does not describe *what the factory actually does, but only what it is capable of doing while working*. The information it provides is analogous to knowing only that Mr. C is a 'civil engineer' ; it describes the potentiality of the factory, whether or not it is working. And like Mr. C, most factories can interrupt and resume work at will. Moreover, because $\wp$ is continuously maintained by the very act of production, the product starts to flow the instant input flows move in. A factory is like a music box which starts playing as soon as it is opened regardless of when this happens. In a factory process, therefore, *there is no time-lag between input and output flows*. And this, we should note, is not an analytical simplification of actuality.

4. To know what a factory actually does, one needs an additional co-ordinate, not included in our table. This co-ordinate is the time length, $\delta \leqslant 1$, the factory works each day. The quantities of material inputs and services consumed and of the product produced *during the period* δ, are then immediately obtained if every co-ordinate of our table is multiplied by δ. However, some fund factors by necessity provide services around the clock even if $\delta < 1$. In the new picture, the corresponding co-ordinates must therefore be multiplied by *one unit of time*. Only these new co-ordinates have a price, whether positive, null, or negative.[2]

---

[1] The matter of user's cost may safely be ignored for our immediate purpose.

[2] The point is of particular importance in cost analysis. In principle, only the daily *variable* cost is proportional to δ, the prices of land and capital services being customarily set for a full day. In actuality, however, there is also a cost of changing shifts and possibly a differential wage rate between shifts. Hence, total cost $TC = A_j + B_j \delta$, j indicating the number of shifts used. Pure competition then determines δ by the condition $TC = (pq + p_w w)\delta$. The reader may find it instructive

A catalogue of all factory processes must nevertheless be based upon the original table. And since all analytical co-ordinates are now *real numbers*, it can be represented by a Dirichlet function :

$$q = G(r, i, m, w ; \; L, K, H, \mathcal{C}, S). \tag{6}$$

It is normal to expect that once the individual items covered by L, K, H, are determined quantitatively and qualitatively, the other elements are also determined through the usual technical constraints represented by limitationality relations. Hence, (6) can be reduced to

$$q = F(L, K, H). \tag{6a}$$

But this reduction should not induce us to ignore — as is frequently done — the other elements of the process not represented in (6a).

5. There is one consequence of the preceding analysis which I have repeatedly found hard to bring home, elementary though the point is. If $Q_0 = \delta q$ is the daily output, from (6a) it immediately follows that

$$Q_0 = \delta F(L, K, H). \tag{7}$$

But this is a far cry from the neo-classical definition of the production function which leads to

$$Q_0 = \emptyset \, (L^o, K^o, H^o), \tag{8}$$

where L^o, K^o, H^o stand for the amounts of services.[1] There seem to be only two fair interpretations of this formula. The first is to read $L^o = \delta L$, $K^o = \delta K$, $H^o = \delta H$. As already explained (*II.* 2), this implies that F is homogeneous of the first degree, a condition met neither by a production unit (factory), nor by every industrial aggregate. The second interpretation is to read $L^o = L \times 1$, $K^o = K \times 1$, $H^o = H \times \delta$. But the basic analytical sin of (8) is still untouched. Let $H^1 = H' \times \delta'$; according to (8), $\emptyset \, (L^o, K^o, H^1) = \emptyset \, (L^o, K^o, H^o)$, which obviously cannot be true in general. The work of one worker for six days is not necessarily equivalent to the work of six workers for one day — as Wicksell once observed — even though the wage bill is the same in both cases. In many mathematical models, the confusion is further increased by treating K^o as a stock

---

to compare these points with Marx's attack on Senior's 'last hour', *Capital*, i, 248 ff.

[1] I take it that the neo-classical definition is that of Boulding and of any other author cited in note 2, p. 499 who cared to describe the variables entering into the function. Actually, it seems difficult to find an author not endorsing the same definition. Stigler, Koopmans and perhaps Pigou are among the very few who view the matter in the same sense as (6a).

capable of being increased by the amount of new capital accumulation ΔK.

Of course, the *numerical*, though not the *dimensional*, discrepancy between the generally used formula $Q_0 = F(L, K, H)$ and (7) disappears if δ is taken as time unit. Q_0 would still represent 'the daily output'. But δ must be constant in order to serve as unit. Perhaps neo-classical economists have proceeded on the assumption that δ is an institutionally determined constant; but if so, they failed to state it explicitly. Be this as it may, treating an important variable as a constant has seriously impaired the value of neo-classical analysis of production — as will be seen presently (*V*. 6).

One has to turn to Marxian economics to find a more adequate approach. In the light of what precedes, Marx was right in allocating a most prominent place to the working day in his economic analysis. And in a certain sense, (7) lends some support to Marx's dearest tenet that labour time, though it has no value, is a measure of value.[1] Unfortunately, for reasons easy to guess, Marx's analytical findings never travelled beyond Marxian economics. All the greater then is the merit of W. H. Nicholls for having broken the neo-classical tradition by introducing the length of the working day as an independent variable in the production function.[2]

V. BALANCE IN ECONOMIC DEVELOPMENT

1. From Marx, however, neo-classical economics has accepted, it seems, two dogmas pertaining to economic development : first that over-population is a bogey, and second, that the same economic laws govern agricultural and industrial production.[3] Yet Engels in 1884,

---

[1] *Capital*, i, 45, 588.

[2] W. H. Nicholls, *Labor Productivity Functions in Meat Packing* (Chicago, 1948). In essence, his analysis is based on the formula $Q_0 = \psi(\tau, L, K, H)$, which for L and K constant leads to $Q_0 = f(\tau, H)$, the notations being ours. None of the simple formulae tried out by Nicholls (pp. 98 ff.) comes near the form $Q_0 = \tau g(H)$, as should follow from (7). There are many reasons why one should not expect this form to fit actual data : the worker's fatigue (cf. Nicholls, p. 25, n. 2), the opening- and closing-effect, etc. Yet none of these reasons work against the separation of the variables as in $Q_0 = h(\tau)g(H)$, with h and g having an S-shape. This form, however, involves some computational difficulties because the simplest S-shape curve is a third degree parabola.

[3] For an analysis of over-population see my essay 'Economic Theory and Agrarian Economics', *Oxford Economic Papers*, xii (1960), pp. 1–40. To straighten out some misinterpretations of that analysis I wish to emphasize that, as ought to be clear, my definition of over-population is a short-run concept. Consequently, even though it might be hard to find in the *changing* actuality a situation where labour marginal productivity is *mathematically* equal to zero, this does not affect the

while quoting approvingly L. H. Morgan's statement that 'Mankind are the only beings who may be said to have gained an absolute control over the production of food', finds it necessary to insert 'almost' to tone down 'absolute'.[1] Were Engels still living, the disturbing evidence accumulated by recent history would perhaps cause him to have many second thoughts on this controversial issue. As it happens, the results obtained in the preceding sections of this paper can throw a great deal of light on it.

2. Let us first draw an analytical picture of the whole economic process, including both production and consumption, at any instant of time. Since by consolidation all flows between production and consumption units must cancel out, the global picture includes only two flow co-ordinates: an input flow from nature and an output flow of waste. As far as the material elements are concerned, the economic process simply transforms natural resources into economically valueless waste.[2]

As it has been repeatedly recognized, man can neither create nor destroy matter and energy. But this is only one-half of the story — the half told by mechanics, that cherished model of most social scientists. However, natural resources do not consist of mere matter and mere energy, but of *matter arranged in some definite structures* and of *free energy*. Mere matter, such as the gold scattered over the bottom of the oceans, has no value for us: we need gold ore where gold is arranged so that we can extract it in useful time. Nor has the immense heat energy contained in the ocean waters any value for us: a sailing ship needs fuel, *i.e.* energy in the free state. All the carbon, oxygen, hydrogen, etc., in the world could not support a human life if they were not arranged in a molecule of sugar, starch or protein.

In the story of nature told by thermodynamics, the laws of which are as inexorable as those of mechanics, the energy-matter constituting natural resources is *qualitatively* different from that forming waste. The energy-matter of natural resources is arranged in some orderly patterns, or as the physicists say, it has a low entropy. In

---

validity of my analysis or policy recommendations for those numerous economies where the *actual* labour productivity is negligible, or only smaller than the prevailing minimum of subsistence. The whole edifice of economic science would collapse if we confuse analytical concepts with evolutionary facts.

[1] F. Engels, *The Origin of the Family, Private Property and the State* (4th edn., New York, 1942), p. 19.

[2] Since this fact might come as a shock to some, I should add that, on the contrary, it reveals a much neglected truth: the actual 'product' of the economic process is not a material flow, but a psychological (or vital) flux: the mere enjoyment of life, as Irving Fisher tried hard to teach us. Real income and labour or, alternatively, real income and leisure, are only the material 'measure' of that flux.

waste we find only disorder, that is, *high entropy*. That is not all. The Second Law of thermodynamics tells us also that the whole universe is subject to a continuous qualitative degradation : entropy increases and the increase is irrevocable. Consequently, *natural resources can pass through the economic process only once : waste is irrevocably waste*. Man cannot defeat this law, any more than he can stop the law of gravitation from working. The economic process, like biological life itself, is *unidirectional*. Money alone moves in a circular flow, because no one throws it away though it is only an artificial token.

3. Natural resources fall into two distinct categories. Some exist as *stocks* in the crust of our planet.[1] These can be used with a speed and rhythm which, in principle, depends only on man's choice. Conceivably, we could exhaust all the known stocks of oil within one year if we wanted to do so and made our plans accordingly. The direct point is that any mine can be operated as a factory process around the clock. *Mining, therefore, does not compel us to keep idle any of the factor funds involved in the process.* The same is true for all manufacturing industries. We should further observe that it is precisely this freedom which man has in using almost at will any mineral deposit — once he has discovered how to use it advantageously — that is responsible for the spectacular progress of technology.

If one would like to brag about man's feats, one should rather choose man's control over inert matter, not over food. The taproot of all food is in photosynthesis ; and photosynthesis requires first of all solar energy. But in contrast with most other stocks of energy, that of the sun is — and may for ever remain — beyond man's control. Solar energy comes to each place on the earth at various epochs of the year in a definite flow rate.

The consequence is that, with a very few exceptions, partial processes in agriculture cannot be arranged *in line*, as in a factory process. They can be arranged only *in parallel*, all beginning at the appropriate phase of the climatic annual cycle in each place. And to pinpoint further the only reason for this necessity : in the island of Bali, for instance, where the climate varies little throughout the year, nothing stands in the way of growing rice *in line*. The same number of buffaloes, ploughs, sickles, flails and villagers could move over the entire field of the village, ploughing, seeding, planting, weeding,

---

[1] Since entropy continuously increases, no natural resource, not even the solar free energy, can be a fund. At the same time, it is now obvious that only energy-matter represents a fund in the strict sense of the term.

harvesting and threshing without interruption. Moreover, the people could then eat each day the rice sown that very day, as it were (*IV*. 3). They would no longer need to wait the long days between ploughing and threshing and, especially, to bear the burden of agricultural loans. Unfortunately, man's condition is such that there are very few spots where this Bali formula can work and, hence, where 'the open-air factory' can become a reality. Elsewhere every kind of agricultural production *inevitably* imposes some idleness on *both* capital and labour over the production period and complete idleness on every fund factor during the rest of the year (as shown in Fig. 1).[1] Although this alone would suffice as proof that industry and agriculture are governed by different laws, the difference is sharper still.

4. An expert on thermodynamics could, no doubt, compute the maximum amount of photosynthesis a certain crop could achieve during one average year on an acre of land in a given location. The size of our catching net — the earth's surface — thus sets a limit to the amount of total photosynthesis each year. To think that one could get a greater yield than this *theoretical* maximum would be absurd. Equally absurd would be to ignore that solar energy is not the only necessary input.

The input that illustrates most conspicuously the fateful work of the Entropy Law is the soil nutrients. The long history of peasant societies may be summarized in a few words : a continuous struggle with the effects of the Entropy Law. Under its pressure, the village economy passed from swidden agriculture, to shifting cultivation, and finally to crop rotation. It also invented, in succession, the caschrom, the scratch-plough, and the ordinary plough ; it also discovered rotation and manuring. These are momentous achievements which contrast with the economic inertia of the contemporary peasant economies. In the contemporary era, however, the peasant economy has come to a crisis that the village alone can no longer solve. The Entropy Law makes the crisis inevitable : the population explosion has only speeded its coming. But leaving aside the population explosion — which is a biological rather than an economic phenomenon — we can easily see that the crisis stems from the scarcity of land — about which we can do rather little — and from the qualitative deterioration of agricultural land through millenary use with manuring only. The tables have turned : it is the turn of the town

---

[1] Clearly, this does not cover the total idleness of the labour fund in excess of the maximum service intensity. But this idleness, if present, is not an immediate consequence of material laws.

now to support the economy of the countryside : the 'manure' must now come from the industrial sector in the form of fertilizers.

5. Undoubtedly, there are many improvements that a sound economic policy can and ought to pursue inside the agricultural sector. We must however decide upon them with the analytical picture of Fig. 1 in mind. Let us ignore the problem of employing villagers outside agriculture ; important though this employment is from the viewpoint of the peasants' income, *it clearly cannot solve the food crisis*. Agronomists, I am sure, can offer many valuable suggestions how to lengthen the service period of land by staggering different crops. A co-operative use of capital equipment could in this case reduce the amount of capital equipment per acre : for, if the whole village cultivates only one crop, everyone needs one plough with its team during the few days of ploughing.

'The horse eats people' is an old peasant saying, which attests the economic awareness of the villagers. And, indeed, the maintenance of draft animals greatly cuts into the net output of a peasant's enterprise. Surprising though it may seem, it is in the agricultural over-populated countries — such as Pakistan, or India, or Indonesia, to name only some — that a replacement of draft animals by mechanical power is most urgent. From this, however, we need not jump to the conclusion that in such — ordinarily low income — countries, the buffaloes must be replaced by heavy tractors like those of the U.S.A. farms. A heavy tractor, in comparison with a small (garden) tractor, certainly saves *labour fund* and *labour time* as well ; but it does not appreciably increase the yield if *ceteris paribus*. If labour is plentiful, too much mechanization is anti-economical ; and if income is low, as is the rule in over-populated countries, heavy machinery is a luxury comparable to that of a splendid villa on the Riviera used for a couple of weeks each year.[1]

We can then say, for the second time, that it is the turn of the town to raise 'buffaloes' and grow 'cattle fodder'.

6. The problem then is how to transfer the modern 'manure', 'buffaloes', and 'fodder' from the town to the country-side. Clearly, in any economic system whatsoever, they must be paid for : the industrial worker must have his income. But the country-side being as desperately poor as we know it is in all over-populated, under-developed countries, we seem to be confronted with the old vicious

---

[1] It is interesting that Gerald K. Boon, *Economic Choice of Human and Physical Factors in Production* (Amsterdam, 1964), pp. 162, 259, arrives at an equivalent conclusion by a careful analysis of cost functions. Boon's study, I may add, is highly valuable in many other respects.

circle (and a vicious thought, as well) that the poor cannot but stay poor. Many may also argue that allocating industrial resources to the production of things needed by agriculture is a fundamentally wrong move : because it slows down industrialization and because salvation can come only from industrialization. But the double dilemma is only apparent, and it is caused by the fact that neo-classical analysis of production has ignored an important variable.

One of the main secrets by which the Western industrialized societies have achieved their spectacular development is — as correctly assessed by Marx and confirmed by (7) — a long working day. This secret solves our dilemma also. For an illustration that would be familiar to all planners, let us take the current industrial bill of goods of a Leontief input-output model. Clearly, if no industry works around the clock, then this bill can be increased by, say, ten per cent immediately with the existent capacity : all we have to do is to lengthen the working day. Industrial development can then go on at the same speed, the real income of the industrial worker can remain untouched, and, at the same time, there would be a surplus available for the industrial needs of agriculture. And to make sure that we have fully grasped the difference between an agricultural and an industrial process, we should note that a lengthening of the working day in agriculture cannot possibly increase output, not even in a Bali formula. It can only release labour power for other possible uses.[1]

I am fully aware of the practical difficulties of all sorts involved in implementing an economic policy based upon these conclusions. But I wish to submit that, in view of our proclaimed economic aims, the

---

[1] A disregard for dimensions is responsible for the fact that the numerous contributions on linear processes have missed the above conclusions. Thus Leontief, *op. cit.*, p. 173, defines X_i, $1 \leqslant i < n$, as 'total net outputs of all various branches of the national economy [during a particular year]'. Clearly, every X_i is a flow, equivalent to our $Q_0 = \delta Q$. But X_n is there defined as 'total employment' measured in number of persons (pp. 173, 179), and in another place (pp. 42, 160) as 'the output of services (by the household industry)'. That is, in one case X_n is the intensity rate of labour services, equivalent to our H, in the other it is the service δH. Then, x_{ik} is also defined as an input flow, equivalent to our δi or δm. The important point, however, is that the coefficients $a_{ik} = x_{ik}/X_i$ have no longer any relation to the time factor, $a_{ik} \backsim i/Q$. The linearity assumption, correctly expressed is that a_{ik} is constant for *all efficient processes* when these are described as in the table of section IV. 2. Indeed, a_{ik} is constant for *any non-agricultural* process if x_{ik}, X_i vary only because δ varies.

Not to depart from Leontief's analytical assumption, let $(1, -a_{21} ; H_1)$, $(-a_{12}, 1 ; H_2)$, represent the analytical co-ordinates of the industrial and agricultural processes, respectively. Let B_1, B_2, be the desired daily (or annual) net output. The standard system then becomes

$$\delta X_1 - a_{12}X_2 = B_1, \quad -\delta a_{21}X_1 + X_2 = B_2,$$

where X_1, X_2, represent the physical scales of the two processes. There are three unknowns, with $\delta \leqslant 1$.

legal rule of a forty- or even forty-eight-hour week constitutes an anachronism for the under-developed countries that possess some industrial potential of a non-parasitary nature. By this remark, a natural conclusion of the analysis presented in this paper, I may have touched a sore spot : the conflict of interest between the town and the country-side. About this conflict Marx said, in passing though, that it constitutes the pivot of all history.[1] Yet, this conflict seems even more important than that upon which he built his doctrine, for no other reason than the fact that its roots lie in an evolutionary law of nature, the Entropy Law.[2]

DISCUSSION OF
PROFESSOR GEORGESCU-ROEGEN'S PAPER

(From this discussion has been omitted Professor Patinkin's criticism of the proposition that the production function must be linearly homogeneous if the definitions given by Boulding and Stigler are to be equivalent (see page 501 of the paper). Professor Patinkin has subsequently become convinced that his criticism is invalid. At his request, it was agreed that there was no point in committing to the record a criticism which was wrong.)

Professor Patinkin said that he felt out of sympathy with Professor Georgescu-Roegen's paper because of its main point, concerning the linear homogeneity of the production function. If it was asked how the argument of the paper affected estimation procedures for production functions, the answer was not to be found in paper. The paper did not seem to lead to operational differences in empirical work on production functions.

Professor Patinkin said that he did not see any fundamental reason for having a different approach to the production function to be used in agriculture as opposed to the production function to be used in other sectors. Professor Georgescu-Roegen had based his argument for a different approach on the distinction between processes that are started in line (characterizing 'factory processes') and processes that have to be started in parallel (characteristic of agricultural processes). Professor Georgescu-Roegen had cited the case of the island of Bali where rather special climatic conditions mean that it is possible to run agricultural processes in line so that a simultaneity in input use and output realization is possible as in a

[1] *Capital*, i, 387.
[2] For additional details on the economic importance of the Entropy Law, see the author's volume *Analytical Economics: Issues and Problems* (Cambridge, Mass., 1965), Part I, Chapter v, sect. 1.

'factory process'. If the agriculture of the whole world is considered as consisting of one large partial process then simultaneity of the same sort is found. It was only by considering partial processes as being confined to one region that Professor Georgescu-Roegen could sustain his distinction successfully. Furthermore the line that had to be drawn between different sectors when they are to be placed in one of the two categories is arbitrary. Any industry that is seriously affected by seasonality encounters the difficulties that Professor Georgescu-Roegen had noted in agriculture. The hotel industry in Rome is clearly a case in point. Also, there are important changes in agricultural methods which overcome some of the seasonal difficulties which lie behind the distinction ; for example, the use, in the United States, of combine teams which move north throughout the season of harvesting.

Professor Georgescu-Roegen said that he believed that much of the mathematics used by economists ignored the real world. His paper was set at a very basic level and asked whether the mathematical formulae commonly employed represent, correctly and with analytical clarity, the phenomena which economists are trying to embrace in their analyses. Great effort has been made by physicists to comprehend phenomena of the physical world within mathematical formulae which would represent as accurately as possible the main characteristics of the phenomena under examination. It is important that economists should also attempt to achieve this goal.

Professor Georgescu-Roegen said that he agreed that there are examples, such as the tourist hotel industry, which show that climate affects the productive rhythm of activities other than agriculture. However, in most such industries climate is a marginal factor whereas it is paramount in agriculture. Even if Professor Patinkin's point, that the agriculture of the whole world may be taken as one, were accepted, it still remains true that the man who ploughs on one day is idle most of the following days. It would be a strange world indeed if the ploughman and his plough could be flown by superjet from one region to another in order to be able to plough also on the following day. The individual equipment and units of labour used today, not to mention the land, are not the same as those used tomorrow in another area. This is the fact which lies at the heart of the over-capitalization of agriculture over the world if taken as whole, or even in part.

Professor Georgescu-Roegen said that the underlying distinctive feature of agriculture, which he had tried to point out in his paper, is the fact that agriculture needs solar energy over which man has no control. Moreover, the economic process as a whole is unidirectional and not, as the textbooks argue, circular. And agriculture reveals this unidirectional process more clearly than most other sectors. Just as, in mining, there is a gradual depletion of some particular resources, so in agriculture there has been a depletion of the nutrient elements of the soil ; the need to re-employ land each year makes it imperative to devise and implement methods to restore the soil fertility. Perspective planning in this direction is therefore called for urgently. The towns will inescapably have to con-tribute 'manure', 'buffaloes' and 'fodder' to the country-side.

Professor Georgescu-Roegen said that he had concentrated on the con-tinuous aspect of the factory process and the relevance of the time element. However, he was aware of the political difficulties raised by the lengthening of the working day. He said that it is particularly in connection with this fundamental co-ordinate of the production function that the analysis of economic production has been improved by his paper.

Professor Mouton said that he spoke as an agronomist and had found the discussion between Professor Georgescu-Roegen and Professor Patinkin rather surprising. In rural economics it is technical agronomists who first prepare production functions and try to formulate relationships between quantities produced and factors of production. Production functions have been prepared for a number of different countries. Those who use these technical relationships to determine price relationships know it is extremely dangerous to extrapolate to a whole region from specific economic condi-tions. Thus, when they start from farm accounts agronomists are careful to specify that the production functions apply only under very specific conditions. It is, for example, carefully remembered that in all agri-cultural production functions time and weather are implicit. In agri-culture, human activity depends on the weather which may render man's will to produce irrelevant.

Professor Delivanis noted that Professor Georgescu-Roegen had written that, 'we need not jump to the conclusion that in ... ordinarily low-income countries, the buffaloes must be replaced by heavy tractors ...' Professor Delivanis said that he wondered if a similar argument applies to industry in these countries. In fact these countries often use very sophisticated and expensive machines because they can be operated by unskilled labour, while less expensive machinery requires a large number of skilled men. Is there a parallel here with cattle-breeding?

Professor Georgescu-Roegen favoured the discarding of a fixed working week in poor countries, and the intensive use of available machinery without legal limit on hours worked per man. Was he suggesting that there should be parallel employment of two or three teams working a standard working week, or had he in mind a completely different organiza-tion of work time?

Professor Bicanic said that he was interested in the economics of waste in the pathology of development. He asked whether Professor Georgescu-Roegen would develop his idea of entropy as applied to economic pro-cesses. What precisely was meant by his statement that 'Waste is irrevocably waste'? Did he include in waste some of the final con-sumption of goods?

Professor Gale Johnson said that there is a minor qualification to be made to Professor Georgescu-Roegen's simple model of corn production. It is true that, under the most simple conditions of agriculture with open-pollination, the maize grain that goes into the production process may be economically identical to part of the output of maize by the farm. However, in more advanced agriculture the development of hybrid corn means that the output is not identical to the input grain.

Professor Gale Johnson said that he had not fully grasped what Professor Georgescu-Roegen had meant by the assertion in his paper (on page 527) : '. . . we should note that a lengthening of the working day in agriculture cannot possibly increase output, . . .' Professor Johnson said that he could imagine cases where the assertion holds true. However, extension of the labour day does not invariably involve other inputs. It is possible to weed a field by hand and so increase output ; and if other factors are permitted to combine with labour the extension of the working day will in many cases, certainly increase output. Intensive activity can increase the yields in spite of the seasonality of farming.

Continuous flow agriculture is perhaps more widespread than Professor Georgescu-Roegen allows. In Singapore and Hong Kong ten or twelve crops per year are realised by intensive methods of seed-bed transplanting, etc. However, Professor Johnson said that he appreciated that whether or not land and machinery is being used the whole time is not relevant to the main distinction that Professor Georgescu-Roegen was making.

Professor Ruttan said that he hoped that Professor Georgescu-Roegen would develop a little further the ideas in the last part of his paper, particularly with reference to the scale of operations in agriculture. In agriculture the typical situation is that of a man or groups of men moving over the materials of production, and having to attune his (their) movement to idiosyncrasies of the materials. In industry the typical situation is where the materials of production are under the straight-forward control of men and where materials move past men. If the problem is approached in this way, does it not follow that mechanization would be expected to have an impact on agriculture, and on the scale of agricultural units, that is analytically different from the impact of mechanization on the scale of operation and the economic organization in industry ?

Professor Dandekar said that he thought that in the last part of Professor Georgescu-Roegen's paper a number of important issues were raised, but that they were so compressed that some points were not at all clear. On page 526 in the first paragraph of section 6, it is argued that in over-populated, under-developed countries 'we seem to be confronted with the old vicious circle . . . that the poor cannot but stay poor.' But at the end of the paper it was proposed that lengthening of the working day would improve the chances of development. What are the interconnecting arguments between these statements ? It was not clear what groups or sectors of workers would be the most important candidates for the lengthened working week which the paper advocated.

Conceptual Problems in Agricultural Development

Professor Dandekar asked whether he would be justified in regarding as a further implication of Professor Georgescu-Roegen's analysis, the idea that as much agricultural processing as possible should be moved into the agricultural areas. This would make it possible to use rural labour more fully, and fill in some of the idle gaps in the use of capital and labour.

In answer to a short question from Professor James about the difference between a stock and a fund, *Professor Georgescu-Roegen* said that he defined a fund as a particular kind of stock which throughout the productive process remains constant in size. A most fitting example of a fund is the total, free and bound, energy in the universe which does not increase or decrease. A stock, on the other hand, can increase and decrease ; stocks of natural resources that could be depleted constitute the most obvious example.

Professor Georgescu-Roegen said that he appreciated Professor Delivanis's intervention in connection with the substitution of machinery for buffaloes and the most appropriate type of machinery to introduce. He wished to stress the fact that in South-east Asia it was often found that it is far better to introduce small (garden-type) tractors rather than the heavy tractors employed in large-scale agriculture elsewhere. He said that he had not felt it necessary for his thesis to generalize on this point to deal with similar kinds of substitution in the industrial sector. But his answer to Professor Delivanis's question was in the affirmative.

In relation to Professor Bićanić's question Professor Georgescu-Roegen said that in a book now in press he dealt with the concept of entropy and with the irrevocability of waste. The idea of entropy is so complicated that it is not clearly understood even by some physicists ; it would take him too long to explain the meaning of the irrevocability of waste.

Professor Georgescu-Roegen said that the remarks of Professor Gale Johnson on his point that agricultural output could not be increased by lengthening the working day, proved that he had been too rash and that he ought to have inserted the words '. . . in most cases'. Counter-examples could certainly be found, but he was dealing with a complex variety of cases and his statement, of a statistical nature, underlined the fact that there is a definite difference in the percentage change in output between the two sectors in response to a longer working day. In industry under normal conditions a 10 per cent increase in length of working day would bring about a 10 per cent increase in output, while in agriculture the increase in crops would be far from 10 per cent.

Professor Georgescu-Roegen said that Professor Gale Johnson was quite right to point out that in agriculture there are numerous processes in which no input is identical with the output. His purpose in using the example of maize-growing, in which the seed is economically identical with part of the output, was to cover a controversial problem in capital theory and make clear with it the analytical difference between fund elements and flow elements in a static production process.

Professor Georgescu-Roegen said that his example of possible continuous agriculture referred to the case of Bali, and that he was not familiar with the situation in Singapore or Hong Kong. But continuous agriculture is certainly a possibility only in a very few and small parts of the world. He said that the example of greater continuity in factor use in agriculture given by Professor Johnson shows that extreme discontinuity could be modified by technical evolution but it is not of sufficiently wide application at the present time to invalidate the distinctions that the paper made.

Professor Georgescu-Roegen said that the problem of full employment of funds (in his sense) in agricultural production is similar to the old question of cottage industries as opposed to factory industry. The idea of cottage industry is to achieve full occupation of the labour force and not necessarily full employment of capital equipment. However, by now, it costs more to produce a yard of cloth by cottage industry even if wages were not included. The technical innovations in the textile industry have made the cottage industry uneconomic. For a parallel, the cost of a horse-drawn carriage today is higher than that of a Cadillac ; the economics of scale existing at the time when horse-drawn carriages were generally in use no longer exist. Technological innovations in agriculture would undoubtedly bring about changes in present patterns of factor idleness, but the natural basis of agriculture is such that a full factory process would not make economic sense except in a few cases. Professor Georgescu-Roegen said that he did not feel able to go into the question of the effects on scale of the different analytical structure of agriculture and industry. He did not think that there was any clear definition of what was meant by scale in such a general context, and he had therefore tried to avoid using the idea at all.

Professor Georgescu-Roegen said that Professor Dandekar was right in thinking that he meant that those who produce the substitutes for traditional agricultural factors should work longer hours. The peasant cultivator has always produced food for others, but himself has gone hungry. The towns have always had the best of the deal. It is high time that the towns should sacrifice some of their welfare and, by working longer hours, increase at least the supply of fertilizers and farm equipment. Referring also to Professor Delivanis's second question, Professor Georgescu-Roegen said that the easiest available way of accelerating the development process in almost every direction is to increase the length of the ordinary work shift. If capital is already continuously used for twenty-four hours per day then obviously this idea does not apply. But in most cases, factories do not work around the clock. Besides, there is shortage of skilled labour. The lengthening of the hours worked by the single (or double) shift is the only logical and simple solution.

Chapter 25

PROBLEMS OF THE RE-STRUCTURING
OF AGRICULTURE IN THE LIGHT OF
THE POLISH EXPERIENCE

BY

J. TEPICHT

Institute of Agricultural Economics, Warsaw

I. INTRODUCTION: THE DIMENSIONS OF
THE PROBLEM

1. The title of this paper could give rise to a misunderstanding about
the author's opinions. It is therefore worth starting by indicating his
disagreement with simplified conceptions of agricultural re-structuring when these conceptions centre on one of the forms of land
concentration. This is not to deny the importance of the problem of
land ownership within the context of the factors determining the
socio-economic and technical-economic structures within this sector.
But in the developed countries, it is a long time since agriculture was
dominated by reciprocal factor substitution of *land* and *labour* both at
the level of production forces and at the level of the relations of
production. It is the substitution of either or both by a third factor,
namely, by man-made and accumulated means of production, which
becomes decisive in determining agricultural structures. This is
manifested in several ways : in the increasing divergence between the
price of land and the theoretical product of the capitalization of the
farming rent that might be derived from the same land as a holding ; in
the diminishing importance of rent in the total cost of production ;
and so on. The old race for new land, even land less favourable to
cultivation, is being replaced by a gradual trend of abandoning
regions which were, until recently, agricultural. The concentration
of investment in certain areas and the discrimination against the
remaining areas is a further result of this trend. These results are to
be found in a number of developed capitalist countries, and in certain
socialist countries as well, in spite of the fact that the level of growth
is at present lower.

2. The new ranking of factors explains why a change in the agrarian structure, *stricto-sensu*, has so little chance, by itself, of resolving the problems which arise, or even of serving as a starting-point for an effective reform in highly or even moderately developed countries. This brings us to the question : What are the principal determinants in bringing about change, the factors internal to agriculture or the factors external to agriculture? For it is not by chance, that, in spite of all the government efforts, the progress of land concentration continues to be so slow in Western Europe, and that conversion schemes are focused on the type of undertakings called family farms, not exceeding some 20–40 hectares which are unable to ensure a rational utilization of the tools and human efforts to the standard found in a modern industrial enterprise. The progress of vertical integration takes away the last appearance of autonomy from these farms ; but without any bias in their favour, one may note that they continue to survive in spite of all hardships, including those caused by the exodus of the young. It is not by chance either, that even the most industrial of the socialist countries, after having reorganized its peasants and lands into collective farms, finds an urgent need for major supplementary investments in agriculture, for an accelerated formation of staff, and for a rapid increase in agricultural incomes. Judging by what is happening in these two geographical areas, it is difficult to avoid the impression that the heavy accent put on the reorganization of ownership and on the *endogenous* factors of agricultural change reflects, in the majority of cases, an impotence with respect to the *exogenous* factors. This impotence may be of a socio-political nature and due to interests of classes or specific influential groups, but it may also follow from the material difficulties of surmounting the barriers to economic growth.

3. The problems of agricultural re-structuring in the majority of highly and moderately developed countries are essentially the difficulties experienced by the peasant economy when it is being integrated within the industrial economy that surrounds it. Depending on whether the industrial set-up of the country is predominantly based upon capitalist or socialist property, the difficulties of peasant agriculture may be comprehended within the first or second of the following questions : How can peasant agriculture be made competitive ; or, how can its rational integration within the planned economy of the country be assured? In either case there is implied by these questions, the basic problem of *the size and nature* of the agricultural enterprise by means of which the insertion of the peasant

535

economy into the industrial economy is to be achieved. The prescription of integrating agriculture within an industrial framework is, of course, unsuitable for the countries of the 'Third World' that are in the process of abandoning post-colonial structures. The endogenous problems in agriculture, especially those which result from antiquated systems of land utilization, most frequently constitute a basic obstacle to the country's overall progress towards a modern economy. In a large number of countries of the 'Third World' it is necessary to clear the way for expansion, sometimes even to stimulate the creation of a peasant agriculture, and not evade the problem by flirting with premature collectivism, or limiting attention to islands of ultra-modern agriculture that seem to deserve preservation under a new social form. It may be specified, then, for clarity, that the present study concerns the structural agricultural problems of a country where the sort of re-structuring commonly called *agrarian reform* was realized twenty years ago, in a rather radical manner. The peasant economy, liberated from the last relics of feudalism by this reform, found itself embedded in a rapidly growing economy which is socialist in its decisive aspects. The problems arising from this co-existence are all the more delicate because peasant agriculture is at the same time a brake and an indispensable cog in the mechanism of growth of the country in question.

II. THE NATURE OF THE PEASANT ECONOMY

4. The astonishing ability to survive (which should not be confused with mere indefinite existence) of the type of farming comprehended here by the expression 'peasant economy' is confirmed by its long history going back into antiquity, by its 'capacity for suffering' [1] through the ups and downs of business cycles in capitalist countries, and even by its slow death, slower than is usually claimed, in a number of leading new-capitalist countries. A supplementary proof of its power of survival may be seen in its semi-latent existence throughout the collective farms of the socialist countries. All this undermines rather than confirms the view of Daniel Thorner,[2] who sees the peasant economy as a socio-economic system and would like

[1] Louis Malassis, *Economie des exploitations agricoles*, Paris, 1958.
[2] Cf. D. Thorner, 'Peasant Economy as a Category in Economic History', Report to the *Second Conference of Economic History of Aix-en-Provence*, 1962, Editions Mouton, Paris-la-Haye, 1965. The same text is to be found in the 'Annals' of the C.N.R.S. and in the VI Section of the E.H.E. (Sorbonne), May–June, 1964, entitled, 'L'Economie paysanne, concept pour l'histoire économique'.

to put it in the place of feudalism in the Marxist succession of socio-economic structures which Thorner criticized as being traced solely from European history. It is worth pointing out, in passing, that Marx himself aimed at making the study of political economy more independent of European standards of measurement. But, without entering here into a debate which has only indirect bearing on our topic, we can assert that the peasant economy is a category that is valuable in analysing more than one of the historical forms of organization of human society. Existing parallel with the market economy, for example, the peasant economy embodies relationships, the formation, apogee and dissolution of which are spread out over periods of time longer than the duration of more transitory structures.

Our effort to provide a definition derives its main inspiration from Polish theoretical research which was made necessary by demands and vicissitudes of practical re-structuring in a straight-forward sense. As a moderately developed country, Poland has an agriculture that *still* provides partial or full employment to about half of its population of families, but which has already achieved a rather high degree of 'intensity' both internally and externally to its peasant enterprises (see the explanation for this terminology on page 538, below). As a socialist state, having central levers of economic command, Poland presents an instructive example of an advanced adaptation of peasant farming to the structures that surround it, and also of the obstacles to complete adaptation. We set out, then, to characterize the peasant economy, well aware of the pitfalls inevitable when using examples taken separately. We will take care to draw upon our knowledge of the forms that the peasant economy takes in other countries to compensate for this shortcoming.

5. At the root of our usage of the adjective 'peasant' lies the unity, or symbiosis, found in the small agricultural undertaking producing for the market while at the same time forming the full domestic economy of the farmers. European statisticians and agricultural economists know well the difficulty of distinguishing one function from the other when it is a question of calculating the effort expended by peasant households, the utilization of farm products, the division of investments into 'productive' and 'non-productive' and so on.

Among the exogeneous factors explaining the persistence of this type of economy, it is necessary to emphasize the still important burden of the agricultural population (*i.e.* the burden of all those who live by working the land) and the very restricted professional mobility of which this population can take advantage. These circumstances are usually accompanied by a state of affairs where the percentage of *persons engaged in agricultural activities* is higher than the percentage of so-called *agricultural population*. This may be taken as a clear sign that there is regular utilization of the *marginal labour* of members of the family which adds generally in a degressive, though disguisedly degressive manner, to the total agricultural income of the family. This leads us to the paradoxical interaction of endogenous factors whereby the development of the relationship between the small farm and the capitalist market has been marked for such a long time by the reinforcement, rather than atrophy, of its peasant characteristics. The marginal resources of the family farm, especially the cheap labour of the members of the household, permit it and, at the same time, force it to become increasingly embedded in this mixed economy based on the productive utilization of its own products, from which state of affairs it is then so difficult to escape. This interior circuit (the use of intermediate products) has become so widespread that a number of countries with peasant farming have adopted the custom, in their economic analyses, of using the category 'global output' which represents the total value of all farm products, including those which are used as raw material or intermediate goods on the farm itself. This value accordingly involves double counting ; for example, fodder is counted once as a crop and a second time in the value of livestock production. While certain countries mistakenly use the concept of global outputs as an index of the growth of agricultural output or as a base for calculating the degree of commercialization, it does prove quite useful for the study of the structure and evolution of peasant farming. Global output and gross output diverge by very little as long as the direct produce of the soil makes up the major part of what is the final product of the farms ; but the two values diverge with the increase in importance of livestock production and, with this, of what may be called 'the internal intensity' or the intermediate productive home-consumption of output on the peasant farm. Another evolution is marked by the transformation from the use of intermediate productive home-grown output to a larger and larger intermediate use of industrial products and by a greater specialization of farm production ; this may be shown in the divergence between gross output and net output (or value added). The growth of 'external intensity', which starts from the commercialization of the product alone and extends to the more elaborate commercialization of the means of production as well, results sooner or later in the disappearance of peasant farming. It

538

results, at the same time, in a weakening of the connection, which is so rigorous in a system of traditional cultivation, between the *surface area of the farms* and the *importance of the agricultural enterprises.*

6. Finally, at the sociological level, the unity between the market and the domestic-oriented activities of the peasant household is characterized by the subordination of the lot of each individual to the general interest of the family unit, and, vice versa, by the dependence of the latter on what may happen to each of its members.

The contribution of effort by each member of the family remains anonymous with respect to the collective income which is at the disposal of the whole family, or more precisely, of the head of the family. Moreover, there is no prospect for personal promotion other than that afforded by the death or retirement of the family head. This situation is eloquently expressed by the grievances of the 'thirty year old minors' heard so often in the French country-side. At the same time, a variety of social and material impediments restrict the freedom of each individual forming the labour force of the family farm, even when their strength and capacities are insufficiently and unprofitably put to use in the particular group to which they belong. The under-employment in which this necessarily results, the labour of children and the old-aged, which would not be counted as labour in other productive branches, the polyvalency of the peasant activity, which runs directly counter to contemporary professional specialization, and the conflict between the vestiges of the patriarchy and the aspirations of a modern man, are all contradictions of the peasant economy in the world of today. These contradictions are only very partially mitigated by the increase in the surface area of the farms ; and for an effective solution, much larger structural changes, both internal and external to agriculture, are necessary.

III. THE POLISH EXAMPLE

7. The Polish experience during the first half of the fifties provides an example of attempts at restructuring starting with massing together of peasant farms, just as they were and without waiting for an appropriate change in the exogenous conditions of the functioning of the agricultural system. Although less than 10 per cent of the peasant families and their land was reorganized into collective farms, the sample is important enough and the study long enough to provide the basis for a thorough analysis ; this is all the more satisfactory

since the simultaneous presence of a mass of individual peasant farms permits the necessary comparisons. At the beginning of the period in question there were approximately 70 thousand families and towards the end there were some 190 thousand families and 1·8 million hectares. The study according to the latest statistics is still being carried on with some 20 thousand families. One of the most important of the observations that may be made is that, after having added up the respective *outputs* of the collective farm and of the private auxiliary economies of its members, almost the same level of output per family and per unit of land is obtained as that from the private farms in similar conditions and located in the same region. A certain superiority in the yield of cereals for the former is compensated by the superiority of livestock production for the latter. These differences are not, however, important enough to hide another fact no less essential for the analysis : the *structure* of the over-all gross output of the co-operative farms, including the members' private auxiliary economies, is strikingly similar to that of individual farms. Rather than something qualitatively new, it is the traditional peasant farm multiplied by the number of assembled farms.

The difference is found essentially in the *division of inputs*. The 'multiplied peasant farm' is divided internally in two sectors : the collective and the auxiliary. From the technical point of view, the former encompasses most of the crop cultivation, including feed grains, while the latter is devoted in a large, if not major, degree to livestock production. From the economic point of view, the former depends on mechanized labour which is for the most part remunerated in a modern manner. We say 'for the most part' because there is often still some payment in kind ; but the remuneration is, as a rule, *proportional* to the effort furnished, as in industry. In contrast, the auxiliary sector depends on the marginal labour of each member's family, and is remunerated in the peasant fashion, that is, concealed in the value of the production sold or consumed by the family. No less important is the putting to profitable use of the 'marginal' equipment of the former private farms such as stables and pigsties, the rapid replacement of which on a national scale by collective structures with modern arrangements would require enormous investments.

8. It is in this manner that the auxiliary sector is able to recover a portion of the traditional, even archaic resources which evaporate when they come into contact with large industrialized agriculture. But by a curious paradox, the auxiliary sector tends to impose a

covert domination over the collective sector. This domination is all the more real because it is based on the production of livestock, the fulcrum of the internal intensity of the peasant economy and its principal contact point with the market. The circuit which is created between the two sectors under discussion — fodder from the collective fields for the private livestock, fertilizer from the latter for the crops of the former, etc. — partially re-establishes the traditional symbiosis between the field and the stable in the peasant economy ; but this is accomplished in a manner that is indirect and heavy with contradictions. In the household, it often means an aggravation of the duties of the wife who, freed from work in the fields, becomes responsible for the stable, pigsty and poultry yards ; it also means a new affirmation of the patriarchal character of the unit's income, especially when this is received from the collective farm in kind. The whole situation is marked by the tendency to subordinate the collective farm, which is treated as a fodder producer, to the so-called supplementary activities of its members. For the country as a whole, the social cost of the agricultural output is increased ; and this social cost is all the more significant because the increasing use of mechanized means of production is not accompanied by a corresponding reduction of the agricultural labour force.

9. In the case of Poland this last point has only a symbolic value. However, recognition of it, as well as recognition of other realities of the 'multiplied peasant farm', is implied by a variety of measures recently taken, as well as by the tendency to maintain as long as possible the inferior types of collective farms (common fields, private stables) in the socialist countries having the more intense peasant agriculture. Usually these measures have two simultaneous objectives : (*a*) revive and develop to whatever extent is practicable what we call here the archaic resources of peasant agriculture ; (*b*) accelerate as much as possible the transformation of the collective farms into typically industrial enterprises with respect to their equipment, the remuneration of labour, their administration and so on. This may appear contradictory ; but it is more likely to be an intelligent manner of looking for ways to attenuate the looming contradictions and to progress toward their solutions.

It is possible, avoiding the effort of analysis, to ascribe the difficulties in question to the socialist economic system. On the other hand, one can honestly recognize here the reflection of the difficulties of reconciling the various aspects of growth. All of the contemporary socialist countries have followed and are following, in one manner or another, two transformations of the situation with which we are here interested : (1) from individual to collective farming ; (2) from a peasant agriculture to one that is industrial. These two transformations are inseparable, but do not both have the same rhythm ; the outcome depends on which is the most rapid. If it is the first, a multiplied peasant economy results. In this case the national economy is laden with additional charges ; it must face the accumulated problems and make up for the delay in the industrialization.

IV. PREREQUISITES FOR RE-STRUCTURING

10. The analysis of past experiences, and the programmes for the recovery and development of collective agriculture, emphasize the need for a considerable increase in investment. This in itself, however, is insufficient unless consideration is also given to the potential of the labour force remaining in agriculture. If this labour force is not reduced sufficiently to permit a radical change in the 'organic composition' of the whole system, the productivity of the combined inputs will decline at the level of the individual farm, as well as at that of the agricultural output of the nation. To these two requirements a third may be added. A large substitution in agriculture of living labour by embodied labour, or 'capital', presupposes a high degree of mobility for the former ; theoretically, full opportunity for rural inhabitants to change occupation, and in changing, to have parity of income with other professions, is required. It is not simply a coincidence that, in the developed capitalist countries of the two sides of the Atlantic, the problem of parity has become so acute in recent times. It is becoming a problem in the most advanced of the socialist countries, Czechoslovakia, where the professional mobility of the rural population is, at the present, surpassing the desired limits.

11. Only when the three exogenous conditions given in the preceding paragraph have been completely fulfilled may one speak of a rational and humane end to the peasant economy, and abandon the marginal resources used for market production and for the needs of the farmer himself. Then only will it be possible to envisage, on the national scale, an organization of agricultural enterprises severed from the households and existing uniquely as a function of the efficiency of advanced techniques and of the individual and collective human capacities for organization and work. Until such conditions

have been brought to fruition, a combination of the past with the future will remain the rule with a growing domination of the latter.

The present state of the economy and technique require this combination ; but that is not to say that they always require it in the same form. The very choice of techniques must vary according to whether one is facing a shortage of agricultural labour or whether one is still obliged to assure agricultural labour fuller employment with an appreciable increase in the value added by such labour. The second situation may be as capital-intensive as the first ; but it implies another type of large agricultural enterprise, at least during the transitory period, and a more gradual disappearance of the small farm.

12. The necessities of technical progress, of product standardization and conditioning, etc., lead to a new division of labour that transforms agriculture more and more into a sort of industrial unit, which is located between other preceding and succeeding units, and by which it is relieved of a variety of its traditional functions. This new division of labour may be realized within the framework of great agricultural-industrial *complexes*, such as those organized in a number of places in the U.S.S.R., Yugoslavia, Czechoslovakia, etc., on the basis of large collective organizations, or, alternatively, around factories processing agricultural output. It may also be effected under the form of *joint production* whereby the activities of co-operative or state establishments, preceding and succeeding the agricultural activity of the peasant, are linked to the latter in a kind of vertical, socialist integration. This is especially the case of Polish production contracts and of the Yugoslavian co-operation contracts, but also to a certain degree, in certain other socialist countries, of the ties between the processing and commercialization enterprises on the one hand, and the auxiliary economies of the collective farm members on the other. Leaving aside the differences of social content, the two new forms of large agricultural organizations, the agro-industrial complex and the joint production enterprise, both have technical antecedents in the developed regions of the capitalist world. Both constitute a proof that the size of the enterprise in this domain can no longer be based, either solely or primarily, on the cultivated surface of a single tenant. The large size of the enterprise is one of the essential factors for an efficient application of existing techniques and, consequently, one of the indispensable aspects of a 'complete' re-structuring of agriculture. But as long as the general level of economic development renders a radical population exodus from

rural areas impossible, or as long as other socio-economic elements necessitate the continuance of peasant farming, priority must be placed on other factors for promoting efficiency. Without neglecting the usefulness of a certain number of specific enterprises where, here and now, an optimal structural combining of all the factors of production including land may be worked for (for example in Poland this might be tried on 12 per cent of the land belonging to state farms), it is advisable on all the rest to try to obtain the most practicable increase of the social productivity of agriculture in general, rather than the maximum technical productivity of a small number of enterprises. A number of structural reforms may contribute to this effort at the same time that they prepare and partially advance a more fundamental re-structuring : (*a*) reforming of the multiple contacts that the peasant farm has with its economic surroundings ; (*b*) reform of those systems of farming that need to be simplified and reconciled with the special services being made available ; (*c*) reform of the territorial distribution of industries in joint production with agriculture with a view to the social and economic stimulation of rural areas ; (*d*) reform of the professional structure of the rural population from which it would be necessary to recruit specialized personnel having knowledge both of business and of agriculture to provide the industrial cadres of a modern agriculture ; (*e*) reform of the rural residential structure the 'topographical nucleus' of the present and prospective changes for the agricultural production units ; (*f*) reform of the administrative structures, especially at the points of contact between the joint production units of which we have been speaking.

13. Last but not least, the democratic character of the administration of the points of contact between the peasant farms and the surrounding socialist sector appears to us as an essential guarantee of the efficacity of the whole system.

The fact that the mass of small agricultural owners have not been ruined, but integrated — and it may be said that in Poland this has happened in an organic fashion to the mutual advantage of all concerned — into a contractual production process, is certainly related to the fact that the different enterprises working at the beginning and the end of this process belong to the same socialist planned sector. It is not, however, simply a question of avoiding the elimination of less competitive units as would be the case with capitalist integration. Economics and humanity make it necessary to avoid, as well, a danger that threatens both worlds, although not to the same

degree : making man the object of the integration of things, transforming him into a passive cog in an immense commercial agriculture, instead of making man himself the subject of a conscious and
direct integration.

Centralized or not, directed by the state or by co-operatives,
the system of contractual production must on the economic plane
adapt itself to the same demands for size that are imposed on many
other institutions in contemporary mass society. In all such cases,
the operation of this imperative has its counterpart which poses a
danger from the social point of view, and also in certain respects from
the economic point of view : the vertical integration of producers
signifies their separation from the decision centres that concern their
productive activity, that is to say, that concern themselves. The
decisions of these centres continue, in spite of everything, to be
partially subjected to the spontaneous regulating forces of market
fluctuations, rather than always avoiding them by means of an
organized feed-back in the form of a practical right of supervision and
questioning on the part of the workers of the land. This requires
that the vertical re-structuring of agriculture generate, step by step,
a new type of horizontal re-structuring.

14. Two conditions must be fulfilled by this type of re-structuring :
(*a*) there must, as an initial basis, be a kind of enterprise, even if only
'partial', to serve as an intermediary between the big enterprise at the
top and the small peasant farms near by ; and this intermediary
organization must be justifiable from the economic point of view
(partial processing of products, transport services, agricultural
machinery) ; (*b*) the perimeter of action of this kind of 'partial'
enterprise must facilitate its direct administration by those concerned
(a village or group of neighbouring villages), at the same time that it
constitutes the territorial framework of a future, modern collective
unit of production.

The preceding suggestions do not entirely resolve the problem ;
they are rather the preliminary outline of the solutions that are being
sought. In the practice of Polish agricultural policy, efforts in this
direction have taken the form of *agricultural circles*, which are
organizations that are at present especially concerned with the maintenance of machinery in the rural areas. In the case of Yugoslavia, it
is the *general agricultural co-operatives* that represent a practical
effort in this sense. In both cases, clarity and consistency are lacking : the Polish circles evolve too easily into a sort of appendix on the general agricultural administration, while the Yugoslav

co-operatives change their objective too easily; rather than enlarging the joint production with the great mass of peasants, their activity is centred on the purchase of peasant land and creation of their own big farms.

It must be hoped that with greater practical experience and with the increasing body of discussions among the exponents and theoreticians of rural socialist re-structuring, the means of realizing this re-structuring will become at once more rational and precise.

V. CONCLUSION: RELEVANCE FOR OTHER COUNTRIES

15. What value may there be in the experience of a particular country belonging to one of the 'Worlds' represented at this Conference ? There is practically none for those who expect to find directly applicable remedies ; but there is a great deal of valuable experience for those who can make indirect use of it in helping to work out what is essential for other contexts.

Considered in this light, a fundamental difference may be observed in the difficulties of agricultural re-structuring between the countries of which contemporary Poland is a partner, and those which are presently the most economically developed. In the former, it is the present level of growth which makes the situation delicate. In the latter, it is especially the institutional obstacles which prevent the objective process of agricultural contraction from being accompanied by an improvement in the conditions of the farmer. In these obstacles may be seen the problem of income disparity which is becoming greater in spite of a higher rate of increase of productivity in agriculture than in industry (at least this is the case in some of those countries). Here one is confronted with the growing discrepancy between the agriculture of regions that have long been favoured and that of regions of traditional under-investment. This discrepancy is seen in the disconcerting inability of a regime dying on the virtues of 'purely technocratic' planning that is to overcome the anarchy of agricultural markets and to replace the reign of oligopolies both upstream and downstream of agriculture by co-operative or state establishments, or even to prevent these oligopolies from reaping the principal benefits of public investment in the agricultural infrastructure.

It has already been emphasized on a general level that the very notion of agricultural re-structuring brings up another combination of problems in the countries of the 'Third World'. Without going into a subject which would require a special study, it seems useful to point out two reasons which frequently explain the manifest tendencies of these countries to copy Europe and North America. One is the understandable desire to progress more quickly ; this, however, leads to an underestimation of the potential resources of a peasant economy which may be further developed and guided by a public industrial sector. It is here that the experience of certain socialist countries may furnish useful inspirations). The other reason lies in the resistance and objective impediments to the realization of agrarian reforms. This is much more difficult in these countries than is the nationalization of the small, modern agricultural sector left over from the time of colonialism.

16. The moral of this study is an appeal for comparisons of which we have tried to show the advantages. Let us hope that it will be heard.

————

DISCUSSION OF PROFESSOR TEPICHT'S PAPER

M. de Farcy said that a central idea in Professor Tepicht's very stimulating paper might be summarized, somewhat naïvely, by saying that in Poland at the present time they are trying to arrive at a socialist form of agriculture which, rather than being centred on the collective, would realize the advantages of vertical integration for a large number of peasant farms, while giving them guarantees against being absorbed by the powerful integrating forces.

Discussion might well centre on two important aspects of the problem : the profitability, or lack of profitability, of collective farms ; and the organization of peasant farms to arm them against the external forces of vertical integration.

As presented by Professor Tepicht, the achievements of the collectives seem to be slight. Even if the production from individual private plots is added, the output of collectives does not seem to be greater than that of comparable private farms. M. de Farcy asked whether Professor Tepicht could analyse the reasons for this poor performance a little more closely than he had been able to do in the compass of his paper. He suggested that the following list of questions would indicate the points on which more clarification would be helpful. Is it largely external conditions that bring about these poor results ? In particular, are the prices of products mainly produced by collectives less favourable than the prices for products grown on private farms ? Are the collectives obliged to incur high costs in purchasing industrial services which they buy in greater quantity than the private farms ? Turning to internal factors : is the quality of work inferior and are the hours that workers are prepared to put into that work shorter than on the private farm ? Is there a lack of incentives to good management ? Of what calibre are the managers of collectives ? (Would it be possible to hear something about the schools or colleges in which the senior management of the collectives are trained ?) There are also basic questions about the sorts of production entrusted to the collectives. Have studies been made to determine the optimum size of enterprise for producing these different lines of product, and to determine the optimum size of production units created within these enterprises to undertake particular parts of the productive process ?

Turning to the question of vertical integration in a peasant framework, M. de Farcy said that Professor Tepicht had emphasised that the peasants would have to arm themselves against the risks of being dominated by more powerful sectors of industry, and had argued that this implies a need for a horizontal defence arranged among themselves ; this defence would need to be of human dimensions which suggests a traditional grouping such as the commune. M. de Farcy said that he fully agreed with this suggestion and cited the case of organizations in the Netherlands which play an analogous role. He said that the representatives of agricultural workers who make up the Council of the Produkschapt du Lait in the Netherlands are designated by the syndicates of agricultural workers themselves and not by government, industry or the large farmers.

M. de Farcy said that he again had a number of particular questions to ask which would amplify what Professor Tepicht had written about this question of communal defence against domination. A number of problems would not be answerable at the local level. How would the local communal groupings be concerted for national purposes ? There would be a need for men who would be able to agitate effectively, but also for men who would be able to discuss skilfully and powerfully on economic matters. How would such men be created from among the peasants ? Does Professor Tepicht think that the peasants would in fact have to entrust their interests to their employers ? On the question of the relationship between the groupings and the enterprises with which they are integrated : How would contracts be determined and agreed upon ? What arrangements would there be for revision of contracts, and what opportunities for appeal against these contracts ?

M. de Farcy asked whether a horizontal organization of the sort Professor Tepicht had suggested would in itself be sufficient. Collectives appear to undertake a number of commercial functions. Would it not be necessary for the groupings to exercise some power over commercial matters as well, and so become involved in activities further forward in the integrated structure ?

Tepicht — Re-structuring of Agriculture

Professor Georgescu-Roegen said that he was sympathetic to Professor Tepicht's dissatisfaction with Daniel Thorner's definition of a peasant economy. But his own dissatisfaction is based on other grounds than those of Professor Tepicht. Professor Thorner defines a peasant economy as one where more than 50 per cent of the national income is produced by 'peasant farms'. This is tantamount to defining a forest country as one in which 50 per cent of the area is covered by forests without explaining what is meant by 'forest'. Thorner does not explain what a peasant is ; he does not seem aware of the difference between a peasant family farm and other agricultural enterprises which operate according to the market price mechanism.

Professor Georgescu-Roegen said that he found that, by contrast, the paper of Professor Tepicht shows great insight in the problem, but he wished to raise one point. According to Professor Tepicht, the character-istic of a peasant family is 'the subordination of the lot of each indi-vidual to the general interest of the family unit'. This subordination, however, is not characteristic of the peasant household alone ; con-sequently, the problem of defining a 'peasant economy' is not solved. Professor Georgescu-Roegen said that in his opinion this problem was satisfactorily solved in 1890 by Baden-Powell. Basing his definition upon a study of the Indian village, Baden-Powell said that a peasant economy consists of fairly isolated production units — the villages — in which the productive activity of every household is connected with that of every other household by indissoluble bonds. Would such a definition be acceptable to Professor Tepicht ?

Professor Tepicht said in answer to Professor Georgescu-Roegen that though he did not accept Thorner's definition of peasant economy, he agreed that it was characterized by durability and a slow rate of obsoles-cence. Professor Tepicht said that he took as his starting-point in defining a peasant economy the requirement of a high percentage of the total population working in agriculture. The peasant economy as he defined it was not necessarily a subsistence economy, but rather an economy where the market activity was not such that agricultural activity was inherently commercialized ; that is, there was a synthesis between commercial and subsistence production.

It is true that the proportions between the two components of total output of the peasant farm — the marketed outputs and the domestically consumed output — vary not only from generation to generation, but also according to the fluctuations in the market prospects. But the two parts of total output are linked by the functioning of the peasant economy which makes the greatest possible use of marginal resources, and particularly of the marginal labour services which work for a return below the 'normal' return for such services as long as this at least means some increases in the family income. It is this characteristic that essentially distinguishes the functioning of the peasant economy from the mechanism of the capitalist economy, in which the entrepreneur expects a rate of profit proportional to his invested capital and in which the workman expects payment proportional to the labour services he expends. This characteristic constitutes, furthermore, the difference between the peasant economy and every sort of industrial economy, including that set within a socialist institutional framework. Professor Tepicht referred to an article by the Soviet economist Shmielev in the review *Voprosy Ekonomiki*, which pointed up the issues raised by the basic characteristics. Replying to those who hoped to see the day come when work time at present devoted by the members of the kolkhoz to their personal plots would be transferred and added to the work that they already perform within the collective, Shmielev predicted that as the kolkhozes became economically sounder, and as income increased, this extra time would not be transferred from work on personal plots, but would simply be taken as leisure time. This point makes beautifully plain the peculiarity of peasant agriculture, in as much as it is truly peasant, whether private or collective.

Professor Dandekar said that Professor Tepicht had referred to the results of a study of agricultural families in Poland which indicates that the gross production of collectives is directly comparable to that of private farms, and that therefore collectivisation has not meant more than putting together a number of traditional peasant farms. Professor Tepicht had gone on to indicate that this situation is to be expected because of the dominance of the peasant sector, and that it would only be changed if the collectives move in the direction of industrialized farming. Professor Dandekar said that he thought it would be profitable to consider another way of looking at the question — a way which is important if the arguments were to be generalized to economies where there is a problem of over-population. Professor Dandekar said that if, after collectivization, the same people still work the same land with the same techniques, collectivization clearly makes no difference. But it would be wrong to discount collectivization simply because this might occur. If the situation prior to collectivization is of a dense population working on scattered and fragmented land, collectivization would make possible a more rational allocation of human labour even if the social, commercial and technical conditions mean that there are no basic changes in the type of agricultural production. By collectivization the surplus labour on family farms would become more mobile, and could be used for other purposes. Professor Dandekar said that Professor Tepicht had argued that labour on collectives should be reduced. In densely populated countries where this is not practicable, at least immediately, collectivization promotes mobility of family labour which could be used for land reclamation and other projects which would not be undertaken by a non-collectivized peasant economy.

Professor Le Bihan said that the industrialized socialist countries are on the eve of radical changes in agriculture because of the need for new arrangements between the different levels of food production under the impact of technical progress. Whether agriculture has been collectivized into large-scale artisan undertakings, or remained uncollectivized, there are important common features in these socialist countries. All upstream and downstream concerns are owned by the state. Because of this fact, it lies in the hands of the state to decide from what level of the whole food-producing industry any vertical integration should be instigated and controlled. Any idea of greater autonomy for the kolkhoz raises the problem of possible contradiction between that autonomy and the relationship with upstream and downstream sections of the industry as greater degrees of integration become necessary. The objectives of the kolkhoz might conflict with the demands of other sections. If this happens the state might find that it faced the conflict between a policy of decentralization and a policy of technical modernization. Professor Le Bihan said that he thought that a number of mistakes might be avoided if the socialist countries examine the experience of Western countries. The capital requirements in the process of integration itself have to be assessed with great care. Professor Le Bihan also suggested that if greater autonomy were to be given to the major units in the industry, and if they were to be called upon to react to active price policy in the movement towards integration, then management skills of a totally new sort would have to be acquired ; those who have been managers under the old arrangements for twenty or thirty years could not become entrepreneurs overnight.

Professor Gale Johnson asked whether in the comparison between the output per family and per unit of land on collective and private farms the measure of output is a gross measure. He asked whether Professor Tepicht had measures of output in terms of value added per worker and per unit of land. Such measures would indicate more exactly what difference there is in the importance of purchased inputs between the two types of farm.

Professor Gale Johnson also asked what was meant by 'large size' as Professor Tepicht used it. Did Professor Tepicht mean units larger than pre-existing peasant farms ? If so, how much larger ? This question is of importance if we are to try to attach some quantitative significance to the statement that there are advantages in large size.

Professor Tepicht said in answer to Professor Gale Johnson that in comparing the different types of farm economy, the study to which he had referred used both global and value-added measures of production. When he referred to the relative equality of output he meant global output. With respect to value added there are considerable differences between the different types of farm. In a comparison between state farms and peasant farms (including co-operatives) the value added by peasant farms is far larger. This partly reflected the fact that the input of marginal labour in these farms is greater than what is indicated by the revenue that is described as wages. In Marxist terms, the organic composition of capital is far higher on state farms which are closer to industrial agriculture.

Professor Tepicht said that these differences in the input-output structure of the different sorts of farms make it impossible to give absolute measures of relative efficiency. Much depends on the historical stage of development, and the existence of pre-industrial and industrial types of structure side by side makes strict comparisons meaningless.

On the question of scale, Professor Tepicht said that he did not mean by large scale the dimensions that are typical in the United States. In the United States the average is about 2·5 hectares per rural inhabitant, in U.S.S.R. it is 2·8 hectares (though with less favourable average national conditions than those in the U.S.A.), in Poland it is 0·6 hectares. All relationships are therefore different. State farms have an average size of 500 hectares and the cost savings from this scale of operation indicate considerable economies of scale. Further economies are being realized by forms of co-operation devised between state enterprises, such as the division and specialization of labour between a number of co-operating farms. The average size of peasant undertaking is between 3 and 10 hectares.

Professor Tepicht said that the nationalized industries have given greater and greater services to agriculture, and that this has inevitably modified the position of the farmer. At the present time farmers are linked with these state sectors by a system of contracts. However, this is regarded as a transitional situation which will last for a period determined by exogenous factors and not by the reorganization of agriculture. Professor Tepicht said that Professor Le Bihan had rightly pointed out some of the potential difficulties in moving towards greater collectivization at a time when the relationship between the farmer and the upstream and downstream sections of the industry had to change. He said that there is little positive that can be said at this stage. How the human element will have to be dealt with depends on the sort of economic solutions that are found. In general, it is clear that both horizontal and vertical integration will be taking place at the same time. At the present stage and with the present growth rate, vertical integration is acting as a stimulant to horizontal integration. The gains to be achieved from horizontal integration are greater where vertical integration has created opportunities for large-scale standardized production. As the technical and manpower capacities of the small peasant farms become inadequate, the idea of horizontal integration will become attractive to the farmer.

Professor Tepicht turned to the points raised by M. de Farcy. The staff who run the Polish production co-operatives are not of poor calibre ; they are mostly chosen from among the most experienced farmers of the village and most of them have taken professional courses. The co-operatives have access to the services of agronomists made available for this job ; beyond a certain size, the co-operative can have an agronomist appointed exclusively for itself by the state. Nor can it be said that the work done by the members of the co-operative is inferior to that which they once devoted to their individual farms. It is simply different. In the first place, the hours worked are less. The farmer who works on the collective farm claims a working day similar to that of an industrial worker in length, in working conditions and in the rate of return for it. The demand for a standard length of working day is in principle met : the second demand, for comparable working conditions, is met in so far as the particular features of the country-side make it possible : the third demand, for parity in wages, is the least realized and this for two main reasons : (1) the modernization of agricultural work by machinery has not been able — because of insufficient expenditure on modernization — to create a large enough out-migration to ensure that the cost of inputs (including the payment of labour) will be competitive with that of low-cost output of the traditional peasant farm ; (2) the mechanization of livestock production is far behind that which has taken place in arable farming, so that it is reasonable and economically sound that members of a co-operative consider it worth while to retain the pre-collective individual system for the greater part of livestock production, thus supplementing family income with returns derived from work of a relatively primitive kind. The appropriate reply to the main point in M. de Farcy's questions is, then, that it is the *external conditions* which cause the labour services of the members of collectives to be divided between collective work in the production of crops and work outside the collective in the production of livestock, and it is this that produces an economic result which is more or less identical to that which held before collectivization took place. Professor Tepicht said that the situation can even be regarded as a success, given the increased total payment to factors of production. However, in talking in terms of the external conditions, we have left on one side the internal conditions that stem from political measures that may be more or less effective according to the time period under analysis. The Polish production co-operatives receive the same prices for their produce as do the private farms, and they are not treated any less favourably in their purchases of inputs — rather they enjoy certain special advantages. Professor Tepicht said that in his view the main external reality for the co-operatives consists of the present difficulty in radically reducing the absolute number of workers on the land. The combination of modern equipment and an unchanged quantity of workers can only increase the cost of production, make a decent level of wages difficult to achieve and give rise to the complications which were dealt with in the paper. Professor Tepicht said that it is his view that the key to the problem lies in the study of what may be called 'the demography of agriculture', and of the role played by this demography in the employment of the population of the country. It would seem to be these indices which are the most necessary for judging to what extent and in what form collectivization in agriculture may be brought to successful maturity.

In this connection, Professor Tepicht noted that Professor Dandekar had correctly stated that the surplus of peasant work force used to improve the local infrastructure (above all in irrigation schemes) might create the basic conditions for a fuller use of the land at the same time as a fuller use of the agricultural labour force. But does this not suggest that there is a need to put peasant holdings on a communal footing straight-away ? Professor Tepicht said that he found it difficult to pass judgement on the problems of a country which he did not know ; however, he could say that the experience of Poland and a number of countries of the east and south of Europe in the first few years after the war showed that such communal effort in the sphere of infrastructure could well succeed, provided that an appeal could be made to peasants who had already been encouraged by the promise of land reform. The redistribution of land, the dispossession of the usurer and profiteer in favour of state-supported co-operative trading may create the psychological conditions necessary for such an appeal to be answered favourably. From this point of view, the collective effort of peasants continuing to be, or rather having finally become, owners of the land and of the produce which they draw from it, constitutes at least a point of departure for a future collective agriculture, though not as yet the achievement of collectivization ; for the way to that achievement is full of pitfalls because of the precocity of the communal activity.

Professor Tepicht turned finally to M. de Farcy's question concerning the power of horizontal self-defence of Polish peasants producing under a contract system. He said that the essential aim is that the peasant shall be able to accept or reject the contract or to choose the one which most suits him. There are in Poland a number of proposals which would replace the present competition between the various industries in the contracting field (for example, the sugar refiners, the brewers, the vegetable oil refiners, etc.) by a co-ordinated plan which would take into account regional needs and potentialities, as well as the different types of farms in the region. There are also divergent opinions on the matter of the level at which the co-ordinated plan ought to be put into effect. One view was that it should be at the level of the central planning of agricultural industries ; another, that it should be the national specialized unions of planters, breeders, etc. ; or, again, at the level of the local sugar refinery, canning factory, or whatever it may be, as opposed to the agricultural groupings within the villages lying within the perimeter of the commercial activity of these plants. The smooth running of the plan, the benefits of 'feed-back' derived from direct confrontation between the interested parties, and finally the value of the system in inculcating a collectivist spirit in the country-side — these are the three objectives that are to be reconciled in making the choice which has not yet been taken.

Chapter 26

TURNING-POINTS IN ECONOMIC DEVELOPMENT AND AGRICULTURAL POLICY

BY

RUDOLF BIĆANIĆ
University of Zagreb, Yugoslavia

I. INTRODUCTION

THE purpose of this paper is threefold : To show that economic development does not follow a single line of an elegant 'take off', but that there are several turning-points in the main trends of development, many pitfalls of structural changes, and those of policy changes to match.

The second task is to find common denominators which would express some underlying conditions of development in the agriculture of all countries, developed ones and those in the process of development ; socialist countries and non-socialist ones. It seems to me that the division of the world into two separate and distinct types of countries is greatly overdone and that the idea of 'one world' in which equal standards should be applied to all human beings, would be more accurate. It would also make it possible to follow the process of development as a continuum. These denominators should be applicable to both socialist and non-socialist countries making it possible to allow international comparisons of both and assess their achievements measured on a comparable basis. We believe that such common measurement is found in counting heads of population, hectares of land and capital investment converted into dollars on the basis of relative prices.

The third purpose — as the main subject of this Conference is to assess the role of agriculture in the process of development — is to measure the relative role of agriculture in a country on a comparative basis, for which purpose percentages were used, taking a country as a whole as 100 per cent.

Conceptual Problems in Agricultural Development

The process of economic development is a complex one during which many of the constituent parts change in the course of time, so that some of them actually reverse their original trends. If this process is enfolded within the global process of growth such changes may easily escape the attention of policy-makers at the right time so that they suddenly find themselves in a situation where policy measures, hitherto quite effective, produce less satisfactory or even adverse results. Such changes in long-term basic trends of development we will call turning-points in economic development.

This is of particular importance in agriculture because of its very complex nature and because the composition of the developmental process is such that some factors operate over long stretches of time and others are of rather short duration.

In this paper we have reduced our observations to three basic factors in agricultural development : labour, land and capital, and have tried to see what the logical consequences for agricultural policy are of turning-points in these three. Changes within these three have not, however, been isolated but have been examined within the framework of the national economy, thus emphasizing the meaning of turning-points for agricultural policy as part of the general economic policy of a country. For analytical purposes these changes have been examined as independent variables.

Since we are trying to study the process on a comparative basis in order to find trends in general developmental policy applicable to very different countries, all in a short paper, the study has necessarily remained on a rather high level of abstraction.

II. LABOUR

As far as labour is concerned a turning-point in agricultural development is reached when the agricultural population stops increasing in absolute figures and begins to decline. Taking into consideration both this point, and the relative importance of total agricultural population in a country, we may divide development into three main stages : one, when the agricultural population increases in absolute terms and decreases relatively ; the second when the population rounds the turning-point, *i.e.* almost stagnates in absolute terms and continues to fall relatively ; and the third when the population in agriculture decreases in both absolute and relative terms. All three stages happen under conditions when the total population of the country is increasing. (See Chart I.) Three stages of agricultural policy rationally correspond to these three stages, that of life parity, of price parity and of income parity.

MOVEMENTS OF AGRICULTURAL POPULATION
ABSOLUTE AND RELATIVE NUMBERS

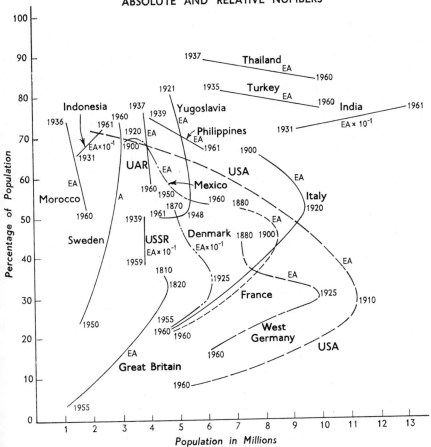

A = Total Agricultural Population   EA = Economically Active Agricultural Population

CHART I

Some examples can be given : in India and Indonesia agricultural population increased both in absolute and relative terms between the 1930's and the 1960's ; therefore they have not yet reached the first stage.

Countries such as Thailand and Turkey show a small decrease in percentages of agricultural population while others, like the Philippines, Morocco and United Arab Republic, show a considerable relative decline in active agricultural population, although the absolute number is still increasing. This could be interpreted as showing that the first group is rather far from the turning-point while the second is approaching it.[1]

Great Britain reached the turning-point in agricultural population as early as 1820, Sweden in 1880, and France in 1900, the U.S.A. in 1910, and Italy in 1920, while Germany did not reach that level of development in agriculturally active population until 1925, the same year as Denmark. Finland developed to this point in 1930, Yugoslavia in 1948, and the Soviet Union as a whole in 1956, with 41 per cent of the active population in agriculture. It is significant that the turning-point was reached in different countries at various levels of percentage of the agricultural population in the total population (varying from 31 to 60 per cent) and at a different national income levels per head (from 132 dollars to over 500 dollars) (see Chart I). This definitely has some connection with development but implies that further research for other variables influencing the developmental process is necessary.

We have said that, *in the first stage*, agricultural population decreases slowly in relative terms but shows an increase in absolute figures. There are an ever-increasing number of men in agriculture, which already suffers from an abundance of labour. There is little opportunity for other work (slow relative change). Income per head from extra agricultural activity is greater than from agriculture, but the opportunities for outside jobs are rather limited. Opportunities for increasing the market for agricultural produce develop slowly. Thus subsistence economy (life parity) is the rule with maximization of physical volume of production per hectare as a common goal. The main function of the organizer of production is to receive supply for the needs of the family.

The price of land is determined far more by population pressure than by yield or income. Dismemberment of family holdings is a process which accompanies the struggle for existence. There is competition for land between the small peasant or landless labourers and big landowners, which usually ends in some kind of land reform.

When there is such agricultural over-population, the use of as much labour as possible is commonly thought to lead to progress ;

[1] On the basis of such figures, some predictions could be made regarding the time when the turning-point in agricultural population would be likely to appear.

thus labour-intensive farming is both an outcome of the situation and is seen as the way to development.

The immobility of labour at this stage results in under-employment or concealed unemployment. Family existence has priority over business calculation, and the peasant way of life over technical progress.

As, during the course of development, population pressure grows less in absolute terms and as the relative decrease of those working on the land gradually opens up markets for agricultural products among the non-agricultural population, the position of agriculture in the national economy changes.

The second stage, rounding the turning-point, occurs when agricultural population stops increasing and begins to decrease in absolute terms but falls relatively at an accelerated pace, so opening the market for agricultural products faster than the decline in agricultural population. At this stage mercantile agriculture prevails. Development depends on the terms under which agricultural products are sold on the market, on the means of production, on the consumption goods bought and on the labour hired. Thus the problem of price scissors (price parity) dominates agricultural policy. The dependence of agricultural output on purchased inputs and hired labour dictates the rate of economic development, and the instability of the relationship between the two determines the pace of progress. The cost-benefit relationship determines the maximization of output per production unit (holding, estate). The operator's main function is that of a commercial manager.

In the third stage, when the agricultural population declines both in absolute and in relative terms, agricultural development is dominated by an ever-increasing shortage of labour which is now more and more finding an outlet in non-agricultural occupations. Both the increase in the agricultural population in the previous stages, and the failure to succeed in commercial farming because of lack of land, capital or managerial ability, lead to the proletarianization of the greater part of the peasants. This speeds up the process of substitution of labour by capital in agriculture. The gradual depopulation of agriculture is linked to a treble change: agriculture is de-proletarianized; there is a change in the quality of labour because of its feminization and senilization, since the able-bodied male labourers and unpaid family members leave agriculture first,[1] and at

---

[1] For changes in the mobility of agricultural population by groups see my paper 'Lack of Institutional Flexibility in Agriculture' in *Proceedings of the Xth Conference of Agricultural Economists: Mysore 1958*, Oxford, 1959.

the same time, there is an increasing technological burden put on those who remain on the land. Thus agricultural labour has to be transformed from a traditional way of life into an occupation which leans heavily on professional training and the ability to introduce innovations. The increased size of the market makes the demand for agricultural products more urgent, and this changes the role of the agricultural operator. He becomes a technical operator rather than a commercial dealer.

Structural changes go hand in hand with the absolute decline of the agricultural population. Those operators who remain on the land get steadily bigger, until a new wave of technical innovations sweeps the relatively smaller ones away.[1] This gradual process splits agriculture into an even smaller number of large capital-intensive commercial farms and a great number of labour-intensive small family farms giving full or part-time occupation to their owners.

Maximization of output per man becomes the main goal of agricultural policy, which gradually makes agricultural development less and less dependent on ever more expensive and scarce human labour.

III. LAND

The most important turning-point regarding land is that when the agricultural area begins to decrease in absolute terms, fewer hectares of land are being worked than before.

We shall distinguish three different situations : one is before the turning-point is reached ; the second occurs around the turning-point ; and the third when the tendency to diminish the cultivated area prevails. This whole analysis concerns the process of development, so that both the increase and decrease of the cultivated area happen under conditions of continuously increasing volume of production.

In the period after 1962 agricultural land was still increasing in some European countries and in many non-European ones : *e.g.* Greece, Spain, U.S.S.R., and in particular Kazakhstan, and to a lesser extent in the Ukraine, also in Canada, India, Japan (see Chart IIa). The turning-point in land increase was reached after the second world

---

[1] In his budget speech 25 January 1965, the American President L. B. Johnson said that of 3·4 million farmers at present in existence in the U.S.A., only 1 million can be expected to gain a decent living from farming, even with continued government assistance — and that in a country where agriculturally active population represents 5 per cent of the total. The aim of French agricultural policy is to reduce agricultural farmers by 1 million.

war and before 1962 in the Netherlands, Switzerland, Finland, U.S.A. and also in Soviet Armenia, Venezuela and the United Arab Republic (see Chart IIb). The turning-point occurred before the second world war, and after 1910 in Belgium, Austria, Italy, Yugo-

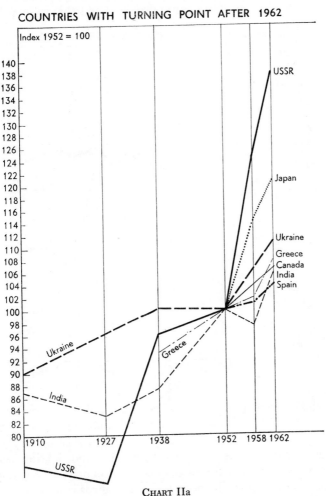

CHART IIa

slavia, Hungary and Estonia (see Chart IIc). Already beyond the turning-point (showing a decrease of agricultural land) in the period 1910 to 1914 were Ireland, Sweden, France, Germany and Czecho-slovakia. The United Kingdom showed a great decline in agricultural area, but it increased again after the second world war, as it did in some other European countries at least during the post-war shortages (see Chart IId).

In the first situation, expansion of agriculture prevails. The main factor of production is land. Extension of the area under cultivation is the main concern of agricultural policy, and this is supposed to

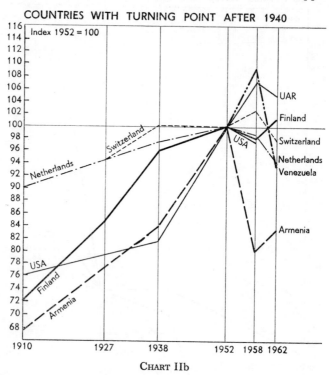

CHART IIb

solve all the main problems of agricultural policy. The cultivated area is increased by bringing under cultivation new, idle or fallow land, by cutting forests, draining marshes, irrigating dry lands, cultivating parts of pastures, making terraces and fighting erosion.

This is the period of the opening up of a country, at individual as well as at social cost; the period of building roads, canals and railways in order to have access to the potentially cultivable area.

The encouragement to expand cultivated area varies from giving land free, distributing communal lands for farming, and creating homesteads, even to the point of elimination of pastures and forests. Land is scarcely a saleable commodity, and only small areas change hands through the market mechanism.

COUNTRIES WITH TURNING POINT BEFORE 1940

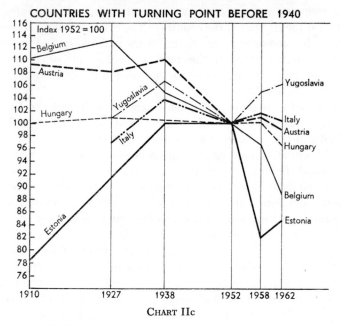

CHART IIc

COUNTRIES WITH TURNING POINT BEFORE 1910

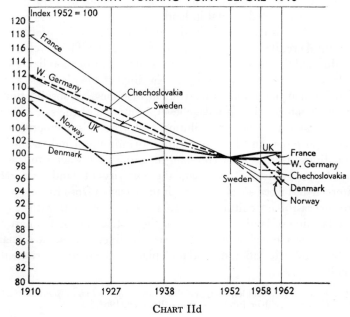

CHART IId

For some countries, the policy of extending the cultivated area did not stop at the frontier of the country. The extension of the domestic territory by acquiring foreign territories and expanding to other countries, either through the market mechanism, or through a system of colonization, were further methods of extending the controlled cultivated area.

The second stage is reached when extension of agricultural lands is no longer possible to any significant extent, when the 'frontier' is reached and the pioneering age is over.

In this situation land redistribution becomes the main concern of agricultural policy. Institutional measures regulating land use become the most important instruments of policy. The problems of economic development are no longer linked to expansion of land but to problems of land redistribution by sale and purchase, by law or by spoils. One of the main concerns of developmental policy is then to satisfy the wants of those who work on land and therefore land gets into the hands of those who, for reasons of thrifty subsistence, of political pressure, or of better husbandry, can gain most from it.

The most common mechanism of land redistribution is the formation of a market for land, so that land is, to an ever greater extent, drawn into commerce. It is bought and sold like any other commodity at commercial prices, which depend in principle on returns from invested capital in land.

Along with this commercialization of land, there arise problems of agricultural credit — especially of mortgage loans. The development of credit facilities forms an important instrument of mobilizing land transfers. Land reform occupies an important place as a real redistributor of land and as instrument of policy. Consolidation of farms with reformation at village level, limitation of the size of holdings and estates by law, prohibition of further dismemberment of agricultural plots, and legal imposition of minimum indivisible size for holdings, are the most common measures. Obstacles to trading in land are gradually removed, the security of land ownership reinforced and freed so as to enable it to be traded from the less to the more successful agricultural operators. Thus contracts with respect to the farming of land are relieved of the restrictions of the past and made as commercial and binding as possible. In some socialist countries collectivization of land took place as the main instrument of agricultural policy.[1]

[1] Cf. R. Bićanić, 'Problems of Socialist Agriculture', *Indian Journal of Agricultural Economics*, Silver Jubilee Number, Bombay, 1964.

In the third situation, in which there is reduction of cultivated area at the same time as an increase in the volume of production, the main policy is that of intensification of agriculture by capital investment. Submarginal land is abandoned because of its low fertility, because of high costs of production or because of inaccessibility or distance. Capital takes over from land as the dominant factor of production and, for reasons of more favourable opportunity costs, becomes the main concern of agricultural policy. Extension of the area under cultivation is no longer pushed ahead, the reallocation of land becomes of secondary importance, because it is not of much use unless there is capital enough to cultivate it properly on a competitive basis. Forest areas are no longer taken to contain reserves for expanding cultivated area, and forests — what is left of them — are carefully conserved.

This general process goes so far that, in some countries, further increase of agricultural area is discouraged by government policy (the use of soil banks, the development of rural tourism, extension of national parks, conservation of forests, guarding of mountain landscapes, etc.).

Thus comparative advantage of investment in agriculture, and the movement of capital from agriculture, are carefully weighed against other competitive investments ; the practice of landless agriculture and livestock breeding is put into operation, hydroponics are carefully studied, etc.

IV. CAPITAL

In dealing with capital in agricultural development we need to emphasize two things : first, that land is not included in capital ; second, that capital is a very heterogeneous category with regard to its productivity and shows partial capital-output ratio coefficients which vary from 0·5 to 33.[1] There are large capital coefficients in land reclamation (irrigation, drainage, terracing), water supply, flood control and erosion, and also in the setting up of vineyards, olive groves and arboriculture. Buildings and construction of roads also

---

[1] M. Cépède gave the following partial macroeconomic capital coefficients for French agriculture : water works 2·7, road building 4·2, farm consolidation 0·5, building construction 33, electricity 7·7, agricultural machinery 12·5, total agriculture 6·4.

For Yugoslavia the partial capital-output ratios in the period 1957–62 were : crop farming 4·7, animal husbandry 2·6, vineyards and arboriculture 12·4, overall agricultural capital coefficient 5·7. (Institute of Agricultural Economics, Belgrade.)

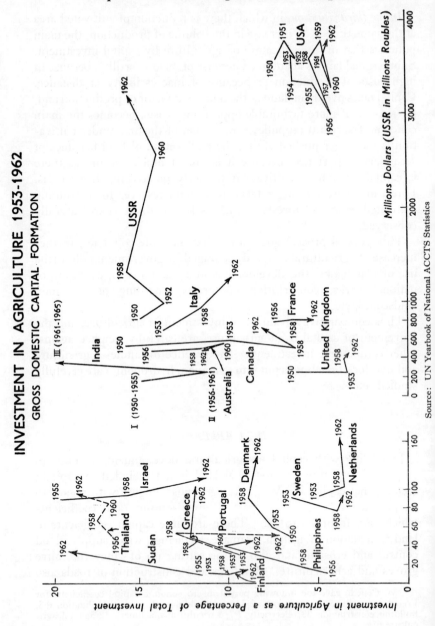

INVESTMENT IN AGRICULTURE 1953-1962
GROSS DOMESTIC CAPITAL FORMATION

CHART III

have heavy capital coefficients. Capital coefficients of medium size are to be found in livestock and machinery investments, and light capital coefficients appear in investment for services, small animals, use of fertilizers, and various kinds of inventory.

Therefore the assessment of the role of capital in agricultural development will not lead us far if we measure it in terms of a homogenized quantity, that is, in financial terms. We have to distinguish clearly its longevity, the various combinations of the physical capital-mix of different partial capital coefficients, and the changes in the quality of capital due to technical progress.

Taking capital as an independent variable we can observe three different stages in agricultural development. The first situation appears when capital is scarce and is the limiting factor in economic development. The second is when capital is available for agriculture at the general rate of interest and is invested according to the profit it makes possible. The third stage is created when capital decreases in financial terms and even begins to decrease in quantity because of the very great influence of technical progress.

(a) Capital as a Limiting Factor

From the capital supply point of view the starting-point in agricultural development is the auto-reproductive process in which internal capital formation prevails over external (seeds from own harvest, animals bred and manure produced from own herd, tools to a great extent home made, and farmhouse built from domestic material by family labour, etc.). Natural risks and capital losses are very considerable (fire, floods, erosion, animal diseases, drought, etc.), insurance against which is practically non-existent and no depreciation allowance calculated. External sources of capital supply play a marginal role.

Starting from this situation agricultural development depends on an increase in external capital goods bought in the market (cattle, machinery, building materials), which can secure a more than proportionate increase in agricultural production for the market. Thus there is a very high elasticity of demand for marginal capital out of money income ; this makes capital very scarce and the hunger for capital very great ; saturation of capital demand depends more on the availability, indivisibility and mobility of capital than on the financial interest rate. Investment in agriculture increases less in millions and more in percentages (*e.g.* Indonesia, Thailand, Philippines, India). At this stage, agriculture can, and actually does, stand an overrated and usurers' rate of interest much above the general rate of interest in the economy.

Thus, in this stage, the rate of interest cannot be used to measure real demand for capital in agriculture. This situation brings about simultaneously a hunger for capital and over-capitalization in production ; over-indebtedness and squandering of windfall gains ; compulsory capital accumulation and exorbitant rates of interest. To counteract this, agricultural policy requires the organization of a specific capital market for agricultural investment (predominantly out of public funds fed by taxing the very same agricultural population ; *i.e.* redistribution by exploitation).

In the course of development the capital coefficient increases and the capital burden becomes heavier. The increasing capital coefficient requires ever more capital investment per unit of product while passing over the threshold of development.[1]

The incremental capital-output ratio (ICOR)[2] in European agriculture in the decade 1949–59 was, in the less-developed countries : Greece 0·9, Spain 1·7, Turkey 1·8, Yugoslavia 2·5, Italy 3·2 and Portugal 3·7. The more developed European countries have the heavier ICOR. In West Germany, the Netherlands, Finland, United Kingdom it was around 7, and in Belgium and Austria it moved between 9 and 20. A heavy capital coefficient operated in the United States (17·6) and Canada (12·2), while in Norway it reaches 31·4.

In the course of general development agricultural investment takes an ever smaller share of the national capital although it increases in absolute quantity (*e.g.* Greece, Denmark, Italy, the Netherlands in Chart III). This increases the possibility of draining some capital back into agriculture by private or public channels. The availability of capital from savings both inside and outside agriculture makes capital-intensive agricultural development possible and brings the rate of interest in agriculture closer to the general rate.

Long-term capital formation in agriculture (secular trend) falls constantly in relative terms which means that other sectors grow faster. From 1800 to 1927 farm capital in the United Kingdom decreased steadily from 19·3 to 2·4 per cent of the national wealth. The net fixed capital formation in Swedish agriculture fell in one

---

[1] Cf. R. Bićanić, 'The Threshold of Economic Growth', *Kyklos*, Basel, 1962, No. 1.
[2] UN-ECE, *Some Factors in Economic Growth in Europe in 1950s*, ch. iii, p. 28.

hundred years (1861–1962) from 33·2 to 3·6 per cent, and in Norway (1865–1962) from 34 to 7·3 per cent. In the U.S.A. in the period 1880–1958 the addition to agricultural product fell from 26·2 to 3·6 per cent. From 1861 to 1880 Australia increased her capital formation in agriculture from 15·6 to 32 per cent ; after that period it fell to 6·2 in 1930–39 while the agricultural share of the national income remained the same (23 per cent). Argentina's gross fixed capital formation in agriculture moved from the beginning to the mid-twentieth century from 23·8 to 5·3 per cent of the total capital formation.[1]

(b) Turning-point

The turning-point in capital investment means that there is a downwards movement in agricultural investment in a relative sense, and almost a stop in the increase of such investment in absolute sums, which means that the turning-point is approaching in Israel, Canada, Australia, France and Finland (see Chart III).

The turning-point in capital occurs when capital supply becomes available at the general rate of interest, and the gain from investment in agriculture nearly proportionate to the amount of capital invested. Income in cash greatly prevails over income in kind and the total income elasticity of demand for capital lies between low coefficients.

Agriculture is treated in the same way as other branches of production and investment criteria are set following the same sort of evaluation. In such a situation reasons for the existence of a special capital market for agriculture have disappeared ; so the special market itself disappears or remains as a historical left-over. Capital is no longer a limiting factor in the development of agriculture.

The financial decision whether or not to invest no longer depends on capital availability at any cost but on the rate of interest — 'Does it pay or not?' Therefore substitutions of land and labour by capital, and above all substitution of capital of a low productivity by capital of a higher productivity, take place. The natural risk of capital loss is decreased but commercial risk is greatly increased due to heavy proportion of fixed assets. The agricultural operator assumes the role of entrepreneur weighing profits and losses, which influence his investment decisions. This means that capital frequently changes form and hands. The entrepreneurial decisions are mainly concerned

[1] S. Kuznets, 'Quantitative Aspects of Long Term Growth of Nations, VI. Long Term Trends in Capital Formation Proportion' *Economic Development and Cultural Change*, July 1961, Part II. Appendix.

to combine factors of production, which will give optimal results in terms of return to global financial capital invested, which also means a reduction in the capital coefficient by a better combination of its component parts. This optimal capital-mix is rather empirical, finding its limits in ecological and technical conditions (crop rotation, animal life cycle, vegetation period, humus preservation). It is also limited by the institutional framework such as family size, property relations, managerial skill, legal regulations, etc.

Managerial skill acquires such importance that it is separated from the providing of capital and treated as a special factor in production (the Marshallian fourth factor of production). Those who cannot invest, or do not do it in a successful way, are doomed and their lands and capital taken over by more skilful 'capital-mixers'.

(c) Capital Decrease

There are signs that in some countries the peak has been reached and we are approaching the stage in development of agriculture where capital will tend to decrease. This decrease is now taking place at an accelerated pace in a relative sense in the fast declining share of agricultural capital in the global national wealth. But one should also expect a decline in the absolute quantity of capital engaged in agriculture in spite of the steadily increasing volume of production.

Yearly gross capital investment in agriculture in the U.S.A., France and Sweden did not increase to any significant extent in the years 1953–1962. It somewhat increased in Sweden and slightly increased in the U.S.A. and France (from 3,600 to 3,800 million dollars in the U.S.A. and in France from 800 million dollars in 1953 to 820 in 1958 and 1962), which means a decline in net capital formation. Capital investment in agriculture fell considerably in relative terms, from 6·6 per cent to 3·6 in Sweden, from 7·1 per cent to 4·4 in the U.S.A., and in France from 7·4 to 5·4 per cent.[1] (Chart III.) This limitation is now manifest in the policy of limitation in agricultural production in several countries due to the saturation of demand or the general low income elasticity of demand for agricultural

---

[1] During the same period Italy, Portugal, Greece, Denmark, Ireland and also the United Kingdom witnessed a considerable increase in capital investment in absolute terms while a decline in relative terms took place (*e.g.* in the United Kingdom from 6·2 in 1950 to 3·6 per cent, in Finland from 16·3 to 8·5, in Italy from 13·6 to 9·8). The absolute sums and the percentages increased in Ireland, Denmark and Greece which still shows a considerable distance from the turning-point. (Computed from the *UN Yearbook of National Accounts Statistics for 1963*; for rates of exchange Table 3 was used.)

products. In others it leads to pressure for an almost complete overhaul of agricultural policy. We interpret these signs in the following way :

(i) The tendency to change the capital-mix in such a way that the average capital coefficient becomes lighter, *i.e.* the productivity of capital per unit of product greater. This is due to the fact that the main capital investment requiring heavy capital coefficient, like land reclamation and building of infrastructure, is already done ; cultivation of submarginal lands abandoned ; building construction of farm operators decreasing because of population decline ; structural investment in machinery largely achieved and only maintenance and replacement required.[1]

(ii) Great technological progress in agriculture is such that the productivity of capital invested is ever increasing. This is particularly so when chemical (fertilizers, insecticides, pesticides) and biotechnical investments are concerned (artificial insemination, antibiotics, hybrid plants, and other uses of genetics, etc.), which again require less than proportionate capital investment per unit of product and give an increasing return.

(iii) Increased specialization of production requires less capital per unit of production than in mixed farming. Over-capitalization is greatly liquidated and idle capital put into operation. There are signs of financial disinvestment in agriculture due to specialization such as cutting of vineyards, orchards and forests, liquidation of some herds and abandonment of stables, and generally less investment in construction than hitherto, because of the greatly decreased number of farms.

(iv) Indirect agricultural investment is directed to extra services which enable the agricultural operator to get activities performed for him. On a micro level he can purchase

---

[1] The yearly increase of tractors in the U.S.A. in 1950 amounted to 8 per cent or 270 thousands and in 1960 only to 0·4 per cent (20 thousands) with the prospect of an annual increase up to 1980 of 50,000 tractors per year. The corresponding numbers for grain combines were 13·2 and 0·5 per cent and for corn pickers 18·5 and 2·6 per cent. Demand moves from structural growth to functional replacement after the saturation point has been reached. This point has not yet been reached in Europe. Between 1950 and 1959 the number of tractors in the EEC countries increased 5 times, of milking machinery 6 times, and combine harvesters 12 times. Landsberg, *et. al. Resources in America's Future*, Johns Hopkins University Press, 1963, pp. 776–777. UN-EEC, *Some Factors in Economic Growth in Europe in 1950s*, 1964. Chapter iii, p. 27.

services which earlier required his own capital investment and keeping greater inventory supplies. On a macroeconomic scale such investments can be supplied by more rational utilization of capital than on an individual basis (*e.g.* fertilizers and oil supply).

Moreover a considerable amount of capital hitherto counted as investment in agriculture has been transferred to services supplied by many branches of industry, commerce and transport and are therefore excluded from the agricultural sector.

The farmers' activity is based to an ever greater extent on scientific research offered to him by public extension services as well as by commercial and other industrial vertical organizations in 'agribusiness'. This decreases his commercial risks, and leads on one side to government protective policy for agriculture and on the other to contract farming. Both factors have greatly rationalized the commercial and financial risk for capital investment previously done on an empirical trial-and-error basis. They also require a lesser amount of capital, and secure a more orderly flow of goods and services. At the same time such development reduces the function of the agricultural entrepreneur to the role of technical manager, and technical parity becomes the main concern of agricultural policy.

Technical progress, jointly with land, labour, capital and managerial activity, becomes the fifth independent variable in agricultural development. Its application increases the interdependence of agriculture in the global economy so that the operation of the market mechanism, and partial uses of a command mechanism, do not provide adequate services. Therefore there is an ever-increasing opening for contract economy and use of the planning mechanism and this all the more as agriculture is increasingly integrated into the process of economic development.

V. CONCLUDING REMARKS

As a further development of this analysis it would be interesting to study the coincidence of the turning-points. For example, the United States reached the population turning-point in 1910, of farmland in 1950 and of capital formation in the sixties. The United Kingdom reached the three turning-points in 1820, around 1910 and the third not yet reached ; France in 1950, and before 1910, etc.

It would also be of considerable interest to pursue the study in more detail to see whether one or several turning-points of land, labour, etc., can occur in long-term development where the most sensitive changes would have to be expected in capital formation. But all this goes beyond the scope of this paper.

―――

DISCUSSION OF PROFESSOR BIĆANIĆ'S PAPER

Professor Neumark said that although Professor Bićanić's study is set at a high level of abstraction, it is not simply making unrealistic models ; its arguments are based on empirical data. The methods Professor Bićanić used are reminiscent of those used by Professor Rostow.

Professor Neumark said that in distinguishing three stages of employment in agriculture, Professor Bićanić characterizes the first as being dominated by the problem of 'life parity'. Professor Neumark asked for the meaning of 'life parity' in this context to be made more clear. What measures does it suggest a government should take within the framework of agricultural policy? Also, at the turning-point of the labour market, Professor Bićanić stated that agricultural policy was dominated by the problem of 'price parity'. Again, the meaning is not clear. Does he mean to indicate a possible policy target, or simply show a policy which follows logically from the fact of rounding of the turning-point? The same question arises over the operational significance of 'income parity' in the third stage, which is characterized by shortage of labour and a growing role of capital.

Professor Neumark referred to Professor Bićanić's discussion of the later stages of land and capital use in agricultural development, and in particular to the assertion that at these stages, agriculture becomes simply one among all other economic activities in which investment is judged according to relative profitability. Professor Neumark said that the author was here heightening the level of abstraction too far. Governments do not treat agriculture in the same way as other sectors, and invariably intervene in the agricultural sector more seriously than in other sectors.

Referring to capital use in the first stage, Professor Neumark asked whether it is in fact possible, under the conditions prevailing in these early stages, to build up a specific capital market. In particular, can governments raise the tax revenue required for this kind of policy? Is it in fact true that at the early stage, agriculture will stand the usurious rates of interest referred to by Professor Bićanić?

Professor Neumark said that although he was interested by the question of the coincidence, or rather the non-coincidence, of the three turning-points, he thought it was more useful to concentrate on the three turning-points in isolation and to explore the accuracy of Professor Bićanić's description of the policy measures corresponding to the different stages of agricultural development.

Professor Dupriez said that he found Professor Bićanić's paper a bold attempt to ascribe to one particular factor of production the main policy problems of a phase of development. The long-run changes in the importance of the various factors are taken to be crucial for policy-makers.

Two illustrations of the usefulness of this historical analysis were given by the enclosure movement in the eighteenth century in Britain, and the contemporary European situation. In the movement of enclosures, the emphasis was on the increasing of output per acre. Land could not be left in the hands of the inefficient ; Ricardo's views were consistent with this principle. In Europe at the present time the concern was with output per production unit. The units were growing in size and reorganizing under conditions of shortage of agricultural labour. In conditions of rapid technical change, European agriculture had to aim at maximizing output per man.

Professor Dupriez said that the analysis needed broadening in certain respects. For example, in many under-developed countries today, the problem is not so much a shortage of land, but rather a surplus of population with respect to the techniques in use. Policy should not so much concern itself with output per cultivated acre, but rather with changing the techniques. Also, in the case of large plantation enterprises (latifundia), and in the case of the kolkhoz, the right economic calculations would be of an industrial type, and output per head remains the dominant consideration.

Professor Tepicht said that he agreed with Professor Bićanić's main conclusion that in the long run there would be at first a relative and then an absolute decline in agricultural population. However, on the conclusion concerning land, he asked if there might not be a confusion here between the undoubted fact that the role of land as a factor of production is declining in importance, and the assertion that the actual physical quantity of land under cultivation will diminish. Such an assertion could not really be proved by the examples of one or two countries at a time when transport facilities and investments in infrastructure are generating, and will continue to generate, forces making for greater world division of labour ; a division which conforms more closely to a rational allocation of resources. There are many economic, social and political obstacles on the road to this goal. However, the contrast between countries of Europe with 0·5 hectares of land per head and other countries of the world, which on the average have five times as much land per head, will one day impose a redistribution of work. This redistribution when it occurs could be read as evidence of a simple contraction if each country were taken in isolation.

To this argument it might be added that in developing countries the demand for a food supply as high as that enjoyed in other countries, as well as the growth of capital per head, will raise, not reduce, the significance of land. Might it not then be premature, at least for a few generations, to discover a turning-point? For it would be deduced from situations limited in their generality, and also constrained by the laws of the capitalist economic game which are at present highly prized but not necessarily eternal?

Professor Tepicht asked if Professor Bićanić would agree that his point about the dismantling of the old diversified agriculture, and its replacement by a series of connected activities (what American economists call 'agribusiness') should be stressed. If so, would it not be right to compare the capital which was used in the old diversified agriculture with the capital invested in the entire complex of 'agribusiness'? The result of such a comparison would be to weaken what the paper argued on this point. From this point of view, we would find a development which is similar to the progress of social division of labour in all the other main sectors of the national economy. If we were to take as an example the large investments which have been announced in the U.S.S.R., both 'downstream and upstream' of farming itself, these imply a change of employment for a large part of the direct farming population of today, even though they would remain closely involved in agriculture. The conclusion that there would be an accompanying contraction of capital would be open to dispute.

Professor Tepicht said that the dangers run by policy-makers when they fail to identify a turning-point are rightly emphasized by Professor Bićanić. However, there is the danger, at the other extreme, of taking so broad a view that the ground under one's feet would be neglected. Thus, if it is true that the specific importance of the agricultural population is decreasing, it is still obvious that it remained of great importance in most of the countries of the world. So the tendency of the policy-makers in certain countries to rush into labour-saving techniques implied strong prejudices, both economic and social, which cause them to neglect one of the essential components of growth. Technical progress, pushed forward heavily, under the banner of 'output per man', is not exactly the same as progress *tout court*, unless we are prepared to accept as progress a disguised unemployment of labour in the country-side. Thus the theoretical model, even though presented at a high level of abstraction, should not disregard crucial realities, but rather be constantly confronted with them.

Professor Pohorille said that he had strong doubts about the notion of attaching 'price parity' policy to the second stage, and 'income parity' to the third stage of the development of agriculture. Parity of prices is not something that can be taken independently, since it is usually designed as an instrument of economic policy which is to lead to parity of incomes. (The fact that policies of price support have not given the results hoped for is another matter altogether.) Agricultural policy cannot really be described adequately in these terms. It would be much better to characterize the agricultural policies in the Western world stressing that before the war they were part of general anti-cyclical policy, and after the war they had been planned within the general context of growth policy and of the problems of the general distribution of national income.

Professor Pohorille wondered whether it is possible at all to examine the tasks of agricultural policy without analysing the concrete economic situations at a particular time. Contemporary policy problems have resulted from disturbances in the international division of labour, rather than from the process of agricultural development itself.

Professor Dandekar said that he thought that the most interesting follow-up to the paper would be to examine the interrelations between the three factors, land, labour and capital, in the different stages. At the beginning of his paper, Professor Bićanić said that the three factors would be treated as independent variables, but they are not really independent. In the second stage it is pretty clear that all three are being treated as working hand-in-hand so that the interrelations between them are very important.

Professor Dandekar noted that towards the end of the second stage, it is argued, there is an absolute decline in agricultural population and with this a decline in the number of small farms. Professor Dandekar asked if Professor Bićanić regarded the resulting pattern of big commercial farms and remaining small farms as essential to this stage.

Professor Dandekar asked why the turning-point in capital — that is, when capital supply became available at the general rate of interest — occurs.

Professor Campbell said that he was disturbed by the symmetry of Professor Bićanić's analysis. The turning-point for labour is stated to occur at various levels of agricultural population, going from 31 per cent of total population down to 60 per cent of total population, and at various levels of national income per head ranging from $132 to $500. However, there are exceptions. Thus in the Australian case the first absolute decline in the rural work force occurred during the second world war, and the first decline in absolute numbers in peace-time was shown in the 1961 census. Whichever date is taken, the agricultural population was less than 20 per cent of the total population, and well below the lower limit given by Professor Bićanić. The national income per head in 1961 was certainly well over $500. Moreover, Australian agriculture had been commercialized over many decades. Stage one in Australia could not be described as one of subsistence agriculture. If Professor Bićanić were to extend the range of his data, the analysis would possibly lose much of its symmetry.

Professor Gulbrandsen asked whether the criteria given in the paper are really usable. A high proportion of 'agricultural population' in an unspecialized economy is only a matter of statistics ; agricultural population under this definition produces many other things apart from food.

Moreover, if a country specializes in agricultural production, it is not axiomatic that the proportion of agricultural population will decrease.

Professor Gulbrandsen said that he also disagreed with the description of the stages. Professor Theodore Schultz had shown that the first stage, that of traditional agriculture, represented a balanced economic system, and that it is not characterized by a labour surplus.

Professor Bićanić said that Professor Neumark had rightly noticed that there are in fact four stages or situations in the population changes. The first of these is the zero or pre-developmental stage when agricultural population increases both absolutely and relatively (this was the case of Mexico in 1920, and of India and Indonesia on Chart I). At the zero stage, agricultural population increases faster than the non-agricultural population; demographic pressure is more important than the pressure of developmental changes on population. Professor Bićanić said he agreed with Professor Gulbrandsen, that there may be some conceptual errors involved in the general statistics, and that the concept of agricultural population is defined in different ways during the course of development. The definition of agriculture itself is a mixed one, containing technological and institutional elements. In fact at earlier stages of development the agricultural population includes many who would at late stages be counted in the secondary and tertiary sectors.

Professor Bićanić said that by 'life parity' he simply meant a policy directed to preventing the population from dying of hunger. This policy is now carried out on an international level as well as on the national level (for example, the Freedom from Hunger Campaign of FAO).

Professor Bićanić said that he agreed that anticipations of unverified changes could be a cause of great errors in agricultural policy, but there is a greater danger of policy-makers lagging behind the turning-points.

Income parity in agricultural policy is necessarily linked to the third stage when agricultural population declines absolutely and relatively. If the rural exodus is faster than technological progress and the substitution of men by machines, there is a grave danger of a shortage of agricultural labour. Here administrative measures are not effective. The policy of income parity is an instrument to retain people on the land. Today, even some socialist countries, which had passed the turning-point, are compelled to introduce a policy of income parity, which a decade ago would have seemed to be quite out of place. The speed of depopulation of agriculture, and also the deterioration of its quality through senilization and feminization, is a general phenomenon throughout the world.

Professor Neumark had raised the question of relative profitability in the third stage of capital development. It is true that there is no country without government intervention in agriculture (which in itself is a significant fact). Professor Bićanić said that he used the term *relative* profitability in the microeconomic sense, in which government measures are taken as constants or as parameters of action, and microeconomic decisions as

Conceptual Problems in Agricultural Development

variables searching for profitability within the given set of data. The turning-point in capital investment is the most controversial and also the most interesting one. He had not expected to find it, but was led to it by empirical research. It sounds incredible that progress in agriculture could occur under conditions of decreasing capital investment. The way to explain it is to distinguish sharply between capital in financial terms and capital in real terms and to measure it by its productivity. Thus the assumption of homogeneous capital cannot be accepted.

With regard to stage one, it has been asked how a specific capital market can be established by means of tax revenue taken from the poorest part of the population. Also some doubt has been expressed whether these poor peasants are able to pay the moneylenders' rate of interest. The exploitation of the peasants is greatest at the boundary between the monetary and the barter sectors, and usury flourishes most in such circumstances. Also, in this first stage there is over-capitalization in one part of agriculture, and hunger for capital in another part. There was, for example, over-capitalization in pre-war Yugoslavia in ploughs. If measured by production units, two of every three peasant holdings were without a plough. But if the optimum size of arable land was taken as a measure, there were twice as many ploughs as necessary.

It should be stressed that government policy is the main instrument of income redistribution in agriculture, be it in the form of primitive capitalist accumulation (in the Marxian sense) or in the form of the primitive socialist accumulation.

Professor Bićanić said that the capital coefficient is an object of particular interest to him. There is some evidence that it is not constant. On the contrary, it changes with economic development, increasing at first very steeply when the infrastructure is being built, and then becoming lighter afterwards, that is, increasing more slowly. This process he called 'creeping over the threshold of economic development'; this threshold represents another type of turning-point.[1]

Professor Bićanić said that he accepted the importance of the question of the interrelation and possible interdependence of the different turning-points. This is material for further research, and he thought it very likely that several turning-points for one variable would be found, instead of simply one. Extra economic forces would be expected to play an important role, particularly at the sensitive period of the turning-point. In answer to Professor Dupriez, Professor Bićanić said that he had not intended to deal with the problems of stagnant agriculture in the predevelopmental stage. However, he thought that at that stage, production per hectare is linearly correlated with production per head of consumers.

[1] Cf. R. Bićanić, 'The Threshold of Economic Growth', *Kyklos*, No. 1, Basel, 1962, and R. Bićanić, 'Social Preconditions and Effects of Moving over the Threshold of Economic Development', *International Social Sciences Journal*, No. 2, Paris, 1964.

578

On the question of latifundia, Professor Bićanić said that they cannot be properly analysed without reference to the adjacent minifundia, which kept the peasants on the land at their own expense and risk. A different role is played by the kolkhozy in a socialist economy ; these provide means of existence to surplus labour.

Professor Bićanić agreed that confusion might arise if the diminishing weight of land as a factor is confounded with an absolute decline of agricultural land. None the less, it is important to recognize that at certain stages of development the quantity of agricultural land decreases, while the volume of production increases, when countries are taken separately as units for observation. This is not only a question of erosion, or of urban development, but of such new changes as the introduction of landless agriculture. The problems associated with this stage had had to be faced in the Soviet Union when the decision had to be taken whether to work the fields of the Ukraine more intensively, or to extend the cultivated area to the virgin lands of Kazakhstan. In fact signs of uncertainties in the agricultural policies of some socialist countries are due to the fact that they had reached a turning-point, so that the impact of a new set of facts was being felt, but the will to adjust was inhibited by preconceived ideas.

Referring to Professor Tepicht's remarks on the dismantling of the old diversified agriculture, Professor Bićanić said that his main concern in the paper had been to demonstrate the interdependence throughout the process of development, between agriculture and other sectors of the economy, and to attempt to measure this interdependence, albeit with crude yardsticks.

Professor Bićanić said that he thought there is a clear difference between the policies of price parity and income parity ; the first is limited to a one-commodity dimension, while the other is based on an institutional concept of the income unit.

Professor Bićanić agreed that large and small farms exist at all stages of development, 'les grandes exploitations et les petites exploitations' of Quesnay. Moreover, labour-intensive small agricultural units and part-time farming are expected to remain for many years to come. These would provide a counterpart and compensation to the industrial mode of production and way of life. Nevertheless, the relationship between large-scale agriculture and small-scale agriculture does change considerably at every stage, and capital is one of the main factors which distinguishes the final stage.

Professor Bićanić said that Professor Campbell was right to criticize the symmetrical approach, which had been used for reasons of theoretical 'beauty' at the high level of abstraction. In other research he had dealt with the asymmetry in agricultural economics which dealt with market and other imperfections. Australia certainly had to be taken as a special case, particularly since the peaks of its two industrial revolutions (urban and agricultural) had largely coincided.

In reply to Professor Gulbrandsen, Professor Bićanić said that the concept of surplus labour is complex. There are at least three different concepts.[1] One of them is related to productivity such that the surplus labour increases with every increase in labour productivity in agriculture. Therefore, the tendency to minimize the role of structural over-population and to emphasize the role of seasonal over-population should be a move in the right direction.

[1] Cf. R. Bićanić, 'Three concepts of Agricultural Overpopulation', *International Explorations of Agricultural Policy*, Iowa University Press, 1964.

Chapter 27

EQUITY AND PRODUCTIVITY ISSUES IN MODERN AGRARIAN REFORM LEGISLATION[1]

BY

VERNON W. RUTTAN

The International Rice Research Institute, Los Banos, Philippines

I. INTRODUCTION

UNDERSTANDING of the economic implications of land tenure systems rests on a dual foundation. First, there is a set of historical generalizations about the consequences of alternative tenure arrangements for economic growth. There is also a set of logical deductions about the effect of alternative tenure arrangements on resource allocation and output levels derived from the neo-classical theory of the firm. Among Western economists, economic history and economic logic have combined to produce a remarkable unity in doctrine to the effect that an agricultural sector organized on an owner-operator pattern (*a*) achieves a more efficient allocation of resources and (*b*) makes a greater contribution to national economic growth than under alternative systems.

Until fairly recently, agricultural productivity has represented only a minor theme in the objectives of most land reform legislation. In this paper, I trace (*a*) the evolution of equity and productivity objectives in Western land tenure reform and (*b*) the emerging role of productivity as a major objective of changes in land tenure in South-east Asia. This is followed by a survey of the empirical evidence on the relationship between land tenure, productivity and agricultural development in South-east Asia. Finally, I suggest that much of this

[1] The author is indebted to P. O. Covar, F. C. Byrnes and A. M. Weisblat for comments on an earlier draft of this paper. The paper draws on material previously published in 'Land Reform and National Economic Development' in G. P. Sicat (ed.), *The Philippine Economy in the 1960's*, U.P., I.E.D.R., Diliman, 1964, pp. 92–119, and 'Equity and Productivity Objectives in Agrarian Reform Legislation : Perspectives on the New Philippine Land Reform Code', *Indian Journal of Agricultural Economics*, Vol. 19, Nos. 3 and 4, July–December, 1964, pp. 115–130.

conflicting empirical evidence can be rationalized by an evolutionary approach to the analytical relationships between land tenure and productivity.

II. EQUITY AND PRODUCTIVITY TRADITIONS IN LAND REFORM

The relative emphasis placed on equity and productivity objectives in land reform policy and programmes has varied widely among countries and over time. In general, political and equity objectives occupied a central role in the land reform movements of the nineteenth and first half of the twentieth century. In recent years, this 'classical' objective has been increasingly complemented by a productivity objective.

(a) The Equity Tradition

In the United States, equity and political considerations have traditionally represented dominant themes in agrarian reform. The confiscation of loyalist estates during and following the Revolution was primarily motivated by equity considerations.[1] Jefferson's political philosophy, which regarded the family farm, owned and operated by the cultivator, as the only sound foundation of social equality and political stability, provided the intellectual foundation for the major U.S. agrarian reform legislation of the nineteenth century and remains an important theme in current agricultural policy discussion.[2]

---

[1] Cf. Irving Mark, *Agrarian Conflicts in Colonial New York, 1711–1775*, Columbia University Press, New York, 1940, p. 16 : 'In the years prior to the American Revolution, there was much agitation against large landholdings. . . . Sometimes the form of struggle was an anti-rent controversy ; sometimes a striving for more secure tenure ; and, in some cases, a "leveling" movement seeking the division of great estates for the benefit of the poor tenants'; and T. F. Marburg, 'Land Tenure Institutions and the Development of Western Society', in Walter Froehlich (ed.), *Land Tenure, Industrialization and Social Stability*, Marquette University Press, Milwaukee 3, Wisconsin, 1961, p. 50 : 'After the break with England . . . the colonial government . . . promptly confiscated estates of British subjects who remained loyal to the Crown. . . . disposal constituted an early step in a sequence of developments which led gradually to the breakup of many of the larger holdings in the East. . . .'

[2] The definitive treatment of the impact of Jeffersonian thought on agrarian policy in the United States is A. Whitney Griswold, *Farming and Democracy*, Harcourt, Brace and Co., New York, 1948. For a recent review of the Griswold interpretation, see J. M. Brewster, 'The Relevance of the Jeffersonian Dream Today', in H. W. Ottoson, *Land Use Policy and Problems in the United States*, University of Nebraska Press, Lincoln, 1963, pp. 86–136.

Agrarian reform legislation during the nineteenth century conceived in the spirit

Other land reforms which appear to have been conceived primarily in an equalitarian frame of reference include the reforms which spread across continental Western Europe beginning with the French Revolution ;[1] the land reforms of Eastern Europe following World War I ;[2] the Latin American land reform movements beginning with the Mexican Revolution of 1910 ;[3] and the Japanese, Korean and Taiwan land reforms following the end of World War II.[4]

The agrarian reform movements which have looked to the liberal-Jeffersonian political philosophy for their intellectual orientation have typically shown little concern for the potential contribution which land reform might make to the growth of agricultural output or to national economic growth. Indeed, leaders of reforms based on this tradition have at times argued that the political and equalitarian objectives were sufficiently important that any disruption of output which might occur as a result of the reforms should be ignored.[5]

In the United States, Western Europe and Japan, this lack of

---

of Jefferson's agrarian democracy includes : (*a*) the Pre-emption Law (1841) which legalized squatting on unsurveyed public domain with the right to purchase up to 160 acres at the minimum price after surveyal ; (*b*) the Homestead Act (1862) which conveyed free title to 160 acres of land after residing on and cultivating the tract for five years. These laws were reinforced by other legislation designed to ensure the economic success of the family farm : (*a*) the Morrill Act (1862) establishing the land grant agricultural colleges ; (*b*) the Hatch Act (1877) providing federal support for the state agricultural experiment stations ; and (*c*) the Smith–Lever Act (1914) creating a federal-state agricultural extension service, Griswold *ibid.*, pp. 139–147.

[1] Commenting on the agrarian legislation of the French Revolution, Griswold points out that 'In one sense these laws were the practical application of the natural rights philosophy inherited from Locke by the French philosophers and passed on by them with the added prestige of the American example, to the leaders of the Revolution. In a more concrete sense they were a rationalization of the existing system of agriculture, stripped of its feudal privileges', p. 95. 'To them (the peasants) the Revolution was the means of preserving the existing agricultural system with themselves in possession of it', p. 98, Griswold, *ibid.* ; for further discussion, see pp. 88–127. See also Gordon Wright, *Rural Revolution in France*, Stanford University Press, Palo Alto, 1964, and C. G. von Dietze, 'Land Tenure Issues in Western Europe Since the French Revolution', in K. H. Parsons (ed.), *Land Tenure*, University of Wisconsin Press, Madison, 1956, pp. 374–383.

[2] David Mitrany, *Marx against the Peasant*, University of North Carolina Press, Chapel Hill, 1951, pp. 87–98, 118–145.

[3] Edmundo Flores, 'Agrarian Reform and Economic Development', in K. H. Parsons, *op. cit.*, pp. 243–246.

[4] M. Kaihara, 'On the Effects of Postwar Land Reform in Japan', in Walter Froehlich, *op. cit.*, pp. 143–156 ; L. I. Hewes, *Japan, Land and Men: An Account of the Japanese Land Reform Program*, Iowa State College Press, Ames, 1955 ; Chen Cheng, *Land Reform in Taiwan*, China Publishing Co., Taiwan, 1961, pp. 90–91 ; and Sidney Kein, *The Pattern of Land Tenure Reform in East Asia after World War II*, Bookman Associates, New York, 1958.

[5] Edmundo Flores, *Land Reform and the Alliance for Progress*, Policy Memorandum No. 27, Center for International Studies, Woodrow Wilson School of Public and International Affairs, Princeton University, 20 May 1963.

concern with the productivity objective reflected the relatively favourable man–land ratios and/or the relatively advanced technology which prevailed at the time of the reform. In Mexico, Bolivia, Korea and Taiwan, where the reforms took place under less favourable technical and economic conditions, the equalitarian orientation of the reforms was used to obtain farmer or peasant loyalty for the national government while development goals in other sectors of the economy were being achieved.[1]

(b) The Productivity Objective

British experience has contributed to the development of a second tradition of land reform which places major emphasis on the achievement of productivity and efficiency in the agricultural sector. Despite the seminal role which Locke's political philosophy played in the thinking of Jefferson and of continental reformers, the owner-cultivator was never regarded, in Britain, as the primary foundation on which to build democratic institutions. The British did not share Jefferson's fears of commerce and industry, and they did not identify democracy with the agrarian way of life. Equity considerations centred around the evolution of more precise definitions of landlord and tenant rights and duties than on transfer of ownership to the cultivator.[2]

The two most striking illustrations of the drive for agricultural productivity as a basis for English agrarian policy are the enclosure movement of the late eighteenth and early nineteenth centuries and the repeal of the Corn Laws in 1846. The enclosure movement complemented the process, begun several centuries earlier, of consolidating the open fields and commons in compact holdings under individual ownership and management and stimulated the first real advances in agricultural practices in Britain since the thirteenth century. The repeal of the Corn Laws represented the triumph of urban interests, both industry and labour, over the same landed classes that had benefited from the enclosure movement, and

---

[1] P. M. Raup, 'The Political Economy of Land Reform', Paper prepared for meeting of the Social Science Research Council — American Farm Economic Association Committee on New Orientation on Research, New York, 30 Nov. 1962.

[2] With the passage of the Landholdings Act of 1923, farm tenants in Great Britain had achieved perhaps the highest degree of security of occupancy possible, short of ownership. For further details, see Griswold, *op. cit.*, pp. 47–85. Also J. J. MacGregor, 'Principles of Tenure in England and Wales', in K. H. Parsons *et al.* (eds.), *op. cit.*, pp. 360–374.

committed Britain to a tradition of free trade in food supplies.[1] The continuing strength of this tradition, which has emphasized efficiency in food production rather than protection for the farm, has been a major obstacle to the entry of the United Kingdom into the European Common Market.[2]

Marxian agrarian policy represents an important heresy, perhaps mutation would be a more appropriate term, of the English tradition.[3] Despite its equalitarian thrust, the welfare of the peasantry has, until recently, remained outside the direct concern of both Marxian theorists and policy-makers. The breaking up of large estates is primarily regarded as an initial step in the Marxian stages of agricultural development. Its primary function is to reduce the political power of the land-owning class preparatory to the 'rationalization' of agricultural production in large-scale units.[4]

Three interrelated factors appear to be involved in the failure of the productivity-oriented Marxian economics to achieve their agricultural productivity objectives : (*a*) over-commitment to large-scale units in agriculture resulting from reliance on ideological rather than pragmatic considerations in the organization of economic activity ; (*b*) lack of confidence in the peasant proprietor's capacity to react rationally when provided with adequate information and market

---

[1] D. G. Barnes, *A History of the English Corn Laws from 1660–1896*, Reprints of Economic Classics, Augustus M. Kelley, New York, 1961 (original edition 1930). 'Only in a brief period in Great Britain were the interests of the manufacturers and consumers identical. Both wanted cheap food although for different reasons, and hence they were united against their temporary common enemy, the agriculturists, and brought in free trade', p. 293. See also Marburg, *op. cit.*

[2] Michael Butterwick, 'British Agricultural Policy and the EEC', *International Journal of Agrarian Affairs*, Vol. 4, No. 2, Apr. 1964, pp. 99–113.

[3] See N. Georgescu-Roegen, 'Economic Theory and Agrarian Economics', Oxford Economic Papers, New Series, Vol. 12, Nos. 1–3 Feb.–Oct. 1960, pp. 1–40. Also D. G. Dalrymple, *Marx and Agriculture: The Soviet Experience*, Mimeo No. 846, Department of Agricultural Economics, Michigan State University, East Lansing, Michigan.

[4] David Mitrany, *Marx against the Peasant*, *op. cit.* Mitrany identifies five development stages in Marxist–Leninist agricultural development policy, as it has evolved in the U.S.S.R. and Eastern Europe. 'The first phase is the class conflict between peasants and landless laborers on the one hand, and landowners and capitalist farmers on the other, that leads to a common front between rural and urban workers in a revolutionary surge. The second phase coincides with the victorious political revolution and ends with the distribution of land to the small peasants and landless laborers. . . . The third phase is a period of transition, with rapid urban and industrial development and the small peasant property organized on a co-operative basis. Finally, when industry can provide the necessary technical equipment . . . the independent peasant class is liquidated and agriculture is concentrated into large collective farms', p. 62. For discussion of Marxian agrarian reform under Asian conditions, see J. Price Gittinger, 'Communist Land Policy in North Vietnam', in *Studies on Land Tenure in Vietnam*, Division of Agriculture and Natural Resources, USOM, Vietnam, Dec. 1959, pp. 30–47. (Reprinted from Far Eastern Survey.)

incentives and (*c*) an effort to achieve increases in surplus production from the agricultural sector by the use of administrative arrangements rather than market incentives.[1]

III. SYNTHESIS OF EQUITY AND PRODUCTIVITY OBJECTIVES IN ASIAN LAND REFORM

Although the increased emphasis given the productivity objective in recent land reform policy has strong roots in English experience and Marxian development theory, it also has an even stronger foundation in the economic environment of most under-developed countries in the 1960's. A striking feature of the land reform movements in the developing countries of South and South-east Asia during the last decade has been the extent to which a dual emphasis on equity and productivity objectives has replaced the almost exclusive concern with equity considerations until at least the mid-1950's.[2]

Four factors of particular importance in accounting for this new synthesis of land reform objectives have been : (*a*) the clash of liberal and Marxist political ideology that has accompanied the withdrawal of colonial authority or the overthrow of domestic authoritarian regimes ; (*b*) the pressure of rising rates of population growth on food supplies which is approaching a level entirely outside the experience of presently developed countries as population growth rates rise to levels between 3 and 4 per cent per year ; (*c*) the demonstration effect of the rapid productivity growth, under conditions of small-scale peasant agriculture, which followed implementation of the Japanese and Taiwan land reform programmes ; (*d*) the apparent power of the modern theory of the firm to identify the productive superiority of an owner-operated agricultural system relative to a share tenure or even a fixed rent leasehold system of agriculture.

The first two factors have contributed to the political motivation for productivity-oriented agrarian reform. Countries such as

---

[1] Lazar Volin, 'Collectivization of Agriculture in Soviet Russia', in K. C. Parsons, *op. cit.*, pp. 407–416. Alec Nove, 'Incentives for Peasants and Administrators', in R. D. Laird, *Soviet Agricultural and Peasant Affairs*, University of Kansas Press, 1963, pp. 51–68.

[2] The same process is also occurring elsewhere. See P. M. Raup, 'The Contribution of Land Reforms to Agricultural Development : An Analytical Framework', *Economic Development and Cultural Change*, Vol. 12, No. 1, Oct. 1963, pp. 1–21. For a review of land reform in South-east Asia prior to the mid-1950's, see E. Jacoby, *Agrarian Unrest in Southeast Asia*, Asia Publishing House, Bombay, 1961.

Malaysia and the Philippines have employed agrarian reform legislation as one element in programmes to satisfy the equalitarian drive of the peasantry. But rural unrest is not the only source of political instability. Political leaders in most South-east Asian countries are sensitive to the demands of a rapidly growing articulate urban population for stable rice prices.[1] Possibility that a land reform programme might make a positive contribution to the problems of both rural and urban unrest makes land reform attractive to many political leaders despite the fact that much of this leadership comes from the land-owning classes.

The other two factors have combined to give a broader 'legitimacy' to the land reform objectives. An intellectual rationale which identifies a positive contribution of land reform to productivity growth has helped gain the support of the national bureaucracy and the commercial and industrial classes committed to rapid economic growth. The net effect has been to provide the liberal intellectual and political élite with a more powerful dialectic with which to reinforce the equalitarian drive of the peasantry.

(a) Philippine Land Reform

This shift from a primary emphasis on equity to a new focus on both equity and productivity objectives and programme instruments is clearly illustrated in Philippine land reform legislation. The Magsaysay land reform legislation of the mid-1950's was directed primarily towards restoring peace and order in the areas of agrarian unrest and to obtaining farmer and peasant loyalty to the newly established national government.[2]

[1] Urban pressures appear to be a major obstacle to rationalization of rice price and marketing programmes in both the rice-exporting and rice-importing countries of South-east Asia. In the rice-exporting countries, rice is typically undervalued in relation to export prices. And in the rice-importing countries, prices paid to farmers frequently exceed import prices while rice is made available to at least part of the urban population at subsidized prices. Urban pressure for lower food prices has also been imparted during the early stages of development in many presently developed countries. See Barnes, *op. cit.*

[2] The Magsaysay land reform legislation includes (a) the Landlord–Tenant Relationships Law of 1954 (Republic Act No. 1199); (b) the National Resettlement and Rehabilitation Administration (NARRA) Act of 1954 (R.A. No. 1160); (c) the Land Reform (Ownership Transfer) Act of 1955 (R.A. No. 1400); (d) the act establishing the Court of Agrarian Relations (R.A. No. 1409). This legislation was designed to reduce the power of the landowners relative to tenants, acquire and redistribute large landed estates and resettle dissident elements on the public domain. For discussions of the Magsaysay and earlier Philippine land tenure legislations, see F. L. Starner, *Magsaysay and the Philippine Peasantry*, University of California Press, Berkeley, 1961 ; H. L. Cooke, 'Land Reform and Development in the Philippines', in Walter Froehlich, *op. cit.*, pp. 168–180 ; J. R. Motheral,

Conceptual Problems in Agricultural Development

The new Philippine Agricultural Land Reform Code of 1963 departs sharply from earlier legislative intent and places important emphasis on both equity and productivity objectives and programme instruments.[1] It bears the unmistakable imprint of a group of young economists and intellectuals who were primarily concerned with the failure of existing agricultural development programmes to generate sufficiently rapid gains in agricultural productivity to match the rapid population growth rate that is now approaching 3·5 per cent per year and may approximate 4·0 per cent in two decades.[2] The productivity orientation of the legislation was particularly useful in gaining the support of the growing industrial classes that would not have been swayed by political appeals for equalitarian justice for the peasantry during a period when organized rural unrest was apparently dormant.

(b) Malayan Land Reform

The evolution of land reform in Malaya parallels, in many respects, the Philippine experience. The 1955 legislation was passed during a period of internal unrest when it was important to obtain the loyalty of the rural Malays. Its primary objective was to increase the security of tenure and to control rents.[3] In contrast, current discussion focuses almost entirely on the modifications in the tenure structure and in related agrarian policy needed to increase incentives

---

'Land Tenure in the Philippines', *Journal of Farm Economics*, Vol. 38, No. 2, May 1956, pp. 465–474 ; R. T. McMillan, 'Land Tenure in the Philippines', *Rural Sociology*, Vol. 20, No. 1, March 1955, pp. 25–33 ; R. S. Hardie, *Land Tenure, Reform Analysis and Recommendations*, Special Technical and Economic Mission, U.S. Mutual Security Agency, Manila, 1952. The Hardie report is particularly valuable for its reproductions of documents of historical significance in the evolution of Philippine land tenure.

[1] Agricultural Land Reform Code (Republic Act No. 3844), Manila, Bureau of Printing, 1963. For a more thorough discussion of the content of R.A. No. 3844 in terms of its equity and productivity emphasis, see O. J. Sacay, 'The Philippine Land Reform Program', *The Philippine Economic Journal*, Vol. ii, No. 2, Second Semester, pp. 69–183 and V. W. Ruttan, *op. cit.*

[2] K. V. Ramachandran, R. A. Almendrala and M. Sivamurthy, 'Population Projections for the Philippines, 1960–1980', *The Philippine Statistician*, Vol. 12, No. 4, Dec. 1963, pp. 145–169.

[3] Federation of Malaya, *The Padi Cultivators (Control of Rent and Security of Tenure) Ordinance*, 1955, Supplement to the Federation of Malaya Government Gazette of April 1955, No. 7, Vol. iii. Notification Federal No. 766. The definitive study of land tenure in the rice-growing areas of Malaya is T. B. Wilson, *The Economics of Padi Production in North Malaya (Land Tenure, Rents, Land Use and Fragmentation)*, Ministry of Agriculture, Federation of Malaya, June 1958. For discussion of tenure on rubber estates, see Ungku Aziz (ed.) *Subdivision of Estates in Malaya, 1951–1960* (3 volumes), Kuala Lumpur : Department of Economics, University of Malaya, 1962. See also G. D. Ness, 'Subdivision of Estates in Malaya, 1951–1960 : A Methodological Critique', *The Malayan Economic Review*, Vol. 9, No. 1, Apr. 1964, pp. 55–62, and G. D. Quirin, 'Estate Subdivision and Economic Development : A Review Article', *ibid.*, pp. 63–79.

to achieve higher levels of productivity in the production of rough rice (padi). Furthermore, this discussion is occurring in a period when there is no evidence of major internal unrest in the rural areas.

(c) Indian Land Reform

The evolution of policy objectives and programme instruments in Indian land reform legislation has been much more complex than in either the Philippines or Malaya. This stems in part from the divergent and, at times, conflicting policies and perspectives of the state and centre governments.

Land reform legislation had been instituted under Congress Party pressure prior to independence (1947). The Bombay Tenancy Act of 1939, for example, restricted eviction and set ceilings on rent in some districts.[1] By the beginning of the First Five-Year Plan (1951), a substantial body of tenancy legislation had been passed by the states and by the end of the Second Five-Year Plan (1960) some form of tenancy legislation existed in every state. It covered abolition of intermediaries, security of tenure for the tenant farmer, regulations of rent, facilities for acquiring ownership rights by the tenant farmer and a ceiling on future acquisitions of landholding.[2]

The concern of the state governments with equity consideration has been in sharp contrast to the productivity orientation of the centre government. The productivity orientation of the centre government and of Congress Party leaders was complicated, however, by ideological commitments and an intellectual tradition which assumed that small-scale peasant proprietorship is incapable of achieving the productivity levels required to meet the development objectives of the Indian economy. The result was insistence on a radical reorganization of the structure of the Indian rural economy to eliminate subsistence farming.[3]

[1] See G. Wunderlich, *Land Reform in Western India: Analysis of Economic Impacts of Tenancy Legislation, 1948–63*. U.S. Department of Agriculture, ERS-Foreign-82, June 1964, for a discussion of the evolution of tenancy legislation in the three states of Gujarat, Maharashtra and Mysore.
[2] V. N. Dandekar, 'From Agrarian Reorganization to Land Reform', *Antha Vijnana*, Vol. 6, No. 1, Mar. 1964, pp. 51–70; M. L. Dantwala, 'Financial Implications of Land Reforms : Zamindari Abolition', *Indian Journal of Agricultural Economics*, Vol. 17, No. 4, Oct.–Dec. 1962, pp. 1–11.
[3] Raj Krishna, 'Some Aspects of Land Reform and Economic Development', in Walter Froehlich, *op. cit.*, p. 234, also pp. 214–254 : '. . . the felt requirements of the food situation, of economic growth, and social justice, ideological commitments to Gandhism and socialism and the reports of Chinese achievements have all combined to strengthen the belief of intellectuals in the necessity of joint farming'. See also the series of papers on 'Land Reform Legislation and Its Implementation

Conceptual Problems in Agricultural Development

The gap between centre policy and state programmes has remained in recent years. The position that complete reorganization of the agrarian structure is a prerequisite for agricultural productivity growth under Indian conditions has been consistently evaded. A more careful assessment of how to supplement existing tenure legislation in order to realize the potential incentive it offers farmers for productivity growth now seems to be in process.[1]

(d) The Policy Dilemma

I am not able to characterize land reform policies and programmes in other countries of South and South-east Asia as definitely as in the Philippines, Malaysia and India. My impression is that in most other countries, the evolution from an equity orientation to a dual equity and productivity orientation is less complete than in these three.[2]

It is clear, however, that the problem which political leaders throughout most of South and South-east Asia face can no longer be cast in terms of choosing between equity and productivity objectives in the formulation of agrarian policy. The social and political unrest among urban consumers which accompanies rising food prices represents at least as important a source of political instability as inequities in the distribution of land ownership and income in rural

in Different States', *Indian Journal of Agricultural Economics*, Vol. 17, No. 1, Jan.–Mar. 1962, pp. 114–195. A. M. Khusro reports in 'Summary of Group Discussion', *ibid.*, pp. 189–195, on an interesting argument among Indian economists as to whether Indian land reform has been primarily productivity or equity oriented.

[1] This shift is particularly apparent in the thought of M. L. Dantwala, one of the intellectual leaders in Indian agrarian policy. In 'The Basic Approach to Land Reforms', *Indian Journal of Agricultural Economics*. Vol. 8, No. 1, Mar. 1953, pp. 95–99, and in other articles written in the early 1950's, he emphasized the primary importance of equity objectives. In his 1960 presidential address, 'Agrarian Structure and Economic Development', *Indian Journal of Agricultural Economics*, Vol. 16, No. 1, Jan.–Mar. 1961, pp. 1–25, Dantwala placed major emphasis on the productivity objectives.

[2] For a discussion of the land tenure situation in a number of Asian countries in the 1950's, see E. H. Jacoby, *Agrarian Unrest in Southeast Asia*, Asia Publishing House, Bombay, 1961, and J. R. Motheral, *Comparative Notes on East Asian Land Tenure Systems*, USOM, ICA, Manila, Philippines, July 1955. In Vietnam, the rent reduction, tenure security and land transfer programmes of the mid-1950's became a casualty of the decline of civil authority even before the demise of the Diem government. See J. P. Gittinger, *op. cit.* For a summary of the Indonesian legislation, see Commercial Advisory Foundation (CAFI), Basic Agrarian Act No. 5, Year 1960, Circular No. 306 (29 Oct. 1960) and 306a (31 Oct. 1960). In Thailand official concern with the economic and political implications of land tenure is only now beginning to emerge. See *Relationship between Land Tenure and Rice Production in Five Central Provinces, Thailand*, 1964, Land Economic Report No. 1. Ministry of National Development, Feb. 1965, and Takeshi Motooka, 'Problems of Land Reform in Thailand with Reference to the Japanese Experience', *Symposium on Japan's Future in Southeast Asia*, Center for Southeast Asian Studies, Kyoto University, Kyoto, 31 May–2 June 1965.

areas. In areas characterized by low agricultural productivity and rapid population growth, achievement of productivity objectives appears to represent a prerequisite for both socio-political stability and equalitarian justice. At the same time, attempts to achieve agricultural productivity objectives by means which fail to satisfy the equalitarian drive of the peasantry do not appear able to provide sufficient motivation for the peasantry to co-operate fully in achieving the productivity and growth objectives.

IV. AN EVOLUTIONARY APPROACH TO THE RELATIONSHIP BETWEEN LAND TENURE AND PRODUCTIVITY

The emergence of productivity as a major objective of change in land tenure systems in South-east Asia imposes a new burden on those engaged in both economic analysis and economic planning. The imperatives of population growth have weakened the ideological commitments of the political élite. They are depending, to an increasing extent, on economists for guidance in the formulation of land tenure policy. A review of the empirical materials available from South-east Asia, however, casts serious doubt on both the precision of our analytical tools and the results of empirical investigations.

(c) Productivity Differentials in Economic Theory and History

The major analytical conclusion which emerges from the modern microeconomic theory of the firm is that 'there is no substitute, from the standpoint of sheer productivity, and irrespective of sociological considerations, for an owner-operated agricultural system'.[1] This generalization is based on formal exploration of the empirical consequences for the equilibrium level of output of the firm of (a) the method of pricing factor inputs and (b) the constraints on decision-making under alternative 'ideal type' tenure arrangements. Deductions based on this theory would clearly lead us to expect that if farms were classified by size, tenure and productivity, that for each size group, a productivity ranking would find the owner-operator at the top, share tenants at the bottom and part owners and lessees in an intermediate position.

[1] L. S. Drake, 'Comparative Productivity of Share and Cash-Rent Systems of Tenure', *Journal of Farm Economics*, Vol. 34, No. 4.

For the evolution of this analysis in U.S. literature, see Reiner Schickele, 'Effect of Tenure Systems on Agricultural Efficiency', *Journal of Farm Economics*, Vol. 23,

The analytical deductions are generally regarded as consistent with historical experience. In Asia the recent Japanese and Taiwan experience has been interpreted as supporting the proposition that an agrarian structure consisting of extremely small owner-cultivator family farms can be viable, reasonably efficient and capable of sustaining rapid increases in agricultural productivity. Analysis of the earlier modifications in tenure arrangements in Japan, beginning with the abolition of feudal privileges and the conversion of the land tax to a cash rather than a commodity basis during the early years of the Meiji Restoration also supports the proposition that the resulting improvements in incentives complemented efforts to introduce technological change in the form of new varieties, higher levels of fertilization and other changes in cultural practices.[1]

Analysis of land tenure data and land reform experience in other Asian countries, however, does not provide the same degree of support for either the logical deductions or the historical generalizations. The growing body of empirical evidence on the relationships among farm size, tenure and productivity is frequently inconsistent with the ordering suggested by the logical deductions. In the smaller size ranges, share tenants typically achieve higher yields than owner-operators. Even in the larger size classes, owner-operators frequently do not exhibit any clear-cut productivity differential relative to other tenure classes.[2]

Furthermore, the new incentives resulting from tenure reform have not yet manifested themselves in the form of differential rates of growth in agricultural productivity or in aggregate farm output. The rate of growth in agricultural output is slower in India, where sustained shifts in tenure relationship have occurred, than in the Philippines and Malaysia where implementation of land tenure legislation has been rather modest. In many areas where 'land to

---

Feb. 1941, pp. 185–207 ; H. O. Heady, 'Economics of Farm Leasing Systems', *Journal of Farm Economics*, Vol. 34, No. 3, Aug. 1947, pp. 650–678 ; D. G. Johnson, 'Resource Allocation Under Share Contracts', *Journal of Political Economy*, Vol. 57, Apr. 1950, pp. 111–123 ; L. S. Drake, *op. cit.*, pp. 535–550.

[1] For Japan, see M. Kaihara, *op. cit.*, L. I. Hewes, *op. cit.*, Takekazu Agura (ed.), *Agricultural Development in Modern Japan*, *Japan FAO Association*, 1–26 Nishigahara, Kita-ku, Tokyo, Japan, 1963, pp. 119–144, 613–677. For Taiwan, see Chen Cheng, *op. cit.*, and S. L. Hsieh and T. H. Lee, 'An Analytical Review of Agricultural Development in Taiwan : An Input–Output and Productivity Approach', Chinese-American Joint Commission on Rural Reconstruction, *Economic Digest Series No. 12*, Taipei, Taiwan, July 1958.

[2] V. W. Ruttan, 'Notes on the Empirical Relationships Between Farm Size, Tenure and Productivity in Southeast Asian Agriculture', IRRI-AE-Staff Memo, May 1964 (Mimeo).

the tiller', or resettlement programmes have been effectively pursued, there has been a tendency for tenancy to reappear in either overt or disguised forms.

It is my impression that a careful review of land reform efforts in South-east Asia, excluding Japan and Taiwan, would not reveal a single example of a successful land reform programme when evaluated in terms of contribution to productivity growth.[1] It appears that neither (*a*) the empirical predictions based on logical deductions from the neo-classical theory of the firm, or (*b*) the historical generalizations from experience in Western Europe, North America and Japan and Taiwan are adequate guides to modern land tenure policy in South-east Asia.

The first step in achieving greater precision in predicting the productivity implications of changes in land tenure arrangements is to reject the assumption that there is any single optimum land tenure system. It seems reasonable to hypothesize that the relationship between land tenure and productivity varies (*a*) with the extent of commercial (or subsistence) production ; (*b*) with the level, rate and direction of technological development ; (*c*) with the extent of diffusion (or concentration) of political and economic power. Each of these categories represents a continuum with undefined origins and termini.

If this hypothesis is valid each rural society can be defined in terms of a plane which intersects each of the several rays. The plane representing a specific country or region at a particular time would occupy a unique position in the defined space. Analysis of the impact of land tenure policies on agricultural productivity should utilize a behaviour model which incorporates the specific economic, technological and socio-political environment of the country for which the analysis is being conducted.

However, a taxonomy consisting of a number of 'ideal type'

---

[1] Apparently this conclusion also applies with some force to land reform efforts in Western and Eastern Europe. Folke Dovring, *Land and Labor in Europe, 1900–1950*, Martinus Nijholf, The Hague, 1960. Dovring indicates that in Europe, 'land policy has more often been a failure than a success. . . . Western countries tried to stop the rural exodus by colonization policy ; the rural exodus continued and was even accelerated. Eastern countries tried to remedy the evils of rural overpopulation by radical land reforms ; population increase continued and overpopulation remained unremedied. Western countries sought to strengthen small and discourage large farms ; farms have continued to grow larger spontaneously, where conditions allowed the trend to manifest itself. The Soviet Union tried to discourage the familistic society by creating large-scale, highly mechanized agriculture ; the response was a drive toward over intensification in the cultivation of the small family gardens and neglect of the collective sector, with overt failure of the livestock industry, and hidden failures in crop production, as the consequence', p. 350.

situations could (*a*) represent a useful device for classifying the particular models and (*b*) serve as a starting-point in the elaboration of particular country or regional models. As a minimum such taxonomy might consist of the four categories outlined below.

(*a*) *Subsistence Agriculture in the Traditional Economy*

In both Western feudal society and in the pre-colonial societies of South and South-east Asia the surplus production of the agricultural sector was typically channelled into the non-agricultural sector through tribute or taxes rendered in the form of labour services or as a share of the produce from the land occupied by the cultivator. The colonial systems adopted by the Spanish, English, Dutch and French in South and South-east Asia continued to employ both methods to extract surplus production in the form of tributes or trade from native agriculture.[1]

With increased monetization of economic activity, extraction of tribute in the form of labour services declined and was replaced by the use of wage labour in public projects and in plantation agriculture. Extraction of tribute in kind, however, did not disappear but instead evolved into the modern share tenure pattern that characterizes much of South and South-east Asian agriculture. Even in areas such as Burma, where the native tax system depended primarily on excise taxes and where the commercial rice-producing areas developed under a system of owner-operation, the share system emerged as the dominant tenure form by 1900.

It is clear that modern share tenure systems have evolved primarily as a response to the monetization and commercialization of economic activity. Colonial administrators and the commercially oriented

---

[1] For the evolution of tenure systems in Western Europe, see B. H. Slicher von Bath, *The Agrarian History of Western Europe*, Edward Arnold Ltd., London, 1963, pp. 29–53, 145–151, 310–324 ; Mogens Boserup, 'Agrarian Structure and Take Off', in W. W. Rostow, *The Economics of Take-Off Into Sustained Growth* (Proceedings of a Conference held by the International Economic Association), Macmillan, London, 1963, pp. 207–215 ; and Folke Dovring, *op. cit.*

For the evolution of colonial economic and political policy in South-east Asia, see K. J. Pelzer, *Pioneer Settlement in the Asiatic Tropics*, American Geographical Society, New York, 1948 ; J. S. Furnivall, *Colonial Policy and Practice*, New York University Press, New York, 1956 ; O. D. Corpuz, *The Bureaucracy in the Philippines*, U.P. Institute of Public Administration, 1957. The Spanish system of colonial administration in the Philippines was very similar, particularly during the early period, to the system employed in the Spanish colonies of Latin America, see David Weeks, 'European Antecedents of Land Tenures and Agrarian Organization of Hispanic America', *Journal of Land and Public Utility Economics*, Vol. 21, No. 1, Feb. 1947, pp. 60–75 ; 'The Agrarian System of the Spanish American Colonies', *ibid.*, Vol. 23, No. 2, May 1947, pp. 154–168 ; 'Land Tenure in Bolivia', *ibid.*, Vol. 23, No. 3, Aug. 1947, pp. 322–336.

indigenous élite have at times used share tenure systems as effective devices to force increased production from a peasant agriculture prior to the extensive monetization of economic activity in rural areas. Owners of estates or haciendas have utilized share tenure arrangements as more effective methods of labour management than use of direct wage labour in the production of crops such as rice which require close personal attention or a high degree of husbandry skill on the part of the cultivator. And peasant farmers have, at times, found that share tenure arrangements exposed them to less income uncertainty from price and yield fluctuations than fixed rent leasehold arrangements or a debt encumbered owner-operatorship. When these factors are considered, along with the historical record of rapid expansion of agricultural output in the colonial areas of South and South-east Asia during the last half of the nineteenth century, one can hardly escape the conclusion that share tenure systems are, at least under some conditions, consistent with rapid growth in agricultural productivity and output.

This conclusion is clearly not consistent with the inferences that are usually drawn from the recent historical record in East Asia or from the standard land tenure models based on the neo-classical theory of the firm. It appears to me, however, that modification of the neo-classical models of the firm to incorporate the dynamic effects of monetization and commercialization of agricultural production in an environment characterized by (*a*) static agricultural technology, (*b*) substantial imperfections in capital and labour markets and (*c*) high rates of growth in the rural labour force would imply more efficient resource allocation and more rapid growth of productivity under share tenure than under fixed rent leasehold or owner-operator systems. The 'subsistence village' or 'household farm' models which have been developed to explore the behaviour of subsistence farming systems represent an important first step in the development of more relevant land tenure models for this type of economy.[1]

(b) Small-Scale Commercial Agriculture in a Developing Economy

From at least 1840 until fairly recently, much of U.S. agriculture, particularly in the North-east and Middle West, has been

[1] E. K. Fisk, 'Planning in a Primitive Economy : Special Problems of Papua–New Guinea', *Economic Record*, Vol. 38, No. 84, Dec. 1962, pp. 462–478. Chihiro Nakajima, 'The Subsistence Farmer in Commercial Economies', A/D/C Seminar on Subsistence and Peasant Economies, East–West Center, Honolulu, 28 Feb.–6 Mar. 1965.

characterized by technically progressive, small-scale, owner-operated commercial farms. These farms operated in an environment (*a*) characterized, but not dominated, by an expanding urban-indusrial sector and (*b*) in an 'open socio-political' structure.[1] Similar conditions have prevailed in many areas of continental Western Europe since the middle of the last century and have become the dominant pattern during the last two decades.[2] A similar pattern has prevailed in Japan, with sharply smaller land area per farm than in the West, at least since World War I.[3]

The historical generalizations concerning the favourable resource allocation and productivity effects of land tenure legislation designed

[1] Griswold, *op. cit.*, pp. 18–46 ; V. W. Ruttan, 'Agriculture in the National Economy', in U.S. Department of Agriculture, *A Place to Live* (The Yearbook of Agriculture, 1963), USGPO, Washington, 1963, pp. 135–138. I. Feller, 'Inventive Activity in Agriculture, 1837–1890', *The Journal of Economic History*, Vol. 22, No. 4, Dec. 1962, pp. 517–560. W. R. Rasmussen, 'The Impact of Technological Change in American Agriculture, 1862–1962', *ibid.*, pp. 578–591. According to Rasmussen, 'two revolutions in American agriculture reflect the impact of technological change in farming during the past century. The first revolution saw the change from manpower to animal power, and centered about the Civil War. The second saw the change from animal power to mechanical power and the adaptation of chemistry to agricultural production. It centered around the post World War II period', *ibid.*, p. 578.

[2] For the evolution of agrarian structure in France, see Gordon Wright, *Rural Revolution in France*, Stanford University Press, 1964. For purposes of tenure analysis, I include under the small-scale commercial agriculture classification some systems which other writers might classify as 'high' or 'late' peasantry. See Daniel Thorner, 'Peasant Economy as a Category in Economic History', *The Economic Weekly*, Special Number, July 1963. Boserup, *op. cit.*, distinguishes four main patterns of tenancy at the beginning of the modern period of rapid economic growth : (1) The 'British' type where the cultivator had been (or was being) restricted and reintegrated as a wage labourer ; and where the entrepreneur was the capitalist tenant. (2) The 'Eastern' type, where the cultivator had become a serf and the entrepreneur was identical with the 'seigneur'. (3) The 'French' type, where the peasant-owner predominated, *i.e.* where the functions of cultivator and entrepreneur were not separated. (4) The 'Mediterranean' type, where the cultivator is a share-cropper, and where there is really no person who can reasonably be described as an entrepreneur (p. 209). In 1831 Richard Jones, *An Essay on the Distribution of Wealth and on the Sources of Taxation: Part I — Rent*, Augustus M. Kelley, New York, 1964, p. 15 (a reprint of the 1831 edition), identified five classes of rent : (*a*) labour or serf rent which he identified as dominant in Eastern Europe ; (*b*) metayer or share rents which he identified as a product of the Greco-Roman system and as dominant in France and Southern Europe ; (*c*) ryot or tax rents, usually transmitted to the sovereign, as sole proprietor, through intermediaries which he identifies as the dominant system in Asia ; (*d*) cottier rents or cash rents which he identifies in Scotland and Ireland ; (*e*) farmers' rents, paid by capitalistic farmers who employed hired labourers and farm rented land. Jones identifies this system as dominant only in England and parts of Scotland, Holland and the Netherlands. He also argues that only in this system are rents determined by economic forces.

The share tenure system which has evolved in the Malay areas of South-east Asia appears to have evolved from a basic serf rent system modified by colonial powers whose main experience had been with metayer or share rent systems.

[3] Tokokazu, Ogura (ed.), *Agricultural Development in Modern Japan*, Japan FAO Association, 1–26, Nishigahara, Kita-ku, Tokyo, 1963.

to transfer a share or lease tenure system to an owner-operator system have been based almost entirely on observations from this category. Similarly, the neo-classical analytical models designed to examine the implications of alternative tenure systems on resource allocation and the growth of productivity have been designed, either implicitly or explicitly, to apply to farms operating in an environment of the type outlined above.[1]

The most significant conclusion to be drawn from the generalizations and the models is that in such an environment, a share tenure system does dampen incentives (*a*) for the adoption of new technology by the tenant and (*b*) investment in capital improvements by the landlord. For societies characterized by small-scale commercial agriculture, technological progress and an expanding urban labour market both the generalizations and the models appear to be consistent with cross-section and time-series observations regarding the relationship between tenure and productivity.

I would anticipate that in agrarian systems which are undergoing a transition from a static subsistence structure to a technically progressive small-scale commercial structure attempts to determine empirical relationships between productivity, tenure and farm size would yield conflicting results. In such situations the direct policy implications cannot be expected to be as clear-cut as policy-makers might desire.[2] Krishna's comment that 'while some land reforms are essential for economic development, economic development is essential for the success of many land reforms',[3] is particularly pertinent during such a transition.

(c) Large-Scale Commercial Agriculture in an Urban-industrial Economy

The emergence of a large-scale commercial agriculture in a technically progressive, rapidly expanding urban-industrial economy in which agriculture represents a relatively small share of total employment and output further modifies the relationship between tenure and productivity. English agriculture has existed in such an environment since the middle of the last century. Much of U.S. agriculture has existed in such an environment since the mid-1920's.

[1] See the articles by R. Schickele, E. O. Heady, D. G. Johnson and L. S. Drake cited in footnote 1 pages 591–2 above.

[2] Peter Dorner, 'Land Tenure, Income Distribution and Productivity Interactions', *Journal of Land Economics*, Vol. 40, No. 3, Aug. 1964, pp. 246–254.

[3] Raj Krishna, 'Some Aspects of Land Reform and Economic Development', in Walter Froehlich, *op. cit.*, p. 223.

Conceptual Problems in Agricultural Development

The most significant feature of an agrarian structure dominated by a progressive urban-industrial economy for the analysis of the relationship between tenure and productivity is the strength of the intersector factor (labour, capital and materials) markets. A strong non-farm labour market increases the bargaining power of the tenant relative to the landlord. A strong non-agricultural financial market modifies the role of land as a store of wealth by providing attractive and dependable non-agricultural investment alternatives. The net effect is to create a bargaining situation between the landlord and the tenant in which both share tenure and leasehold arrangements are, at least, as consistent with efficiency and rapid technological change as owner-operatorship.[1]

In the dynamic environment in which U.S. agriculture has operated in the last several decades, land rental has played an important role in facilitating rapid change in farm size. In the areas with the most advanced production techniques, particularly in the great grain-producing areas of the Middle West, which account for almost half of national agricultural output, the proportion of leased- or tenant-operated land is higher than in other regions and is continuing to rise.[2]

(d) Plantation, Latifundia and Collective Agriculture

There is a fourth general category of agriculture in which production is carried on in large-scale units described by terms such as

[1] In 1940 T. W. Schultz, in 'Capital Rationing, Uncertainty and Farm Tenancy Reform', *Journal of Political Economy*, Vol. 48, No. 3, pp. 309–324, pointed out that in an agriculture characterized by high levels of capital input per worker, 'changing tenant farmers over to encumbered owners limits measurably the returns of farmers who have limited assets', p. 323. D. G. Johnson, *op. cit.*, develops the implications of share tenure leases for allocative efficiency when the tenant has effective alternative employment opportunities for his labour and the landlord has alternative investment opportunities for his capital.

[2] In a study sponsored by the North Central Land Tenure Research Committee, W. G. Miller, W. E. Chryst and H. W. Ottoson, *Relative Efficiencies of Farm Tenure Classes in Intrafirm Resource Allocation*, Iowa Agricultural and Home Economics Experiment Station, Research Bulletin 461, Iowa State College, Ames, Nov. 1958, failed to find significant differences in the marginal productivity of resources related to tenure.

J. O. Bray, 'Farm Tenancy and Productivity in Agriculture : The Case of the United States', *Food Research Institute Studies*, Vol. 4, No. 1, 1963, pp. 25–38. According to Bray, 'The most impressive development of recent years has been the persistent increase in the percentage of land in the hands of part owners. These operators generally own more land than do full owners in the same region, and, in addition, rent an acreage similar to that farmed by the average tenant. Thus, they farm larger acreages than either full owners or tenants. The increasing proportion of all land in farms held by part owners suggests that this group contains some of the most successful owners who have enlarged their operations by renting, as well as successful tenants who have bought a farm but continue to rent other land', p. 36.

plantation agriculture, latifundia and collective or co-operative farms. Although there are great differences in the organization of such farms, the farms included in this general category are typically characterized by the use of labour which is attached to the farm unit through some sort of contractual relationship rather than through classical labour market relations. The labour force typically has use rights to cultivate small areas from which they obtain the entire product although their labour is usually available to the larger unit at the discretion of the management of the large unit.[1]

Such farms have typically been most viable under conditions (*a*) of new land development in the tropics ; (*b*) in situations where the technology requires rather large scale capital investment in certain operations but where mechanization is not available or is relatively inefficient for other operations and (*c*) where there are substantial advantages in vertically integrating farm operations with transportation or processing activities.[2] It is generally presumed that such operations make relatively inefficient use of labour resources. Land reforms which have attempted to convert such farms to small-scale owner-operator units have typically been accompanied by reductions in output per hectare and per worker. At the same time, attempts to consolidate small units into large-scale units to take advantage of presumed scale economies have had great difficulty in organizing incentives to maintain the level of productivity achieved on smaller farms.

IV. CONCLUSIONS

In concluding this paper, I would like to emphasize three points :

1. It is no longer feasible to frame the land reform issue in South and South-east Asia in terms of a choice between equity and productivity objectives. The imperatives of political stability and economic growth are such that both objectives must be achieved simultaneously.

2. The historical generalizations and analytical constructs that

---

[1] David Weeks, *op. cit.*, and Alec Nove, *op. cit.* The functional organization of Soviet agriculture is similar in many respects to the pattern that prevailed prior to 1861. For the evolution of tenure relationships in Russian agriculture, see Jerome Blum, *Lord and Peasant in Russia from the Ninth to the Nineteenth Century*, Princeton University Press, 1961.

[2] J. M. Brewster, 'The Machine Process in Agriculture and Industry', *Journal of Farm Economics*, Vol. 32, No. 1, Feb. 1950, pp. 69–81. Brewster argues that the introduction of machine methods does not represent a significant factor in determining whether farms are organized on a family or larger than family basis.

have been employed to predict the effects of alternative land tenure systems on productivity, in economies characterized by small-scale commercial farmers operating in an environment characterized by technological progress, an expanding urban-industrial economy and a relatively 'open' socio-political structure, do not provide a reliable guide to the impact of tenure changes in agrarian systems which are organized along different lines.

3. The first step in achieving greater precision in predicting the productivity implications of changes in land tenure arrangements is to reject the assumption that there is any single optimum land tenure system. It is hypothesized in this paper that the relationship between land tenure and productivity varies with (*a*) the extent of commercial (or subsistence) production ; (*b*) with the level, rate and direction of technological development ; and (*c*) with the extent of diffusion (or concentration) of political power.

DISCUSSION OF PROFESSOR RUTTAN'S PAPER

Professor Hsia said that it was not clear what Professor Ruttan intended to include under land reform legislation. For example, mention was made of the repeal of the Corn Laws in England and this seemed to have only remote relevance to the cases under discussion. As a result of this lack of clear delineation, Professor Ruttan seemed to vacillate between land tenure legislation and measures intended simply to improve agricultural productivity.

Professor Hsia said that he regretted that Professor Ruttan had not tried to analyse a specific case of agrarian reform legislation in detail in order to pinpoint the equity and productivity issues in it.

The paper spoke of 'the almost exclusive concern with equity considerations until at least the mid-1950's' in South-east Asia. Professor Hsia said that he was not convinced that productivity aims were absent at that stage. In any case, the distinction between equity and productivity is not as sharp as Professor Ruttan was there implying.

On the question of the divergence in policy objectives between federal and state governments in India, Professor Hsia said that such a divergence applies to only one or two out of the fifteen states, and it seems scarcely justifiable to generalize from these exceptions.

Professor Hsia asked how, precisely, the addition of the productivity objective affects land reform programmes, and the subsequent pattern of productivity. He also asked whether Professor Ruttan intended the 'evolutionary approach' as a policy recommendation for future land reform legislation.

Professor Obolenski said that in his paper Professor Ruttan has simplified Marxist theory and ascribed to it aspects which are not and never have been part of a Marxist programme. Thus, for example, it was incorrectly stated that until recently the welfare of the peasantry was outside the direct concern of Marxist theoreticians and politicians. From its inception, Marxism proclaimed the raising of the living standards of all who live by their labour as one of its fundamental aims.

With reference to the supposed inability of socialist agriculture to achieve the aims it has set itself, Professor Obolenski said that this is refuted by the course of the economic life of the Soviet Union. In the U.S.S.R. the gross agricultural production in 1964 was 241 per cent of pre-revolutionary production, and the gross national income was 335 per cent of the figure in 1950.

Professor Ruttan argued that the Soviet Union laid too great a stress on large-scale agricultural enterprises ; yet it is plain that the capitalist countries themselves follow a path leading to concentration of agricultural production.

Professor Obolenski said that Professor Ruttan's assertion that socialist countries strive for greater agricultural production only by means of administrative measures is clearly contradicted by the facts. Certainly mistakes have been made during the great task of socialist reconstruction of agriculture ; these were readily admitted. Thus, for example, some farms were enlarged to a point where they became unmanageable. However, one must not turn the exceptions into a rule.

All measures in the U.S.S.R. are designed to provide material and moral stimuli to production on a broad economic base in order to raise the standard of living. Professor Obolenski said he based this statement on his own experience as a Director of the Institute of Agricultural Economics where research is mainly concentrated on the problem of raising the productivity of labour as a priority. He said that in the paper he was presenting to the Conference, these problems were dealt with in greater detail.

Professor Georgescu-Roegen said that with respect to the equity-productivity issue he would have liked to see the problem divided into two parts, reflecting the two broad categories found in the world today. On the one hand there are the agrarian problems of the over-populated countries, which include most of the far eastern countries ; on the other hand there are the agrarian problems of the countries where land is so plentiful that anyone could obtain up to 1,000 hectares by simply applying to the government and paying only the necessary stamp duty. An example of such a country is Colombia. But in India, for instance, land is a limitative factor ; in this case, a useful concept of 'productivity' is different from that which would be appropriate for a country of the second type. Professor Georgescu-Roegen said that he wished only to express his hope that Professor Ruttan's further research would include consideration of the fundamental problem of measuring labour productivity without which there can be no clear understanding of the various solutions of equity problems and of productivity issues.

Professor Dandekar said that he did not think that Professor Ruttan implied that there is a conflict between the two objectives of productivity and equity. Many economists think that there is this inherent conflict in land reform. Thus there is a school of thought in India which states that equity is a luxury, and that productivity problems should come before questions of distribution. He did not think this was true. At a low standard of living, equity is a prerequisite of productivity. When we talk in terms of investment in human beings it is clear that food should come before education, and equity is required to provide the essential 12 ounces of cereals. Such equity is a precondition of productivity. Looked at this way, land reform need not immediately lead to increased productivity of land, but rather increased productivity of labour.

On tenurial arrangements Professor Dandekar said that the owner-cultivator system is one important basis of tenure reforms. But in India it has not led to increased production. Even before the major land reform in India, 80 per cent of the land was under owner-cultivators, so that the problem was restricted to only 20 per cent of the land. In turning tenants into owner-cultivators it is wrong to expect an immediate increase in production ; what we should expect in the first instance is an improvement of man.

Professor Dandekar referred to Professor Ruttan's argument that in some conditions share-cropping is consistent with rapid growth in output, even with static agricultural technology. He did not think that Professor Ruttan was right in his attempt to single out conditions under which share-cropping is good. The system dampens down the technological improvement without which there can be no improvement of productivity.

Professor Bićanić said that parts of Yugoslavia have experienced many of the problems with which Professor Ruttan's paper dealt, having had five land reforms in the last 200 years.

Professor Ruttan's paper dealt with land reform techniques from the point of view of the pure economist. However, land reforms involve social reform, not just the production of crops and animals.

There are two main types of land reform : those that aim at the abolition of rent and those that involve a real redistribution of land. They should be considered separately.

On the question of the productivity consequences of reform, Professor Bićanić said that it is essential to introduce a time dimension into the discussion explicitly. For example, in Yugoslavia after the post first world war land reform, agricultural production fell by 20 per cent and it took five years to restore the original level of production and to surpass it. However, the initial 'decreased' production in fact fed more people on the spot than before the land reform. It is important to know whether we mean productivity in the market sector or total productivity in terms of preserving and improving the living conditions of agricultural producers.

In general, land reforms often take place after great wars or other major upheavals. They are pregnant with political implications. There are certainly important economic considerations, but what was meant by 'economic' in this context is unclear.

Professor Zemborain said he wished to emphasize that there was a great difference between the idea of land reform and, simply, increases in agricultural productivity. The latter only involved a better use of labour and capital, whereas land reform represented a change in the social structure such that peasants had a better opportunity to take part in the political life of the country. To be successful, in most of the countries of the world, land reform had to be accompanied by an increase in industrial employment opportunities, and was closely bound up with the creation of a middle class. Land reform was, in this view, part of the change from a peasant to an industrial society.

Professor Campbell referred to the following passage in Professor Ruttan's paper : 'The agrarian reform movements which have looked to the liberal-Jeffersonian political philosophy . . . have typically shown little concern for the potential contribution which land reform might make to the growth of agricultural output. . . .' In newer countries like Australia where there have been land reform policies which established family farms by the breaking up of extensive holdings, equity considerations had been very important. However, it is not true that they paid little regard to the growth of agricultural output and of the national economy. There has usually been a strong concern to increase output per acre and the aggregate agricultural production of the country, and, in fact, to develop the country-side as a whole. Land policies had achieved this in the latter part of the nineteenth century and the early twentieth century. More recently, as a result of the upsurge of agricultural productivity since the war, there has been increasing recognition that the family farm as originally conceived might be too small to take full advantage of modern technology and that consequently larger farms are necessary to achieve efficient production.

Professor Ashton said that the issues dealt with in agrarian reform policies generally involve major conflicts of personal interests. Under these circumstances there is not necessarily any economic logic in these policies, and other motives have to be sought. Frequently, the policies are justified on the grounds of equity. But it is necessary to ask the question : equity for whom? The wider public interest is not necessarily considered. To connect land reform with questions of productivity could therefore be useful in helping to decide whether the benefit accrues to the public at large or only to the sector or region in question. If reform policies are to be based on productivity criteria, it suggests that a wider interpretation of equity is accepted.

In the United Kingdom, where there is a highly developed tenancy system, there is no evidence which would support the proposition that productivity is higher on owner-occupied farms.

Professor Ruttan said he would emphasize that in general land reform by itself does not directly affect growth of agricultural output. Land reform is simply a major factor in changing the incentives to achieve the existing potential. The evolutionary approach to the problem could provide a framework within which to start thinking, but it would only have direct bearing on policy decisions to a limited extent. Policy has to be considered in greater detail in relation to each individual country.

Professor Ruttan agreed that productivity objectives have become an increasingly important element in recent land reform legislation and referred to his discussion in a recent issue of the *Indian Journal of Agricultural Economics*.[1] The recent land reform measures in the Philippines are a clear-cut example of a productivity-oriented land reform. In these measures there were three elements : large plantation agriculture using direct labour was excluded from the measures ; they applied to farms operated on share tenure system ; and they were accompanied by agricultural credit and extension activity in the different land reform districts.

Referring to Professor Obolenski's objections, Professor Ruttan accepted that it had been in the earlier stages that the heaviest reliance on administrative arrangements in the U.S.S.R. had been seen. He also accepted that Marxian theorists did pay attention to the position of the peasants, but they took a long time deciding what they thought about them, and what they should do. Lenin had been more astute in sizing up the political role of the peasants than his Marxist colleagues in the East European countries.

Professor Ruttan agreed with Professor Georgescu-Roegen on the distinction between the predicaments in over-populated and under-populated countries. This is very important for the role of equity. Earlier land reform movements had taken place in under-populated areas. In the over-populated countries of South-east Asia, politicians are not free to emphasize either the equity objective or the productivity objective to the exclusion of the other. A politician will simply not remain in office if he accepts a decline in productivity over a five-year period. It is essential to operate on the incentive and equity side in such a way that productivity is boosted. Professor Ruttan said that in this connection he accepted the general point that land reform movements are designed to reform society as well as reorganize the agricultural sector.

On the question of the correct measure of productivity for parts of the world with different pressures of population, Professor Ruttan said that in

[1] V. W. Ruttan, 'Equity and Productivity Objectives in Agrarian Reform Legislation : Perspectives on the New Philippine Land Reform Code', *Indian Journal of Agricultural Economics*, Vol. xix, Nos. 3 and 4, July–Dec. 1964, pp. 114–130.

the Far East gross agricultural output per hectare is the most important productivity measure in the short run. A high marginal productivity of labour cannot have the same importance as in the developed Western countries. Indeed, in the very short run it might be an appropriate aim to force the marginal productivity of labour even lower.

In answer to Professor Bićanić, Professor Ruttan said that the two types of land reform — abolition of rent and land redistribution — can often be set in an evolutionary context. From an original share tenure system the easiest shift is to one of lease tenure which gives the greatest incentive to utilize current inputs. However, as managerial capacity increases it will become necessary to move on to the next step. Thus, changes that were made at the turn of the century in Japan are now being made in many South-east Asian countries.

In answer to Professor Campbell, Professor Ruttan said that in the Jeffersonian-liberal types of reform productivity played only a minor role. It is by other policies than land reform that the productivity objective is furthered.

In reply to Professor Dandekar, Professor Ruttan cited the cases of the Philippines and Burma in the nineteenth century, where through the introduction of commercialism and the opening of the Suez Canal, the share tenure system sustained great increases in productivity. Professor Ruttan said that while he did not defend the sort of exploitation that this involved, it had undoubtedly resulted in increased marketable surplus in these economies.

To Professor Ashton, Professor Ruttan said that the tenure modifications in Britain in fact showed what one would expect with increases in owner-operatorship elsewhere. Tenure arrangements in Britain have been developed which provide production incentives similar to those achieved elsewhere under an owner-operatorship system. Thus the discussion should not be cast in terms of tenancy versus owner-operatorship but rather an effective balancing of equity and productivity objectives at any point in a nation's economic and political development.

Chapter 28

THE NEW SYNTHESIS OF
RURAL AND URBAN SOCIETY
IN THE UNITED STATES

BY

KARL A. FOX

Iowa State University, U.S.A.

What a man sees depends both upon what he looks at
and also upon what his previous visual-conceptual experience has taught him to see.—Thomas S. Kuhn [1]

I. INTRODUCTION — THE BACKGROUND OF THE INQUIRY

CHANGES in United States agriculture during the past quarter of a
century have been so thoroughly documented that at first glance it
seems difficult to illuminate them further. However, I believe I have
found a new way of looking at these changes which has far-reaching
implications.

The research which led me to the concept of the *functional
economic area* began as an exercise in pure science. However, there
was much evidence that a new organizing concept of this general
type would have important consequences for policy.

(a) Signs of Stress in Rural Institutions

By the mid-1950's it was generally recognized that the farm
population of the United States was declining and that commercial
farms were growing larger. Capital was being substituted for labour
on a relatively fixed land area. Nevertheless the 1959 Census of
Agriculture indicated a more rapid exodus of people from farms than
most economists had anticipated.[2]

---

[1] Thomas S. Kuhn, *The Structure of Scientific Revolutions* (University of Chicago
Press, Chicago), 1962, p. 112.
[2] For a discussion of these and related changes through 1960 see Karl A. Fox,
'Commercial Agriculture : Perspectives and Prospects', in *Farming, Farmers, and
Markets for Farm Goods: Essays on the Problems and the Potentials of American
Agriculture*, Supplementary Paper No. 15, published by the Committee for Eco-
nomic Development, New York, Nov. 1962, pp. 7–72.

There were many signs of stress in rural areas. These were perceived by some to be related to the changes in commercial agriculture, although the connections were not clearly formulated. Thus, it was recognized that villages and small towns (under 1,500 residents) in agricultural regions were tending to lose population unless they were located within 25 miles or so of cities. Towns with more than 1,500 residents in 1950 generally showed increases in the 1960 Census, and the percentage rates of increase were most marked for the larger towns and small cities. These phenomena were noted by quantitative geographers and rural sociologists, though not in general by agricultural economists.

Political scientists were aware of the high cost and backward methods of county governments in the agricultural regions. Residents of the rapidly growing cities were making more and more insistent demands for the reapportionment of state legislatures on the basis of population rather than area. After 1960, these demands for reapportionment received increasing support from decisions of the U.S. Supreme Court and other federal courts. Migration from farms to cities led to substantial decreases in the populations of rural legislative districts and supplemented the natural increase of the urban populations.

There were also signs of stress in the Land Grant Universities. They had been created in the nineteenth century to serve a population which was then predominantly agricultural and almost exclusively rural. During the 1950's and early 1960's enrolments in the colleges of arts and sciences, business and engineering in such universities increased rapidly while enrolments in colleges of agriculture remained fairly stable. Curricula within the colleges of agriculture gave increasing attention to training students for the food-processing and farm supply industries. Undergraduate training in colleges of agriculture drew closer to that of colleges of engineering and business administration. At both graduate and undergraduate levels, colleges of agriculture were under pressure to include more basic science in their curricula and reduce the number of highly applied farm-oriented courses.

The research programmes of the colleges of agriculture made some accommodation to the emergence of new clienteles, particularly in the food-processing and farm supply industries. Agricultural extension services had difficulty in establishing new clienteles. The increasing sophistication and educational attainments of commercial farmers posed a problem, as leading farmers began to ignore the county agent and to go directly to research workers on the state university campuses for technical advice.

(b) Efforts of the Federal Government to Deal with Structural Unemployment and Depressed Areas

During the late 1950's and early 1960's increasing concern was expressed about the existence of 'depressed areas', some of them in agricultural regions and others in coal-mining areas and the New England textile towns. During 1960–61 there was active debate among presidential advisers and before the Joint Economic Committee of Congress concerning the existence and extent of *structural* as distinct from *cyclical* unemployment. Cyclical unemployment should, by definition, respond to fiscal and monetary instruments if these were applied with sufficient expertness and vigour. Whatever unemployment remained after these measures had begun to generate a price inflation would presumably be structural in nature.

The existence of structural unemployment in certain areas was sufficiently evident to bring them special attention from the federal government. However, the federal programmes were typically organized on a county basis even though the depressed areas and their labour markets often covered a number of contiguous counties. Some federal agencies were specialized to the agricultural sector, so the available palliatives were applied to fractions of individual counties rather than to the entire economies of multi-county labour market areas.

In 1963 and after, increasing emphasis was placed on vocational education and retraining to cope with structural unemployment. This opened the way for new approaches and conceptualizations of the problems of 'depressed areas'.

(c) Signs of Stress in the National System of Economic and Demographic Data

Producers of economic data in the United States are in general well motivated and relatively sophisticated. Bodies of data such as the national income accounts and time-series on production, prices and employment are intended to reflect important economic realities.

It is easy to underestimate the influence of data systems upon policy. For example, official time-series dividing national income and employment into farm and non-farm components tend to suggest standards of equity and goals for public policy to citizens, organization leaders and legislators. What is *perceived* to be true is true in its consequences for action.

By the late 1950's, federal statisticians had decided that it was no longer meaningful or possible to maintain the traditional rural and urban dichotomies in their vital statistics. Rural babies were being born in urban hospitals and rural accident victims were dying in them. A new Census definition of a farm, introduced in 1959, brought with it a 20 per cent reduction in the official estimate of farm population.

The publication of 1950 and 1960 Census data in terms of Standardized Metropolitan Statistical Areas (SMSA's) and State Economic Areas (SEA's) brought mixed reactions from social scientists. Each SMSA consisted of one or more contiguous counties organized around a central city of at least 50,000 people ; more than 200 such areas were recognized as of 1960. The SMSA's included nearly two-thirds of the population of the United States. They were used extensively by economists and demographers.

The State Economic Areas proved to be of little or no use to social scientists. Together the SMSA's and SEA's completely covered the map of the United States. The SMSA's included all the important metropolitan areas and there were strong pressures from local groups to add additional counties to many of them. The SEA's failed to attract a significant clientele, suggesting that they did not succeed in partitioning the non-metropolitan areas of the United States in a manner which was useful to either public or private decision-makers.

In the early 1960's the national data system was confronted with several new challenges. An input-output model of about 87 sectors was imposed upon and conformed with the national income and product accounts. A group of 15 or 20 leading economists collaborated on a new econometric model of the United States, based on quarterly data for the 1947–60 period. Meanwhile, several groups of urban economists and city planners were experimenting with models of metropolitan economies. Another group was focusing its efforts on regional income and product accounts.

These lines of endeavour disclosed important gaps in economic data for local areas as well as opportunities for relating economic models of cities or small regions to economic models of the nation as a whole. Attendance at a national conference on regional accounts led me (December 1962) to write a short paper 'On the Current Lack of Policy Orientation in Regional Accounting'. This was followed by others on 'the major problems of rural society', 'economic models for area development research', and 'integrating national and regional models for economic stabilization and growth'.

(d) Some Sources of Stress

Agricultural economists have given much attention to the changes in factor proportions in commercial agriculture itself. They have also studied the joint effects of transportation costs and economies of size upon the optimal location of plants processing agricultural products.

Much less attention has been given to the changing spatial distribution of consumer-oriented services in agricultural regions. Business and labour economists have recognized that the almost universal ownership of passenger automobiles has greatly increased the shopping radii of farm and village families and has facilitated part-time farming by persons holding full-time jobs in factories.

There has been pressure from school authorities at the state level to enlarge rural school districts and increase the minimum sizes of elementary and secondary schools. Food supermarkets in the larger towns have been displacing small grocery stores in villages several miles distant. Department stores and apparel stores with wider product lines have been established in the larger towns and small cities in agricultural regions. The passenger automobile has been a major factor in these developments. Higher real incomes of farm and non-farm people alike have stimulated a demand for greater variety and breadth in consumer goods and services, recreational activities, public school curricula and medical services — all favourable to the introduction of larger and more complex economic units. Rural churches have encountered similar pressures.

(e) The Need for a New Organizing Concept

With so many different and apparently unrelated symptoms of stress, it may seem unreasonable to expect that a single organizing concept can bring them all into a common analytical framework. The concept must explain the stresses on rural institutions and give focus to efforts for relieving them. It must produce a satisfactory delineation of areas for dealing with structural unemployment and conducting other 'people-oriented' activities such as local government, vocational education and training and public education up to the junior or community college level. It must clarify the tasks of extension education and area economic development. The concept should also help to link data on income, employment and production for local areas to the corresponding national data systems for the more effective regionalization of national policies for economic stabilization and growth.

II. THE FUNCTIONAL ECONOMIC AREA

The concept of the functional economic area seems to emerge at a weak point near the intersection of several specialized fields — geography, regional economics, agricultural economics, labour economics, urban economics, macroeconomics, demography and perhaps other specialities. It is basically an ecological concept.

(a) The Ecology of a Four-Wheeled Species

My illustrations in this paper refer to the American Mid-west. The sizes and shapes of the areas delineated depend upon the almost universal ownership of passenger automobiles and a road system of sufficient quality that persons driving at legal speeds can travel 50 miles in 60 minutes.

It costs money to operate an automobile. Persons with low incomes may feel that their mobility is limited by these operating costs. For others, the scarce factor is time. If we allow ten hours a day for sleeping, eating and personal maintenance and eight hours for regular employment, all other activities must be compressed into six hours a day. If more than two hours a day are used up in travel between home and work, other activities are curtailed.

For most Americans time rather than money appears to be the limiting factor in determining the length of the home-to-work commuting trip. We may think of each family as living at the centre of a set of concentric circles of *minutes* of driving time. For the great majority of families, a circle of 60 minutes' radius will include the places at which the members work and the places at which they obtain nearly all of their goods and services. Their schools, churches and most frequently visited friends are usually within much less than a 60-minute radius.

There is much evidence that cities of 25,000 population or more play a special role as labour market centres and as shopping centres for specialized goods and services. If we are interested in population clusters rather than individual families, we may put the point of our compasses on a city of 25,000 or more people and draw our circles of *minutes of commuting time* around it as a centre. In Iowa, if we draw a 60-minute commuting 'circle' around each city of 25,000 or more (excluding small cities which are near to, and apparently satellites of, larger cities), the neighbouring 60-minute 'circles' tend to touch or overlap each other. There are a few gaps in the western part of Iowa — that is, areas which are more than 60 minutes' driving time from any city of 25,000 people or more.

It is important to note that circles of minutes around a central city do *not* translate themselves into circles on an ordinary map. We shall see shortly that a rectangular road system in relatively homogeneous agricultural country translates concentric circles of minutes into nesting squares of miles.

(b) *The Revolutionary Effect of a Rectangular Road Grid*

Shortly after the American Revolution of 1776–83, settlers began moving westward over the Appalachian Mountains and on to the prairies. About 1785, the (pre-constitutional) congress decided that the Western lands should be identified and measured off according to the quadrangular survey principle. Thus, the greater portion of the United States was surveyed into 'sections' (one-mile squares); townships six miles on an edge containing 36 sections; and counties which in most cases included integral numbers of townships. Settlers or homesteaders usually bought some fraction of a section of land; the 'quarter section' or 160-acre tract became the modal size.

In the flat or rolling country of the Mid-west it was quite logical to build public roads along the edges of the sections. The sides of the sections were universally oriented in the four cardinal compass directions, so the section roads ran either east and west or north and south. Although the airline distance between the south-east and the north-west corners of a particular section is only 1·414 miles, a wagon would have to travel two miles in getting from one point to the other — one mile west and one mile north if it started at the south-east corner.

Starting from any crossroads, a 'radius' of two miles would take us two miles in each of the four cardinal directions. The other feasible two-mile trips would take us one mile west and one mile north, one mile east and one mile north, one mile east and one mile south or one mile west and one mile south. Hence, a perimeter connecting the eight possible terminal points of two-mile wagon trips will be a square with its *corners* pointing in the four cardinal compass directions! This 'rotated' square pattern will characterize any set of points which are equidistant by road from some given point of intersection on the rectangular road grid. The angle of rotation is 45 degrees.

Most counties in the American Mid-west include from 12 to 16 townships. A 16-township county would measure 24 miles on each edge. If the county seat town and principal shopping centre were located at the exact centre of such a county, persons living near one of the corners of the county would be nearly 24 miles from the county seat while persons living near the county line but directly east, west, north or south of the county seat would be only 12 miles away. However, so long as the prevailing modes of travel (foot, horseback, or horses and wagons) involved speeds of five miles an hour or thereabouts, shopping and labour market areas tended to be considerably smaller than counties. The incongruities of 'rotated' shopping areas and compass-oriented county boundaries was evidently not much of a problem.

The advent of the automobile did not increase speeds from five miles to 50 miles an hour overnight. In the 1920's most rural roads were quite bad and were vulnerable to wet weather. Many of the automobiles in operation during the 1920's could not cruise comfortably, even on good roads, at more than 35 miles an hour.

By 1940, most of the passenger automobiles in service could cruise at 60 miles an hour or more on good highways. By this time also, the Mid-western states had a fairly dense grid of all-weather roads. In Iowa the so-called primary roads (connecting towns of 1,000 people or more) averaged about 12 to 15 miles apart. Any farm resident who wished to travel 40 or 50 miles to a sizeable town could travel all but six or seven miles of the way on all-weather roads.

Since 1940, the vast majority of Mid-western families, urban and rural, have possessed the capacity to travel about over the predominantly rectangular road grid at average speeds of around 50 highway miles per hour. From any given point, they have been able to travel 50 miles east, north, west or south in 60 minutes. But 50 miles over the other feasible routes have terminated at points such as the following : 40 miles west and 10 miles north, 30 miles west and 20 miles north, 25 miles west and 25 miles north, and so on. These terminal points of 50-mile trips lie on a 'rotated' square containing an area of 5,000 square miles! This area is equivalent to nearly nine counties of 16 townships each ($9 \times 576 = 5,184$). Hence, some counties would lie completely inside such a 50-mile square, but others would be cut diagonally by its sides.

It is most useful to consider a 50-mile square centred on a city of 25,000 people or more. In Iowa, this is a good first approximation to a one-hour commuting or labour market area. In comparison with such an area, the county becomes hopelessly small in size. At the same time, the 45-degree rotation of the boundaries of the labour market area (given a rectangular road grid) means that many counties are divided among two or more labour markets or functional economic areas.

(c) Some Visual Evidence of the Sizes and (Conditional) Shapes of Functional Economic Areas in the American Mid-west

Figure I shows 50-mile commuting perimeters centred on the business districts of seven Iowa cities which had 1960 populations of 50,000 people or more. The counties containing such cities are recognized as Standard Metropolitan Statistical Areas (SMSA's) in the official data system of the United States. The seven SMSA's include only about 900,000 people. If we add to these the other persons living within 50 miles' commuting distance of the central cities, we increase this total by several hundred thousand. Figure II shows similar 50-mile squares centred on some Iowa cities of 25,000 to 35,000 population. (The square centred on Clay County has as its centre a city of less than 10,000 people which is growing rapidly and has an unusually large volume of retail sales, suggesting that the region is in the process of becoming a functional economic area.)

Figure III combines the 50-mile squares shown on both of the preceding figures. The regularity of the pattern is highly suggestive of some important underlying phenomenon. It appears that the rotation of the squares at a 45-degree angle to the county lines and the four cardinal points of the compass is an essential feature of this phenomenon. The pattern would look much less impressive if every square were drawn with its sides paralleling the county lines.

Figure IV calls attention to the gaps between some of the 50-mile squares. Several of these gaps have considerable value in diagnosing the economic growth potentials of the trading centres in them. Also, the recent and prospective completions of four-lane super highways (parts of the so-called inter-state highway system) will narrow or close some of the apparent gaps.

The gap centring on Taylor County appears to be of a more ominous nature. The total population of Taylor County declined 17·2 per cent from 1950 to 1960. The number of adult males aged

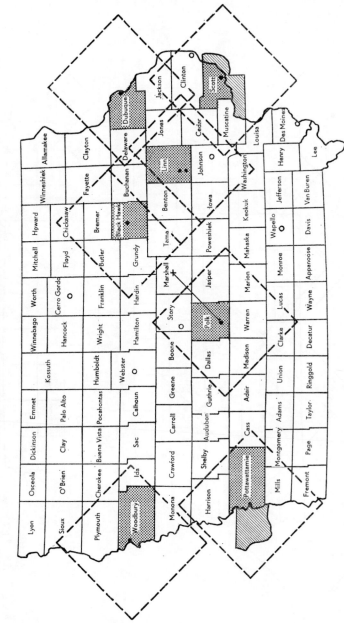

50-MILE COMMUTING DISTANCES FROM THE CENTRAL BUSINESS
DISTRICTS OF IOWA SMSA CENTRAL CITIES*

Fig. I

*Central cities of 50,000 people or more in 1960. Each shaded county or pair
of shaded contiguous counties are SMSA's

50-MILE COMMUTING DISTANCES FROM THE CENTRAL BUSINESS DISTRICTS OF
IOWA FEA CENTRAL CITIES WITH LESS THAN 50,000 POPULATION*

*1960 Census

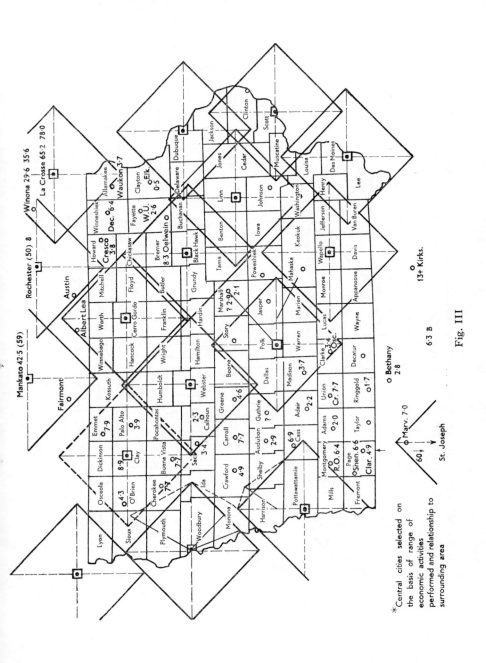

Fig. III

*Central cities selected on the basis of range of economic activities performed and relationship to surrounding area

GAPS BETWEEN 50-MILE COMMUTING PERIMETERS AND PERCENT CHANGES IN TOTAL POPULATIONS OF IOWA COUNTIES NEAR GAPS, 1950-1960

25 to 34 declined by 41 per cent! Taylor County is nearly two hours' driving time from either Des Moines or Omaha–Council Bluffs. It is simply not feasible for residents of Taylor County to commute every day to one of these places, which are the nearest places of rapid economic growth and diversified job opportunities. Hence, the most plausible hypothesis is that many young men from Taylor County have migrated to other areas, often to Des Moines or Omaha–Council Bluffs.

Figure V shows per cent changes in the total populations of Iowa counties from 1950 to 1960. Within each of the 50-mile squares we see represented a process which might be called 'creeping urbanization'. In every case, population in the county containing the central city of the area has increased more rapidly, or decreased less rapidly, than in the peripheral counties. Most of the peripheral counties lost population during the 1950's. With the exception of Wapello County, the population of each county containing an FEA central city increased quite significantly.

There is evidence that this centripetal process has been going on since about 1920. Farm population has been declining, but many of the young people who could not find adequate opportunities on farms have moved into the central city of the same functional economic area. Towns which had 3,000 people or more in 1920 have in most cases shown substantial growth, while villages of 1,000 population or less as of 1920 have in most cases remained stable or declined in population.

The durability of houses and the desire of many retired farm couples for low-rent housing have lent a somewhat specious stability to the populations of Iowa villages and small towns. The percentage of people over 65 years of age in such villages has increased very rapidly and is higher than in any other size group of populated places. Main Street enterprises in most villages of less than 1,000 population have suffered severe declines as a result of competition from supermarkets and other enterprises in larger towns. The total number of job opportunities in villages of less than 1,000 population has been supported in part by the increasing volume of purchased farm inputs handled in them, including commercial mixed feeds, oil and gasoline, seeds and fertilizers, farm machinery, replacement parts, repair services and the like.

Figure VI shows the distribution of town population sizes in the Fort Dodge area. The population of Fort Dodge itself (about 30,000) suggests that it is a city of a different hierarchical order and

PER CENT CHANGES IN TOTAL POPULATIONS OF IOWA COUNTIES, 1950-1960*

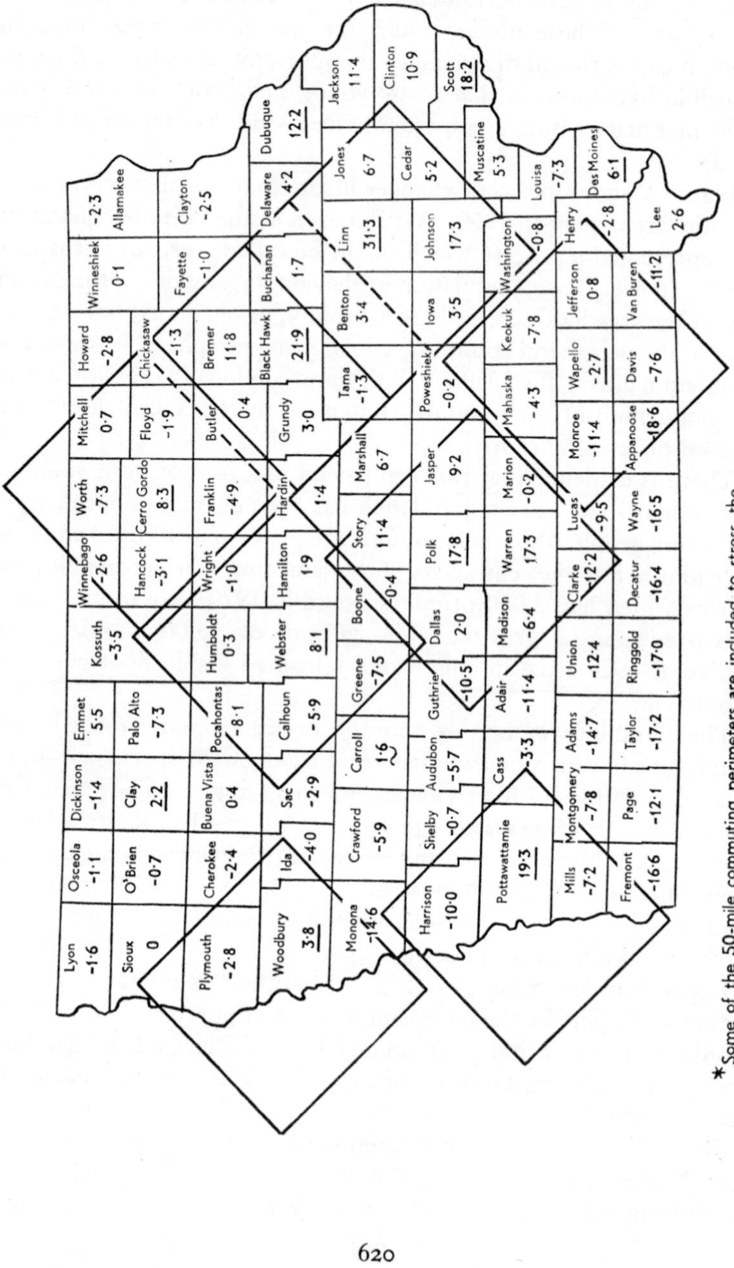

*Some of the 50-mile commuting perimeters are included to stress the redistribution of population occurring within functional economic areas

Fig. V

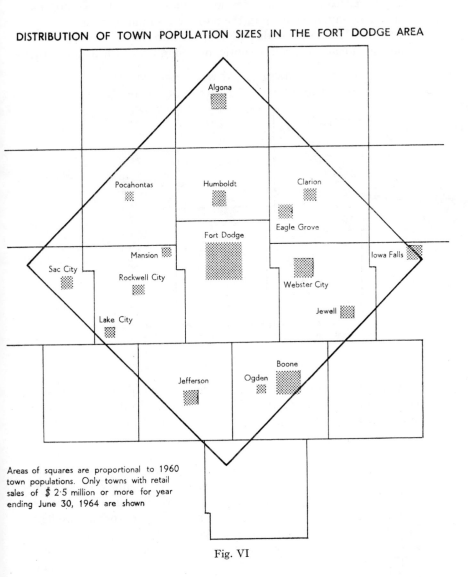

DISTRIBUTION OF TOWN POPULATION SIZES IN THE FORT DODGE AREA

Areas of squares are proportional to 1960 town populations. Only towns with retail sales of $ 2·5 million or more for year ending June 30, 1964 are shown

Fig. VI

function than the other towns within 50 highway miles of it. Visits to Fort Dodge and the other towns strongly confirm this impression. Many specialized goods and services found in Fort Dodge are not available at any other town in the 50-mile commuting area.

Figure VII shows the distribution of town population sizes in a 'non-functional' area of six contiguous counties which includes the

DISTRIBUTION OF TOWN POPULATION
SIZES IN SIX CONTIGUOUS COUNTIES

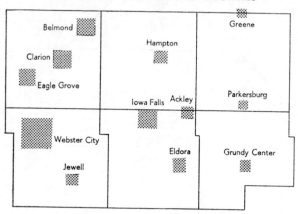

Areas of squares are proportional to 1960 town populations. Only towns with retail sales of $ 2·5 million or more for year ending June 30, 1964 are shown

Fig. VII

east-central portion of the Fort Dodge functional area. The six-county area has no central city of the Fort Dodge class. The configuration of town sizes immediately creates misgivings as to the usefulness or viability of this area for purposes of administration or legislative representation.

Figure VIII shows the home-to-work commuting pattern in the Fort Dodge area as of 1960. Each arrow represents workers who reside in a particular township and who work in some county other than that in which the township is located. Where workers resident in a township work in two or more other counties the arrow is a vector sum of the two or more real home-to-work commuting movements.

The arrows indicate that the vast majority of workers living inside of the 50-mile square also work within its boundaries. At the same time, most workers living just outside the 50-mile boundary also work outside that boundary.

COMMUTING PATTERN IN THE FORT DODGE AREA

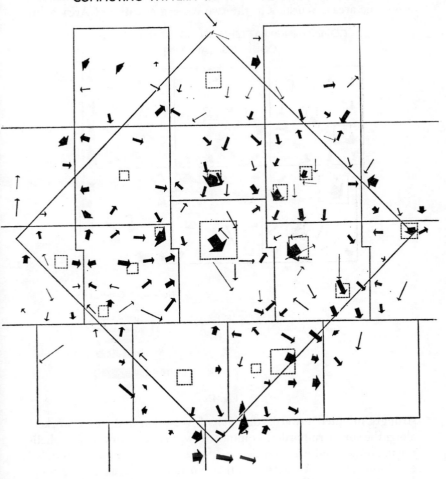

Number of Commuters:

0-5
6-10
11-25
26-50
51-100
101-250

Fig. VIII

Figure IX shows that the commuting pattern in the six contiguous counties (Area N) contrasts sharply with that in the functional economic area. Workers in the two eastern counties of Area N are

COMMUTING PATTERN IN SIX CONTIGUOUS COUNTIES (AREA N)

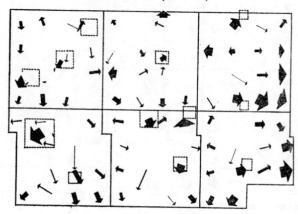

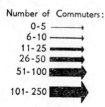

Number of Commuters:
0-5
6-10
11-25
26-50
51-100

101-250

Fig. IX

strongly oriented towards the city of Waterloo and many workers along the southern borders commute towards other towns (Marshalltown, Nevada and Ames) which are part of the Des Moines functional economic area. Workers in other portions of Area N commute towards Fort Dodge or Mason City. Area N, then, appears to be pulled in several different directions. It would be difficult to create a sense of common interest among the residents of such an area.

III. SOME IMPLICATIONS

Space does not permit extensive treatment of the implications of the functional economic area concept. The commuting patterns, and evidence based on retail sales, suggest that the people of the Mid-west have *de facto* grouped themselves and many of their daily economic and social concerns into functional economic areas of about the sizes shown. Each functional economic area is relatively bounded, closed or self-contained as a labour market and with respect to a cluster of residentiary (consumer-oriented) activities.

We will relate our discussion of implications to the areas of stress to which we referred in our opening section.

(a) Relieving Stresses on Rural Institutions

If we visualize a functional economic area as a low-density city whose export-oriented workers are engaged in agriculture, food-processing and farm supply activities, perhaps the simplest and most direct approach is to suggest that functional economic areas be given municipal or quasi-municipal government responsibilities.

We may ask, then, which functions of local government could be better organized on a multi-county functional economic area basis than on a single county or single small-town basis? I would suggest that the following functions be considered in this light :

(*a*) school districts including public junior colleges and four-year colleges ;

(*b*) centres for vocational education, training and retraining ;

(*c*) university-wide extension programmes, including extension programmes for farmers and those engaged in farm supply or farm product processing and storing activities ;

(*d*) police and fire protection ;

(*e*) public health services ;

(*f*) social welfare services ;

(*g*) the maintenance and construction of local streets and roads, as distinct from those connecting major population centres and maintained by the State Highway Commission ;

(*h*) urban and rural or regional zoning ; and

(*i*) public library services.

Further, we might suggest that legislative districts be related to functional economic areas. The populations of functional economic areas in the American Mid-west will generally run from 150,000 to 500,000 people. The population of Iowa is 2,800,000 people. A state legislature of 100 persons might be composed of five or six representatives elected from each of the less populous functional economic areas and as many as 15 to 18 from the most populous areas. (The very large cities, such as Chicago, in some states would present a different problem and require larger numbers of legislators.) If legislators were elected from sub-districts within functional economic areas (FEA's), they could represent local constituents on narrow issues and co-operate with their colleagues on an FEA basis on issues of broader concern.

United States congressional districts have populations of at least 400,000. A congressional district could be made up of two or three of the less populous functional economic areas. Some states are separated by rivers and, for historical reasons, some of the FEA central cities are located on these river boundaries. As congressional districts must lie within state lines, the river cities provide half-FEA's which can be combined with one or more adjacent FEA's to make up a congressional district. Congressmen from adjoining states could co-operate (as at present) on issues which affected the halves of FEA's which they represented on their respective sides of the river.

We might also ask which functions of state or federal governments could be better organized and implemented on an FEA basis? Here we may list for consideration programmes of the Bureau of Employment Security ; of the post office ; of area economic development ; of state planning for outdoor recreation facilities ; and of state planning and operation of mental health and medical facilities to the extent that these are publicly supported.

We have already noted that declining farm populations and increasing city populations make a great deal of ecological sense when viewed in an FEA framework. We could conceptualize and perhaps measure, within each FEA, continuous surfaces of house rentals, values of houses, rates of construction of new houses, and expected wage rates and income distributions reflecting costs of home-to-work commuting travel to the central city from each point in the area. To some extent, perhaps, the economic forces operating in an FEA turn it into a centrifuge such that certain age-groups and labour force categories tend to locate in concentric rings of commuting *minutes* around the central city. This insight should be explored in detail at a later date.

(*b*) *Improving the National Data System as a Basis for Anticipating the Regional Impacts of National Economic Policies*

Under this heading I would suggest the following :

1. We should develop income and product accounts and data on labour force and employment by occupation and industry for each

member of a complete set of approximately 400 functional economic areas, which would include the entire territory of the United States and add to United States totals for the respective variables and income components.

2. We should develop 'impact models' of typical functional economic areas and some individual (actual) FEA's which are compatible with the new Brookings–SSRC Quarterly Econometric Model of the United States. The residentiary or consumer-oriented activities will tend to be similar for different FEA's in fairly broad regions of the United States.

3. Perhaps we should look forward to the development of a complete set of some 400 FEA models integrated with (or into) an econometric model of the United States as a whole.

4. Serious thought should be given to the development of data or informational systems relating to FEA's from the standpoints both of policy-makers within the FEA and of national policy-makers. Some portions of the data systems needed to appraise the performance of an FEA's economy should be compatible with the national data system, as indicated above. However, it might be that other components of the information system needed by the mayor or other officials of an FEA could be independent of the national data system and also of the data systems of other FEA's. At the moment this is only a hypothesis. It is perhaps more likely that standardization of the local components of FEA data systems from one FEA to another would be highly desirable, as the governmental and other residentiary functions carried on within different FEA's would probably show great similarities.

If there are to be some policy-making bodies or officials concerned with problems of FEA-wide scope, these officials will have needs for data systems which cover both rural and urban parts of the FEA in a consistent and systematic manner. Users of such data would include municipal or quasi-municipal FEA governments, zoning commissions, city and regional planning commissions, school boards and others. Existing or potential business firms would also be interested in measures of economic performance and future potential for the FEA.

(c) *Implications for the Land Grant Universities*

We shall limit ourselves here to a few brief observations :

1. There is a great deal of inertia in adjusting government programmes and public agencies, including the land grant universities,

to the new synthesis of urban and rural society embodied in the functional economic area concept. I have argued that 'the major problem of rural society in the United States is our belief that a rural society exists and can be dealt with independently of the society as a whole'. I refer here only to United States society as it exists in the 1960's; there may be highly significant rural–urban dichotomies in other countries, and I suspect that the sharpness of the dichotomy varies widely between regions in countries characterized by 'dual economies'.

2. Considering the current structure of the United States economy and society, it seems difficult to justify the administrative separation of teaching programmes in rural sociology from those in sociology or of teaching programmes in agricultural economics from those in economics. The changing organization and integration of production and distribution and the changing patterns of labour force requirements evidently call for related adjustments of curricula within our land grant universities.

3. The extension and applied research programmes of our land grant universities should evidently be reorganized to facilitate co-operation across the traditional lines between colleges. In essence, the 'people's college' must turn and face the people.

During the early 1900's, two enterprising companies built up tremendous volumes of mail order retail business with farm people. Farmers and their families ordered items by mail on the basis of pictures and descriptions in a printed catalogue. Since 1945 these same firms have built many department stores in urban locations.

Recently an executive of one of these mail order houses summarized his company's policy as follows: 'We found that the farmers were moving into town so we decided to move in with them.' Perhaps the problem facing the colleges of agriculture is as simple as that.

––––––

DISCUSSION OF PROFESSOR FOX'S PAPER

Dr. Odhner said that his main criticism was that the approach in the paper was almost purely analytical and not instrumental. Thus there was little aid to planning in the concepts used, and there were no forecasts of future development in the techniques of agricultural production or of communications, nor of the economic and social structures of the future. Underlying this was the absence of an attempt to find what future social behaviour will be. Rather than try to plan for future administrative and social needs on the basis of the present situation, we need to try to find out what the future structure will be so that administration and policy can be adapted to it. For example, there is the question of whether commuting is a normal function of people. Dr. Odhner considered that it is more likely to be a transitory stage ; people are obliged to endure two hours in the car because structural change in the community means that they are not able to change their jobs. The congestion of cities will also influence where people will want to live in the future. Dr. Odhner asked whether a concept of one central town might not be replaced by a pattern of a central city with subsidiary towns fulfilling some of the main functions.

Mr. Allen said that Professor Fox may have detected a permanent phenomenon. An article by Colin Clark in the *Journal of the Town Planning Institute* analysed the spread of cities over the centuries and claimed that a clear pattern emerged. A significant proportion of people are prepared to travel fifty minutes a day each way, thus setting limits to the suburbs, which expand with the speed of transport. It is possible that there is an in-built tendency for some of us to want to spend one hour a day, each way, in commuting? It should be remembered that in the United States, if weekends and holidays are included, the commuter would only commute for two-thirds of the year.

Mr. Allen also asked if Professor Fox had related his study to the hierarchical social pattern studied by other experts such as Professor Lösch.

There were a number of questions concerning the shape of functional economic areas. *Dr. Odhner* asked how a functional economic area would be planned in the absence of the very special road grid system of the American Middle West. *Professor Ruttan* asked if Professor Fox had considered the linear region as a useful concept for areas with topographical patterns which differed from Iowa. *Professor Renborg* said that mountains or other natural divisions would disrupt any simple pattern of commuter time and wished to know how much emphasis Professor Fox would put on this criterion and what other criteria he would employ. *Professor Robinson* said that the uniformity of the quality of land in Iowa is not widely paralleled and certainly not in Europe. *Mr. Allen* said that the development of super highways would change the patterns of movement. Would not this alter the system of squares ? Inter-state highways would tend to elongate the Iowa rectangles.

Professor Zemborain asked what effect the past and present transportation systems would have upon Professor Fox's concept. He cited the case of Argentina where the building of large cities had been intimately connected with the construction of the railway system. The distances between villages and the railway system had been affected by the change from horse-drawn wagons to trucks. However, trade centres, made redundant by new transport arrangements, died very slowly.

Professor Robinson asked what repercussions on the organization of agriculture were implied by the desire of people to have access to the facilities and services of big towns. In some respects we seem to have come a full circle. An old pattern, for example in Southern Italy, was that of the agricultural work force living in large towns and walking to work. On the other hand, planning tends to create small holdings in the countryside on the assumption that it is more economical to live on the farm. Is this now a mistaken policy?

Professor Dandekar said that the weakness of the dichotomy between country and town found by statisticians is perhaps near the truth ; we are in fact considering overlapping low-density cities. Referring to Taylor County and those areas which do not form a functional economic area, he was surprised that Professor Fox did not suggest lack of job opportunities as the main reason for depopulation. The important question is why there are no job opportunities there. If people wish to stay where they are, and if there are gaps in the functional economic network, they will have to be filled by the redistribution of job opportunities.

Professor Fox's study is based on consumer market areas, whereas in India it is necessary to deal with agricultural produce market areas. People do not commute but farmers come to market with their produce. A similar study based on producer market areas may therefore be of interest.

Professor Gulbrandsen said that he viewed the problem of transport as a question of potential external economies for agriculture. The relevant question is : Should we localize cities to give agriculture the necessary external economies, or should we leave land idle if it is too far from the cities?

Professor Fox replied to Dr. Odhner's criticism that his approach was in the first instance analytical and only secondarily policy-oriented. The boundary of the functional economic area serves to indicate the geographical limit within which certain kinds of policies can be carried out more efficiently than by the present system of local (county) government in the U.S.A. which is adapted to a horse-and-wagon society. On the question of the structure of society several decades in the future, he said that he expected further improvements in automobile transport would reduce the average time taken to travel fifty miles. Expected improvements in transport in Iowa would not change boundary lines very much. Interstate highways would tend to elongate some areas a few miles in the directions of the highways and fill in or diminish some of the gaps in Figure I. Of profoundly radical changes, such as individual helicopters, we cannot know the effect but functional economic areas of about the present size would be ecologically important during the era of the passenger automobile. Agricultural extension work and economic development planning on a functional economic area basis are being tried in Iowa and are proving successful.

Fox — Synthesis of Rural and Urban Society

In answer to questions about the boundaries of the functional economic areas, Professor Fox said that the basic concept was of a circle of *minutes* of commuting (travel) time. Thus, geographic boundaries of the areas may be regarded as transformations of a circle with a radius of 60 minutes. In a mountain or desert region with roads many miles apart and with feasible automobile speeds of (say) 20 miles an hour on some roads and 40 miles an hour on others, a 60-minute circle around a city would lead to a geographical boundary of very irregular shape and averaging between 20 and 40 road miles from the central city. Even in the case of Salt Lake City where on one side there are mountains and on the other desert, one can draw a geographical boundary connecting all points which can be reached in passenger automobiles by driving 60 minutes outward from the city. Few workers will commute daily into Salt Lake City from points outside this boundary.

On the question of central place hierarchies, Professor Fox said that he had made the relation of his work to this clear in another paper.[1] Borchert and Adams (1963), Brian J. L. Berry (1964) and others had described and empirically illustrated the existence of a hierarchy of central places in Iowa and the upper Mid-west.[2] Philbrick (1957) and others had put forward similar hierarchical schemes.[3] Quoting from another paper of his own Professor Fox said : 'We shall assume here that the villages, towns and cities in an agricultural region covering one or more States can be allocated among four categories. Each category is defined by means of a specified array of retail goods and consumer oriented services. Let us characterise these trade-hierarchic categories from smallest to largest, in terms of volume of retail sales and typical population size of the town, into convenience centres, partial shopping centres, complete retail shopping centres and central cities of functional economic areas.' The desirable sizes of retail stores, schools, etc., in each of the four kinds of consumer centres will vary from one country to another. Clearly, economies of scale in public institutions played a role in the size of the consumer centres and so in the structure of the functional economic areas.

Professor Fox said that he would put the population of a convenience centre in Iowa at about 1,000, that of a partial shopping centre at between

[1] Karl A. Fox and T. Krishna Kumar, 'Delineating Functional Economic Areas', Chapter 1, pp. 13–55, in *Research and Education for Regional and Area Development* (Iowa State University Press, Ames), 1966. 287 pp. (The 'author' of the book is the Iowa State University Center for Agricultural and Economic Development.)

[2] John R. Borchert and Russell B. Adams, *Trade Centers and Trade Areas of the Upper Midwest*, Upper Midwest Economic Study, Urban Report No. 3, Sept. 1966. 44 pp. (multilithed). (Enquiries for copies should be addressed to Professor Borchert, Department of Geography, University of Minnesota, Minneapolis.)

Brian J. L. Berry, H. Gardiner Barnum, and Robert J. Tennant, 'Retail Location and Consumer Behavior', *Papers and Proceedings of the Regional Science Association*, 9, 1962, pp. 65–106.

[3] A. K. Philbrick, in 'Principles of Areal Functional Organization in Regional Human Geography', *Economic Geography*, Vol. 33 (Oct. 1957).

2,500 and 5,000 and a complete retail shopping centre at from 6,000 to 20,000 or more. The central city of the functional economic area may range from 25,000 to as many as a million people. The pattern of areas shown in Fig. I probably does not have to be modified to take account of cities even as large as Minneapolis. This system only requires major modification in metropolitan areas such as Chicago, New York, Los Angeles and a few others. The functional economic area is primarily a labour market area — we may think of the U.S.A. as divided into about 400 labour market areas with the commuting of workers across boundaries at a minimum. This labour market area approach fits well into location theory, central place theory and the theories of quantitive geographers.

In reply to Professor Dandekar, Professor Fox said that he would emphasize that the functional economic area is a low-density city, which is essentially homogeneous as a labour market and supplier of consumer services, and in which town and open-country dwellers have essentially the same living standards and value systems. In another recent paper he had said : 'The image of the traditional [rural-urban] dichotomy lingers in the minds of many older people, rural and urban alike, and contributes to much confusion concerning appropriate solutions for the economic and educational problems of rural people. The greatest problem of rural society in the United States is the belief that a rural society still exists and can be manipulated successfully apart from the society as a whole. For better or for worse, the city as an economic and cultural entity has surrounded the country.' Professor Fox said that his emphasis was as Professor Dandekar had stated.

Professor Fox said that the desirability of purposefully creating non-agricultural job opportunities in a farming area more than fifty miles from a central city would vary from one such area to another. If Iowa were as crowded as England it might be reasonable to plan fully and to locate a city of 50,000 in Taylor County or similar areas undergoing rural depopulation. In the actual circumstances in Iowa it would seem more appropriate to develop outdoor recreation facilities in such areas which would appeal seasonally or periodically to city residents in Iowa and in the more populous states to the east.

Professor Fox said that Professor Gulbrandsen's question as to whether we should retire land which is more than a certain distance from the centre of a functional economic area was an interesting one. In principle, in each functional economic area we should be able to construct a house-rental surface which is highest at the centre of the central city, and which declines as it radiates outwards (perhaps to zero at a distance of more than fifty miles). This rental surface, roughly speaking, would reflect the rental value of dwellings at different locations from the standpoint of persons who work in the central city. Thus, as farms increase in size and diminish in number, 'surplus' farm houses more than fifty miles from the central city may be abandoned for lack of interested tenants, while those within twenty miles of the central city may command very good rentals from persons who work in the city. On the other hand, rentals of agricultural land depend on distances from the relevant agricultural market, so that where that market lies outside the functional economic area the rental surface of agricultural land may slope in the opposite direction from that for dwellings.

>

Chapter 29

THE TERMS OF TRADE OF
AGRICULTURE IN CONTEXT OF
ECONOMIC GROWTH [1]

BY

YAIR MUNDLAK

The Hebrew University of Jerusalem, Israel

I. INTRODUCTION

IN a competitive economy, where the value marginal productivities
of the factors of production are equal in all sectors, the terms of trade
of one sector are meaningfully measured by the rent realized by
factors specific to this sector. If there are no specific factors, there is
no problem of terms of trade.

Of course, if the economy moves with friction, there are short-run
discrepancies in the remuneration of the various factors. Such discrepancies may be important, and of great concern to policy-makers.
However, the economy is, sooner or later, likely to iron them out.
Since this process takes time, and since with time some of the data
change, the original discrepancies also change. Thus, if one is
interested in the short-run adjustments, he may do well by starting
with the study of the long-run dynamic equilibrium path of the
economy. By long run we mean the competitive solution of the
economy when all adjustments have been made, and with dynamic
variables held constant. But for the story to be revealing, the
dynamic variables should be accounted for. This is 'the context of
economic growth' which dominates much of this paper.

The analysis will involve a division of a closed economy into two
sectors : agriculture and non-agriculture. The agricultural sector
employs a specific factor, say land. Thus, our ultimate objective
will be to trace the changes which take place in the rent of land. One

[1] This work was supported by the Maurice Falk Institute for Economic Re-
search in Israel, and by the Ford Foundation grant for International Studies in
Agricultural Development at the University of Chicago. I am indebted to Ron
Mosenson for valuable comments on an earlier draft.

of the interesting aspects of such an investigation is that it provides a measure of the changes in the value marginal productivity of the specific factor. If one introduces a supply function for this factor, he can also obtain the optimum level for its employment.

Such an analysis can be applied to evaluate the desirability of undertaking heavy investment in desalination of water for agricultural production. Perhaps, in order to obtain results for such practical questions, it will be necessary to extend the model to an open economy.

Finally, by way of reservation, it should be mentioned that this paper is not the result of long study, but rather my introduction to the subject. For a start, I have sought simplified assumptions.

The works on two-sector models listed in the bibliography have been very useful in this analysis.[1] The illuminating survey by Hahn and Matthews contains, among many other things, references to other works on two-sector models.

II. ASSUMPTIONS

We deal with a closed economy which consists of two sectors : agriculture and non-agriculture, to which we refer as sectors 1 and 2 respectively. Sector 1 employs labour (L_1), capital (K_1) and 'Land' (A), whereas sector 2 employs labour (L_2) and capital (K_2). Labour and capital are homogeneous, transferable and fully employed. The last assumption implies :

$$L_1 + L_2 = L \qquad (2.1)$$

$$K_1 + K_2 = K, \qquad (2.2)$$

where L and K are the total supply at a given time.

The production function of 2 is linear homogeneous in L and K, whereas that of 1 is linear homogeneous in L, K and A, separable with respect to A and homogeneous of degree $\mu < 1$ in K and L. Furthermore, there are Hicks-neutral technical changes in the two sectors designated as γ_j, $j = 1, 2$.
Specifically

$$Y_1 = F_1(L_1, K_1, A)e^{\gamma_1^* t} \qquad (2.3)$$

F_1 can be separated into two components :

$$Y_1 = \bar{F}_1(A)\bar{\bar{F}}_1(L_1, K_1)e^{\gamma_1^* t}.$$

[1] In particular, the analysis by Takayama (5). In the revision I also benefited from the exposition by Ronald Jones (2).

Conceptual Problems in Agricultural Development

Let us assume that $A(t) = A(o)e^{\hat{A}t}$, and let $\bar{F}_1(Ao)\bar{\bar{F}}_1(L_1K_1)$ $= H(L_1K_1)$ so that

$$Y_1 = H(L_1, K_1)e^{\gamma_1 t}.$$

By homogeneity of degree μ

$$Y_1 = L_1^\mu \, h(k_1)e^{\gamma_1 t},$$

where $\qquad k_1 = \dfrac{K_1}{L_1}$ and $\gamma_1 = \gamma_1^* + (1-\mu)\hat{A}.$

Throughout, the sign $\wedge$ will imply rates of change with respect to time :

$$\hat{A} = \frac{\dot{A}}{A} = \frac{dA}{dt\,A}$$

For sector 2 we have :

$$Y_2 = F_2(L_2K_2)e^{\gamma_2 t} \qquad (2.4)$$
$$= L_2 f(k_2)e^{\gamma_2 t},$$

where $\qquad k_2 = \dfrac{K_2}{L_2}.$

It is assumed that the production functions behave properly : $g(k_j) > 0$, $g'(k_j) > 0$, $g''(k_j) < 0$, for $k_j > 0$ and $g(0) = 0$, $g(\infty) = \infty$, $g'(0) = \infty$, $g'(\infty) = 0$, where $g = h, f$ and g', g'' are respectively the first and second derivatives with respect to k_j ($i = 1, 2$). We also assume that the isoquants are convex to the origin.[1]

Let P be the price of the agricultural products in terms of the non-agricultural product $\left(P = \dfrac{P_1}{P_2}\right)$. By the assumption of competition :

$$w = PF_{1L} = F_{2L} \qquad (2.5)$$

$$r = PF_{1K} = F_{2K}, \qquad (2.6)$$

where w and r are the wage and rental (on capital) respectively and F_{jL}, F_{jK} are respectively the marginal productivities of labour and capital in the jth sector.

Equations (2.1)–(2.6) specify the supply side of the model for given K and L. They could be easily described by an Edgeworth-Bowley Box diagram and summarized by a transformation curve between Y_1 and Y_2. (See next section.) To determine the composi-

---

[1] Convexity constitutes an additional assumption only for the non-linear function.

tion of output, we need a demand specification. For simplicity, and without much sacrifice of reality, we assume that the price elasticity is zero and that there is an Engel-like curve with constant elasticity $0 < \eta < 1$ for agricultural products.[1] Real income is measured by the transformation curve, whose co-ordinates are Y_1 and Y_2 and hence the demand equation is summarized by :

$$y_1 = \eta_0 y_2^\eta, \tag{2.7}$$

where $y_j = \dfrac{Y_j}{L}$ is the *per capita* demand. Y_1 is fully consumed, whereas Y_2 consists also of capital goods.

III. EQUILIBRIUM IN THE SHORT RUN

This section deals with the equilibrium determination for a given endowment of resources and a given technology. We start by ex-

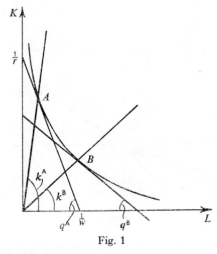

Fig. 1

pressing some known relationships in a convenient form for our discussion. The relation between factor intensity and wage rental

[1] The reference to a zero price elasticity is in a demand relation of the form $y_1 = d(y_2, P)$. We can start with a more conventional demand relation $y_1 = D(y, P)$ where y is *per capita* income. Total differentiation yields :

$$\hat{y}_1 = \frac{1}{1 - E_{1y}\pi_1} [(E_{1y}\pi_1 + E_{1p})\hat{P} + \pi_2 E_{1y}\hat{y}_2],$$

where E_{1y} and E_{1p} are the income and price elasticities respectively and π_j is the proportion of total income spent on the jth product. Our demand relation implies that the substitution and income effects of a change in price always offset each other. In terms of this relation the price elasticity is negative.

ratio, $q = \dfrac{w}{r}$, is shown in Fig. 1. Let A be a point of tangency with the unit isocost. The wage–rental ratio determines uniquely the capital–labour ratio, k_j^A, in the jth sector. This is an outcome of the convexity of the isoquants. By the homogeneity assumption, the marginal rates of substitution of the isoquants on the ray that goes through A are all equal to q. As q declines, a new tangency solution, B, is obtained. Note that $k_j^B < k_j^A$. That is, $\dfrac{dk_j}{dq}$ is positive everywhere. So is also the elasticity of substitution defined as :

$$\sigma_j = \frac{dk_j}{dq} \cdot \frac{q}{k_j} = \frac{\hat{k}_j}{\hat{q}} > 0. \tag{3.1}$$

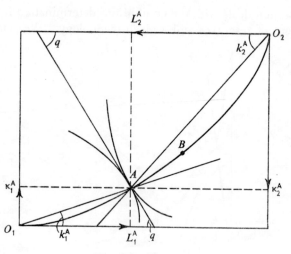

Fig. 2

Much of the analysis can be pictured by an Edgeworth-Bowley box diagram. It is drawn in Fig. 2, where 0_1 and 0_2 are the origins for sectors 1 and 2 respectively. The curve $0_1 A 0_2$ is the contract curve for the economy. Since we deal with a competitive market, only one q exists at any particular moment. This q determines the capital-labour ratio in each sector. At point A, we have k_1^A and k_2^A. The figure is drawn for a case where agriculture is more labour intensive and industry is more capital intensive. That is, $k_2 > k_1$. q also determines the allocation of resources to the two sectors. Thus for point A we have L_1^A, L_2^A, K_1^A and K_2^A. Given the allocation of re-

638

sources, and the production functions, the outputs of the two sectors are also given. To determine q we need the demand equation. Before considering it, let us note the relationships between q and the other variables. Suppose that the economy moves from A to B. We can immediately see that capital intensity in both sectors is higher in B than in A. Hence, from (3.1) we know that $q^B > q^A$. We also note the following relationships:

$$
\begin{array}{ll}
Y_1^B > Y_1^A & Y_2^B < Y_2^A \\
L_1^B > L_1^A & L_2^B < L_2^A \\
K_1^B > K_1^A & K_2^B < K_2^A \\
k_1^B > k_1^A & k_2^B > k_2^A.
\end{array}
$$

That is, when resources are given, an increase in the wage–rental ratio can increase capital–labour ratio in the two sectors by moving resources from the more capital- to the more labour-intensive sector. Consequently, output of the more labour-intensive sector increases and that of the more capital-intensive sector decreases. If we define the proportions of labour and capital allocated to the two sectors as:

$$
l_1 = \frac{L_1}{L} \qquad l_2 = 1 - l_1 \tag{3.2}
$$

$$
\rho_1 = \frac{K_1}{K} \qquad \rho_2 = 1 - \rho_1
$$

we can note the following: At the origin 0_1 all resources are given to sector 2. Hence $l_1 = \rho_1 = 0$. At that point, we get the lowest possible capital–labour ratio for sector 2, and hence the lowest possible q for this economy. Let this value be $\underline{q}$. Similarly, at the other origin, 0_2, we have $l_1 = \rho_1 = 1$. Furthermore, at this point we get the highest possible k_1 and hence the highest q for this economy. Let it be $\bar{q}$.

Let total product be

$$
Y = PY_1 + Y_2 \tag{3.3}
$$

and the proportion of sector 1 in total product is

$$
\pi_1 = \frac{PY_1}{Y}. \tag{3.4}
$$

We note that since at $\underline{q}$ no Y_1 is produced, π_1 is also zero. On the other hand, we have $\pi_1(\bar{q}) = 1$. In fact, π_1 is related to l_1 and ρ_1 by [1]

$$\pi_1 = \rho_1 \frac{\alpha_K}{\mu} + l_1 \frac{\alpha_L}{\mu},\qquad(3.5)$$

where α_K and α_L are the capital and labour shares in total product respectively. Thus, π_1 is almost a weighted average of l_1 and ρ_1. In

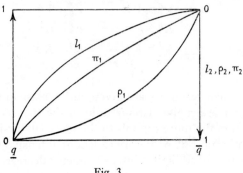

Fig. 3

Fig. 3 the relationships between the three ratios, l_1, ρ_1, π_1 and q are traced for the case where agriculture is the labour-intensive sector.

In order to show the short-run equilibrium we draw the transformation curve in Fig. 4. The axes represent *per capita* production, y_j. The supply price always changes in the same direction as y_1 and hence is increasing with q when sector 1 is more labour intensive. Total product, y, measured in terms of Y_2 is shown on the horizontal axis.

Finally, the demand relationship is drawn. It intersects with the transformation curve at point N, and the momentary equilibrium

[1]
$$\pi_1 = \frac{PY_1}{Y} = \frac{1}{Y}[rK_1 + WL_1 + R],$$

where R is the rent on land and can be written as $R = PY_1(1 - \mu)$ (see 4.16 below). Let

$$\alpha_K = \frac{rK}{Y} \qquad \alpha_L = \frac{WL}{Y}$$

we can write

$$\pi_1 = \rho_1 \alpha_K + l_1 \alpha_L + \pi_1(1 - \mu)$$

$$= \rho_1 \frac{\alpha_K}{\mu} + l_1 \frac{\alpha_L}{\mu}.$$

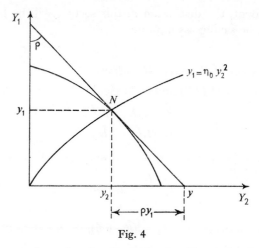

Fig. 4

in our economy is uniquely determined. From the foregoing discussion it is clear that given the point N, q is determined and hence the allocation of resources and hence the wage and the rental rates.

IV. DISPLACEMENT FROM EQUILIBRIUM

If γ_1 and γ_2 are zero and if K and L do not change, then the above system will freeze at the equilibrium point. Beside allowing for technical change, the following assumptions are made to put the system in motion. Labour grows geometrically at rate $\hat{L}$:

$$L(t) = L_0 e^{\hat{L}t}. \tag{4.1}$$

Gross saving is proportional to income. Capital depreciates at rate λ. Hence capital accumulation is:

$$\hat{K} = \frac{sY}{K} - \lambda. \tag{4.2}$$

Since capital goods are produced by sector 2, we also have

$$sY \leqslant Y_2$$

which implies

$$\frac{s}{1-s} PY_1 \leqslant Y_2 \tag{4.3}$$

or

$$\frac{s}{1-s} \leqslant \frac{\pi_2}{\pi_1}.$$

We now trace the effect of the above assumptions. As it will become evident, the discussion in this section will be largely independent of the saving assumption.

(a) Displacement of Resource Allocation

Differentiation of (2.2) and (2.1) yield :

$$\hat{K} = \rho_1\hat{K}_1 + \rho_2\hat{K}_2 \tag{4.4}$$

$$\hat{L} = l_1\hat{L}_1 + l_2\hat{L}_2. \tag{4.5}$$

Using (3.1), (4.4) and (4.5) we can write (see Appendix A.1)

$$\hat{l}_1 = \frac{l_2}{i}(-\hat{k} + \bar{\sigma}_\rho\hat{q}) \tag{4.6}$$

$$\hat{l}_2 = \frac{l_1}{i}(\hat{k} - \bar{\sigma}_\rho\hat{q}), \tag{4.7}$$

where $\qquad\qquad \bar{\sigma}_\rho = \rho_1\sigma_1 + \rho_2\sigma_2$

and $i = l_1 - \rho_1$ is an intensity parameter. It represents the difference between the proportions of sector 1 in total labour and capital respectively. Thus, i is positive (negative) when sector 1 is more labour (capital) intensive.

If the overall capital–labour ratio remains constant, we see that the proportion of total employment in the labour (capital) intensive sector grows (declines) with q. This just repeats what has already been shown in Fig. 3. Furthermore, it can be shown that (Appendix A.1)

$$\rho_j = \sigma_j\hat{q} + \hat{l}_j - \hat{k}. \tag{4.8}$$

Again, let 1 be the more labour-intensive sector. An increase in q, with $\hat{k} = 0$, will increase l_1 and hence ρ_1. Consequently, ρ_2 must decline. This repeats the drawing in Fig. 3.

In order to evaluate the net effect of a change in the overall capital–labour ratio we have to evaluate its effect on q. This we do below. However, it should be noted that a change in k with q held constant acts in opposite direction to a change in q. An increase in the overall capital–labour ratio decreases the proportion of labour and capital employed in the labour-intensive sector.

(b) Displacement of Supply

As shown in Appendix A.2, the rate of change of *per capita* production can be written as follows

$$\hat{y}_1 = (\mu - \alpha_1)\hat{k}_1 + \mu\hat{l}_1 + [\gamma_1 - (1 - \mu)\hat{L}] \qquad (4.9)$$

$$\hat{y}_2 = (1 - \alpha_2)\hat{k}_2 + \hat{l}_2 + \gamma_2, \qquad (4.10)$$

where α_j is the labour share in the jth sector defined as $\alpha_j = \dfrac{WL_j}{P_j Y_j}$ ($P_2 = 1$).

If labour allocation and capital intensity remain unchanged, the *per capita* production in agriculture depends on the last terms in (4.9). In the absence of technical change in agriculture, *per capita* production will decline at the rate $(1 - \mu)\hat{L}$. This is a direct result of the fact that the size of land remains constant and intensification of labour and capital results in diminishing returns. The diminishing returns can be offset by technical change. Thus, even though the labour force grows at the rate $\hat{L}$, the economy would remain at its initial position if it so happens that $\gamma_1 = (1 - \mu)\hat{L}$ and $\gamma_2 = 0$. In some cases below it will be convenient to treat the last term in (4.9) together, so we define

$$\gamma_1' = [\gamma_1 - (1 - \mu)\hat{L}]. \qquad (4.11)$$

The more general situation to be examined is that where resource allocation is allowed to change. To do this, we substitute for $\hat{k}_j$ and $\hat{l}_j$ (Appendix A.2) to get:

$$\hat{y}_1 = -\mu\frac{l_2}{i}\hat{k} + E_{1q}\hat{q} + \gamma'_1 \qquad (4.12)$$

$$\hat{y}_2 = \frac{l_1}{i}\hat{k} + E_{2q}\hat{q} + \gamma_2, \qquad (4.13)$$

where E_{jq} is the (partial) elasticity of output in j^{th} sector with respect to q. It is shown in Appendix A.2 that

$$E_{1q}i > 0 \qquad E_{2q}i < 0. \qquad (4.14)$$

Output of the labour (capital) intensive sector increases (decreases) with q, other things held constant. This is a restatement of the relation between movements along a given transformation curve and changes in q.

Suppose now that $\gamma_1' = \gamma_2 = 0$, and q is constant. An increase in the overall capital–labour ratio ($\hat{k} > 0$) will decrease (increase) output of the labour (capital) intensive sector. This effect is shown diagrammatically in Fig. 5 where the pertinent part of Fig. 2 is redrawn. The increase in k is represented by moving the origin of sector 2 from 0_2

to $\bar{O}_2$. A ray parallel to O_2A is drawn from $\bar{O}_2$ and intersects with the ray O_1A at point B. At B the capital–labour ratios in the two sectors are the same as at point A and since the production functions are homogeneous, q is the same at the two points. The net effect is therefore to decrease output in 1 (labour intensive) and increase output in 2 (capital intensive).

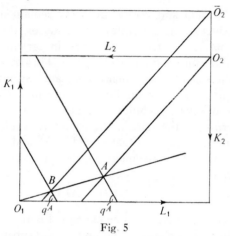

Fig. 5

(c) Displacement of Price and Rent

The supply price can be obtained from the equilibrium conditions, (2.5) or (2.6). To get the displacement, the conditions are differentiated (Appendix A.3) and the following is obtained :

$$\hat{P} = I\hat{q} + \frac{(1-\mu)}{\mu}\hat{Y}_1 - \frac{\gamma_1}{\mu} + \gamma_2, \qquad (4.15)$$

where $I = \dfrac{\alpha_1}{\mu} - \alpha_2$ is the difference between the labour share in payments to capital and labour in sector 1 and the labour share in sector 2. I has the same sign as i. Hence, when $I > 0$, an increase in q is associated with an increase of the price of the labour-intensive product. We can think of this effect as coming from a change in factor proportions. Since, however, Y_1 is not linear, we also have the scale effect which is represented by the second term in (4.15). As we recall, an increase in q results in an increase in Y_1 and that again leads to an increase in P. The last two terms in (4.15) show the impact of technical change on $\hat{P}$. If technical change in agriculture is relatively large, $\dfrac{\gamma_1}{\mu} > \gamma_2$, P will decline.

Rent on land is that part of the value of agricultural product that is not paid to labour and capital. By Euler's theorem it can be expressed as :

$$R = (1 - \mu)PY_1. \qquad (4.16)$$

Hence

$$\hat{R} = \hat{P} + \hat{Y}_1. \qquad (4.17)$$

Using (4.15)

$$\hat{R} = I\hat{q} + \frac{1}{\mu}\hat{Y}_1 - \frac{\gamma_1}{\mu} + \gamma_2. \qquad (4.18)$$

We shall show below that q is expected to increase with time, so $\hat{q} > 0$. Also, total agricultural product increases, $\hat{Y}_1 > 0$. Under these circumstances, if agriculture is more labour intensive, rent can decline with time, only if technical change in agriculture is sufficiently large to offset the other terms in (4.18). However, if agriculture is more capital intensive, then for rent to decline we need

$$\left| I\hat{q} + \frac{\gamma_1}{\mu} \right| > \frac{1}{\mu}\hat{Y}_1 + \gamma_2. \qquad (4.19)$$

To get some notion of the relative magnitudes of the various quantities in question, we should proceed with our analysis. The discussion to this point has not utilized the demand relationships, and therefore applies to any demand specification. We turn now to analyse the demand relation used in this paper.

(d) Displacement of Wage–Rental Ratio

The demand relation (2.7) imposes the following constraint on the displacement of the two products :

$$-\hat{y}_1 + \eta\hat{y}_2 = 0. \qquad (4.20)$$

The displacement of the system under the equality of supply and demand is obtained by using (4.12) and (4.13) in (4.20). The result is a relation between changes in resources and technology and between the wage–rental ratio :

$$\hat{q} = \sigma^{-1}\hat{k} + \frac{E_o}{E_q}, \qquad (4.21)$$

where σ^{-1} is the inverse of the overall elasticity of substitution between capital and labour :

$$\sigma^{-1} = -\frac{l}{iE_q} \qquad (4.22)$$

$$E_q = -(E_{1q} - \eta E_{2q}) \tag{4.23}$$

$$E_o = \gamma_1' - \eta\gamma_2 = \gamma_1 - (1 - \mu)\hat{L} - \eta\gamma_2 \tag{4.24}$$

$$l = \mu l_2 + \eta l_1 = \mu - (\mu - \eta)l_1 \tag{4.25}$$

$$iE_q < 0, \; \sigma > 0$$

Let us first examine the sign of $\hat{q}$ under the assumption $\hat{k} = 0$. For $\mu = 1$, changes in labour and capital affect q only to the extent that k changes. But, when $\mu < 1$, not only k matters but also the scale, and this is reflected in the term $-(1 - \mu)\hat{L}$ in E_o. In addition, E_o represents the effects of change in technology in the two sectors. Since η is relatively small, the effect of γ_2 is depressed and hence E_o is likely to be positive. In what follows we assume that it is positive. Consequently $\dfrac{E_o i}{E_q} < 0$, which means that q decreases (increases) when agriculture is more labour (capital) intensive. That is, when technical change in agriculture is sufficiently rapid to make E_o positive, resources move out of agriculture. We already know that a shift of resources from a capital-intensive to a labour-intensive sector increases q.

The change in q caused by E_o depends on the difference in capital intensities in the two sectors. Obviously, for given elasticities of substitution, when there is only a small difference in capital intensity, a shift of resources between sectors will not change much the capital intensity in the two sectors and hence only a small change in q will result. The effect of E_o on q should therefore be directly related to the absolute value of i. What also matters is the relative magnitude of the transferred resources. A given value of E_o mobilizes more resources the larger is agriculture relative to the non-agricultural sector. From (4.25) we see that for $\mu > \eta$, l is a decreasing function of l_1 and is bounded by μ and η. Thus, the larger is l, the smaller is l_1 and hence the smaller should be the effect of E_o on q. To show it, insert (A.2.5) and (A.2.7) in E_q and write :

$$E_q = -\left(\frac{l}{i}\bar{\sigma}_\rho + D\right), \tag{4.26}$$

where

$$D = \sigma_1(\mu - \alpha_1) - \sigma_2(1 - \alpha_2)\eta. \tag{4.27}$$

We thus see that

$$\lim_{|i| \to 0} \left| \frac{E_o}{E_q} \right| = 0.$$

The smaller is l_1, the larger is l and hence, the smaller is $\left|\dfrac{E_o}{E_q}\right|$.

Also, the smaller is l_1 the smaller is i likely to be and this again weakens the impact of E_0. All this holds true under the assumption that D remains unchanged. However, as we see below, changes in D are not large and in particular when $i=0$, D is $(\mu\,\sigma_1-\sigma_2\eta)\,(1-a_2)$.

We turn now to examine the impact of $\hat{k}$ on $\hat{q}$. Using (4.26) in (4.22), we can write :

$$\sigma = \bar{\sigma}_p + \frac{i}{l}D = \bar{\sigma}_p + \delta. \tag{4.28}$$

From (A.3.9) we see that $i<0$ implies $\dfrac{\alpha_1}{\mu}>\alpha_2$ and therefore

$$(\mu-\alpha_1)<\frac{1}{\mu}(\mu-\alpha_1)<(1-\alpha_2). \tag{4.29}$$

If σ_1 were close to $\sigma_2\eta$, we would have $Di<0$ and hence $\delta<0$. When σ_2 is not much larger than σ_1, η has to be close to 1 to produce negative δ. For small values of η and $\dfrac{\sigma_2}{\sigma_1}$ not much larger than 1, D is likely to be positive and hence the sign of δ will be the same as that of i. To get some boundaries on the magnitude of δ one can write :

$$-\frac{|\,i\,|}{l}(1-\alpha_2)\sigma_2\eta<\frac{|\,i\,|\,D}{l}<\frac{|\,i\,|}{l}(\mu-\alpha_1)\sigma_1. \tag{4.30}$$

For $\mu>\eta$, l is smaller than μ, and for $\mu=\cdot9$ is likely to be approximately $\cdot7$–$\cdot8$, depending on η and l_1. The share of capital in agriculture, $\mu-\alpha_1$, is likely to be smaller. Hence, it is likely that $\dfrac{\mu-\alpha_1}{l}<1$. The share of capital in the non-agricultural sector, for all we know, is likely to be in the range of $\cdot25$–$\cdot40$ and hence we can use for the lower limit $\eta\dfrac{1-\alpha_2}{l}<\cdot5\eta$.

Insert a value of $\eta=\cdot3$ and get the boundaries :

$$-\cdot15\,|\,i\,|\,\frac{\sigma_2}{\bar{\sigma}_p}<\frac{\delta}{\bar{\sigma}_p}<|\,i\,|\,\frac{\sigma_1}{\bar{\sigma}_p}.$$

When σ_1 and σ_2 are approximately equal, we have

$$-\cdot15\,|\,i\,|<\frac{\delta}{\bar{\sigma}_p}<|\,i\,|.$$

A high value for $|i|$ is 0 2. Thus, for

$$i>0 \qquad -\cdot03 < \frac{\delta}{\bar{\sigma}_p} < \cdot2 \qquad (4.31)$$

$$i<0 \qquad \cdot03 > \frac{\delta}{\bar{\sigma}_p} > -\cdot2.$$

But if we incorporate our previous result that δ is likely to have the same sign as i, we can further narrow down the boundaries

$$i>0 \qquad 0 < \frac{\delta}{\bar{\sigma}_p} < \cdot2$$

$$i<0 \qquad 0 > \frac{\delta}{\bar{\sigma}_p} > -\cdot2. \qquad (4.32)$$

Incorporating (4.32) in (4.28), we can state that it is likely that

$$\cdot8\bar{\sigma}_p \leqslant \sigma \leqslant 1\cdot2\bar{\sigma}_p$$

and more specifically,

$$\text{when } i>0 \qquad \bar{\sigma}_p \leqslant \sigma \leqslant 1\cdot2\bar{\sigma}_p$$

$$\text{when } i<0 \qquad \cdot8\bar{\sigma}_p \leqslant \sigma \leqslant \bar{\sigma}_p. \qquad (4.33)$$

The larger are the elasticities of substitutions in the two sectors, the larger is the overall elasticity of substitution and the smaller is the impact of capital accumulation on q. Furthermore, on the basis of (4.33), we can have $\sigma^{-1}>1$

$$\text{when} \qquad i>0 \quad \text{and} \quad \bar{\sigma}_p \leqslant \cdot83$$

$$i<0 \quad \text{and} \quad \bar{\sigma}_p \leqslant 1.$$

From our previous discussion we also have

$$\left(\frac{E_o}{E_q}\right) i < 0.$$

Hence, when agriculture is more capital intensive, the differential rate of technical change, E_o, contributes to the increase in q and therefore $\bar{\sigma}_p \leqslant 1$ is a sufficient condition for q to grow faster than k. On the other hand, when agriculture is more labour intensive, the two terms of (4.21) have opposite signs and a sufficient condition for q to grow slower than k is that $\bar{\sigma}_p \geqslant \cdot83$. Furthermore, when $k>0$, we cannot have $\hat{q}<0$ for very long. For $E_o>0$, we can have $\hat{q}<0$ only when $i>0$. Here the differential change in technology moves resources out of the labour intensive sector and thus depresses q. But as the contributing sector declines enough, i starts to decline and the effect of this process weakens and tends to disappear.

(e) The Response of Agriculture to Changes in Resources and Technology

After having determined how q is affected by changes in technology and growth of resources, we can now determine how the allocation of resources and rent are determined by those changes. It is convenient to start with l_1. Substitute (4.21) in (4.6) to obtain :

$$l_1 = \frac{l_2}{i}\left[\left(\frac{\bar{\sigma}_\rho}{\sigma} - 1\right)\hat{k} + \frac{\bar{\sigma}_\rho}{E_p}E_o\right]. \tag{4.34}$$

Recalling (4.28) we write :

$$\frac{\bar{\sigma}_\rho}{\sigma} - 1 = -\frac{\delta}{\sigma} = -\frac{i}{l}D\frac{1}{\sigma} \tag{4.35}$$

and using (4.22) :

$$\frac{\bar{\sigma}_\rho}{E_q} = -\frac{i\bar{\sigma}_\rho}{l\sigma}. \tag{4.36}$$

Combining (4.34), (4.35) and (4.36) :

$$l_1 = -\frac{l_2}{l\sigma}[D\hat{k} + \bar{\sigma}_\rho E_o]. \tag{4.37}$$

This is an exact result and independent of the empirical approximations mentioned above. However, on the basis of the foregoing discussion we know that E_o and D are likely to be positive and hence we can conclude that changes in technology and labour force, as reflected in E_o, and the increase in capital labour ratio, both lead to a decline in the proportion of the agricultural labour force. We further note that, other things being equal, the decline is proportional to l_2; therefore, it is relatively larger the smaller is the proportion of labour in agriculture. It is all related to the fact that our demand constraint requires a more rapid development of the non-agricultural sector. If it does not come about from a relatively rapid technical change in that sector, it can only be done by increasing its proportion in total resources. It is therefore clear that by increasing the value of η this process can reverse its direction.

The question is whether it is possible to have a decline in the total labour force in agriculture, that is, $\hat{L}_1 < 0$. Since our previous findings hold for $\hat{L} = 0$, in this case we have $0 < \hat{l}_1 = \hat{L}_1$. It is then clear that for small values of $\hat{L}$, there will be a decline in L_1. However, how small should $\hat{L}$ be depends on the parameters in (4.37).

We turn now to examine the displacement of rent. In an expanding economy the demand relation imposes growth on the two sectors. Thus, a necessary condition for rent to decline is that P declines. It is likely that not only P must decline but so also must L_1. This can be seen by using (4.37) and (A.2.3) in (4.18) and simplifying to obtain :

$$\hat{R} = \left[\left(\frac{\alpha_1}{\mu} - \alpha_2\right) + \left(1 - \frac{\alpha_1}{\mu}\right)\sigma_1\right]\hat{q} + \gamma_2 + \hat{L}_1. \qquad (4.38)$$

For non-negative q, $R<0$ can come about either from $\hat{L}_1<0$ or from a negative coefficient for $\hat{q}$. The coefficient is positive for $i>0$ and except for small values of σ_1 it is also positive for $i<0$.

To get the relation between $\hat{R}$ and changes in resources and technology we can substitute the expressions for $\hat{L}_1$ and $\hat{q}$ in (4.38). This is basically what is done in Appendix A.4, where we use a somewhat different route. The end result is of course the same. One way to present it is :

$$\hat{R} = A_1\hat{k} + A_o\hat{E}_o + \gamma_2 + \hat{L}, \qquad (4.39)$$

where

$$A_o = -\frac{i}{l\sigma}\left(I + \right)\frac{E_{1q}}{\mu} < 0$$

$$A_1 = -\frac{1}{i}(l_2 + lA_o).$$

It is shown in the Appendix A.4 that when $i>0$ $A_1>0$ and when $i<0$ the sign of A_1 is undetermined. As we have already stated, when agriculture is more labour intensive, R can decline only by having a relatively large value for γ_1. This will produce a large value for E_o. In an economy with no technical change in industry, with stable population and with no capital accumulation, rent will decline when there is any technical change in agriculture. In an economy which displays growth in capital-labour ratio, in population and in industrial productivity, rent can still decline if the productivity in agriculture increases fast enough. (4.39) enables us to evaluate how fast it must increase for rent to decline for any set of values of the parameters in question. When agriculture is the capital-intensive industry, it is also possible that the increase in capital intensity in the economy will lead to a decline in rent. However, this effect is likely to be relatively unimportant.

On the other hand, when demand for agricultural products, as presented by γ_2 and $\hat{L}$, increases more rapidly than technology in agriculture, rent will increase.

For small i the results simplify greatly. In fact, the following limiting values are easily obtained :

$$\lim_{i=0} \sigma = \bar{\sigma}_\rho$$

$$\lim_{i \to 0} A_0 = \frac{l_2}{l} \frac{\bar{\sigma}_\rho}{\sigma} = -\frac{l_2}{l}$$

$$\lim_{i \to 0} A_1 = \frac{\eta(1 - a_2)}{l}.$$

Thus, for small i, the following approximate expression replaces (4.39) :

$$R \approx \frac{-l_2}{l}\gamma_1 + [1 + \frac{l_2}{l}(1 - \mu)]\hat{L} + (1 + \frac{l_2}{l}\eta)\gamma_2 + \frac{\eta}{l}(1 - a_2)\hat{k}. \quad (4.40)$$

For small i, the rate of change can be approximated in terms of $\hat{L}$, γ_1, γ_2, η, a_2 and μ.[1]

As we recall, in section 2, γ_1 was defined as technical change proper $(\gamma_1{}^*)$ plus the change in production due to changes in land $[(1 - \mu)A]$:

$$\gamma_1 = \gamma_1{}^* + (1 - \mu)\hat{A}.$$

This relation can be incorporated in (4.39) to yield the dynamic relations between rent and changes in land.

---

[1] To get some idea on the size of i, let us note that we can write

$$\frac{k_1}{k_2} = \frac{\rho_1}{1 - \rho_1} \frac{1 - l_1}{l_1}$$

Upon simplification we get

$$i = \frac{(1 - m)(1 - l_1)}{\frac{1}{l_1} - (1 - m)} \qquad \text{where } m = \frac{k_1}{k_2}$$

The following table gives values of i computed for selected values of m and l_1 :

	\multicolumn{4}{c}{$\frac{k_1}{k_2} = m =$}
l_1	2	4	$\frac{1}{2}$	$\frac{1}{4}$
0·05	−0·045	−0·012	0·02	0·04
0·1	−0·08	−0·020	0·05	0·07
0·2	−0·013	−0·30	0·09	0·14
0·5	−0·17	−0·30	0·17	0·25

V. CONCLUDING REMARKS

The foregoing discussion traced the response of the system to changes in resources and technology. Such changes were taken as given. Growth theory treats changes in capital stock, and sometimes changes in population and technology, as endogenous. So, our next step should be to investigate what happens to the capital stock in our economy. To this effect, the saving relation was introduced and it is only in this context that the particular saving behaviour becomes important.

The question of capital accumulation is reserved for another analysis. However, it should be noted that the results of such analysis can be easily utilized in our discussion, for all it requires is to substitute for $\dot{k}$ the values which the analysis will yield.

Of the assumptions made in this paper, perhaps the most restrictive one is that of the demand relation. It is restrictive in two respects : (1) a constant elasticity η is used and (2) demand is unaffected by price. The first restriction is more apparent than real, for the elasticity can be made to depend on the level of *per* capita consumption. For instance, it is likely to be smaller when *per capita* consumption is high. However, in a local analysis, such as conducted above, taking it as fixed involves little loss in generality. This remark goes also for the other parameters in question such as the factor shares, including the share of land. The second restriction is of a different nature. To assume that price does not affect demand may not be a bad assumption in this case, where we deal with very broad commodities, one of which is food (agriculture). It should be noted that in our formulation an increase in the price of agricultural product increases income. It is possible that the income and substitution effects do not offset each other, as assumed here.

An extension of the demand relation is of particular interest if this framework is used for analysing other sector classifications which may have stronger substitution effects. Such extension is also reserved for another treatment. However, we can conjecture the impact of such an extension on some of our results. For instance, we recall that a necessary condition for a decline in rent is that the price of agricultural product will go down. If the demand relation contained a negative price elasticity, then the decline in price would have increased consumption of agricultural product and the decline in resources employed in agriculture would have been smaller than what our analysis shows. Consequently, the decline in rent would be smaller. The reverse is also true; an increase in price would reduce consumption and rent would increase by less than is shown by our results.

REFERENCES

1. Hahn, F. H., and Matthews, R. C. O., 'The Theory of Economic Growth: A Survey', *The Economic Journal*, lxxiv (December 1964), 779–902.
2. Jones, R. W., 'The Structure of Simple General Equilibrium Models', *The Journal of Political Economy*, lxxiii (December 1965), 557–572.
3. Meade, J. E., *A Neo-Classical Theory of Economic Growth* (London, Allen and Unwin, 1961).
4. Solow, R. M., 'Note on Uzawa's Two Sector Model of Economic Growth', *Review of Economic Studies*, xxix (October 1961).
5. Takayama, A., 'On a Two-Sector Model of Economic Growth: A Comparative Statics Analysis', *Review of Economic Studies*, xxx (June 1963).
6. Uzawa, H., 'On a Two-Sector Model of Economic Growth: I', *Review of Economic Studies*, xxix (October 1961).
7. Uzawa, H., 'On a Two-Sector Model of Economic Growth: II', *Review of Economic Studies*, xxx (June 1963).

APPENDIX

(A.1.) Displacement of Labour and Capital Allocation

Rewrite (4.4), (4.5) and (3.1).

$$\hat{K} = \rho_1 \hat{K}_1 + \rho_2 \hat{K}_2 \qquad (A.1.1)$$

$$\hat{L} = l_1 \hat{L}_1 + l_2 \hat{L}_2 \qquad (A.1.2)$$

$$l_1 + l_2 = 1 \qquad \rho_1 + \rho_2 = 1$$

$$\sigma_j = \frac{\hat{k}_j}{\hat{q}}. \qquad (A.1.3)$$

Subtracting (A.1.2) from (A.1.3) and rearranging terms:

$$\hat{k} = \rho_1 \hat{k}_1 + \rho_2 \hat{k}_2 + i[\hat{L}_2 - L_1], \qquad (A.1.4)$$

where $i = l_1 - \rho_1$.

Note that $l_j = L_j - L$ and $l_1 l_1 + l_2 l_2 = 0$.

$$k = \sigma_\rho q + i(l_2 - l_1) \qquad (A.1.5)$$

$$= \bar{\sigma}_\rho q - i \frac{1}{l_2},$$

where $\bar{\sigma}_\rho = \rho_1 \sigma_1 + \rho_2 \sigma_2$ is a weighted sum of the elasticities of substitution in the two sectors with ρ_j being the weights.

Rearranging we get :

$$l_1 = \frac{l_2}{i}(-k + \bar{\sigma}_\rho q) \tag{A.1.6}$$

$$l_2 = \frac{l_1}{i}(k - \bar{\sigma}_\rho q). \tag{A.1.7}$$

To explore the connection between change in l_j and in ρ_j note that (A.1.3) can be written as :

$$\sigma_j q = (K_j - K) - (L_j - L) + (K - L) \tag{A.1.8}$$

$$= \rho_j - l_j + k.$$

Hence

$$\rho_j = \sigma_j q + l_j - k \tag{A.1.9}$$

(A.2.) Displacement of Supply

Differentiation of 2.3 yields :

$$Y_1 = \alpha_1 L_1 + (\mu - \alpha_1)K_1 + \gamma_1. \tag{A.2.1}$$

Subtracting L from both sides and rearranging :

$$y_1 = Y_1 - L = (\mu - \alpha_1)k_1 + \mu l_1 + [\gamma_1 - (1 - \mu)L] \tag{A.2.2}$$

Using (A.1.3)

$$y_1 = (\mu - \alpha_1)\sigma_1 q + \mu l_1 + [\gamma_1 - (1 - \mu)L]. \tag{A.2.3}$$

Using (A.1.6)

$$y_1 = - \mu \frac{l_2}{i}k + E_{1q}q + [\gamma_1 - (1 - \mu)L], \tag{A.2.4}$$

where

$$E_{1q} = \frac{1}{i}[\sigma_1(\mu - \alpha_1)i + \mu l_2 \sigma_\rho] = \frac{B_1}{i}. \tag{A.2.5}$$

A similar development for Y_2, noting that $\mu = 1$ yields :

$$y_2 = \frac{l_1}{i}k + E_{2q}q + \gamma_2, \tag{A.2.6}$$

where

$$E_{2q} = \frac{1}{i}[(1 - \alpha_2)\sigma_2 i - l_1 \bar{\sigma}_\rho] = \frac{B_2}{i}. \tag{A.2.7}$$

We turn now to examine the signs of E_{1q} and E_{2q}. First it should be noted that if $i > 0$, then B_1 is a sum of positive terms and hence must be positive. Suppose, therefore, that i is negative. We then write :

$$\frac{B_1}{\mu} = i\sigma_1\left(1 - \frac{\alpha_1}{\mu}\right) + l_2 \bar{\sigma}_\rho$$

$$\frac{B_1}{\mu} = - \frac{\alpha_1}{\mu}\sigma_1 i + l_1 \sigma_1 - \rho_1 \sigma_1 + (1 - l_1)\rho_1 \sigma_1 + l_2 \rho_2 \sigma_2$$

$$\frac{B_1}{\mu} = -\frac{\alpha_1}{\mu}\sigma_1 i + l_1\sigma_1(1 - l_1) + l_2\rho_2\sigma_2 \tag{A.2.8}$$

and again B_1 is a sum of positive terms.

Similarly we examine the sign of B_2. When i is negative, B_2 is a sum of negative terms and hence negative.

For positive i we write :

$$B_2 = -\alpha_2\sigma_2 i + \sigma_2 i - l_1\rho_1\sigma_1 - l_1\rho_2\sigma_2$$
$$= \alpha_2\sigma_2 i - l_1\rho_1\sigma_1 + \sigma_2[l_1 - \rho_1 - l_1\rho_2] \tag{A.2.9}$$

and the last term is

$$l_1(1 - \rho_2) - \rho_1 = (l_1 - 1)\rho_1 \qquad < 0.$$

Hence B_2 consists of negative terms only.

We then conclude :

$$B_1 = E_{1q}i > 0 \qquad (E_{1q} \text{ has same sign as } i)$$
$$B_2 = E_{2q}i < 0 \qquad (E_{2q} \text{ has opposite sign to } i). \tag{A.2.10}$$

(A.3.) Displacement of Price

Differentiation of the equilibrium conditions (2.5) and (2.6) yield :

$$w = P + F_{1L} = F_{2L} \tag{A.3.1}$$
$$r = P + F_{1K} = F_{2K}. \tag{A.3.2}$$

Since Y_1 is homogeneous of degree μ in K and L, we have from Euler's theorem :

$$\mu Y_1 = F_{1L}L_1 + F_{1K}K_1. \tag{A.3.3}$$

Differentiation of (A.3.3) yields

$$\mu Y_1 = \alpha_1 F_{1L} + (\mu - \alpha_1)F_{1K} + \alpha_1 L_1 + (\mu - \alpha_1)K_1. \tag{A.3.4}$$

Using (A.2.1), (A.3.4) is rewritten :

$$(\mu - 1)Y_1 = \alpha_1 F_{1L} + (\mu - \alpha_1)F_{1K} - \gamma_1. \tag{A.3.5}$$

Using (A.3.1) and (A.3.2) in (A.3.5) and rearranging terms we get

$$P = \frac{\alpha_1}{\mu}w + \left(1 - \frac{\alpha_1}{\mu}\right)r + \left(\frac{1 = \mu}{\mu}\right)Y_1 - \frac{\gamma_1}{\mu}. \tag{A.3.6}$$

Similar development for sector 2 yields :

$$0 = \alpha_2 w + (1 - \alpha_2)r \qquad - \gamma_2. \tag{A.3.7}$$

Subtracting (A.3.7) from (A.3.6) :

$$P = Iq + \left(\frac{1 - \mu}{\mu}\right)Y_1 - \frac{\gamma_1}{\mu} + \gamma_2, \tag{A.3.8}$$

where $I = \frac{\alpha_1}{\mu} - \alpha_2$ is related to the capital intensity. It can be shown that

$$\frac{k_2}{k_1} = \frac{\alpha_1(1 - \alpha_2)}{\alpha_2(\mu - \alpha_1)}. \tag{A.3.9}$$

Hence, as $i>0$, $k_2>k_1$ and so $\dfrac{\alpha_1}{\mu}>\alpha_2$, so $I>0$.

(A.4.) Displacement of Rent

Combine (4.21) and (A.2.4) and note that $y_1 = Y - L$:

$$Y_1 = -\mu\frac{l_2}{i}k + E_{1q}\left(\sigma^{-1}k - E_0\frac{i}{\sigma l}\right) + \gamma_1 + \mu L. \qquad (A.4.1)$$

Using (4.21) in (4.15):

$$P = \frac{1-\mu}{\mu}Y_1 + I\left(\sigma^{-1}k - \frac{i}{l\sigma}E_0\right) + \gamma_2 - \frac{\gamma_1}{\mu}. \qquad (A.4.2)$$

We can now rewrite (4.18)

$$R = \left[-\frac{l_2}{i} + \sigma^{-1}\left(\frac{E_{1q}}{\mu} + I\right)\right]k - \left(\frac{E_{1q}}{\mu} + I\right)\frac{iE_0}{\sigma l} + \gamma_2 + L. \qquad (A.4.3)$$

Using an obvious notation for the coefficients in (A.4.3):

$$R = A_1\hat{k} + A_oE_o + \gamma_2 + L$$

note

$$\left(\frac{E_{1q}}{\mu} + I\right)\sigma^{-1} = -A_o\frac{l}{i}.$$

Hence

$$A_1 = -\frac{1}{i}(l_2 + lA_o). \qquad (A.4.4)$$

Since $E_{1q}i>0$ and $Ii>0$ we have $A_o<0$ independently of the sign of i. The sign of A_1 depends on whether the absolute value of lA_o is larger or smaller than l_2 and on i. To examine it write:

$$-lA_o = \frac{i}{\sigma}\left[\left(\frac{\alpha_1}{\mu} - \alpha_o\right) + \left(1 - \frac{\alpha_1}{\mu}\right)\sigma_1 + \frac{l_2}{i}\bar{\sigma}_\rho\right]. \qquad (A.4.5)$$

First, take up the case where $i>0$.

$$-lA_o = \left[\left(\frac{\alpha_1}{\mu} - \alpha_2\right)\frac{i}{\sigma} + \frac{l_2}{\sigma}\left(\bar{\sigma}_\rho + (\mu - \alpha_1)\frac{\sigma_1 i}{l}\frac{l}{l_2\mu}\right)\right]. \qquad (A.4.6)$$

Since $\dfrac{l}{l_2\mu} = \dfrac{l}{l - \eta l_1} > 1$

$$\bar{\sigma}_\rho + (\mu - \alpha_1)\frac{\sigma_1 i}{l}\frac{l}{l_2\mu} > \bar{\sigma}_\rho + (\mu - \alpha_1)\frac{\sigma_1 i}{l} > \sigma$$

and since $\left(\dfrac{\alpha_1}{\mu} - \alpha_2\right)i>0$ we have for $i>0$

$$-lA_o > l_2 \qquad (A.4.7)$$

and therefore $A_1>0$.

On the other hand, for $i < 0$, the inequality is reversed and we have

$$\bar{\sigma}_p + (\mu - \alpha_1)\frac{\sigma_1 i}{l}\frac{l}{l_2\mu} < \bar{\sigma}_p + (\mu - \alpha_1)\frac{\sigma_1 i}{l} < \sigma.$$

However, the first term is still positive and hence we cannot say whether $-lA_o$ is smaller or larger than l_2, and hence the sign of A_1 is undetermined for this case.

Finally, note that it is possible to trace back the effects of the various variables on R by writing :

$$R = \left[\frac{E_{1k}}{\mu} + E_{qk}\left(\frac{E_{1q}}{\mu} + E_{pq}\right)\right]k + E_{qt}\left(E_{pq} + \frac{E_{1q}}{\mu}\right)E_o + \gamma_2 + L, \quad (A.4.8)$$

where the symbol E_{sm} refers to the elasticity of s with respect to m and in particular :

$$E_{1k} = -\mu\frac{l_2}{i} \qquad E_{1k}i < 0 \qquad E_{1q}i > 0$$

$$E_{qk} = \sigma^{-1} \qquad E_{qk} > 0$$

$$E_{qT} = \frac{-i}{l\sigma} \qquad E_{qT}i < 0$$

$$E_{pq} = I \qquad E_{pq}i > 0.$$

DISCUSSION OF PROFESSOR MUNDLAK'S PAPER

(The chapter above is a revised version of the paper which Professor Mundlak gave at the Conference. The revisions are predominantly of a sort to improve the exposition rather than to alter the sense. The discussion has therefore only had to be modified by changing the numbers of equations referred to by contributors to the discussion and by excluding some of Professor Mundlak's clarifying argument which is now embodied in the text of the chapter.)

Professor Haberler said that it would have helped in the elucidation of Professor Mundlak's paper if the author had attempted to formulate the results in non-mathematical terms. Professor Haberler said that he had received the paper late, and had not been able to study the full implications of the model used by Professor Mundlak. For this reason he hoped that Professor Mundlak would, for example, explain whether it would make much difference to the main conclusions of his paper if there were more than two factors assumed in the model and if there were more than two industries ; or, to take another example, whether it would affect the conclusions if Professor Mundlak had assumed other than Hicks-neutral technical change in the process of growth. Professor Haberler said that if this special nature of technical change were not assumed then the result that technical change in agriculture reduces the rent to land would not necessarily follow from the model.

Professor Haberler said that most of the results of the paper are not new. For example, the fact that technical change and increases in population in the non-agricultural sector would be expected to bring about an increase in land rent and so an improvement in the terms of trade for agriculture is a Ricardian conclusion. Professor Haberler said that it is intuitively plausible that if rent to land is to remain constant, technical change in agriculture would have to exceed that in the non-agricultural sector when population is constant. Professor Haberler said that although the paper was stated to be an introduction to the subject, and, as such, was rigorous and clear, it would have been very useful if Professor Mundlak had attempted to indicate the results of relaxing the very tight assumptions upon which the model is based, and had considered the short-run aspects of the problem.

Professor Haberler said that his first comments above suggested some of the immediate limitations of Professor Mundlak's paper. However, since the paper dealt with the problems of terms of trade between two sectors, it was, perhaps, its remoteness from the main body of literature on the theory of the terms of trade that deserved most comment. Again, in this field, the paper was a very useful introduction but it was a pity that Professor Mundlak had not even tried to speculate what the effects on his model would have been of introducing some of the major complications normally treated in the literature on the terms of trade.

Professor Haberler said that it is possible to distinguish two main types of discussion in the literature on the terms of trade. There is the discussion conducted at a low level of abstraction by United Nations economists such as Prebisch and Singer, which conclude that the terms of trade move inexorably against less-developed countries ; and there is the discussion at a more abstract level which centres on the famous article by Chenery in which trade theory and growth theory are contrasted. In both these sets of discussion the arguments turn on such problems as the complementarities in production, the rigidities of prices and wages, the degrees of monopoly, the existence of external economies in both a static and a dynamic sense and the essential question of the unemployment or underemployment of resources. Professor Haberler said that the work by Chenery is particularly important for the present discussion. Chenery made a sharp distinction between trade theory and growth theory. While trade theory made a large number of simplifying assumptions and then discussed the effects of introducing one or two of the complications, growth theory does not start with the same simplifying assumptions and allows, for example, for complementarities between factors, and for external economies. Growth theories also typically employ production functions which are homogeneous to a higher degree than one. Professor Haberler said that Chenery overdrew the differences between trade theory and growth theory, but the problems he raised can scarcely be avoided in any serious discussion of the terms of trade. It would have been interesting to see, within a closed model, an attempt to assimilate these complications.

Professor Haberler said that it might have been better for many growth theorists if they started with a simple and rigorous model of the sort presented by Professor Mundlak before introducing the many complexities they discuss. However, he thought that it would have been useful if Professor Mundlak had attempted to show the connections between his work and the points raised in the literature on the subjects he was treating.

Professor Mundlak said that his main motivation in constructing the model in his paper was to show the relationship in a growth model between those major aspects of the sectoral problem which empirical studies all agree are important. On many of the complicating issues the literature shows diverse opinions, but it is generally accepted that the income elasticity of demand for food is less than one, that in the process of growth there is a shift of labour from agriculture to industry and that land is the significant specific factor in agriculture and is not of the same significance in industry. His paper was an attempt to put major empirical aspects of the problem into a formal framework.

Professor Mundlak said that he certainly regarded himself as an empirical economist, and that the apparent level of abstraction of the model in his paper did not mean that it was not of direct empirical use. In Israel new but costly methods of desalination of water have been considered, and in a country which suffers from lack of water the question has naturally arisen whether it would pay to desalinate water in order to bring new land into cultivation. At present the cost of producing water is at least twice as great as the marginal value product of that water. However, in the future it may well prove profitable to undertake the operation. A model to determine under what conditions it would pay to undertake such a project might well rest on assumptions no less restricted than those in the model under discussion. Rather than discard, *a priori*, a neo-classical model of the type he had put forward, it is better to see what predictive value it has. Professor Mundlak said that in such centrally planned projects as the one he had given as an example it would seem that such a model has important applications. Also it is not enough to criticize the model on the ground that it rests upon unrealistic assumptions of long-run equilibrium, if it has predictive value in immediate problems. However, Professor Mundlak accepted that ultimate tests of the model would be best made by using very long-run data since he abstracted from the business cycle and other shorter-run complications.

Professor Mundlak said that he had not simply wished to restate Ricardo, and that he would point out that equation (4.6) in the paper tells us more than the bald Ricardian results. It is the rent function on the long-run equilibrium growth path, and states what parameters are relevant to the determination of long-run rent. The five relevant parameters are the rate of growth of the labour force (n), the rates of technical advance in the two sectors (γ_1 and γ_2), the share of rent in total agricultural production $(1-\mu)$, and the income elasticity of demand (η). These five parameters are as much open to direct empirical observation as any parameters used in economics. The model gives insight into how an economy with two sectors moves, and can help to answer such questions as what costs in terms of other resources can justifiably be incurred in reproducing the specific factor in one sector.

Professor Haberler said that he wanted to comment on two of the points raised by Professor Mundlak's reply. Professor Mundlak seemed to regard the complications which he [Professor Haberler] had raised which are normally dealt with by growth theorists, and which are not even alluded to by Professor Mundlak, as short-run complications only. Thus Professor Mundlak feels that the long-run assumptions he uses mean that he does not have to discuss these complications. Professor Haberler said that this very classical approach is attractive, but he wondered whether, in the context of growth and trade theory, it is really permissible.

Professor Haberler said that on the methodological problem, Professor Mundlak had appealed to Friedman's argument that the realism of assumptions are irrelevant and that a theory stands or falls by the testing of its conclusions. This is a convenient idea and one which is plausible in the case of simple comparative static models. However, Professor Haberler said that he had not meant that Professor Mundlak's model should not be tested by the validity of its conclusions, but rather that there seems to be a need for further theoretical preparation before tests would be of much interest. For example, a theoretical exploration of the results of including more than one specific factor is called for.

Professor Robinson said that in order to illuminate policy decisions it is of interest to try to predict the terms of trade between agricultural products and other products over a 15- to 20-year period. In attempting this it is important to recognize that agriculture is more likely to be out of equilibrium than industry. To expand the output of fertilizer by 30 per cent is a very different problem from trying to expand the output of agriculture by the same percentage. To expand the output of fertilizer would not require changing the structure of the industry greatly, but the expansion of agricultural output would require the reshaping of millions of farms. Moreover, agriculture could remain out of equilibrium for long periods of time ; the present European situation bears this out. For output to contract because of low prices would require many years of sustained low prices. The converse applies in countries such as India where the supply response to high prices is slow and where, during the period of adjustment, acute shortages and inflations are likely to occur.

Professor Robinson said that Professor Mundlak's model is not helpful for middle-term forecasts of this type where the balance of trends in the world as a whole are the problem. Professor Robinson said that mathematicians would be of more help if they evolved more econometric models which take account of the length of periods of adjustment, and the difficulties arising from different sectoral reactions to disequilibrium.

Professor Nicholls said that in the under-developed countries which are not over-populated, especially in parts of Latin America and Africa, a simple two-sector model is inadequate for consideration of the terms of trade between agriculture and industry. On the agricultural side it is particularly important to distinguish between crops and livestock products. For example, in Brazil during the last decade the terms of trade between crop prices and industrial prices have turned against agriculture, reflecting the relative ease with which supply responds to temporary rises in prices since land is not yet a limiting factor. This is in spite of the fact that most techniques remain relatively primitive. On the other hand, again in Brazil, terms of trade between prices of livestock products and industrial produce have turned in favour of agriculture. In this case, supply response appears to be relatively low, reflecting the low technical level of livestock production and the more complex managerial skills required to expand livestock as against crop production. Thus, in crop production in Brazil, with a highly elastic supply curve and lower income elasticity of demand, rises in relative prices have been prevented. In livestock production, however, with a highly inelastic supply curve, which is also difficult to shift downward, and with high income elasticity of demand, terms of trade have become more favourable to this sector of agriculture.

Professor Nicholls said that since he suspected that such important differences in crop and livestock sectors of agriculture are relatively widespread phenomena, he would suggest that, at least for the basis of policy formation, more complex models which deal directly with at least these two sub-sectors of agriculture as a whole, are essential in the consideration of inter-sectoral terms of trade.

Professor Patinkin said that the assumption in the model, as laid out in Professor Mundlak's paper, that the amount of land is constant, means that it is hard to apply the results to many countries in the process of development. Account should be taken of the fact that in normal development the quantity of land employed by agriculture may change markedly. The United States and Israeli economies are important instances.

Professor Patinkin said that he thought that Professor Mundlak was taking an unduly optimistic view of the ease with which equation (4.40) in his paper could be used for empirical purposes. Such differences as arose between intuitive results and the results in the paper are due to the different quantitative estimates of factors. Professor Mundlak's estimates would depend upon being able to estimate the quantitative effects of the parameters γ_1 and γ_2 which are measures of technical advance in the two sectors. If the particular assumptions made about the neutral nature of technical change are not realized, and if the quantitative estimates are inaccurate, the equation is not going to be accurate.

Dr. Stipetić said that the implications of equation (4.40) should be set against such knowledge as we have about the long-run changes in land rents. There seems to be a general long-run tendency for land rent to decline ; at least this is true for most developed countries. This is usually taken to reflect the declining role of land as a factor of production. Is this result consistent with the behaviour of the parameters in Professor Mundlak's equation? The facts from long-run analyses do not seem to support the equation. Professor Stipetić said that he felt cautious about the model presented since it does not include many variables which on *a priori* grounds he thought ought to be included.

Professor Mundlak said that if he had given the impression that all short-run growth models should be superseded by long-run models he wanted to remove it completely. Short-run problems dictate much of economic policy, and no politician would be able to stay in power if he only considers the long run. Professor Mundlak said that he had not been arguing for the discarding of short-run models, but only that great attention should be paid to long-run considerations as well.

Professor Mundlak said that even if he had the time he would not at this stage feel able to speculate on the effect of introducing the complications to which Professor Haberler had referred. It is clearly necessary that short-run problems should be grafted onto the model but he had not yet set about doing so. However, it is clear that the model, even as now presented, is consistent with a number of typical short-run phenomena. For instance, suppose the long-run growth path of the economy is such that labour should steadily move out of agriculture. If this move took place without friction, then wages would be the same in agriculture and in industry. However, such a move means occupational and geographical mobility and it is not frictionless. Since adjustment takes time, it is very likely that labour in agriculture will always exceed the long-run equilibrium quantity, the result being lower wages in agriculture. This is often considered as unfavourable for the terms of trade of agriculture. This is a short-run problem which is not explicitly taken up by the model, but one which is easily predictable from it. In fact, if the long-run solution would call for labour movement into agriculture, then friction would favour the terms of trade of agriculture. The process is completely reversed. He agreed with Professor Robinson that there is a need for other models to supplement his own.

Professor Mundlak said that the adoption of the assumption of Hicksian neutral technical change was used in this first attempt, and he agreed that in later stages of developing the model it would be necessary to introduce Harrodian assumptions about labour-augmenting technical change which is very important in the agricultural sector. In this matter, as in many others, growth theory is at such a stage of development that it is essential to begin by making very simple assumptions in order to make any progress. The field is wide open for improvements.

Professor Mundlak said that in arguing that empirical predictions and not assumptions are important he had not meant to be taken too seriously. It would of course be best if models could be built that were free of all restricting assumptions, but at the present stage it is necessary to employ simplifying assumptions and to test the models that are generated from them by the conclusions they produce. The assumptions in his model are open to general discussion and if better predictive results are achieved with different assumptions he would recommend their adoption.

To Professor Nicholls, Professor Mundlak answered that the concept of terms of trade between the two major sectors followed from the simple assumptions of his model. He recognized that for many purposes the simplicity of the model would render it uninteresting, and that there are sectoral difficulties within each sector. Professor Mundlak said that in other questions also it had been implied that there is a need for models more elaborate and complex than his own. But this depends on the questions that the model is to be expected to answer. He suggested that there are a number of questions which can be answered with models as simple as his own.

Professor Mundlak said that it is not essential to the model that land be treated as constant. There are two ways that a change in the amount of land can arise and be dealt with in the model. Either it may change exogenously, in which case it could be treated in the same way as the role of technical change in agriculture, or it may change through the use of capital and labour to reproduce it, in which case this would be made implicit in a supply function for reproducing land.

In answer to Dr. Stipetic, Professor Mundlak said that if the decline in rent which he asserted to be a widespread phenomenon were not predicted by the parameters in his equation when correctly specified, then his model would have to be modified. However, no test had yet been made and it is not sufficient criticism to say that plausible variables have been omitted.

INDEX

Page references in the Index in bold type under the Names of Participants in the Conference indicate their Papers or Discussions of their Papers. Page references in italic indicate Contributions by Participants to the Discussions.

Index

Index

Index

Index

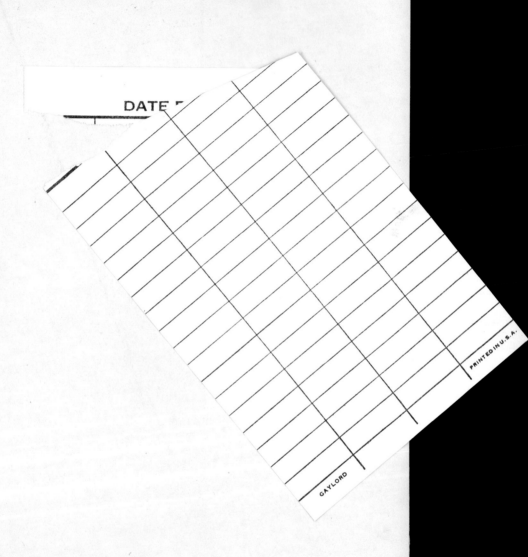